VOLUME FIVE HUNDRED AND FORTY TWO

METHODS IN ENZYMOLOGY

Conceptual Background and Bioenergetic/Mitochondrial Aspects of Oncometabolism

METHODS IN ENZYMOLOGY

Editors-in-Chief

JOHN N. ABELSON and MELVIN I. SIMON
Division of Biology
California Institute of Technology
Pasadena, California

ANNA MARIE PYLE
Departments of Molecular, Cellular and Developmental Biology and Department of Chemistry Investigator
Howard Hughes Medical Institute
Yale University

Founding Editors

SIDNEY P. COLOWICK and NATHAN O. KAPLAN

VOLUME FIVE HUNDRED AND FORTY TWO

METHODS IN ENZYMOLOGY

Conceptual Background and Bioenergetic/Mitochondrial Aspects of Oncometabolism

Edited by

LORENZO GALLUZZI

Université Paris Descartes;
Institut Gustave Roussy, France

GUIDO KROEMER

INSERM UMRS 872; Cordeliers Research Cancer Paris; Hôpital Européen Georges Pompidou; Université Paris Descartes, France

ELSEVIER

AMSTERDAM • BOSTON • HEIDELBERG • LONDON
NEW YORK • OXFORD • PARIS • SAN DIEGO
SAN FRANCISCO • SINGAPORE • SYDNEY • TOKYO
Academic Press is an imprint of Elsevier

Academic Press is an imprint of Elsevier
525 B Street, Suite 1800, San Diego, CA 92101-4495, USA
225 Wyman Street, Waltham, MA 02451, USA
Radarweg 29, PO Box 211, 1000 AE Amsterdam, The Netherlands
The Boulevard, Langford Lane, Kidlington, Oxford, OX5 1GB, UK
32 Jamestown Road, London NW1 7BY, UK

First edition 2014

For information on all Academic Press publications visit our website at store.elsevier.com

ISBN: 978-0-12-416618-9
ISSN: 0076-6879

14 15 16 17 11 10 9 8 7 6 5 4 3 2 1

CONTENTS

CONTRIBUTORS

Woo Suk Ahn
Department of Chemical Engineering, Massachusetts Institute of Technology, Cambridge, Massachusetts, USA

Federica Baldassari
Section of Pathology, Oncology and Experimental Biology, Laboratory for Technologies of Advanced Therapies (LTTA), Department of Morphology, Surgery and Experimental Medicine, Ferrara, Italy

Leah K. Billingham
Autoimmunity Branch, Immunoregulation Section, National Institutes of Arthritis, Musculoskeletal and Skin Diseases (NIAMS), National Institutes of Health (NIH), Bethesda, Maryland, USA

Massimo Bonora
Department of Experimental and Diagnostic Medicine, Section of General Pathology, Interdisciplinary Centre for the Study of Inflammation (ICSI), BioPharmaNet, University of Ferrara, Ferrara, Italy

Jing Chen
Department of Pathology, Immunology and Laboratory Medicine, The University of Florida College of Medicine, Gainesville, Florida, USA

Christos Chinopoulos
Department of Medical Biochemistry, Semmelweis University, Budapest, Hungary

Wai Kan Chiu
Department of Pharmacology & Chemical Biology, and Pittsburgh Institute for Neurodegenerative Diseases, University of Pittsburgh School of Medicine, Pittsburgh, Pennsylvania, USA

Elena De Marchi
Section of Pathology, Oncology and Experimental Biology, Laboratory for Technologies of Advanced Therapies (LTTA), Department of Morphology, Surgery and Experimental Medicine, Ferrara, Italy

Ralph J. DeBerardinis
Children's Medical Center Research Institute, University of Texas Southwestern Medical Center, Dallas, Texas, USA

Jayanta Debnath
Department of Pathology, Helen Diller Family Comprehensive Cancer Center and Biomedical Science Graduate Program, University of California, San Francisco, California, USA

Ruslan I. Dmitriev
Biochemistry Department, University College Cork, Cork, Ireland

Jerzy Duszynski
Nencki Institute of Experimental Biology, Warsaw, Poland

Andrea Eigentler
Daniel Swarovski Research Laboratory, Department of Visceral, Transplant and Thoracic Surgery, Medical University of Innsbruck, Innsbruck, Austria

Mario Fasching
OROBOROS INSTRUMENTS, Innsbruck, Austria

Christian Frezza
Medical Research Council Cancer Unit, University of Cambridge, Hutchison/MRC Research Centre, Cambridge, United Kingdom

Keigo Fukano
Department of Biomedical Sciences, Graduate School of Veterinary Medicine, Hokkaido University, Sapporo, Japan

Simone Fulda
Institute for Experimental Cancer Research in Pediatrics, Goethe-University, Frankfurt, Germany

Paulo A. Gameiro
Department of Chemical Engineering, Massachusetts Institute of Technology, Cambridge, Massachusetts, USA

Edoardo Gaude
Medical Research Council Cancer Unit, University of Cambridge, Hutchison/MRC Research Centre, Cambridge, United Kingdom

Erich Gnaiger
OROBOROS INSTRUMENTS, Innsbruck, and Daniel Swarovski Research Laboratory, Department of Visceral, Transplant and Thoracic Surgery, Medical University of Innsbruck, Innsbruck, Austria

Juliet Goldsmith
Department of Pathology, Helen Diller Family Comprehensive Cancer Center and Biomedical Science Graduate Program, University of California, San Francisco, California, USA

Michelle F. Green
Duke Molecular Physiology Institute, Duke University Medical Center, Durham, North Carolina, USA

Mika Hilvo
Medical Research Council Cancer Unit, University of Cambridge, Hutchison/MRC Research Centre, Cambridge, United Kingdom, and Biotechnology for Health and Well-Being, VTT Technical Research Centre of Finland, Espoo, Finland

Matthew D. Hirschey
Duke Molecular Physiology Institute; Sarah W. Stedman Nutrition and Metabolism Center; Department of Medicine, and Department of Pharmacology and Cancer Biology, Duke University Medical Center, Durham, North Carolina, USA

Frank K. Huynh
Duke Molecular Physiology Institute, Duke University Medical Center, Durham, North Carolina, USA

Paul M. Hwang
Center for Molecular Medicine, National Heart, Lung, and Blood Institute, National Institutes of Health, Bethesda, Maryland, USA

James Hynes
Biochemistry Department, University College Cork, Cork, Ireland

Raffaella Iurlaro
Cell Death Regulation Group, Bellvitge Biomedical Research Institute (IDIBELL), Barcelona, Spain

Lei Jiang
Children's Medical Center Research Institute, University of Texas Southwestern Medical Center, Dallas, Texas, USA

Yang Jin
Department of Biosciences, University of Oslo, Oslo, Norway

Ju-Gyeong Kang
Center for Molecular Medicine, National Heart, Lung, and Blood Institute, National Institutes of Health, Bethesda, Maryland, USA

Agnieszka Karkucinska-Wieckowska
Department of Pathology, The Children's Memorial Health Institute, Warsaw, Poland

Hibiki Kawamata
Brain and Mind Research Institute, Weill Medical College of Cornell University, New York, USA

Mark A. Keibler
Department of Chemical Engineering, Massachusetts Institute of Technology, Cambridge, Massachusetts, USA

Kazuhiro Kimura
Department of Biomedical Sciences, Graduate School of Veterinary Medicine, Hokkaido University, Sapporo, Japan

Gergely Kiss
Department of Medical Biochemistry, Semmelweis University, Budapest, Hungary

Timothy R. Koves
Duke Molecular Physiology Institute; Sarah W. Stedman Nutrition and Metabolism Center, and Department of Medicine, Duke University Medical Center, Durham, North Carolina, USA

Gerhard Krumschnabel
OROBOROS INSTRUMENTS, Innsbruck, Austria

Clara Lucía León-Annicchiarico
Cell Death Regulation Group, Bellvitge Biomedical Research Institute (IDIBELL), Barcelona, Spain

Beth Levine
Departments of Internal Medicine and Microbiology, Center for Autophagy Research, and Howard Hughes Medical Institute, University of Texas Southwestern Medical Center, Dallas, Texas, USA

Joseph Loscalzo
Cardiovascular Division, Department of Medicine, Brigham and Women's Hospital, Harvard Medical School, Boston, Massachusetts, USA

Dominika Malinska
Nencki Institute of Experimental Biology, Warsaw, Poland

Clayton E. Mathews
Department of Pathology, Immunology and Laboratory Medicine, The University of Florida College of Medicine, Gainesville, Florida, USA

Cristina Muñoz-Pinedo
Cell Death Regulation Group, Bellvitge Biomedical Research Institute (IDIBELL), Barcelona, Spain

Michael J. Palladino
Department of Pharmacology & Chemical Biology, and Pittsburgh Institute for Neurodegenerative Diseases, University of Pittsburgh School of Medicine, Pittsburgh, Pennsylvania, USA

Dmitri B. Papkovsky
Biochemistry Department, University College Cork, Cork, Ireland

Simone Patergnani
Section of Pathology, Oncology and Experimental Biology, Laboratory for Technologies of Advanced Therapies (LTTA), Department of Morphology, Surgery and Experimental Medicine, Ferrara, Italy

Martin Pelletier
Autoimmunity Branch, Immunoregulation Section, National Institutes of Arthritis, Musculoskeletal and Skin Diseases (NIAMS), National Institutes of Health (NIH), Bethesda, Maryland, USA

Paolo Pinton
Department of Experimental and Diagnostic Medicine, Section of General Pathology, Interdisciplinary Centre for the Study of Inflammation (ICSI), BioPharmaNet, University of Ferrara, and Section of Pathology, Oncology and Experimental Biology, Laboratory for Technologies of Advanced Therapies (LTTA), Department of Morphology, Surgery and Experimental Medicine, Ferrara, Italy

Jacques Pouysségur
Institute of Research on Cancer and Aging (IRCAN) CNRS-Inserm-UNS UMR 7284, U 1081 Centre A. Lacassagne, Nice, France, and Centre Scientifique de Monaco (CSM), Monaco

Madhu Ramaswamy
Autoimmunity Branch, Immunoregulation Section, National Institutes of Arthritis, Musculoskeletal and Skin Diseases (NIAMS), National Institutes of Health (NIH), Bethesda, Maryland, USA

Guy A. Rutter
Section of Cell Biology, Department of Medicine, Imperial College London, London, United Kingdom

Fahri Saatcioglu
Department of Biosciences, University of Oslo, and Institute for Cancer Genetics and Informatics, Oslo University Hospital, Oslo, Norway

Marco Sciacovelli
Medical Research Council Cancer Unit, University of Cambridge, Hutchison/MRC Research Centre, Cambridge, United Kingdom

Gregg L. Semenza
Vascular Program, Institute for Cell Engineering; Departments of Pediatrics, Medicine, Oncology, Radiation Oncology, and Biological Chemistry, and McKusick-Nathans Institute of Genetic Medicine, Johns Hopkins University School of Medicine, Baltimore, Maryland, USA

Larissa A. Shimoda
Division of Pulmonary and Critical Care Medicine, Johns Hopkins University School of Medicine, Baltimore, Maryland, USA

Richard M. Siegel
Autoimmunity Branch, Immunoregulation Section, National Institutes of Arthritis, Musculoskeletal and Skin Diseases (NIAMS), National Institutes of Health (NIH), Bethesda, Maryland, USA

Jørgen Sikkeland
Department of Biosciences, University of Oslo, Oslo, Norway

Hélène Simonnet
Centre de Recherche en Cancérologie de Lyon UMR INSERM U1052/CNRS 5286 Centre Léon Bérard, Lyon, France

Anatoly A. Starkov
Brain and Mind Research Institute, Weill Medical College of Cornell University, New York, USA

Gregory Stephanopoulos
Department of Chemical Engineering, Massachusetts Institute of Technology, Cambridge, Massachusetts, USA

Andrei I. Tarasov
Oxford Centre for Diabetes, Endocrinology and Metabolism, University of Oxford, Oxford, United Kingdom

Michael A. Teitell
Molecular Biology Institute; Department of Pathology and Laboratory Medicine; Department of Bioengineering; Department of Pediatrics; California NanoSystems Institute; Jonsson Comprehensive Cancer Center, and Broad Center of Regenerative Medicine and Stem Cell Research, UCLA, Los Angeles, California, USA

Tara TeSlaa
Molecular Biology Institute, UCLA, Los Angeles, California, USA

Atif Towheed
Department of Pharmacology & Chemical Biology, and Pittsburgh Institute for Neurodegenerative Diseases, University of Pittsburgh School of Medicine, Pittsburgh, Pennsylvania, USA

Arnaud Vigneron
Centre de Recherche en Cancérologie de Lyon UMR INSERM U1052/CNRS 5286 Centre Léon Bérard, Lyon, France

Ping-yuan Wang
Center for Molecular Medicine, National Heart, Lung, and Blood Institute, National Institutes of Health, Bethesda, Maryland, USA

Mariusz R. Wieckowski
Department of Biochemistry, Nencki Institute of Experimental Biology, Warsaw, Poland

Aleksandra Wojtala
Nencki Institute of Experimental Biology, Warsaw, Poland

Chendong Yang
Children's Medical Center Research Institute, University of Texas Southwestern Medical Center, Dallas, Texas, USA

Yi Yang
Synthetic Biology and Biotechnology Laboratory, State Key Laboratory of Bioreactor Engineering, School of Pharmacy, East China University of Science and Technology, Shanghai, China

Huafeng Zhang
School of Life Science, University of Science and Technology of China, Hefei, PR China

Jie Zhang
Department of Chemical Engineering, Massachusetts Institute of Technology, Cambridge, Massachusetts, USA

Zhe Zhang
Department of Chemical Engineering, Massachusetts Institute of Technology, Cambridge, Massachusetts, USA

Yuzheng Zhao
Synthetic Biology and Biotechnology Laboratory, State Key Laboratory of Bioreactor Engineering, School of Pharmacy, East China University of Science and Technology, Shanghai, China

Alexander V. Zhdanov
Biochemistry Department, University College Cork, Cork, Ireland

PREFACE: ONCOMETABOLISM: A NEW FIELD OF RESEARCH WITH PROFOUND THERAPEUTIC IMPLICATIONS

In 1924, the German physiologist Otto Heinrich Warburg was the first to report the propensity of neoplastic cells to metabolize glucose via aerobic glycolysis rather than via the citric acid cycle (also known as Krebs cycle) as a fuel for mitochondrial respiration (Koppenol, Bounds, & Dang, 2011; Warburg, 1924). For a long time since then, however, much greater attention has been attracted by the discovery of the genetic and epigenetic alterations that characterize cancer cells than by their metabolic profile. Such a tendency was so pronounced that, at the end of the twentieth century, several chemotherapeutic agents specifically targeting oncogene addiction, i.e., the process whereby cancer cells rely for their survival and growth on the constitutive activation of oncogenic signaling pathways and/or on the permanent inactivation of oncosuppressive mechanisms, were licensed for use in cancer patients (Luo, Solimini, & Elledge, 2009). Conversely, no chemotherapeutic agent specifically devised to target the metabolism of cancer cells was available then, nor it is now, although (1) several widely employed and effective chemotherapeutics including methotrexate, 5-fluorouracil, gemcitabine, and many others (which are cumulatively known as antimetabolites), *de facto* exert antineoplastic effects by operating as metabolic inhibitors (but were discovered and developed based on empirical, as opposed to mechanistic, grounds); and (2) the safety and therapeutic profile of many of these agents are being evaluated in a growing number of clinical trials (Chabner & Roberts, 2005; Galluzzi, Kepp, Vander Heiden, & Kroemer, 2013). Indeed, it is only over the past decade that the complexity and prominent therapeutic implications of oncometabolism, which can be defined as the ensemble of metabolic rearrangements that accompany oncogenesis and tumor progression, have been fully recognized as a central aspect of malignant transformation (Hanahan & Weinberg, 2011). As a result of such a refocus in the interest of researchers and clinicians, the metabolic rewiring of cancer cells is now viewed as a rich source of targets for the development of novel chemotherapeutic agents, and an intense wave of investigation currently explores this possibility (Galluzzi et al., 2013; Vander Heiden, 2011).

At odds with long-standing beliefs, it is now clear that the so-called Warburg effect represents only the tip of the iceberg of metabolic alterations associated with oncogenesis, which also encompass an increased flux through the pentose phosphate pathway, elevated rates of lipid biosynthesis, intense glutamine consumption, an improved control of redox homeostasis, and (at least in the initial steps of malignant transformation) decreased levels of macroautophagy (Schulze & Harris, 2012; Vander Heiden, Cantley, & Thompson, 2009; White, 2012). A few other common misconceptions about oncometabolism are in the process of being reconsidered based on preclinical and clinical findings from several laboratories worldwide.

First, the metabolic rewiring of neoplastic cells should not be considered as a self-standing hallmark of malignancy, but rather as a phenomenon that intimately accompanies, allows for and cannot be mechanistically separated from many, if not all, aspects of oncogenesis (Galluzzi et al., 2013; Locasale & Cantley, 2011; Wellen & Thompson, 2012). Accumulating evidence indicates indeed that (1) several metabolic intermediates such as ATP, acetyl-CoA, α-ketoglutarate, and reactive oxygen species play a major role in cell-intrinsic as well as cell-extrinsic signaling pathways (Galluzzi, Kepp, & Kroemer, 2012; Locasale & Cantley, 2011; Wellen & Thompson, 2012); (2) multiple proteins with prominent metabolic functions such as cytochrome *c* (which operates as an electron shuttle in the mitochondrial respiratory chain) and the M2 isoform of pyruvate kinase (PKM2, which catalyzes the last step glycolysis) participate in signal transduction (Galluzzi, Kepp, & Kroemer, 2012; Galluzzi, Kepp, Trojel-Hansen, & Kroemer, 2012; Gao, Wang, Yang, Liu, & Liu, 2012; Luo et al., 2011; Yang et al., 2011); and (3) several proteins initially viewed as "pure" signal transducers including (but not limited to) the antiapoptotic Bcl-2 family members BCL-X_L and MCL1 also impact on metabolic functions such as the handling of Ca^{2+} ions at the endoplasmic reticulum and the enzymatic activity of the F_1F_0 ATP synthase (Alavian et al., 2011; Perciavalle et al., 2012; Rong & Distelhorst, 2008).

Second, the metabolic changes linked to malignant transformation should not be considered as a general property shared by all types of cancer. It has indeed been clearly demonstrated that several variables including tissue type and oncogenic driver (and presumably many others) determine the metabolic profile of developing tumors (Yuneva et al., 2012). This has obvious implications for the use of metabolic inhibitors in cancer therapy.

Third, it should be kept in mind that the metabolic profile of neoplastic cells is far less specific than previously thought, but (with some exceptions) resemble that of highly proliferating nontransformed cells (Altman & Dang, 2012; Michalek & Rathmell, 2010). This notion is corroborated by the fact that the most severe side effects of antimetabolites involve highly proliferating normal tissues, such as the intestinal epithelium and bone marrow (Chabner & Roberts, 2005'). Nonetheless, the clinical success of these widely employed chemotherapeutic agents points to the existence of a therapeutic window for the use of metabolic inhibitors in cancer patients (Galluzzi et al., 2013; Vander Heiden, 2011). As it stands, multiple facets of oncometabolism can be considered as forms of "nononcogene addiction," a term referring to the fact that the survival of malignant cells relies not only on the constitutive activation of oncogenes and/or the permanent inactivation of oncosuppressive mechanisms but also on a wide array of genes and functions that are not inherently tumorigenic (Luo et al., 2009).

Finally, tumors (in particular solid neoplasms but also hematological malignancies) should no longer be considered as homogenous entities predominantly composed of malignant cells. Indeed, it is now clear that neoplastic lesions contain a large amount of nontransformed cells, including endothelial, stromal, and immune cells, and that oncogenesis takes place in the context of a complex network of physical and functional interactions among the malignant and nonmalignant components of the tumor microenvironment (Nagaraj & Gabrilovich, 2010; Pietras & Ostman, 2010). These interactions, part of which have direct metabolic implications (Nieman et al., 2011; Sonveaux et al., 2008; Whitaker-Menezes et al., 2011), are also attracting attention as targets for the development of novel antineoplastic agents (Galluzzi, Senovilla, Zitvogel, & Kroemer, 2012; Zitvogel, Galluzzi, Smyth, & Kroemer, 2013).

In *OncoMetabolism*, a thematic collection covering two volumes of the successful *Methods in Enzymology* series, leading researchers summarize the current state of the field from both a conceptual and methodological standpoint. The first volume, entitled "*Conceptual background and bioenergetic/mitochondrial aspects of oncometabolism*," provides a robust theoretical background on cancer-associated metabolic alterations, discussing how these relate to other aspects of oncogenesis such as the relentless proliferation and resistance to death exhibited by neoplastic cells. Thereafter, this volume offers a collection of techniques that can be employed to study the major bioenergetic and mitochondrial aspects of oncometabolism, including (but not limited to) alterations in glycolysis and oxidative phosphorylation. The second volume, entitled "*Cell-wide metabolic alterations associated with malignancy*," proposes a series of methods for the investigation of global facets of oncometabolism, including (but not limited to) deregulations in Ca^{2+} fluxes and autophagy, as well as (malignant) cell- or tissue-wide metabolomic alterations.

OncoMetabolism is expected to provide a comprehensive and reliable methodological guide to beginners and experts in this exciting and rapidly expanding area of cancer research.

ACKNOWLEDGMENTS

Lorenzo Galluzzi and Guido Kroemer are supported by the Ligue contre le Cancer (équipe labellisée), Agence National de la Recherche (ANR), AXA Chair for Longevity Research, ARC, Cancéropôle Ile-de-France, Institut National du Cancer (INCa), Fondation Bettencourt-Schueller, Fondation de France, Fondation pour la Recherche Médicale (FRM), the European Commission (ArtForce), the European Research Council (ERC), the LabEx Immuno-Oncology, the SIRIC Stratified Oncology Cell DNA Repair and Tumor Immune Elimination (SOCRATE), the SIRIC Cancer Research and Personalized Medicine (CARPEM), and the Paris Alliance of Cancer Research Institutes (PACRI).

REFERENCES

Alavian, K. N., Li, H., Collis, L., Bonanni, L., Zeng, L., Sacchetti, S., et al. (2011). Bcl-xL regulates metabolic efficiency of neurons through interaction with the mitochondrial F1FO ATP synthase. *Nature Cell Biology*, *13*, 1224–1233.

Altman, B. J., & Dang, C. V. (2012). Normal and cancer cell metabolism: Lymphocytes and lymphoma. *The FEBS Journal*, *279*, 2598–2609.

Chabner, B. A., & Roberts, T. G., Jr. (2005). Timeline: Chemotherapy and the war on cancer. *Nature Reviews. Cancer*, *5*, 65–72.

Galluzzi, L., Kepp, O., & Kroemer, G. (2012). Mitochondria: Master regulators of danger signalling. *Nature Reviews. Molecular Cell Biology*, *13*, 780–788.

Galluzzi, L., Kepp, O., Trojel-Hansen, C., & Kroemer, G. (2012). Non-apoptotic functions of apoptosis-regulatory proteins. *EMBO Reports*, *13*, 322–330.

Galluzzi, L., Kepp, O., Vander Heiden, M. G., & Kroemer, G. (2013). Metabolic targets for cancer therapy. *Nature Reviews. Drug Discovery*, *12*, 829–846.

Galluzzi, L., Senovilla, L., Zitvogel, L., & Kroemer, G. (2012). The secret ally: Immunostimulation by anticancer drugs. *Nature Reviews. Drug Discovery*, *11*, 215–233.

Gao, X., Wang, H., Yang, J. J., Liu, X., & Liu, Z. R. (2012). Pyruvate kinase M2 regulates gene transcription by acting as a protein kinase. *Molecular Cell*, *45*, 598–609.

Hanahan, D., & Weinberg, R. A. (2011). Hallmarks of cancer: The next generation. *Cell*, *144*, 646–674.

Koppenol, W. H., Bounds, P. L., & Dang, C. V. (2011). Otto Warburg's contributions to current concepts of cancer metabolism. *Nature Reviews. Cancer*, *11*, 325–337.

Locasale, J. W., & Cantley, L. C. (2011). Metabolic flux and the regulation of mammalian cell growth. *Cell Metabolism*, *14*, 443–451.

Luo, J., Solimini, N. L., & Elledge, S. J. (2009). Principles of cancer therapy: Oncogene and non-oncogene addiction. *Cell*, *136*, 823–837.

Luo, W., Hu, H., Chang, R., Zhong, J., Knabel, M., O'Meally, R., et al. (2011). Pyruvate kinase M2 is a PHD3-stimulated coactivator for hypoxia-inducible factor 1. *Cell*, *145*, 732–744.

Michalek, R. D., & Rathmell, J. C. (2010). The metabolic life and times of a T-cell. *Immunological Reviews*, *236*, 190–202.

Nagaraj, S., & Gabrilovich, D. I. (2010). Myeloid-derived suppressor cells in human cancer. *Cancer Journal*, *16*, 348–353.

Nieman, K. M., Kenny, H. A., Penicka, C. V., Ladanyi, A., Buell-Gutbrod, R., Zillhardt, M. R., et al. (2011). Adipocytes promote ovarian cancer metastasis and provide energy for rapid tumor growth. *Nature Medicine*, *17*, 1498–1503.
Perciavalle, R. M., Stewart, D. P., Koss, B., Lynch, J., Milasta, S., Bathina, M., et al. (2012). Anti-apoptotic MCL-1 localizes to the mitochondrial matrix and couples mitochondrial fusion to respiration. *Nature Cell Biology*, *14*, 575–583.
Pietras, K., & Ostman, A. (2010). Hallmarks of cancer: Interactions with the tumor stroma. *Experimental Cell Research*, *316*, 1324–1331.
Rong, Y., & Distelhorst, C. W. (2008). Bcl-2 protein family members: Versatile regulators of calcium signaling in cell survival and apoptosis. *Annual Review of Physiology*, *70*, 73–91.
Schulze, A., & Harris, A. L. (2012). How cancer metabolism is tuned for proliferation and vulnerable to disruption. *Nature*, *491*, 364–373.
Sonveaux, P., Vegran, F., Schroeder, T., Wergin, M. C., Verrax, J., Rabbani, Z. N., et al. (2008). Targeting lactate-fueled respiration selectively kills hypoxic tumor cells in mice. *The Journal of Clinical Investigation*, *118*, 3930–3942.
Vander Heiden, M. G. (2011). Targeting cancer metabolism: A therapeutic window opens. *Nature Reviews. Drug Discovery*, *10*, 671–684.
Vander Heiden, M. G., Cantley, L. C., & Thompson, C. B. (2009). Understanding the Warburg effect: The metabolic requirements of cell proliferation. *Science*, *324*, 1029–1033.
Warburg, O. (1924). Über den Stoffwechsel der Carcinomzelle. *Biochemische Zeitschrift*, *152*, 309–344.
Wellen, K. E., & Thompson, C. B. (2012). A two-way street: Reciprocal regulation of metabolism and signalling. *Nature Reviews. Molecular Cell Biology*, *13*, 270–276.
Whitaker-Menezes, D., Martinez-Outschoorn, U. E., Lin, Z., Ertel, A., Flomenberg, N., Witkiewicz, A. K., et al. (2011). Evidence for a stromal-epithelial "lactate shuttle" in human tumors: MCT4 is a marker of oxidative stress in cancer-associated fibroblasts. *Cell Cycle*, *10*, 1772–1783.
White, E. (2012). Deconvoluting the context-dependent role for autophagy in cancer. *Nature Reviews Cancer*, *12*, 401–410.
Yang, W., Xia, Y., Ji, H., Zheng, Y., Liang, J., Huang, W., et al. (2011). Nuclear PKM2 regulates beta-catenin transactivation upon EGFR activation. *Nature*, *480*, 118–122.
Yuneva, M. O., Fan, T. W., Allen, T. D., Higashi, R. M., Ferraris, D. V., Tsukamoto, T., et al. (2012). The metabolic profile of tumors depends on both the responsible genetic lesion and tissue type. *Cell Metabolism*, *15*, 157–170.
Zitvogel, L., Galluzzi, L., Smyth, M. J., & Kroemer, G. (2013). Mechanism of action of conventional and targeted anticancer therapies: Reinstating immunosurveillance. *Immunity*, *39*, 74–88.

Lorenzo Galluzzi
Guido Kroemer

CHAPTER ONE

The Metabolic Alterations of Cancer Cells

Marco Sciacovelli[*,1], **Edoardo Gaude**[*,1], **Mika Hilvo**[*,†,1], **Christian Frezza**[*,2]

[*]Medical Research Council Cancer Unit, University of Cambridge, Hutchison/MRC Research Centre, Cambridge, United Kingdom
[†]Biotechnology for Health and Well-Being, VTT Technical Research Centre of Finland, Espoo, Finland
[1]These authors contributed equally to this work
[2]Corresponding author: e-mail address: cf366@mrc-cu.cam.ac.uk

Contents

Abstract

Cancer cells exhibit profound metabolic alterations, allowing them to fulfill the metabolic needs that come with increased proliferation and additional facets of malignancy. Such a metabolic transformation is orchestrated by the genetic changes that drive tumorigenesis, that is, the activation of oncogenes and/or the loss of oncosuppressor genes, and further shaped by environmental cues, such as oxygen concentration and nutrient availability. Understanding this metabolic rewiring is essential to elucidate the fundamental mechanisms of tumorigenesis as well as to find novel, therapeutically exploitable liabilities of malignant cells. Here, we describe key features of the metabolic transformation of cancer cells, which frequently include the switch to aerobic glycolysis, a profound mitochondrial reprogramming, and the deregulation of lipid metabolism, highlighting the notion that these pathways are not independent but rather cooperate to sustain proliferation. Finally, we hypothesize that only those

Methods in Enzymology, Volume 542
ISSN 0076-6879
http://dx.doi.org/10.1016/B978-0-12-416618-9.00001-7

genetic defects that effectively support anabolism are selected in the course of tumor progression, implying that cancer-associated mutations may undergo a metabolically convergent evolution.

1. INTRODUCTION

Cancer is not a single disease but a set of diseases with distinct histological and genetic features. Nevertheless, all cancers are characterized by common biological traits among which unrestrained proliferation is the most fundamental (Hanahan & Weinberg, 2011). Several lines of evidence indicate that cancer cells undergo a complex metabolic reprogramming to satisfy the increased demands of macromolecules and energy for proliferation. Although these observations have been experimentally validated only recently, thanks to the development of appropriate analytical techniques and cancer models, the hypothesis that the metabolism of cancer cells is altered is not new. In fact, it was at the beginning of the twentieth century, thanks to the blossoming field of biochemistry, that the energy requirements of cell proliferation were initially taken into consideration. The first to make this link was the German scientist von Wassermann, a renowned bacteriologist, who hypothesized that cancer cells increase oxygen consumption due to their accelerated proliferation rate. Following this hypothesis, he set out to target cancer cell using a derivative of selenium (selenium–eosin), a drug that was thought to disrupt cell respiration (Wassermann, Keysser, & Wassermann, 1911). Despite very promising trials in rodents, this drug was found to be toxic in humans, and further investigations revealed that selenium was not as effective in killing cancer cells as initially thought, making Wassermann's observations circumstantial (Sugiura & Benedict, 1929). A more systematic study of cancer metabolism started few years later, through the painstaking work of Otto Warburg. In his early career, Warburg (1924) observed that cancer cells consume large amounts of glucose even in the presence of oxygen, when normal cells predominantly use respiration as energy source. After the discovery that respiration is carried out by the mitochondria, Warburg (1956) went on to propose that cancer transformation was caused by inherent defects in mitochondrial function. Despite debated, Warburg's observation stimulated several lines of research devoted to the understanding of the biochemical determinants of cancer transformation that lead to the discovery of several metabolic liabilities of cancer cells (Weinhouse, 1976). The elucidation of the structure of

DNA and the discovery of the role of oncogenes and tumor suppressor genes in cancer in the late 1970s momentarily diverted the interest of scientists to the emerging field of cancer genetics. Cancer metabolism had been partially neglected until the late 1990s, when the laboratory of Chi Van Dang demonstrated that the oncogene c-*Myc* actively contributes to aerobic glycolysis by regulating the expression of the glycolytic enzyme lactate dehydrogenase A (Shim et al., 1997), revamping the field of cancer metabolism. At this stage, however, altered metabolism was seen nothing more than an epiphenomenon of cell transformation, a by-product of the oncogene-induced reprogramming. The link between metabolism and cancer was substantiated in the early 2000s by the discovery that mutations of the housekeeping metabolic enzymes succinate dehydrogenase (SDH) (Baysal et al., 2000) and fumarate hydratase (FH) (Tomlinson et al., 2002) were associated with hereditary forms of cancer, highlighting the possibility that, in some circumstances, altered metabolism could be the cause, not the effect, of cancer transformation (Gottlieb & Tomlinson, 2005). These observations inspired a renaissance of the field of cancer metabolism, which combined the biochemical knowledge of cell metabolism developed in the pregenomic era with novel high-throughput techniques such as transcriptomics and metabolomics. This multidisciplinary effort culminated with the discovery that a number of oncogenes and tumor suppressors are implicated in the regulation of cancer metabolism and that key metabolic enzymes, if mutated, predispose to cancer. It is now clear that metabolic reprogramming is an obligate step of cancer transformation required to sustain unrestrained proliferation triggered by the activation of oncogenic signaling cascades (Fig. 1.1), turning cancer metabolism from anecdotal evidence to a hallmark of cancer (Hanahan & Weinberg, 2011). In this chapter, we will discuss three major components of the metabolic transformation of cancer cells, aerobic glycolysis, mitochondrial reprogramming, and deregulated lipid metabolism (Fig. 1.2) and how they contribute to biomass generation for cancer growth and proliferation (Fig. 1.3).

2. AEROBIC GLYCOLYSIS: A HALLMARK OF CANCER CELL METABOLISM

Aerobic glycolysis is the metabolic process that converts glucose into lactate even under aerobic conditions, when normal cells fully oxidize glucose in the mitochondria (Fig. 1.2). This metabolic feature of cancer cells, first noticed by Warburg (1924) in the first decades of the twentieth century, was initially thought to be caused by inherent mitochondrial dysfunction

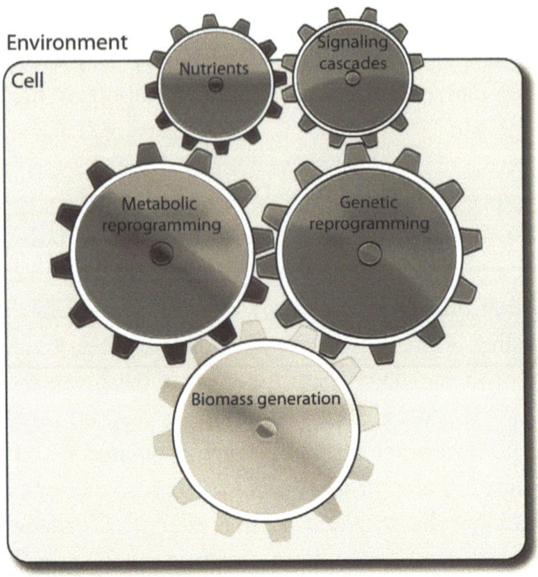

Figure 1.1 *The metabolic and genetic reprogramming in cancer cells is devoted to biomass generation.* The tight interconnection between the genetic reprogramming and the metabolic reprogramming is depicted here as a clockwork system. The metabolic reprogramming that supports biomass generation is orchestrated by oncogenes and tumor suppressors but at the same time is linked to nutrient availability, which also controls the activity of oncogenic signaling cascades. This figure shows that proliferation is the result of a combination of cell signaling and metabolism.

Warburg (1956). However, as we will discuss in detail later, the oxidative capacity of mitochondria is intact in cancer cells. We now know that aerobic glycolysis is genetically determined (see review from Chapter 3 and Levine & Puzio-Kuter, 2010), and its role in supporting proliferation goes beyond cancer cells. Indeed, aerobic glycolysis has been observed in nontransformed cells, including activated lymphocytes (reviewed in Wang & Green, 2012) and embryonic stem cells (Zhang, Nuebel, Daley, Koehler, & Teitell, 2012). These observations indicate that aerobic glycolysis can be considered as a universal metabolic phenotype (*metabotype*) of proliferating cells. However, the reasons why cancer cells switch to aerobic glycolysis are still debated.

2.1. Aerobic glycolysis and the energy requirements of proliferation

Glycolysis, the process that converts glucose into pyruvate, generates 2 molecules of ATP per molecule of glucose, whereas the full oxidation of glucose

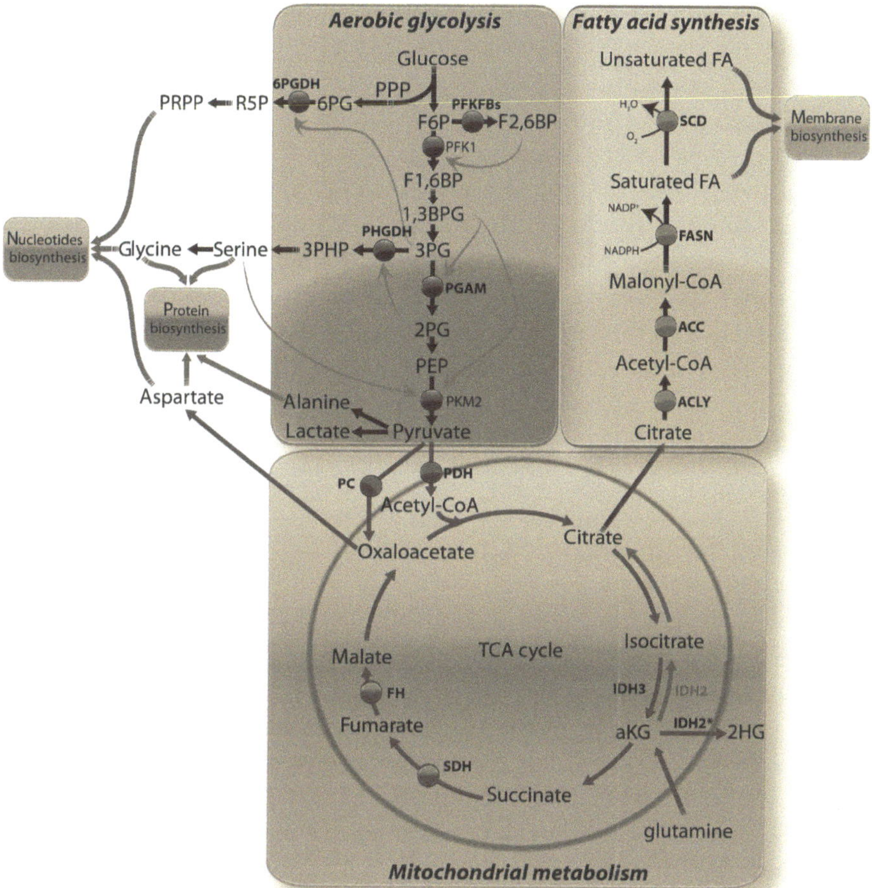

Figure 1.2 *The metabolic alterations of cancer cells.* Three major components of the metabolic transformation of cancer cells are depicted: aerobic glycolysis, mitochondrial metabolism, and lipid synthesis. Glucose branching to biosynthetic processes is regulated by various glycolytic enzymes, indicated as blue circles. The activity of these enzymes is subjected to allosteric control and posttranslational modification, indicated by the orange arrows. Mitochondrial metabolism is fueled by glucose and glutamine. PDH (red circle) is a mitochondrial gatekeeper and its activity is tightly regulated by genetic and environmental cues. When PDH is inactive, glutamine becomes the major carbon source for TCA cycle intermediates and anabolic precursors such as oxaloacetate and citrate. Both oxidative (gray arrows) and reductive (red arrows) metabolism of glutamine is used by cancer cells. PC could also provide glucose-derived carbons to the mitochondria, when PDH is inactive. Fatty acid biosynthesis requires acetyl-CoA obtained from citrate, which can be synthesized from glucose or glutamine. Several regulated steps (green circles) lead to the synthesis of palmitate, which is further converted into unsaturated lipids by the oxygen-dependent enzyme SCD. Dotted arrows indicate multiple

(*Continued*)

through mitochondria generates 31 molecules of ATP (Salway, 2004). Based on pure stoichiometric calculations, aerobic glycolysis is an inefficient pathway for ATP generation, and the switch to this *metabotype* seems a paradox for cancer cells. However, this argument fails on at least two points. First, it has been proposed that fast proliferation does not increase the demand of ATP for the cell. In fact, in a seminal work, Kilburn and colleagues calculated that the majority of ATP in the cell is used by processes involved in cell homeostasis, such as the maintenance of electrochemical gradients and protein turnover, and not by those devoted to the generation of biosynthetic precursors required for cell division (Kilburn, Lilly, & Webb, 1969). In line with this view, a more recent study calculated that ATP demand is not increased for cells that divide in the order of days, such as those in tumor tissues (Locasale & Cantley, 2011). Second, glycolysis is not inefficient at providing cells with ATP when kinetic aspects are taken into consideration. In fact, although glycolysis generates ATP with lower efficiency than oxidative phosphorylation, it produces ATP at a faster rate (Pfeiffer, Schuster, & Bonhoeffer, 2001). Furthermore, as shown in early experiments of Warburg, the glycolytic flux is so enhanced in cancer cells that by the time that one molecule of glucose is fully oxidized via respiration, 24 additional ATPs are generated via aerobic glycolysis, demonstrating that aerobic glycolysis could still supply 2/3 of the ATP, if respiration fails (Koppenol, Bounds, & Dang, 2011). Given the large amount of glucose taken up by cancer cells in the unit time, it has been speculated that the amount of ATP generated by aerobic glycolysis might even exceed the ATP requirements of cancer cells. Importantly, without sufficient ATP turnover, ADP and phosphate would become a limiting factor for glycolysis, and the accumulation of ATP would cause the allosteric inhibition of key metabolic enzymes, such as phosphofructokinase, blocking the entire glycolytic flux. Therefore, it

Figure 1.2—Cont'd enzymatic steps. Abbreviations: PPP, pentose phosphate pathway; PFK1, phosphofructokinase 1, PFKFBs, 6-phosphofructo-2-kinase/fructose-2,6-biphosphatases; F6P, fructose-6-phosphate; F1,6BP, fructose-1,6-bisphosphate; F2,6BP, fructose-2,6-bisphosphate; 1,3BPG, 1,3 bisphosphoglycerate; 3PG/2PG, 3/2 phosphoglycerate; PGAM, phosphoglycerate mutase; 6PGDH, 6-phosphogluconate dehydrogenase; 6PG, 6-phosphogluconate; R5P, ribulose-5-phosphate; PRPP, phosphoribosyl pyrophosphate; PHGDH, phosphoglycerate dehydrogenase; PEP, phosphoenolpyruvate; PKM2, pyruvate kinase variant M2; PDH, pyruvate dehydrogenase; PC, pyruvate carboxylase; FH, fumarate hydratase; SDH, succinate dehydrogenase; IDH, isocitrate dehydrogenase; IDH*, mutant IDH; aKG, alpha-ketoglutarate; 2HG, 2 hydroxyglutarate; ACLY, ATP-citrate lyase; ACC, acetyl-CoA carboxylase; FASN, fatty acid synthase; FA, fatty acid, SCD, stearoyl-CoA desaturase. (See the color plate.)

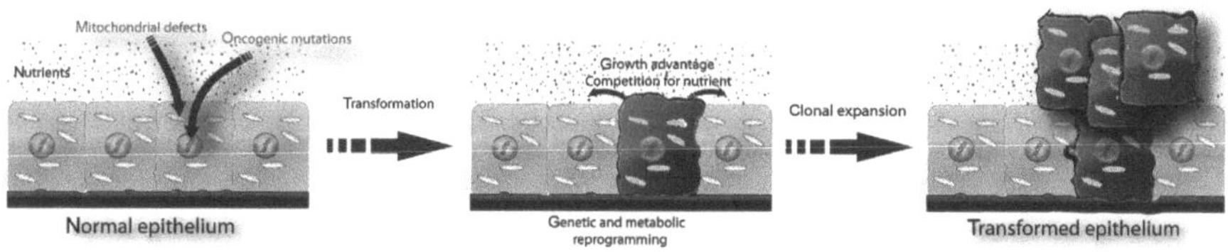

Figure 1.3 *Metabolic evolution of cancer cells.* We propose a general model of tumor formation and evolution: cancer transformation occurs upon mutations of key nuclear or mitochondrial enzymes. These mutations lead to a profound genetic and metabolic reprogramming that supports proliferation. Cancer-relevant mutations increase the fitness of cancer cells in the context of competition for nutrients and grant cancer cells a selective growth advantage. Only the mutations that support this feature will be selected in the process of clonal expansion. We propose that promotion of anabolism is a general characteristic of the most common oncogenic lesions. (See the color plate.)

has been proposed that aerobic glycolysis must be coupled to, or even driven by, aberrant ATP-consuming reactions present in cancer cells (Racker, 1976). A defective Na^+/K^+ plasma membrane ATPase has been initially proposed to be the major sink of ATP in cancer cells (Fagan & Racker, 1978). More recently, other examples of cancer-specific ATP-consuming reactions have been reported. The first example entails the phosphoinositide 3-kinase (PI3K)-mediated upregulation of the endoplasmic reticulum enzyme ENTPD5, which, together with cytidine monophosphate kinase-1 and adenylate kinase-1, engages into an ATP-hydrolyzing futile cycle that results in a compensatory increase in aerobic glycolysis (Fang et al., 2010). The second example involves the expression of the cancer-specific M2 variant of pyruvate kinase (PKM2), which was shown to uncouple pyruvate production from ATP generation by stimulating the transfer of the high-energy phosphate of phosphoenolpyruvate to the glycolytic enzyme phosphoglycerate mutase (PGAM) instead of ADP (Vander Heiden et al., 2010). These data suggest that (1) cancer cell proliferation is not associated with increased ATP demands, (2) high rates of aerobic glycolysis can produce sufficient amounts of ATP, and (3) high rates of aerobic glycolysis are possible thanks to the presence of ATP-depleting reactions.

Besides generating ATP, glycolysis is embedded in a complex network of metabolic reactions involved in the biosynthesis of anabolic precursors required for cell growth and proliferation, such as proteins and nucleotides (Fig. 1.2; Vander Heiden, Cantley, & Thompson, 2009). For instance, glucose supplies the carbon backbone for the nonessential amino acids cysteine, glycine, serine, and alanine (Salway, 2004). Furthermore, by generating 5-phosphoribosyl-α-pyrophosphate (PRPP) through the pentose phosphate pathway (PPP), glycolysis actively supports the *de novo* synthesis of purines and pyrimidines required for DNA biosynthesis (Salway, 2004). Accordingly, several reports showed that the expression of glycolytic enzymes and the activation of the PPP are strictly coupled to cell cycle progression and controlled by the cell cycle machinery (Buchakjian & Kornbluth, 2010; Vizan et al., 2009). Thanks to the branching to the oxidative arm of the PPP, glycolysis also contributes to the regeneration of NADPH, a key player in maintaining the redox homeostasis (Kondoh, Lleonart, Bernard, & Gil, 2007) and key cofactor involved in several biochemical pathways, including lipid biosynthesis.

2.2. The branching of glycolysis is a tightly controlled process

From the observations earlier, it is clear that the proper distribution of glucose-derived carbons towards anabolic pathways is of critical importance to support proliferation. In the following subsections, we will discuss key metabolic nodes that regulate glucose branching, which have been recently discovered (Fig. 1.2).

2.2.1 Phosphofructokinase 1 and 6-phosphofructo-2-kinase/fructose-2, 6-biphosphatases

Phosphofructokinase 1 (PFK1) catalyzes the irreversible conversion of fructose-6-phosphate (F6P) and ATP into fructose-1,6-bisphosphate (F1,6BP) and ADP. PFK is a highly regulated enzyme and a key branching point of glycolysis (Ros & Schulze, 2013; Fig. 1.2). Several mechanisms of regulation of PFK activity have been proposed, including regulation of its expression by oncogenes (described in Chapter 3), allosteric regulation, and posttranslational modifications. The most potent allosteric activator of PFK1 is fructose-2,6-bisphosphate (F2,6BP), a metabolite produced by the group of bifunctional enzymes 6-phosphofructo-2-kinase/fructose-2,6-biphosphatases (PFKFBs) from F6P (Yalcin, Telang, Clem, & Chesney, 2009). Several reports demonstrated that the steady-state levels of F2,6BP are increased in cancer cells (Hue & Rousseau, 1993; Nissler, Petermann, Wenz, & Brox, 1995), in part thanks to the oncogene-mediated overexpression of PFKFBs (Ros & Schulze, 2013), consistent with an allosterically activated PFK1. However, more recent data challenged this view. Yi and colleagues (2012) demonstrated that glycosylation, a cancer-specific posttranslational modification of PFK1, inhibits PFK1 enzymatic activity and redirects glucose through the PPP. These results are apparently at odds with the general view that cancer cells require a high PFK1 activity to maintain aerobic glycolysis and, on the contrary, seem to suggest that low PFK is advantageous for cancer cells. It is therefore possible that PFK1 activity is context type-dependent and cancer type-dependent, and more studies are required to solve this apparent conundrum.

2.2.2 Pyruvate kinase

Pyruvate kinase (PK) catalyzes the conversion of phosphoenolpyruvate (PEP) into pyruvate and is the final step of glycolysis. It has been shown that cancer cells express the M2 variant of PK, also known as PKM2 (reviewed in Chaneton & Gottlieb, 2012b). Unlike the constitutively active isoenzyme PKM1, PKM2 is a tightly regulated enzyme, whose activity is controlled by its oligomerization state, phosphorylation (Christofk, Vander Heiden, Wu, Asara, & Cantley, 2008; Hitosugi et al., 2009), acetylation (Lv et al., 2011), and various allosteric regulators, such as fructose-1,6-bisphosphate (Salway, 2004) and serine (Chaneton et al., 2012). Moreover, the reversible oxidation of key cysteine residues triggered by oxidative stress further controls its enzymatic activity (Anastasiou et al., 2011). It was shown that when PKM2 is inactive due to genetic, metabolic, or environmental cues, glycolytic intermediates upstream PEP accumulate and are diverted to the PPP (Kruger et al., 2011) or to the serine biosynthesis (Chaneton et al., 2012). Therefore, it has been proposed that the selective expression of PKM2 appears advantageous for cancer cells, since it would allow a fine tuning of glycolysis to match the requirements of anabolic precursors and antioxidant molecules (Chaneton & Gottlieb, 2012b). A recent paper from the laboratory of Vander Heiden shed some light on this issue. By using an elegant transgenic mouse model, it was shown that PKM2 deletion accelerates tumor formation in a BRCA1-deficient breast cancer mouse model, suggesting that PKM2 is not necessary for tumor cell proliferation. Furthermore, histology analyses revealed that the highly proliferative PKM2-deficient cancer cells decrease also the expression of the M1 variant of PK, PKM1, establishing an anticorrelation between PK activity and proliferation (Israelsen et al., 2013). Together, these results seem to suggest that PKM2 expression in tumors facilitates low PK activity, which is needed for cancer cells growth and proliferation.

2.2.3 3-Phosphoglycerate dehydrogenase

The diversion of glucose-derived carbons to serine biosynthesis is emerging as an important branch point for cancer cells (Locasale, 2013). PHGDH is the first enzyme in the serine biosynthesis pathway, which converts the glycolytic intermediate 3-phosphoglycerate (3PG) into 3-phosphohydroxypyruvate (3PHP) (Locasale, 2013; Fig. 1.2). It was recently demonstrated that PHGDH amplification strongly supports cancer cell proliferation by promoting the diversion of glucose into the serine biosynthesis pathway, making PHGDH a bona fide oncogene, at least in breast and skin cancers (Locasale et al., 2011; Possemato et al., 2011). However, how enhanced flux through PHGDH drives tumorigenesis remains unclear and other functions of PHGDH, which are independent from its metabolic activity, might play a role in this process.

2.2.4 Phosphoglycerate mutase

Another important enzyme recently found to regulate glycolytic branching is PGAM (Hitosugi et al., 2012). PGAM catalyzes the conversion of 3PG into 2-phosphoglycerate (2PG) (Fig. 1.2). These metabolites have been recently shown to modulate two important biosynthetic branches of glycolysis: the oxidative arm of the PPP and the serine biosynthesis pathway. In detail, 3PG was found to inhibit the PPP enzyme 6-phosphogluconate dehydrogenase (6PGDH) and 2PG to activate PHGDH (Hitosugi et al., 2012). Hence, by modulating the ratio between these two metabolites, PGAM could coordinate biosynthetic processes that branch from glycolysis (Fig. 1.2) (Hitosugi et al., 2012). The finding that PGAM is overexpressed in a variety of human cancers and regulated by the tumor suppressor p53 suggests that its expression might offer a striking metabolic flexibility to cancer cells (Chaneton & Gottlieb, 2012a).

2.2.5 1,3-Bisphosphoglycerate regulates glycolytic flux

The regulation of enzyme activity by small molecule metabolites is emerging as a novel mechanism of control of central carbon metabolism, beyond the classical allosteric regulation. In an elegant proteomics/metabolomics study, Moellering and Cravatt (2013) demonstrated that the reactive glycolytic intermediate 1,3-bisphosphoglycerate (BPG) could inactivate several glycolytic enzymes by binding to functional lysine residues, leading to the redirection of glycolytic intermediates to biosynthetic pathway (Fig. 1.2). This interesting observation may represent a self-regulatory mechanism of glycolysis that diverts glucose-derived carbons to branches of glycolysis when glycolytic flux is high.

In summary, these data demonstrate that aerobic glycolysis is not a mere consequence of mitochondrial dysfunction and is far from being an inefficient metabolic alternative for cancer cells. Instead, high rates of glycolysis coupled to the appropriate distribution of carbons towards anabolic pathways for biomass generation allow cancer cell to fine-tune the metabolic needs of proliferation.

3. MITOCHONDRIAL METABOLISM IN CANCER CELLS

Despite the initial hypothesis of Warburg (1956) that mitochondrial function was impaired in cancer, several lines of evidence indicate that cancer cells do not exhibit overt defects in oxidative metabolism. Indeed, as demonstrated by Zu and Guppy (2004), mitochondria are still the major contributors of ATP generation in cancer cells, whereas glycolysis only contributes to an average of 17% of the total cellular ATP. As we will see in the following sections, mitochondria not only contribute to ATP synthesis but also cooperate with aerobic glycolysis to support anabolic reactions.

3.1. Mitochondria and the anabolic requirements of cancer cells

Mitochondria are important metabolic hubs of the cell: they orchestrate the biosynthesis of lipids, proteins, and nucleic acids and maintain redox homeostasis via the generation of NADH and NADPH (DeBerardinis et al., 2007). Importantly, these reactions are tightly controlled by oncogenes and tumor suppressors to perfectly match the metabolic need of cancer cells (Frezza & Gottlieb, 2009). A key biosynthetic node in the mitochondria is the tricarboxylic acid (TCA) cycle (Salway, 2004). Although in normal cells the TCA cycle is mostly involved in glucose oxidation, the mitochondrial metabolism of cancer cells is generally uncoupled from glycolysis. In fact, due to genetic and environmental cues, such as hypoxia, the enzyme responsible for the entry of glucose-derived carbons in the mitochondria, the pyruvate dehydrogenase complex (PDH), is frequently inactivated in cancer (Sutendra & Michelakis, 2013) (Fig. 1.2). To maintain their function, mitochondria can bypass PDH inactivation in at least two ways: The first mechanism entails a diversion of glucose-derived pyruvate to oxaloacetate via the mitochondrial enzyme pyruvate carboxylase (PC). This reaction has been shown to play a key role in the growth of cancer cells, especially in glutamine-deprived conditions (Cheng et al., 2011). The second mechanism to overcome PDH inhibition is to utilize an alternative carbon source, glutamine. Glutamine oxidation in the mitochondria (glutaminolysis) is particularly important for cancer cells and several cell lines, including human glioma and HeLa cells, required glutamine for survival and proliferation (reviewed in DeBerardinis et al., 2007).

Glutamine oxidation in the mitochondria is a major means of ATP production in cancer cells (Fan, Kamphorst, Mathew et al., 2013), and it contributes to the synthesis of anabolic precursors, such as oxaloacetate and aspartate. Furthermore, by the reversal of the NADPH-dependent isocitrate dehydrogenase (IDH2), glutamine-derived alpha-ketoglutarate (aKG) could also contribute to the biosynthesis of citrate in a process called reductive carboxylation (Fig. 1.2). This process seems to occur only under specific circumstances, such as hypoxia, and overt mitochondrial dysfunction (Gameiro, Laviolette, Kelleher, Iliopoulos, & Stephanopoulos, 2013; Gameiro, Yang et al., 2013; Metallo et al., 2012; Mullen et al., 2012; Wise et al., 2011); however, its quantitative contribution to citrate biosynthesis has been recently challenged (Fan, Kamphorst, Rabinowitz, & Shlomi, 2013). These results indicate that, while glucose contributes to the anabolic reactions that branch from glycolysis, glutamine is the major source of carbons for the biosynthetic and bioenergetic processes that take place in the mitochondria of cancer cells.

3.2. Mitochondrial tumor suppressors

Given the crucial role of mitochondria in supporting cell proliferation, the discovery that the mitochondrial enzymes—FH (Tomlinson et al., 2002) and SDH (Baysal et al., 2000)—are mutated in hereditary cancer syndromes came as a bombshell. How was it possible that dysfunctions of housekeeping genes, once thought to be incompatible with cell survival, were the cause of cancer? Elegant work from the laboratory of Eyal Gottlieb revealed that the inactivation of SDH and the subsequent accumulation of succinate lead to the competitive inhibition of the aKG-dependent prolyl hydroxylases domain (PHD) proteins (Selak et al., 2005). Since PHDs are involved in the oxygen-dependent hydroxylation and following degradation of the hypoxia-inducible transcription factors (HIFs), the accumulation of succinate indirectly causes the stabilization of HIFs, important players in tumorigenesis (Gottlieb & Tomlinson, 2005). The disruption of PHDs activity was also documented in FH-deficient tumors and FH-deficient cells, where the striking accumulation of fumarate was associated with a potent stabilization of HIF (Isaacs et al., 2005; O'Flaherty et al., 2010), even though HIF seems not be required for tumor formation in Fh1-deficient animals (Adam et al., 2011). Other HIF-independent mechanisms have been recently proposed to explain the tumorigenesis in FH-deficient patients. Given its mild electrophilic nature, fumarate was shown to bind to reactive thiol residues of the Kelch-like ECH-associated protein 1 (Keap1), the negative regulator of the transcription factor nuclear factor erythroid 2-related factor (NRF2), a master regulator of the antioxidant response of the cell (Kansanen, Kuosmanen, Leinonen, & Levonen, 2013). Hence, when accumulated, fumarate abrogates the interaction of Keap1 with NRF2 and elicits a potent antioxidant response (Adam et al., 2011; Ooi et al., 2011). Several lines of evidence suggest that the activation of the NRF2-mediated antioxidant defense pathway may promote tumorigenesis; however, its role in FH-deficient tumors remains to be addressed (Ooi & Furge, 2012). Another oncogenic mechanism of fumarate, in part shared with succinate, entails the competitive inhibition of aKG-dependent dioxygenases involved in histone and DNA demethylation, which results in epigenetic alterations that affect the expression of genes involved in malignant transformation (Letouze et al., 2013; Xiao et al., 2012).

Although these findings helped to elucidate the mechanism of tumorigenesis in these cancer syndromes, it was still unclear how FH- and SDH-deficient cells could survive without a functional TCA cycle. Recent experiments on FH-deficient cells shed some light on this conundrum. It was shown that, in order to preserve their mitochondrial function, FH-deficient cells engage in a linear metabolic pathway that starts from glutamine oxidation and ends with the excretion of bilirubin, involving the biosynthesis and degradation of heme. Importantly, this pathway enables FH-deficient cells to generate mitochondrial NADH and to use the accumulated TCA cycle metabolites (Frezza et al., 2011). Of note, the inhibition of components of the heme biosynthesis and degradation pathway causes mitochondrial dysfunction and selectively kills FH-deficient cells, suggesting that this metabolic escape valve is critical to maintain mitochondrial function (Frezza et al., 2011). We and others have characterized several other metabolic consequences of the loss of FH and the ensuing accumulation of fumarate. For instance, high levels of fumarate trigger the reversal of the fumarate-producing enzyme argininosuccinate lyase. This reaction causes the accumulation of the small molecule metabolite argininosuccinate, which is accumulated in the cell and retrieved in the urine of Fh1-deficient animals (Adam et al., 2013; Zheng et al., 2013). Furthermore, in mouse Fh1-deficient fibroblasts, fumarate was shown to bind and inactivate mitochondrial aconitase, inhibition of which helps to explain the lack of reductive carboxylation in these cells (Ternette et al., 2013). Finally, fumarate was recently shown to bind to the cysteine residue of glutathione, leading to glutathione depletion and to increased oxidative stress in human FH-deficient cells (Sullivan et al., 2013). These results show that, upon the loss of FH, cells undergo a profound metabolic and (epi)genetic rewiring, which are required to support mitochondrial function and to buffer the striking accumulation of fumarate. We propose that it is only in those cells where these metabolic adaptations are allowed that fumarate can accumulate and activate oncogenic signaling cascades. In this scenario, the metabolic transformation in FH-deficient cells can be seen as a prerequisite, rather than a consequence, of cancer transformation.

3.3. Isocitrate dehydrogenase (IDH): A bona fide mitochondrial (and cytosolic) oncogene

Recent whole-genome sequencing efforts revealed that another enzyme of the TCA cycle, IDH2, together with its cytosolic variant IDH1, is mutated in gliomas (Parsons et al., 2008), leukemia (Mardis et al., 2009), and chondrosarcomas (Amary et al., 2011). Unlike the examples of SDH and FH, IDH mutations do not cause the functional inactivation of the enzyme. In fact, it was demonstrated that mutant IDH acquires a neomorphic activity, and instead of converting isocitrate into aKG, it catalyzes the reduction of aKG into the R enantiomer of 2-hydroxyglutarate (2HG), a poorly characterized metabolite present only in traces in normal cells (Fig. 1.2) (Dang et al., 2009). Similarly to fumarate and succinate, 2HG acts as a competitive inhibitor of aKG-dependent dioxygenases. However, its role as PHDs inhibitor has been debated and evidence suggests that while the *S* enantiomer inhibits PHDs, R-2HG could activate PHDs and promote HIF degradation (Koivunen et al., 2012). Additionally, several lines of evidence indicate that the oncogenic effect of 2HG depends, at least in part, on the competitive inhibition of aKG-dependent histone (Chowdhury et al., 2011; Lu et al., 2012) and DNA demethylases (Figueroa et al., 2010; Turcan et al., 2012; Xu et al., 2011). Importantly, it was shown that the incubation with 2HG is sufficient to transform cells and promote leukemogenesis (Losman et al., 2013) and that the targeted inhibition of mutant IDH in leukemia (Wang et al., 2013) and glioblastoma (Rohle et al., 2013) cancer cells induces differentiation.

Overall, this brief overview suggests that, contrary to the generally held views, mitochondria function is well preserved in cancer cells, and, even when inherently damaged, such as in the germline mutations of FH and SDH, they can still accommodate the metabolic needs of the cells thanks to their striking flexibility. Furthermore, it is emerging that small molecule metabolites produced by the mitochondria, such as fumarate, succinate, and 2HG, can induce profound genetic and epigenetic changes that support and, in some cases, drive tumorigenesis.

4. LIPID METABOLISM IN CANCER CELLS

Investigations on the lipid metabolism of malignant cells started in the early 1950s thanks to pioneering work of Sidney Weinhouse who demonstrated for the first time that carbons from glucose and acetate are actively incorporated into lipids in tumor tissues (Medes, Thomas, & Weinhouse, 1953). However, the first molecular evidence that lipid metabolism is deregulated in cancer cells traces back to the 1990s, when Kuhajda and colleagues (1994) discovered that fatty acid synthase (FASN) was overexpressed in breast cancer patients. Thereafter, several lines of evidence confirmed the important role of FASN in cancer, leading to the hypothesis that this enzyme could behave as a bona fide oncogene, at least in some types of cancer such as prostate and breast (Baron, Migita, Tang, & Loda, 2004; Menendez & Lupu, 2007). Other enzymes of fatty acid biosynthesis have been recently found to be deregulated in cancer, including ATP-dependent citrate lyase (ACLY) (Hatzivassiliou et al., 2005) and acetyl-CoA carboxylase (ACC) (Brusselmans, De Schrijver, Verhoeven, & Swinnen, 2005; Chajes, Cambot, Moreau, Lenoir, & Joulin, 2006), further underlining the relevance of this metabolic pathway for cancer cell proliferation.

4.1. Fatty acid biosynthesis supports cancer cell proliferation

In normal conditions, *de novo* fatty acid biosynthesis utilizes carbon units derived from glucose. This process requires first the breakdown of glucose into citrate in the TCA cycle. Then, citrate is exported in the cytosol and converted into acetyl-coenzyme A (acetyl-CoA) and oxaloacetate by the enzyme ACLY. Acetyl-CoA is activated by ACC, an enzyme that catalyzes the irreversible carboxylation of acetyl-CoA to produce malonyl-CoA. Finally, the acetyl and malonyl groups are repeatedly assembled by the multifunctional enzyme FASN until a 16-carbon saturated fatty acid, palmitate, is obtained. As indicated earlier, citrate can also be synthesized from glutamine, especially under hypoxia, when carbons from glucose are excluded from the mitochondria (Metallo et al., 2012) (Fig. 1.2). Since hypoxia is a common feature of solid tumors (Vaupel, 2004), glutamine could be a significant contributor to lipid biosynthesis in cancer cells.

De novo-synthesized palmitate is the building block for the generation of mono- and polyunsaturated fatty acids, and stearoyl-CoA desaturase (SCD) is the oxygen-dependent enzyme in charge of this process

(Paton & Ntambi, 2009) (Fig. 1.2). SCD converts palmitate and stearate into the monounsaturated fatty acids palmitoleic and oleic acid, respectively. Recent evidence suggests that SCD expression is regulated by the PI3K–Akt–mTOR pathway (Chang, Wang, Lu, Thewke, & Mason, 2005), and it is frequently overexpressed in cancer (Igal, 2010). However, SCD is an oxygen-dependent enzyme; therefore, under hypoxic conditions, its enzymatic activity could be greatly reduced. In line with this observation, it was recently shown that under hypoxia, the amount of unsaturated fatty acids generated from glutamine is decreased, indicating a possible SCD inactivation. Importantly, hypoxic cells become dependent on extracellular unsaturated lysophospholipids to overcome SCD deregulation (Kamphorst et al., 2013). These results might explain why some aggressive cancers rely on exogenous lipids for survival and why adipocytes might attract metastatic cancer cells (Nieman et al., 2011). The degree of lipid saturation is an important factor for the physicochemical properties of cell membranes. For instance, increased levels of saturated fatty acids make cell membranes less susceptible to oxidation, suggesting that increased palmitate biosynthesis uncoupled to desaturation might lead to resistance to oxidative stress (Rysman et al., 2010). Furthermore, high degree of lipid saturation correlates with increased vertical and horizontal movement of lipids in the membrane, leading to reduced uptake of chemotherapeutic agents into cells (Warmoes et al., 2013).

Fatty acid biosynthesis is strictly dependent on NADPH availability and, as a consequence, this pathway is coupled to the availability of this cofactor. Interestingly, it has been demonstrated that under stress conditions, cancer cells maintain NADPH levels by reducing fatty acid biosynthesis and increasing NADPH generation through fatty acid oxidation. Of note, this metabolic rearrangement, mostly orchestrated by the AMP-dependent protein kinase (AMPK; see Chapter 3), allows cell survival under energy stress conditions, such as glucose starvation and anchorage-independent growth (Jeon, Chandel, & Hay, 2012). These results indicate that fatty acid oxidation is as important as fatty acid biosynthesis for cancer cells, and recent evidence suggests that these processes might coexist in the cell (Carracedo, Cantley, & Pandolfi, 2013). In fact, it has been recently found that the inhibition of fatty acid oxidation and *de novo* lipid biosynthesis similarly sensitizes leukemia cells to apoptosis, proposing the intriguing hypothesis that fatty acid biosynthesis might provide the substrates for fatty acid oxidation and that the two processes might occur simultaneously (Samudio et al., 2010). More studies are needed to understand the relation of fatty acid synthesis and oxidation and how these processes occur in different cancers or cancer subtypes as in different phases of tumorigenesis.

5. FUTURE PERSPECTIVES: ALTERED METABOLISM SHAPES TUMOR EVOLUTION

Several lines of evidence suggest that cancers evolve from benign to malignant lesions by acquiring a series of mutations that can "drive" tumorigenesis and support abnormal cell proliferation (Vogelstein et al., 2013). Several of these driving mutations, such as *Myc*, *Ras*, and *Raf*, have a clear role in directing the metabolic transformation of cancer cells (see Chapter 3 in the accompanying review and Levine & Puzio-Kuter, 2010). Therefore, it is tempting to speculate that these mutations are selected based on their "anabolic efficiency," that is, how much they control key anabolic pathways. Recent evidence seems to support this hypothesis. In the attempt to investigate how environmental conditions select for specific genetic alterations, Yun et al. (2009) demonstrated that long-term glucose starvation leads to the selection of clones of cells that express oncogenic mutations, such as *K-Ras* and *B-Raf*. Surprisingly, even though the genetic lesions acquired were diverse and heterogeneous, they all converged to a similar *metabotype*, that is, increased uptake of glucose, a key advantage for growth in glucose-depleted conditions (Yun et al., 2009). These results are consistent with the idea of a convergent metabolic evolution of cancer mutations, which are selected for their capacity to support anabolic reactions. In this chapter, we would like to propose that tumor initiation might be driven by clones of cells that successfully switched their metabolism towards anabolic programs and increased their abilities to snatch nutrients from the tissue microenvironment, resulting in a selective growth advantage on normal cells (Fig. 1.3). Furthermore, during tumor evolution, only those clones that maintain a sufficient metabolic flexibility to adapt to the harsh environmental conditions would be selected. Therefore, understanding the metabolic needs of cancer during its evolution might open novel perspectives not only for targeting tumor progression but also for preventing tumor initiation.

6. CONCLUSIONS

Vogelstein and colleagues (2004) described cancer as a genetic disease and proposed that one of the challenges of future cancer research would be the "delineation of the pathways through which these genes act and the basis for the varying actions in specific cell types." It is now evident that most of the previously characterized oncogenes and tumor suppressors control key steps of cell metabolism, and several metabolic enzymes are now considered oncogenes or tumor suppressors. We now know that the metabolic changes instigated by these mutated genes dramatically alter the genotype and phenotype of the cell and significantly contribute to the evolutionary success of the tumor. We have provided evidence that the definition of cancer as a genetic disease is, at least, simplistic. Indeed, genetic alterations and metabolic reprogramming are two sides of the same story: how cancer arises and evolves.

REFERENCES

Adam, J., Hatipoglu, E., O'Flaherty, L., Ternette, N., Sahgal, N., Lockstone, H., et al. (2011). Renal cyst formation in Fh1-deficient mice is independent of the Hif/Phd pathway: Roles for fumarate in KEAP1 succination and Nrf2 signaling. *Cancer Cell, 20*(4), 524–537.

Adam, J., Yang, M., Bauerschmidt, C., Kitagawa, M., O'Flaherty, L., Maheswaran, P., et al. (2013). A role for cytosolic fumarate hydratase in urea cycle metabolism and renal neoplasia. *Cell Reports, 3*(5), 1440–1448.

Amary, M. F., Bacsi, K., Maggiani, F., Damato, S., Halai, D., Berisha, F., et al. (2011). IDH1 and IDH2 mutations are frequent events in central chondrosarcoma and central and periosteal chondromas but not in other mesenchymal tumours. *Journal of Pathology, 224*(3), 334–343.

Anastasiou, D., Poulogiannis, G., Asara, J. M., Boxer, M. B., Jiang, J. K., Shen, M., et al. (2011). Inhibition of pyruvate kinase M2 by reactive oxygen species contributes to cellular antioxidant responses. *Science, 334*(6060), 1278–1283.

Baron, A., Migita, T., Tang, D., & Loda, M. (2004). Fatty acid synthase: A metabolic oncogene in prostate cancer? *Journal of Cellular Biochemistry, 91*(1), 47–53.

Baysal, B. E., Ferrell, R. E., Willett-Brozick, J. E., Lawrence, E. C., Myssiorek, D., Bosch, A., et al. (2000). Mutations in SDHD, a mitochondrial complex II gene, in hereditary paraganglioma. *Science, 287*(5454), 848–851.

Brusselmans, K., De Schrijver, E., Verhoeven, G., & Swinnen, J. V. (2005). RNA interference-mediated silencing of the acetyl-CoA-carboxylase-alpha gene induces growth inhibition and apoptosis of prostate cancer cells. *Cancer Research, 65*(15), 6719–6725.

Buchakjian, M. R., & Kornbluth, S. (2010). The engine driving the ship: Metabolic steering of cell proliferation and death. *Nature Reviews. Molecular Cell Biology, 11*(10), 715–727.

Carracedo, A., Cantley, L. C., & Pandolfi, P. P. (2013). Cancer metabolism: Fatty acid oxidation in the limelight. *Nature Reviews. Cancer, 13*(4), 227–232.

Chajes, V., Cambot, M., Moreau, K., Lenoir, G. M., & Joulin, V. (2006). Acetyl-CoA carboxylase alpha is essential to breast cancer cell survival. *Cancer Research, 66*(10), 5287–5294.

Chaneton, B., & Gottlieb, E. (2012a). PGAMgnam style: A glycolytic switch controls biosynthesis. *Cancer Cell, 22*(5), 565–566.

Chaneton, B., & Gottlieb, E. (2012b). Rocking cell metabolism: Revised functions of the key glycolytic regulator PKM2 in cancer. *Trends in Biochemical Sciences, 37*(8), 309–316.

Chaneton, B., Hillmann, P., Zheng, L., Martin, A. C., Maddocks, O. D., Chokkathukalam, A., et al. (2012). Serine is a natural ligand and allosteric activator of pyruvate kinase M2. *Nature, 491*(7424), 458–462.

Chang, Y., Wang, J., Lu, X., Thewke, D. P., & Mason, R. J. (2005). KGF induces lipogenic genes through a PI3K and JNK/SREBP-1 pathway in H292 cells. *Journal of Lipid Research, 46*(12), 2624–2635.

Cheng, T., Sudderth, J., Yang, C., Mullen, A. R., Jin, E. S., Mates, J. M., et al. (2011). Pyruvate carboxylase is required for glutamine-independent growth of tumor cells. *Proceedings of the National Academy of Sciences of the United States of America, 108*(21), 8674–8679.

Chowdhury, R., Yeoh, K. K., Tian, Y. M., Hillringhaus, L., Bagg, E. A., Rose, N. R., et al. (2011). The oncometabolite 2-hydroxyglutarate inhibits histone lysine demethylases. *EMBO Reports, 12*(5), 463–469.

Christofk, H. R., Vander Heiden, M. G., Wu, N., Asara, J. M., & Cantley, L. C. (2008). Pyruvate kinase M2 is a phosphotyrosine-binding protein. *Nature, 452*(7184), 181–186.

Dang, L., White, D. W., Gross, S., Bennett, B. D., Bittinger, M. A., Driggers, E. M., et al. (2009). Cancer-associated IDH1 mutations produce 2-hydroxyglutarate. *Nature, 462*(7274), 739–744.

DeBerardinis, R. J., Mancuso, A., Daikhin, E., Nissim, I., Yudkoff, M., Wehrli, S., et al. (2007). Beyond aerobic glycolysis: Transformed cells can engage in glutamine metabolism that exceeds the requirement for protein and nucleotide synthesis. *Proceedings of the National Academy of Sciences of the United States of America, 104*(49), 19345–19350.

Fagan, J. B., & Racker, E. (1978). Determinants of glycolytic rate in normal and transformed chick embryo fibroblasts. *Cancer Research, 38*(3), 749–758.

Fan, J., Kamphorst, J. J., Mathew, R., Chung, M. K., White, E., Shlomi, T., et al. (2013). Glutamine-driven oxidative phosphorylation is a major ATP source in transformed mammalian cells in both normoxia and hypoxia. *Molecular Systems Biology, 9*, 712.

Fan, J., Kamphorst, J. J., Rabinowitz, J. D., & Shlomi, T. (2013). Fatty acid labeling from glutamine in hypoxia can be explained by isotope exchange without net reductive IDH flux. *Journal of Biological Chemistry, 288*(43), 31363–31369.

Fang, M., Shen, Z., Huang, S., Zhao, L., Chen, S., Mak, T. W., et al. (2010). The ER UDPase ENTPD5 promotes protein N-glycosylation, the Warburg effect, and proliferation in the PTEN pathway. *Cell, 143*(5), 711–724.

Figueroa, M. E., Abdel-Wahab, O., Lu, C., Ward, P. S., Patel, J., Shih, A., et al. (2010). Leukemic IDH1 and IDH2 mutations result in a hypermethylation phenotype, disrupt TET2 function, and impair hematopoietic differentiation. *Cancer Cell, 18*(6), 553–567.

Frezza, C., & Gottlieb, E. (2009). Mitochondria in cancer: Not just innocent bystanders. *Seminars in Cancer Biology, 19*(1), 4–11.

Frezza, C., Zheng, L., Folger, O., Rajagopalan, K. N., MacKenzie, E. D., Jerby, L., et al. (2011). Haem oxygenase is synthetically lethal with the tumour suppressor fumarate hydratase. *Nature, 477*(7363), 225–228.

Gameiro, P. A., Laviolette, L. A., Kelleher, J. K., Iliopoulos, O., & Stephanopoulos, G. (2013). Cofactor balance by nicotinamide nucleotide transhydrogenase (NNT) coordinates reductive carboxylation and glucose catabolism in the tricarboxylic acid (TCA) cycle. *Journal of Biological Chemistry, 288*(18), 12967–12977.

Gameiro, P. A., Yang, J., Metelo, A. M., Perez-Carro, R., Baker, R., Wang, Z., et al. (2013). In vivo HIF-mediated reductive carboxylation is regulated by citrate levels and sensitizes VHL-deficient cells to glutamine deprivation. *Cell Metabolism, 17*(3), 372–385.

Gottlieb, E., & Tomlinson, I. P. (2005). Mitochondrial tumour suppressors: A genetic and biochemical update. *Nature Reviews. Cancer, 5*(11), 857–866.

Hanahan, D., & Weinberg, R. A. (2011). Hallmarks of cancer: The next generation. *Cell, 144*(5), 646–674.

Hatzivassiliou, G., Zhao, F., Bauer, D. E., Andreadis, C., Shaw, A. N., Dhanak, D., et al. (2005). ATP citrate lyase inhibition can suppress tumor cell growth. *Cancer Cell*, *8*(4), 311–321.

Hitosugi, T., Kang, S., Vander Heiden, M. G., Chung, T. W., Elf, S., Lythgoe, K., et al. (2009). Tyrosine phosphorylation inhibits PKM2 to promote the Warburg effect and tumor growth. *Science Signaling*, *2*(97), ra73.

Hitosugi, T., Zhou, L., Elf, S., Fan, J., Kang, H. B., Seo, J. H., et al. (2012). Phosphoglycerate mutase 1 coordinates glycolysis and biosynthesis to promote tumor growth. *Cancer Cell*, *22*(5), 585–600.

Hue, L., & Rousseau, G. G. (1993). Fructose 2,6-bisphosphate and the control of glycolysis by growth factors, tumor promoters and oncogenes. *Advances in Enzyme Regulation*, *33*, 97–110.

Igal, R. A. (2010). Stearoyl-CoA desaturase-1: A novel key player in the mechanisms of cell proliferation, programmed cell death and transformation to cancer. *Carcinogenesis*, *31*(9), 1509–1515.

Isaacs, J. S., Jung, Y. J., Mole, D. R., Lee, S., Torres-Cabala, C., Chung, Y. L., et al. (2005). HIF overexpression correlates with biallelic loss of fumarate hydratase in renal cancer: Novel role of fumarate in regulation of HIF stability. *Cancer Cell*, *8*(2), 143–153.

Israelsen, W. J., Dayton, T. L., Davidson, S. M., Fiske, B. P., Hosios, A. M., Bellinger, G., et al. (2013). PKM2 isoform-specific deletion reveals a differential requirement for pyruvate kinase in tumor cells. *Cell*, *155*(2), 397–409.

Jeon, S. M., Chandel, N. S., & Hay, N. (2012). AMPK regulates NADPH homeostasis to promote tumour cell survival during energy stress. *Nature*, *485*(7400), 661–665.

Kamphorst, J. J., Cross, J. R., Fan, J., de Stanchina, E., Mathew, R., White, E. P., et al. (2013). Hypoxic and Ras-transformed cells support growth by scavenging unsaturated fatty acids from lysophospholipids. *Proceedings of the National Academy of Sciences of the United States of America*, *110*(22), 8882–8887.

Kansanen, E., Kuosmanen, S. M., Leinonen, H., & Levonen, A. L. (2013). The Keap1-Nrf2 pathway: Mechanisms of activation and dysregulation in cancer. *Redox Biology*, *1*(1), 45–49.

Kilburn, D. G., Lilly, M. D., & Webb, F. C. (1969). The energetics of mammalian cell growth. *Journal of Cell Science*, *4*(3), 645–654.

Koivunen, P., Lee, S., Duncan, C. G., Lopez, G., Lu, G., Ramkissoon, S., et al. (2012). Transformation by the (R)-enantiomer of 2-hydroxyglutarate linked to EGLN activation. *Nature*, *483*(7390), 484–488.

Kondoh, H., Lleonart, M. E., Bernard, D., & Gil, J. (2007). Protection from oxidative stress by enhanced glycolysis; a possible mechanism of cellular immortalization. *Histology and Histopathology*, *22*(1), 85–90.

Koppenol, W. H., Bounds, P. L., & Dang, C. V. (2011). Otto Warburg's contributions to current concepts of cancer metabolism. *Nature Reviews. Cancer*, *11*(5), 325–337.

Kruger, A., Gruning, N. M., Wamelink, M. M., Kerick, M., Kirpy, A., Parkhomchuk, D., et al. (2011). The pentose phosphate pathway is a metabolic redox sensor and regulates transcription during the antioxidant response. *Antioxidants & Redox Signaling*, *15*(2), 311–324.

Kuhajda, F. P., Jenner, K., Wood, F. D., Hennigar, R. A., Jacobs, L. B., Dick, J. D., et al. (1994). Fatty acid synthesis: A potential selective target for antineoplastic therapy. *Proceedings of the National Academy of Sciences of the United States of America*, *91*(14), 6379–6383.

Letouze, E., Martinelli, C., Loriot, C., Burnichon, N., Abermil, N., Ottolenghi, C., et al. (2013). SDH mutations establish a hypermethylator phenotype in paraganglioma. *Cancer Cell*, *23*(6), 739–752.

Levine, A. J., & Puzio-Kuter, A. M. (2010). The control of the metabolic switch in cancers by oncogenes and tumor suppressor genes. *Science, 330*(6009), 1340–1344.

Locasale, J. W. (2013). Serine, glycine and one-carbon units: Cancer metabolism in full circle. *Nature Reviews. Cancer, 13*(8), 572–583.

Locasale, J. W., & Cantley, L. C. (2011). Metabolic flux and the regulation of mammalian cell growth. *Cell Metabolism, 14*(4), 443–451.

Locasale, J. W., Grassian, A. R., Melman, T., Lyssiotis, C. A., Mattaini, K. R., Bass, A. J., et al. (2011). Phosphoglycerate dehydrogenase diverts glycolytic flux and contributes to oncogenesis. *Nature Genetics, 43*(9), 869–874.

Losman, J. A., Looper, R. E., Koivunen, P., Lee, S., Schneider, R. K., McMahon, C., et al. (2013). (R)-2-hydroxyglutarate is sufficient to promote leukemogenesis and its effects are reversible. *Science, 339*(6127), 1621–1625.

Lu, C., Ward, P. S., Kapoor, G. S., Rohle, D., Turcan, S., Abdel-Wahab, O., et al. (2012). IDH mutation impairs histone demethylation and results in a block to cell differentiation. *Nature, 483*(7390), 474–478.

Lv, L., Li, D., Zhao, D., Lin, R., Chu, Y., Zhang, H., et al. (2011). Acetylation targets the M2 isoform of pyruvate kinase for degradation through chaperone-mediated autophagy and promotes tumor growth. *Molecular Cell, 42*(6), 719–730.

Mardis, E. R., Ding, L., Dooling, D. J., Larson, D. E., McLellan, M. D., Chen, K., et al. (2009). Recurring mutations found by sequencing an acute myeloid leukemia genome. *New England Journal of Medicine, 361*(11), 1058–1066.

Medes, G., Thomas, A., & Weinhouse, S. (1953). Metabolism of neoplastic tissue. IV. A study of lipid synthesis in neoplastic tissue slices in vitro. *Cancer Research, 13*(1), 27–29.

Menendez, J. A., & Lupu, R. (2007). Fatty acid synthase and the lipogenic phenotype in cancer pathogenesis. *Nature Reviews. Cancer*, 7(10), 763–777.

Metallo, C. M., Gameiro, P. A., Bell, E. L., Mattaini, K. R., Yang, J., Hiller, K., et al. (2012). Reductive glutamine metabolism by IDH1 mediates lipogenesis under hypoxia. *Nature, 481*(7381), 380–384.

Moellering, R. E., & Cravatt, B. F. (2013). Functional lysine modification by an intrinsically reactive primary glycolytic metabolite. *Science, 341*(6145), 549–553.

Mullen, A. R., Wheaton, W. W., Jin, E. S., Chen, P. H., Sullivan, L. B., Cheng, T., et al. (2012). Reductive carboxylation supports growth in tumour cells with defective mitochondria. *Nature, 481*(7381), 385–388.

Nieman, K. M., Kenny, H. A., Penicka, C. V., Ladanyi, A., Buell-Gutbrod, R., Zillhardt, M. R., et al. (2011). Adipocytes promote ovarian cancer metastasis and provide energy for rapid tumor growth. *Nature Medicine, 17*(11), 1498–1503.

Nissler, K., Petermann, H., Wenz, I., & Brox, D. (1995). Fructose 2,6-bisphosphate metabolism in Ehrlich ascites tumour cells. *Journal of Cancer Research and Clinical Oncology, 121*(12), 739–745.

O'Flaherty, L., Adam, J., Heather, L. C., Zhdanov, A. V., Chung, Y. L., Miranda, M. X., et al. (2010). Dysregulation of hypoxia pathways in fumarate hydratase-deficient cells is independent of defective mitochondrial metabolism. *Human Molecular Genetics, 19*(19), 3844–3851.

Ooi, A., & Furge, K. A. (2012). Fumarate hydratase inactivation in renal tumors: HIF1alpha, NRF2, and "cryptic targets" of transcription factors. *Chinese Journal of Cancer, 31*(9), 413–420.

Ooi, A., Wong, J. C., Petillo, D., Roossien, D., Perrier-Trudova, V., Whitten, D., et al. (2011). An antioxidant response phenotype shared between hereditary and sporadic type 2 papillary renal cell carcinoma. *Cancer Cell, 20*(4), 511–523.

Parsons, D. W., Jones, S., Zhang, X., Lin, J. C., Leary, R. J., Angenendt, P., et al. (2008). An integrated genomic analysis of human glioblastoma multiforme. *Science, 321*(5897), 1807–1812.

Paton, C. M., & Ntambi, J. M. (2009). Biochemical and physiological function of stearoyl-CoA desaturase. *American Journal of Physiology, Endocrinology and Metabolism, 297*(1), E28–E37.
Pfeiffer, T., Schuster, S., & Bonhoeffer, S. (2001). Cooperation and competition in the evolution of ATP-producing pathways. *Science, 292*(5516), 504–507.
Possemato, R., Marks, K. M., Shaul, Y. D., Pacold, M. E., Kim, D., Birsoy, K., et al. (2011). Functional genomics reveal that the serine synthesis pathway is essential in breast cancer. *Nature, 476*(7360), 346–350.
Racker, E. (1976). Why do tumor cells have a high aerobic glycolysis? *Journal of Cellular Physiology, 89*(4), 697–700.
Rohle, D., Popovici-Muller, J., Palaskas, N., Turcan, S., Grommes, C., Campos, C., et al. (2013). An inhibitor of mutant IDH1 delays growth and promotes differentiation of glioma cells. *Science, 340*(6132), 626–630.
Ros, S., & Schulze, A. (2013). Balancing glycolytic flux: The role of 6-phosphofructo-2-kinase/fructose 2,6-bisphosphatases in cancer metabolism. *Cancer Metab, 1*(1), 8.
Rysman, E., Brusselmans, K., Scheys, K., Timmermans, L., Derua, R., Munck, S., et al. (2010). De novo lipogenesis protects cancer cells from free radicals and chemotherapeutics by promoting membrane lipid saturation. *Cancer Research, 70*(20), 8117–8126.
Salway, J. G. (2004). *Metabolism at a glance.* Oxford, U.K: Wiley-Blackwell.
Samudio, I., Harmancey, R., Fiegl, M., Kantarjian, H., Konopleva, M., Korchin, B., et al. (2010). Pharmacologic inhibition of fatty acid oxidation sensitizes human leukemia cells to apoptosis induction. *Journal of Clinical Investigation, 120*(1), 142–156.
Selak, M. A., Armour, S. M., MacKenzie, E. D., Boulahbel, H., Watson, D. G., Mansfield, K. D., et al. (2005). Succinate links TCA cycle dysfunction to oncogenesis by inhibiting HIF-alpha prolyl hydroxylase. *Cancer Cell,* 7(1), 77–85.
Shim, H., Dolde, C., Lewis, B. C., Wu, C. S., Dang, G., Jungmann, R. A., et al. (1997). c-Myc transactivation of LDH-a: Implications for tumor metabolism and growth. *Proceedings of the National Academy of Sciences of the United States of America, 94*(13), 6658–6663.
Sugiura, Kanematsu, & Benedict, Stanley R. (1929). The action of certain dyestuffs on the growth of transplantable tumors. *The Journal of Cancer Research, 13*(4), 340–358.
Sullivan, L. B., Martinez-Garcia, E., Nguyen, H., Mullen, A. R., Dufour, E., Sudarshan, S., et al. (2013). The proto-oncometabolite fumarate binds glutathione to amplify ROS-dependent signaling. *Molecular Cell, 51*(2), 236–248.
Sutendra, G., & Michelakis, E. D. (2013). Pyruvate dehydrogenase kinase as a novel therapeutic target in oncology. *Front Oncol, 3,* 38.
Ternette, N., Yang, M., Laroyia, M., Kitagawa, M., O'Flaherty, L., Wolhulter, K., et al. (2013). Inhibition of mitochondrial aconitase by succination in fumarate hydratase deficiency. *Cell Reports, 3*(3), 689–700.
Tomlinson, I. P. M., Alam, N. A., Rowan, A. J., Barclay, E., Jaeger, E. E. M., Kelsell, D., et al. (2002). Germline mutations in FH predispose to dominantly inherited uterine fibroids, skin leiomyomata and papillary renal cell cancer. *Nature Genetics, 30*(4), 406–410.
Turcan, S., Rohle, D., Goenka, A., Walsh, L. A., Fang, F., Yilmaz, E., et al. (2012). IDH1 mutation is sufficient to establish the glioma hypermethylator phenotype. *Nature, 483*(7390), 479–483.
Vander Heiden, M. G., Cantley, L. C., & Thompson, C. B. (2009). Understanding the Warburg effect: The metabolic requirements of cell proliferation. *Science, 324*(5930), 1029–1033.
Vander Heiden, M. G., Locasale, J. W., Swanson, K. D., Sharfi, H., Heffron, G. J., Amador-Noguez, D., et al. (2010). Evidence for an alternative glycolytic pathway in rapidly proliferating cells. *Science, 329*(5998), 1492–1499.
Vaupel, P. (2004). The role of hypoxia-induced factors in tumor progression. *The Oncologist, 9*(Suppl 5), 10–17.
Vizan, P., Alcarraz-Vizan, G., Diaz-Moralli, S., Solovjeva, O. N., Frederiks, W. M., & Cascante, M. (2009). Modulation of pentose phosphate pathway during cell cycle

progression in human colon adenocarcinoma cell line HT29. *International Journal of Cancer, 124*(12), 2789–2796.
Vogelstein, B., & Kinzler, K. W. (2004). Cancer genes and the pathways they control. *Nature Medicine, 10*(8), 789–799.
Vogelstein, B., Papadopoulos, N., Velculescu, V. E., Zhou, S., Diaz, L. A., Jr., & Kinzler, K. W. (2013). Cancer genome landscapes. *Science, 339*(6127), 1546–1558.
Wang, R., & Green, D. R. (2012). Metabolic checkpoints in activated T cells. *Nature Immunology, 13*(10), 907–915.
Wang, F., Travins, J., DeLaBarre, B., Penard-Lacronique, V., Schalm, S., Hansen, E., et al. (2013). Targeted inhibition of mutant IDH2 in leukemia cells induces cellular differentiation. *Science, 340*(6132), 622–626.
Warburg, O. (1924). On the metabolism of cancer cells. *Naturwissenschaften, 12*, 1131–1137.
Warburg, O. (1956). On the origin of cancer cells. *Science, 123*(3191), 309–314.
Warmoes, M., Jaspers, J. E., Xu, G., Sampadi, B. K., Pham, T. V., Knol, J. C., et al. (2013). Proteomics of genetically engineered mouse mammary tumors identifies fatty acid metabolism members as potential predictive markers for cisplatin resistance. *Molecular & Cellular Proteomics, 12*(5), 1319–1334.
Wassermann, A. V., Keysser, F., & Wassermann, M. (1911). Beiträge zum problem: Geschwülste von der blutbahn aus therapeutisch zu beeinflussen. *Deutsche Medizinische Wochenschrift, 37*(51), 2389–2391.
Weinhouse, S. (1976). Warburg hypothesis 50 years later. *Zeitschrift Fur Krebsforschung Und Klinische Onkologie, 87*(2), 115–126.
Wise, D. R., Ward, P. S., Shay, J. E., Cross, J. R., Gruber, J. J., Sachdeva, U. M., et al. (2011). Hypoxia promotes isocitrate dehydrogenase-dependent carboxylation of alpha-ketoglutarate to citrate to support cell growth and viability. *Proceedings of the National Academy of Sciences of the United States of America, 108*(49), 19611–19616.
Xiao, M. T., Yang, H., Xu, W., Ma, S. H., Lin, H. P., Zhu, H. G., et al. (2012). Inhibition of alpha-KG-dependent histone and DNA demethylases by fumarate and succinate that are accumulated in mutations of FH and SDH tumor suppressors. *Genes & Development, 26*(12), 1326–1338.
Xu, W., Yang, H., Liu, Y., Yang, Y., Wang, P., Kim, S. H., et al. (2011). Oncometabolite 2-hydroxyglutarate is a competitive inhibitor of alpha-ketoglutarate-dependent dioxygenases. *Cancer Cell, 19*(1), 17–30.
Yalcin, A., Telang, S., Clem, B., & Chesney, J. (2009). Regulation of glucose metabolism by 6-phosphofructo-2-kinase/fructose-2,6-bisphosphatases in cancer. *Experimental and Molecular Pathology, 86*(3), 174–179.
Yi, W., Clark, P. M., Mason, D. E., Keenan, M. C., Hill, C., Goddard, W. A., 3rd., et al. (2012). Phosphofructokinase 1 glycosylation regulates cell growth and metabolism. *Science, 337*(6097), 975–980.
Yun, J., Rago, C., Cheong, I., Pagliarini, R., Angenendt, P., Rajagopalan, H., et al. (2009). Glucose deprivation contributes to the development of KRAS pathway mutations in tumor cells. *Science, 325*(5947), 1555–1559.
Zhang, J., Nuebel, E., Daley, G. Q., Koehler, C. M., & Teitell, M. A. (2012). Metabolic regulation in pluripotent stem cells during reprogramming and self-renewal. *Cell Stem Cell, 11*(5), 589–595.
Zheng, L., Mackenzie, E. D., Karim, S. A., Hedley, A., Blyth, K., Kalna, G., et al. (2013). Reversed argininosuccinate lyase activity in fumarate hydratase-deficient cancer cells. *Cancer & Metabolism, 1*(1), 12.
Zu, X. L., & Guppy, M. (2004). Cancer metabolism: Facts, fantasy, and fiction. *Biochemical and Biophysical Research Communications, 313*(3), 459–465.

CHAPTER TWO

Autophagy and Cancer Metabolism

Juliet Goldsmith*, Beth Levine†, Jayanta Debnath*,[1]

*Department of Pathology, Helen Diller Family Comprehensive Cancer Center and Biomedical Science Graduate Program, University of California, San Francisco, California, USA

†Departments of Internal Medicine and Microbiology, Center for Autophagy Research, and Howard Hughes Medical Institute, University of Texas Southwestern Medical Center, Dallas, Texas, USA

[1]Corresponding author: e-mail address: jayanta.debnath@ucsf.edu

Contents

Methods in Enzymology, Volume 542
ISSN 0076-6879
http://dx.doi.org/10.1016/B978-0-12-416618-9.00002-9

Abstract

The metabolism of malignant cells is profoundly altered in order to maintain their survival and proliferation in adverse microenvironmental conditions. Autophagy is an intracellular recycling process that maintains basal levels of metabolites and biosynthetic intermediates under starvation or other forms of stress, hence serving as an important mechanism for metabolic adaptation in cancer cells. Although it is widely acknowledged that autophagy sustains metabolism in neoplastic cells under duress, many questions remain with regard to the mutual relationship between autophagy and metabolism in cancer. Importantly, autophagy has often been described as a "double-edged sword" that can either impede or promote cancer initiation and progression. Here, we overview such a dual function of autophagy in tumorigenesis and our current understanding of the coordinated regulation of autophagy and cancer cell metabolism in the control of tumor growth, progression, and resistance to therapy.

1. INTRODUCTION

The ability of cells to adapt to stress requires diverse changes in cellular metabolism. One of the principal pathways contributing to this metabolic adaptive response is macroautophagy (commonly termed autophagy), a tightly regulated lysosomal digestion process. Because degradation through autophagy allows recycling of nutrients, autophagy serves as an important survival and fitness pathway induced by a wide array of stresses including nutrient deprivation, growth factor withdrawal, oxidative stress, infection, and hypoxia (Avivar-Valderas et al., 2011; Boya, Reggiori, & Codogno, 2013; Lum et al., 2005; Yin, Kharbanda, & Kufe, 2009). In addition to its role in the stress-induced response, autophagy plays an essential homeostatic function by selectively removing damaged or nonfunctional proteins and organelles. These quality control functions have been demonstrated to be especially crucial in certain cell types: Liver cells are reliant on autophagy for the breakdown of stored metabolites, pancreatic β-cells utilize autophagy to manage high levels of endoplasmic reticulum (ER) stress, and postmitotic neurons require autophagy to remove potentially damaging proteins that cannot be diluted by cell division (Hara et al., 2006; Jung et al., 2008; Komatsu et al., 2005, 2006; Wu et al., 2009).

Given these key functions of autophagy in normal cells and tissues, it is not surprising that disruptions in autophagy have been implicated in numerous human diseases, including neurodegeneration, liver disease, inflammation, type 2 diabetes, and cancer (Debnath, 2011; Levine & Kroemer, 2008; Murrow & Debnath, 2013). While autophagy has been demonstrated to improve disease outcome in many cases by facilitating stress-induced metabolic adaptation or cellular homeostasis, the role of autophagy is more complex in cancer. Autophagy serves as an important tumor suppressor mechanism that impedes cancer initiation; at the same time, autophagy can promote the survival of tumor cells in response to diverse microenvironmental and therapeutic stresses and support anabolic capacity in fast-replicating, metabolically stressed tumor cells (Kimmelman, 2011; Levine & Kroemer, 2008; Rabinowitz & White, 2010). Despite this widely accepted notion that autophagy critically fuels metabolism in tumor cells under duress, many questions remain with regard to the interrelationships between autophagy and metabolism in cancer. This chapter focuses on the coordinated regulation of autophagy and cancer cell metabolism, controlling tumor growth, progression, and resistance.

2. OVERVIEW OF THE AUTOPHAGY MACHINERY

The process of macroautophagy occurs in a series of distinct steps: (1) initiation of the isolation membrane; (2) nucleation; (3) elongation of the double-membrane structure to form the autophagosome; and (4) fusion to the lysosome to form an autolysosome, in which the contents are degraded (Fig. 2.1). Studies in yeast have revealed over 30 autophagy related genes and proteins (ATGs and Atgs respectively) involved in the autophagic trafficking process, many of whose mammalian orthologues have also been identified (Nakatogawa, Suzuki, Kamada, & Ohsumi, 2009). This section provides an overview of the key molecular complexes that comprise the autophagy machinery in mammalian cells—more detailed reviews can be found elsewhere (Klionsky, 2013; Klionsky & Emr, 2000; Yang & Klionsky, 2010).

2.1. Initiation and the ULK complex

In mammals, autophagosome initiation requires the ULK complex, which consists of ULK1/2 (orthologous to yeast Atg1) associated with ATG13, FIP200, and ATG101 (Mizushima, 2010; Fig. 2.1A). At least three different ULK proteins are involved in different aspects of autophagy, among which ULK1/2 bear the highest similarity to yeast Atg1. Under nutrient-rich conditions, the ULK complex interacts with mTORC1 and remains inactivated

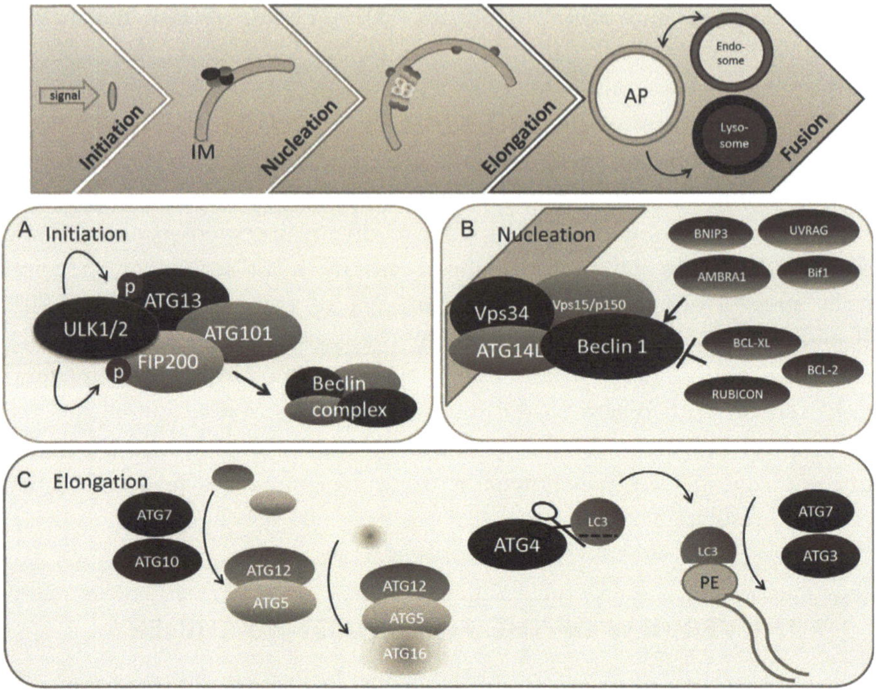

Figure 2.1 *Overview of the autophagy trafficking process.* The process of macroautophagy occurs in a series of distinct steps: (1) initiation of the isolation membrane (IM); (2) nucleation; (3) elongation of the double-membrane structure to form the autophagosome (AP); and (4) fusion to endosomes and lysosomes ultimately results in the formation of an autolysosome, in which the contents are degraded. (A) Initiation is mediated by the ULK complex. Activation of ULK activity leads to the phosphorylation of FIP200 and ATG13 and initiates nucleation via interaction with the Beclin 1 complex. (B) Beclin 1/VPS34/ATG14/PIK3R4 (p150) complex interacts with multiple interacting partners that positively and negatively regulate Beclin 1/VPS34 lipid kinase activity, resulting in the fine-tuning of autophagosome nucleation. (C) Elongation requires two ubiquitin-like conjugation pathways that form the ATG12–ATG5/ATG16 complex and phosphatidylethanolamine (PE)-conjugated LC3. (See the color plate.)

by mTORC1-mediated phosphorylation. However, upon nutrient deprivation, mTORC1 dissociates from the complex resulting in the dephosphorylation of inhibitory sites and concomitant autophosphorylation of activating sites in ULK1/2 (Chan, 2009). The kinase activation of ULK1/2 then leads to the phosphorylation and activation of ATG13 and FIP200 (Jung et al., 2009). The active complex then initiates nucleation by interaction with the Beclin 1/ATG14/VPS34 complex.

2.2. Nucleation and Beclin 1/ATG14/VPS34 complex

The formation of autophagosomes requires the activity of class III phosphatidylinositol 3-kinase (PI3K) VPS34, which is essential for phosphatidylinositol 3-phosphate production during the early stages of phagophore nucleation. VPS34 forms a complex with the yeast Atg6 orthologue Beclin 1, ATG14L, and VPS15/PIK3R4 (p150) (Itakura, Kishi, Inoue, & Mizushima, 2008; Zhong et al., 2009). Various binding partners of Beclin 1 have been identified (Fig. 2.1B), including UV irradiation resistance-associated gene (UVRAG) (Itakura et al., 2008; Liang et al., 2006), ATG14L/Barkor (Matsunaga et al., 2009; Zhong et al., 2009), and AMBRA1 (Fimia et al., 2007), all of which positively regulate Beclin 1 activity. Notably, ATG14L plays a critical role in specifying the site of the VPS34 complex relocation and therefore phagophore nucleation (Matsunaga et al., 2009). UVRAG also interacts with SH3GLB1/Bif-1 (an N-BAR domain protein), which potentially leads to phagophore membrane curvature, and expedites autophagosome–lysosome fusion (Liang et al., 2008; Takahashi et al., 2007). In addition to these positive regulators, other Beclin 1-interacting partners, including BCL-2, BCL-xL, Rubicon (RUN domain and cysteine-rich domain containing, Beclin 1-interacting protein), AKT, and EGFR, are negative regulators of the Beclin 1/VPS34 autophagy-promoting complex (Matsunaga et al., 2009; Pattingre et al., 2005; Wang et al., 2012; Wei et al., 2013; Zhong et al., 2009). Overall, these studies indicate that multiple class III PI3K complexes exist concurrently within the cell, suggesting that these proteins can exquisitely tune the level of autophagy. Notably, several proteins in this complex have tumor-suppressive or antiproliferative effects, which are discussed in detail in the succeeding text.

2.3. Elongation and the ATG12/ATG8 conjugation systems

The elongation of the phagophore membrane requires two ubiquitin-like conjugation systems. In the first, ATG7 and ATG10 (E1- and E2-like enzymes, respectively) conjugate ATG12 to ATG5. The ATG5–ATG12 complex binds ATG16 and forms a large multimeric complex called the ATG16L complex, which is essential for the elongation of the nascent phagophore (Fig. 2.1C). The second conjugation system involves cleavage of the ubiquitin-like molecule, ATG8, by the protease ATG4 to expose a C-terminal glycine residue required for subsequent activation and conjugation reactions. Several mammalian orthologues of yeast Atg8 have been identified, of which the best characterized is microtubule-associated protein 1 light chain 3 (LC3) alpha (MAP1LC3A) (Weidberg, Shpilka, Shvets, Shinder, & Elazar, 2010). Atg4 also has four mammalian isoforms, although the specificities are entirely known (Li et al., 2011; Shpilka, Weidberg, Pietrokovski, & Elazar, 2011). Ultimately, LC3 is conjugated to the lipid phosphatidylethanolamine (PE) via ATG7 and E2-like ATG3 and is subsequently recruited to both the outer and inner surfaces of the autophagosomal membrane (Fig. 2.1C). LC3 and other Atg8 family members can mediate membrane tethering and hemifusion, which may be important in the fusion of the ends of the phagophore membrane into a closed autophagosome (Weidberg et al., 2011).

In addition, LC3 is an important mediator for selectively targeting cargo for autophagic degradation. Several ubiquitin-binding proteins have been identified as cargo receptors for autophagy substrates (Johansen & Lamark, 2011), including p62/SQSTM1 (Bjørkøy et al., 2005), NBR1 (Kirkin et al., 2009), NDP52 (Thurston, Ryzhakov, Bloor, von Muhlinen, & Randow, 2009), and OPTN (Wild et al., 2011). These cargo receptors contain a well-conserved linear amino acid motif called the LIR (LC3-interacting region) that is necessary for specific targeting to the autophagosome. Interestingly, the LIR consensus sequence has been identified in a number of proteins, suggesting that the repertoire of LC3-interacting proteins acting as cargo receptors for selective autophagy may be expansive. In support of this, a large-scale proteomic study demonstrated that the mammalian ATG8 family has 67 high-confidence interactions with other cellular proteins (Berhends, Sowa, Gygi, & Harper, 2010).

2.4. Fusion

After an autophagosome forms, it fuses with the endosome or lysosome where the engulfed components may be recycled (Fig. 2.1E). Autophagosomes travel along microtubules, pushed by dynein, to lysosomes. Fusion requires ESCRT, SNAREs—specifically syntaxin 17 (Itakura, Kishi-Itakura, & Mizushima, 2012)—VPS family proteins, and RAB7. Fusion to the lysosome is the last step in the degradation of the intracompartmental components, and impaired lysosome function prevents complete autophagic flux. Hence, lysosomotrophic agents such as hydroxychloroquine (HCQ) are used experimentally to inhibit autophagy. These lysosomal inhibitors are proposed to impair autophagosome maturation and flux by altering the pH of the lysosome; nonetheless, it is important to recognize that these compounds impact a broad array of processes other than autophagy.

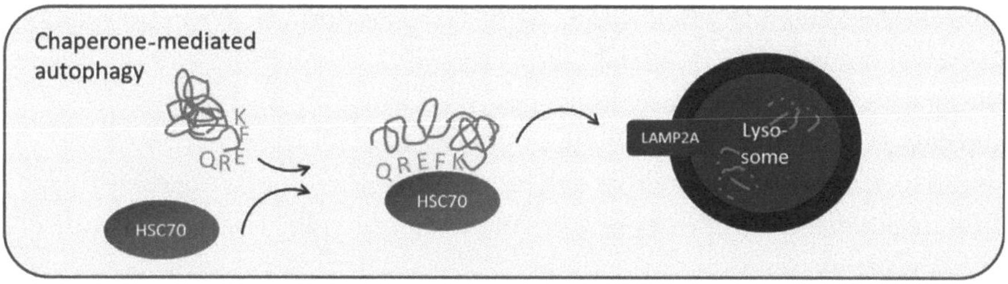

Figure 2.2 *Chaperone-mediated autophagy*. Chaperone-mediated autophagy is an additional route by which proteins are degraded in the lysosome. HSC70 binds to proteins with a KFERQ pentapeptide motif, assists in protein unfolding, and delivers the targeted protein directly to the lysosome for degradation via interaction with lysosome-associated membrane protein 2A (LAMP2A).

2.5. Chaperone-mediated autophagy

Although this chapter principally focuses on macroautophagy, it is important to recognize that multiple routes of autophagic degradation exist, including microautophagy and chaperone-mediated autophagy (CMA) (Mizushima, 2007). CMA warrants special attention because of its emerging role in cancer (Kon et al., 2011; Lv et al., 2011; Vakifahmetoglu-Norberg et al., 2013). CMA is a highly selective form of autophagy in which specific proteins are targeted to the lysosome via their interaction with a cytosolic chaperone protein—HSC70—that recognizes and binds to a specific pentapeptide motif, the KFERQ sequence. This interaction leads to binding to the lysosome via a variant of the lysosome-associated membrane protein 2A (LAMP2A), and after some unfolding, the targeted protein is directly delivered into the lysosome for degradation (Fig. 2.2; Cuervo, Terleckyh, Dice, & Knecht, 1994; Dice, Chiang, Spencer, & Backer, 1986; Koga, Martinez-vicente, Macian, Verkhusha, & Cuervo, 2011). Interestingly, CMA can be induced in mammalian cells when macroautophagy is inhibited and vice versa, indicating that a switch in one type of autophagy can compensate for a deficiency in the other (Massey, Kaushik, Sovak, Kiffin, & Cuervo, 2006; Wang et al., 2008).

3. METABOLIC STIMULI REGULATING AUTOPHAGY

Metabolic stresses often occur in solid tumors and the tumor microenvironment—rapidly multiplying tumor cells and tumors that have yet to initiate angiogenic programs and often cannot maintain nutrient supply and quickly become hypoxic. To forestall senescence or death, tumor cells metabolically reprogram and engage autophagy to survive in the hostile tumor microenvironment (DeBerardinis, Lum, Hatzivassiliou, & Thompson, 2008; Lozy & Karantza, 2012). Metabolites, oxygen concentration, and oncogenes all regulate the initiation of autophagosome formation, and the regulation of autophagy is finely balanced by the integration of all these signals (Fig. 2.3). In this section, we provide an overview of the regulation of autophagy by specific metabolites and metabolic stressors in tumor cells, focusing on cancer-relevant pathways.

3.1. Nutrient starvation

Autophagy is strongly induced in response to nutrient starvation, which is primarily controlled by mammalian target of rapamycin (mTOR). mTOR

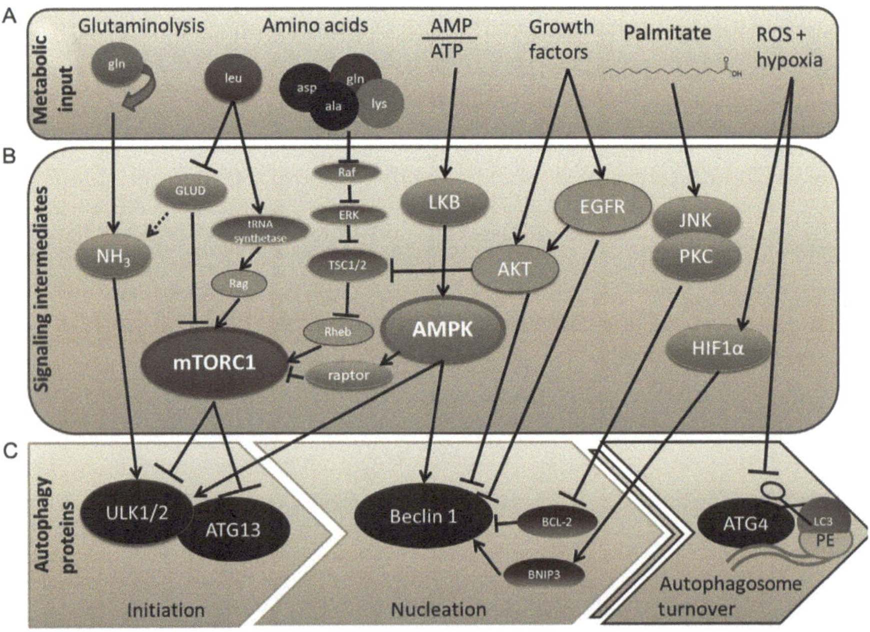

Figure 2.3 *AMPK and mTORC1 as metabolic regulators of autophagy.* The control of autophagosome formation and turnover is tightly controlled by many upstream metabolic stimuli. Metabolic input (A), such as concentrations of ammonia, general and specific amino acid levels, ATP to ADP ratio, and signals of growth and stress such as growth factors, reactive oxygen species, and palmitate, signals to initiate autophagosome formation and inhibits turnover either through signaling intermediates (B) or by directly inhibiting or activating key autophagy-related proteins (C). AMPK1 and mTOR are the main signaling integrators and modulators of autophagy—they sense glucose and amino acid levels and act on ULK1/2, ATG13, and Beclin 1 to inhibit autophagy in times of plenty and promote autophagy under energy-lean circumstances. (See the color plate.)

was initially identified as a key negative regulator of autophagy in yeast and has been confirmed to function as a major regulator of mammalian autophagy (Kroemer, Mariño, & Levine, 2010). mTOR acts as a master sensor of metabolic state; signals from growth factors, amino acids, oxidative stress, and DNA damage alter mTOR interactions with binding partners, thereby regulating mTOR activity. Active mTORC1 under nutrient conditions modulates the rates of translation, lipid synthesis, and mitochondrial proliferation and phosphorylates ULK1/2 and ATG13 to block autophagy. Under nutrient deprivation, ATG13 and ULK1/2 are dephosphorylated by an unknown phosphatase, leading to autophagosome formation (Jung et al., 2009; Jung, Ro, Cao, Otto, & Kim, 2010; Neufeld, 2010; Zoncu, Efeyan, & Sabatini, 2011).

3.2. Glucose

As noted by Otto Warburg in 1924, cancer cells preferentially utilize glycolysis over oxidative phosphorylation as a source of energy in aerobic conditions. Glycolysis is thought to provide a growth advantage by maintaining intracellular pools of metabolites for anabolism (Vander Heiden, Cantley, & Thompson, 2009). Therefore, cancer cells are more sensitive to low levels of glucose than nontransformed cells. Low glucose levels induce autophagy in a wide variety of mammalian cell types, and this regulation appears to be partially dependent on the activation of AMPK (Williams, Forsberg, Viollet, & Brenman, 2009). AMPK is activated by a high ratio of AMP to ATP (Kroemer et al., 2010). Under conditions of low intracellular energy, activated AMPK induces autophagy both by phosphorylating ULK1, resulting in its activation, and by inhibiting mTORC1 via phosphorylation of Raptor (Kim, Kundu, Viollet, & Guan, 2011; Mihaylova & Shaw, 2011). During glucose deprivation, AMPK-dependent Beclin 1 phosphorylation activates the proautophagy Beclin 1/VPS34 complex (Kim et al., 2013). However, the balance of nutrient availability is crucial for autophagy induction, especially since autophagy is an ATP-consuming process. Under starvation conditions, the addition of glucose (up to a threshold) promotes autophagy via a p38 MAPK-dependent pathway (Moruno-Manchón, Pérez-Jiménez, & Knecht, 2013).

3.3. Amino acids

Autophagy is inhibited in an mTORC1-dependent manner based on the levels of amino acids in the cytoplasm. Amino acids activate Rag GTPases, which promote translocation of mTORC1 to the lysosomal surface, resulting in mTORC1 activation and inhibition of autophagy via ULK1/2. Intralysosomal amino acid levels also regulate mTORC1 activity in a vacuolar ATPase-dependent manner, which may function as a means of feedback inhibition of the autophagic process (Sancak et al., 2008, 2010; Zoncu, Bar-Peled, et al., 2011). Amino acid levels also alter the signaling of the RAS/RAF1/ERK1/2 pathway, which regulates autophagy induction. High amino acid levels inhibit the activation of RAF1, which prevents ERK1/2-dependent phosphorylation of Gα-interacting protein, resulting in decreased stimulus-induced autophagy in HT-29 intestinal cells (Gozuacik & Kimchi, 2004; Ogier-Denis, Pattingre, El Benna, & Codogno, 2000; Pattingre, Bauvy, & Codogno, 2003). Specific amino acids also have distinct effects on autophagy inhibition. Leucine has the strongest inhibitory effect on autophagy. Leucyl-tRNA synthetase, an intracellular leucine sensor, binds to and regulates Rag GTPase interaction with mTORC1, leading to autophagy inhibition (Han et al., 2012).

3.4. Glutamine

When glucose levels are low, cells commonly shift to glutaminolysis to maintain tricarboxylic acid (TCA) cycle ATP and NADPH production. Ammonia produced during glutaminolysis increases autophagic flux by an mTORC1-independent pathway (Cheong, Lindsten, & Thompson, 2012; Eng, Yu, Lucas, White, & Abraham, 2010). Moreover, leucine levels regulate glutamate dehydrogenase 1 (GLUD1) activity that promotes autophagy by inhibiting mTORC1 activity and modulating reactive oxygen species (ROS) levels (Lorin et al., 2013). Accordingly, the production of ammonia by GLUD1-mediated oxidative deamination of glutamate to alpha ketoglutarate may also regulate autophagy in a similar fashion to ammonia generated from glutaminolysis; however, this intriguing hypothesis requires further testing. Moreover, it is important to recognize that glutaminolysis may not always promote autophagy. Indeed, glutamine and leucine together have been reported to activate mTORC1 and therefore inhibit autophagy in a glutaminolysis-dependent manner (Durán et al., 2012), indicating that the regulation of autophagy by glutamine is sensitive to metabolic context. Glutamine depletion was reported to decrease mRNA levels of *Atg5* in wild-type MEFs, supporting the finding that glutaminolysis may promote autophagy (Lin et al., 2012).

3.5. Lipids and free fatty acids

In cancer cells, the impact of altered lipid metabolism on autophagy regulation is not as well defined as that of glucose and glutamine metabolism. Fatty acid synthesis is generally restricted to specific tissues, but is often upregulated in cancers (Santos & Schulze, 2012). Palmitate, the simplest and most abundant fatty acid and the product in fatty acid synthesis, stimulates autophagy in the muscle, liver, neurons, and pancreatic cells. Palmitate-induced autophagy is mediated by JNK1 activity and PKC activity and is independent of mTOR (Komiya et al., 2010; Martino et al., 2012; Tan et al., 2012). However, the induction of autophagy may not increase autophagic flux in pancreatic cells (Las, Serada, Wikstrom, Twig, & Shirihai, 2011), although there are conflicting data about the turnover of long-lived proteins. Further experiments, such as using GFP and mCherry-tagged LC3, will clarify this point. Additionally, the autophagic response to fatty acids may be highly tissue-specific. In hepatocytes, palmitate was found to promote apoptosis instead of autophagy, while oleate—the most abundant monounsaturated fatty acid—was found to promote autophagy via increasing ROS levels (Mei et al., 2011). The synthetic fatty acid 2-hydroxyoleic acid induced ER stress and autophagy in glioma cell lines but not a control fibroblast cell line and resulted in glioma cell differentiation (Marcilla-Etxenike et al., 2012; Terés et al., 2012). How 2-hydroxyoleic acid induces ER stress remains unknown. However, it has been shown that excess lipid storage in nonadipose tissue can cause ER stress, which increases autophagy via MTOR, JNK, and increased transcription of autophagy genes (B'chir et al., 2013; Ogata et al., 2006; Qin, Wang, Tao, & Wang, 2010; Tomohiro & Klionsky, 2007).

3.6. Hypoxia and ROS

Hypoxia and ROS, often found in the poorly vascularized tumor microenvironment, have been shown to increase autophagic flux via several mechanisms. Most directly, ROS inhibit ATG4 autophagosome turnover activity, allowing for the maintenance of lipidated LC3 necessary for autophagosome formation (Scherz-Shouval et al., 2007). Autophagy is upregulated during hypoxia via hypoxia-inducible factor 1α (HIF1α) induction of BNIP3 and BNIP3L, which binds to Beclin 1 to promote autophagy (Bellot et al., 2009). AMPK promotes autophagy independently of HIF in response to severe hypoxia. While BNIP3-regulated autophagy protects cells from death, AMPK-induced autophagy promotes cell death, pointing to the influence of cellular context on the outcome of autophagy (Papandreou, Lim, Laderoute, & Denko, 2008).

ROS-mediated damage also likely controls autophagy. ROS damages DNA, proteins, and organelles (Wellen & Thompson, 2010), and accumulated damage and subsequent metabolic stress activate autophagic programs. In addition to starvation, JNK-mediated autophagy induction is often associated with oxidative stress (Eisenberg-Lerner & Kimchi, 2007; Kang, Zeh, Lotze, & Tang, 2011). Stress-activated JNK results in the phosphorylation of BCL-2, an antiapoptotic protein that binds to and inhibits Beclin 1 (Pattingre et al., 2005), causing the release of Beclin 1 and autophagy induction (Wei, Pattingre, Sinha, Bassik, & Levine, 2008). In another indirect mechanism, low-oxygen concentrations lead to the acidification of the environment, and autophagy is upregulated in response to low pH independent of oxygen concentration (Wojtkowiak et al., 2012).

4. AUTOPHAGY AND TUMOR SUPPRESSION

Scientific evidence supports both tumor-promoting and tumor-suppressive functions for autophagy and the exact role of autophagy during cancer progression depends on tumor type, context, and stage. Here, we discuss the genetic evidence supporting the role of autophagy-related proteins (ATGs) as tumor suppressors and review the potential mechanisms through which autophagy impairs tumor initiation and progression (Fig. 2.4).

4.1. ATGs as tumor suppressors

Genetic evidence that autophagy can prevent tumor formation first emerged through studies of *beclin 1* (Liang et al., 1999), which was found to be monoallelically deleted in 40–75% of cases of sporadic human breast, ovarian, and prostate cancer. Furthermore, mice lacking a single copy of *beclin 1* developed spontaneous lymphoma, hepatocellular carcinoma, and lung adenocarcinomas (Qu et al., 2003; Yue, Jin, Yang, Levine, & Heintz, 2003). Notably, the second allele of *beclin 1* was not lost in these tumors, further corroborating that *beclin 1* functioned as a haploinsufficient tumor suppressor. In addition, multiple Beclin 1-interacting partners have been implicated as tumor suppressors. UVRAG, a Beclin 1-interacting protein that positively regulates autophagy, is allelically deleted in human colon carcinoma (Liang et al., 2006; Liang, Feng, Ku, & Oh, 2007). Moreover, frameshift mutations in the polyadenine tract of the UVRAG gene resulting in decreased

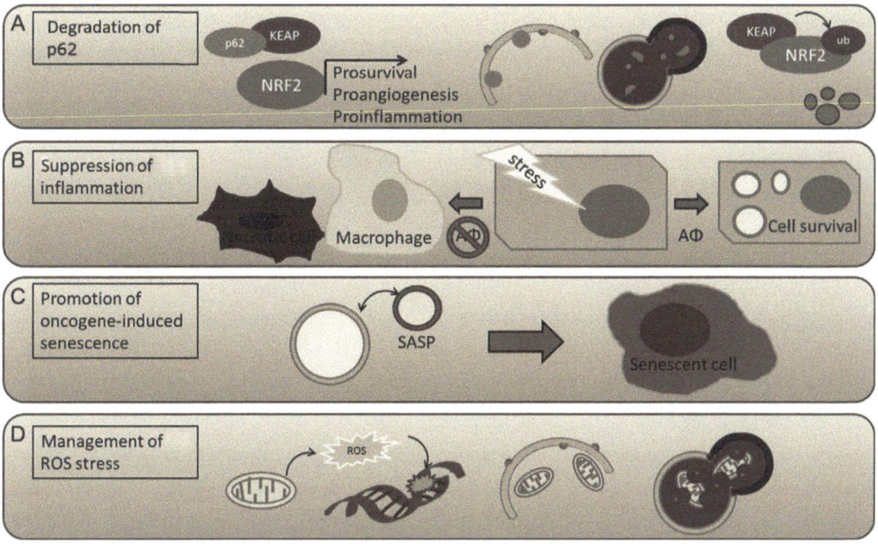

Figure 2.4 *Tumor-suppressive roles of autophagy.* (A) Autophagy prevents p62/NRF2 pathway prosurvival, proangiogenic, and proinflammatory signaling. During hypoxia, p62 binds to and sequesters KEAP, thereby preventing ubiquitination of NRF2. NRF2 can then promote transcription of prosurvival, proangiogenic, and proinflammatory genes that enhance tumor growth. When autophagy is active, p62 is degraded by sequestration within the autophagolysosome, thereby allowing ubiquitination and degradation of NRF2. (B) Autophagy prevents necrosis and inflammation by promoting survival in stressed cells. Tumor cells with decreased autophagy (AΦ) are more prone to necrosis following stress, which recruits macrophages, promotes inflammation, and fuels tumor growth. (C) Autophagy promotes oncogene-induced senescence by enhancing the senescence-associated secretory phenotype (SASP). (D) Autophagy suppresses reactive oxygen species (ROS) accumulation and genomic damage, which helps prevent genomic instability—an important driver of tumorigenesis. The mitochondria produce ROS under normal metabolic conditions and may increase the production of ROS when damaged. Mitophagy is upregulated in response to ROS and clears excess and damaged mitochondria, which then mitigates ROS production. (See the color plate.)

autophagy are present in gastric carcinomas (Kim et al., 2008). Mice lacking SH3GLB1/Bif-1, which interacts with Beclin 1 via UVRAG, exhibit a significantly higher rate of spontaneous tumors (Takahashi et al., 2007), and reduced SH3GLB1/Bif-1 expression, which correlates with decreased autophagy, is observed in gastric carcinoma (Lee et al., 2006). Besides the well-characterized oncoprotein BCL-2 interaction with Beclin 1, two other oncoproteins have been more recently shown to interact with Beclin 1 leading to autophagy suppression and oncogenesis. AKT-mediated Beclin 1 serine phosphorylation enhances its interaction with vimentin and decreases autophagy. Depletion of vimentin or expression of a nonphosphorylatable Beclin 1 mutant in AKT-overexpressing cells increases autophagy and inhibits transformation, supporting the hypothesis that autophagy suppresses tumor initiation in AKT-driven tumors (Wang et al., 2012). EGFR-mediated Beclin 1 tyrosine phosphorylation suppresses the formation of the proautophagy Beclin 1/VPS34 complex, which may contribute to tumor progression and chemoresistance in human non-small cell lung cancer xenografts harboring oncogenic EGFR mutations (Wei et al., 2013).

In addition to Beclin 1 and its associated proteins, other ATGs have been implicated as suppressors of spontaneous tumorigenesis. Mice with systemic mosaic deletion of *Atg5* and liver-specific *Atg7*$^{-/-}$ mice develop liver adenomas (Inami et al., 2011; Takamura et al., 2011). *Atg4C* knockout mice exhibit increased susceptibility to fibrosarcomas in a chemical carcinogen model (Mariño et al., 2007). Mice with hematopoietic stem cell deletion of *Atg7* develop an atypical myeloproliferation resembling human myelodysplastic syndrome and acute myeloid leukemia (Mortensen et al., 2011). Frameshift mutations in *ATG2B*, *ATG5*, and *ATG9B* have been reported in gastric and colorectal carcinomas, further suggesting that the components of the core autophagic machinery act as tumor suppressors in human cancers (Kang et al., 2009).

4.2. Autophagy-dependent degradation of p62/SQSTM1

The accumulation of p62/SQSTM1, an autophagy cargo receptor, promotes tumorigenesis: Liver tumor size is reduced in *Atg7*$^{-/-}$ mice by simultaneous *p62* deletion (Takamura et al., 2011), *p62* gene targeting reduces anchorage-independent growth of human hepatocellular carcinoma cells (Inami et al., 2011), *p62*$^{-/-}$ mice fail to develop RAS-induced lung carcinomas (Duran et al., 2008), and *p62*-null cells have impaired RAS transformation (Guo et al., 2011). In KRAS-driven tumor cells, p62 activates Nrf2 and NF-κB, which stimulate proangiogenic and proinflammatory responses, respectively, thereby contributing to aggressive tumor progression. Thus, increased autophagy enhances p62 degradation, leading to diminished angiogenic and inflammatory responses (Duran et al., 2008; Kim, Hur, et al., 2011; Mathew et al., 2009).

p62/SQSTM1 activation of the Nrf2 pathway in autophagy-deficient cells is especially important in tumor progression (Komatsu et al., 2010). Notably, the Nrf2 pathway, due to inactivating somatic mutations in the E3 ubiquitin ligase Keap1, has been implicated as a survival pathway in non-small cell lung carcinomas (Singh et al., 2006). The transcription factor Nrf2 (nuclear regulatory factor 2) regulates the expression of a wide range of genes that promote angiogenesis and facilitate cell survival. Keap1 ubiquitinates Nrf2 resulting in its degradation under normal conditions. Accumulated p62/SQSTM1 in autophagy-deficient cells directly binds to Keap1, disrupting Keap1-mediated degradation of Nrf2 and promoting aberrant Nrf2-mediated transcription (Komatsu et al., 2010). Thus, aberrant regulation of Nrf2 in autophagy-deficient cells may be an important pathway in tumor cell survival (Fig. 2.3A). Indeed, this pathway has been implicated in the spontaneous tumorigenesis of autophagy-defective liver cells (Inami et al., 2011; Takamura et al., 2011) and in the early growth acceleration of BRAF-driven lung cancers lacking Atg7 (Strohecker et al., 2013).

4.3. Autophagy prevents protumor inflammation and facilitates senescence

Because autophagy promotes tumor cell adaptation and survival during hypoxic and metabolic stress, it may suppress tumor progression by inhibiting necrosis. In solid tumors, necrotic cell death causes macrophage infiltration and proinflammatory cytokine production, and chronic inflammation generally favors cancer growth and progression (DeNardo, Johansson, & Coussens, 2008). Thus, by limiting necrosis, autophagy may actually suppress tumor growth by preventing leukocyte infiltration of the primary tumor site (Fig. 2.3B). Indeed, this ability of autophagy to restrict necrosis prevented macrophage-associated tumor inflammation and inhibited primary tumor growth in apoptosis-resistant cells (Degenhardt et al., 2006). Additionally, autophagy can facilitate the transition to senescence (Fig. 2.3C), which also prevents immune activation due to necrosis, and can lead to the elimination of premalignant cells by senescence-mediated surveillance (Kang, Yevsa, et al., 2011; Narita et al., 2011; Young et al., 2009). Autophagy allows the cancer cells to quietly survive but helps to restrict proliferation by facilitating senescence, thereby overall suppressing tumor growth.

4.4. Autophagy clears dysfunctional mitochondria

Autophagy is an important mechanism for the clearance of damaged mitochondria, a process termed mitophagy. Mitochondrial number may indirectly regulate tumor progression as the mitochondria produce ROS, which can promote tumor progression via damage to proteins or DNA causing chromosomal instability (Ishikawa et al., 2008). In response to ROS, mitophagy is upregulated to remove excess mitochondria and mitigate ROS production (Fig. 2.3D). Increased ROS production from increased metabolic rate can damage the mitochondria, which in turn can increase metabolic stress in the cell. Accordingly, in autophagy-defective cells, metabolic stress induces more DNA damage, increased genomic instability, and increased accumulation of damaged mitochondria than in wild-type control cells (Belaid et al., 2013; Mathew et al., 2009). By clearing damaged mitochondria and controlling intracellular ROS levels, autophagy may exert a tumor suppressor function.

5. TUMOR-PROMOTING FUNCTIONS OF AUTOPHAGY

Although reduced autophagy can promote tumor development, autophagy provides cancer cells with certain selective advantages to cope with stress and promote metabolic adaptation. Hence, a basal level of autophagy appears to be necessary for the optimal survival and fitness of cancer cells. The following section provides an overview of several potential mechanisms by which autophagy may promote tumor progression (Fig. 2.5).

5.1. Autophagy and metabolic adaptation in cancer

5.1.1 Autophagy and oxidative mitochondrial metabolism

Strong oncogenic insults like RAS activation lead to increased autophagy. In pancreatic ductal adenocarcinoma (PDAC), where activating KRAS mutations are present in greater than 90% of tumors, elevated autophagy is found in both primary PDAC tumors and cell lines. Genetic inhibition of autophagy in PDAC cells potently suppresses proliferation *in vitro* and elicits robust tumor regression and prolonged survival in pancreatic cancer xenografts and genetic mouse models (Yang et al., 2011). Because RAS activation is marked by profound metabolic alterations that promote energy production and support the biosynthesis of macromolecules needed for rapid proliferation, it has been hypothesized that autophagy maintains key metabolic pathways in RAS-transformed cells. In support, the loss of autophagy during RAS transformation is associated with reduced mitochondrial oxygen consumption and decreased levels of TCA cycle intermediates (Guo et al., 2011; Yang et al., 2011). This requirement for autophagy to maintain oxidative mitochondrial metabolism of RAS-transformed cells indicates that the protumor effects of autophagy are not limited to survival functions in

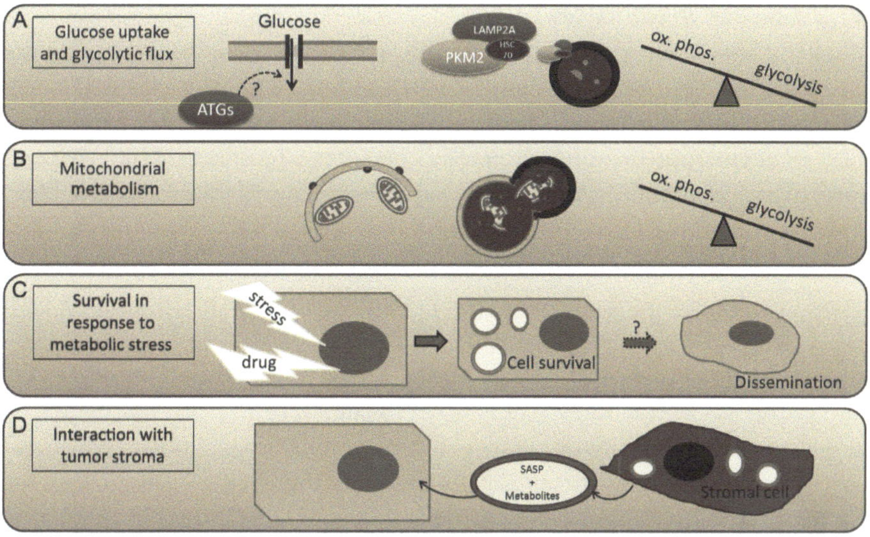

Figure 2.5 *Tumor-promoting roles of autophagy.* (A) Autophagy promotes glucose uptake and glycolytic flux. Autophagy has been shown to promote glucose uptake in cancer cells, although the mechanism remains to be elucidated. Additionally, increased chaperone-mediated autophagy (CMA) promotes degradation of PKM2, a rate-limiting glycolytic enzyme. Thus, CMA can control the rate of flux through glycolysis and whether glycolytic intermediates are used for energy production or anaplerosis. (B) Autophagy selectively degrades the mitochondria and therefore the machinery necessary for fatty acid oxidation and oxidative phosphorylation. This enhances the shift to glycolysis, which is characteristic of cancer cells. (C) Autophagy promotes survival in response to metabolic stress such as growth factor deprivation, acidic environment, and ER stress by recycling cytoplasmic material in order to maintain the basal energy state and clear damaged, misfolded proteins. This process may also be important for survival during tumor dissemination and metastasis. (D) Autophagy in stromal cells induced by the hypoxic and acidic tumor microenvironment promotes the secretion of metabolites and growth signals via senescence-associated secretion phenotype (SASP) that enhances tumor cell growth. (See the color plate.)

response to external stresses. Rather, autophagy contributes to the metabolic fitness of the entire tumor population. Remarkably, this requirement for autophagy may be oncogene-dependent, as autophagy has been demonstrated to restrict, rather than promote, proliferation driven by oncogenic PI3K in a three-dimensional mammary culture model (Chen, Eritja, Lock, & Debnath, 2013). As RAS is one of the few oncogenes that stimulate—rather than suppress—autophagy, it will be interesting to determine whether this requirement for autophagy is conserved in other oncogenic contexts.

5.1.2 Glucose metabolism

Many tumors preferentially use aerobic glycolysis, which allows for the accumulation of metabolic intermediates required for anabolism (Hsu & Sabatini, 2008). CMA and selective macroautophagy both play important roles in regulating the shift to aerobic glycolysis in cancer cells. CMA is upregulated in diverse tumor types and is necessary for tumor growth and metastasis in lung cancer cells, and inhibition of CMA decreases the rate of glycolysis characteristic of tumor growth (Kon et al., 2011). More specifically, CMA controls the levels of the metabolic enzyme PKM2 (Fig. 2.4A), which is often upregulated in many tumor types and particularly glioblastoma. The PKM2 isoform of pyruvate kinase is slower at metabolically converting phosphoenolpyruvate to pyruvate than the M1 isoform; this causes glycolytic intermediates to accumulate and drives tumor cell proliferation and growth by promoting key biosynthetic side reactions in the glycolytic pathway. CMA can selectively degrade PKM2, thereby regulating the levels of the metabolic intermediates, glucose-6-phosphate and fructose-1,6-bisphosphate, and the levels of ATP (Lv et al., 2011). Recently, PKM2-specific deletion was shown to have increased mammary tumor formation driven by *Brca-1* deletion (Israelsen et al., 2013), consistent with the notion that cancer cells prefer low pyruvate kinase activity. Therefore, the degradation of PKM2 by CMA may promote tumor progression.

The number of mitochondria present also regulates the shift to anaerobic metabolism. BRAF-driven melanoma cells decrease the rate of mitochondrial biogenesis in order to shift from oxidative phosphorylation to glycolysis (Haq et al., 2013; Ho et al., 2012; Vazquez et al., 2013). If mitophagy is aberrantly activated, decreased numbers of mitochondria shift the cells to glycolysis in a similar mechanism to BRAF regulation of mitochondrial biogenesis (Fig. 2.4B). RCAN1-1L, whose expression is increased in response to oxidative stress, can open the MPT pore and decrease ATP levels. This inhibits mTOR signaling via AMPK, resulting in increased mitophagy and a shift to glycolysis (Ermak et al., 2012). In addition to shifting the metabolic pathways to preferentially use glucose, autophagy also facilitates glucose uptake (Fig. 2.4A) and glycolytic flux in RAS-transformed cells, which is required for adhesion-independent proliferation (Lock & Debnath, 2011; Lock et al., 2011).

5.1.3 Amino acids

In addition to glucose, amino acids are necessary for cancer cell growth. In yeast, autophagic breakdown of proteins during starvation generates cytosolic amino acid pools crucial for survival (Onodera & Ohsumi, 2005). Amino acids feed into cataplerotic pathways and can be used to maintain biosynthetic capacity in rapidly dividing cancer cells. Glutamine, the most abundant amino acid in mammalian cells, is important in cancer progression as a metabolic intermediate (DeBerardinis et al., 2008; Gaglio et al., 2011). As glycolytic rates increase, tumor cells rely increasingly on glutamine to replenish the TCA cycle and maintain ATP production (Burgess, 2013). In pancreatic cancer, glutamine feeds into glutaminolysis, utilizing steps in the TCA cycle to generate NADPH, maintain the cellular redox state, and provide metabolites for anaplerosis (Son et al., 2013). In wild-type MEFs, loss of autophagy was also found to decrease the levels of intracellular glutamine and also mimic the metabolic changes associated with glutamine depletion, indicating that autophagy normally helps to maintain intracellular stores of glutamine. However, in the same study, glutamine deprivation did not increase the levels of autophagy, and the *Atg5* mRNA level decreased (Lin et al., 2012). Therefore, how autophagy may increase specific amino acids during deprivation remains to be defined.

5.1.4 Lipids

Lipid metabolism is altered in cancer—tumor cells reactivate *de novo* lipid synthesis, ATP-citrate lyase is required for transformation *in vitro*, cholesterol synthesis in prostate cancer is increased, and fatty acid oxidation is an important source of energy for prostate cancer cells (Santos & Schulze, 2012). Autophagy in the specific form of lipophagy is important for the degradation of lipid droplets in the adipose tissue (Singh & Cuervo, 2012), and autophagy regulates lipid metabolism in hepatocytes as triglyceride hydrolysis is impaired in $Atg5^{-/-}$ cells (Singh et al., 2009). Whether these processes affect tumor lipid metabolism requires further study.

Additionally, autophagy impacts lipid metabolism by altering the mitochondrial number. *Atg7* deleted, p53 mutant cells in a KRAS-driven NSCLC model have intracellular lipid accumulation because of increased dysfunctional mitochondria that compromises fatty acid oxidation, suggesting that autophagy is crucial to maintain lipid metabolism in KRAS and p53 mutant cells. This prevents the efficient growth of tumor cells and turns them into lipid cysts instead of tumors (Guo et al., 2013).

5.2. Autophagy promotes cell survival under metabolic stress

As discussed earlier in the text, autophagy is strongly activated under periods of oxidative and metabolic stress, and depending on the extent and severity of stress, autophagy serves to prolong cell survival in the primary tumor and possibly also during tumor dissemination and metastasis (Fig. 2.4C). In melanoma cells driven by oncogenic Ras or MEK, the removal of leucine does not induce autophagy to the same extent as nontransformed, immortalized melanocytes. The aberrant activation of mTOR via Ras prevents autophagy induction and the cells are sensitized to apoptosis, presumably because translation continues although the lost leucine is not recycled intracellularly (Sheen, Zoncu, Kim, & Sabatini, 2011). Following growth factor withdrawal, autophagy is essential for maintaining cell survival in apoptosis-deficient hematopoietic cells and can sustain viability for several weeks. IL3-deprived cells become less glycolytic and use autophagy as a catabolic process to maintain mitochondrial respiration and levels of ATP (Lum et al., 2005). Increased autophagy regulated by the PI3K/AKT/mTOR pathway prolongs cancer cell survival under acidic environment stress produced by glycolysis (Wojtkowiak et al., 2012). Autophagy also prevents ER stress-induced cell death during protein overproduction (e.g., induced by oncogenes such as Myc) by clearing excess and misfolded proteins (Tomohiro & Klionsky, 2007). Indeed, Myc-driven tumors have increased cell growth, ER stress, and metabolic rate, and autophagy inhibition enhances therapy-induced apoptosis in a Myc-driven model of lymphoma (Amaravadi, Yu, & Lum, 2007; Dang, 1999; Miller, Thomas, Islam, Muench, & Sedoris, 2012).

5.3. Autophagy in the tumor stroma

Autophagy prolongs tumor cell survival under stressful conditions. It should be noted that the acidic, hypoxic, or nutrient-starved environment also induces autophagy in the surrounding stromal cells, which promotes tumor growth (Fig. 2.4D). Serum-deprived mesenchymal stem cells induce autophagy and support MCF7 growth in xenograft models by secreting growth factors and antiapoptotic factors (Sanchez et al., 2011). While autophagy-induced senescence in cancer cells limits growth, autophagy-induced senescence in the tumor stroma may promote cancer by enhancing the senescence-associated secretory phenotype (SASP) and promoting the secretion of growth factors and cytokines that enhance tumor progression (Capparelli, Chiavarina, et al., 2012; Capparelli, Guido, et al., 2012; Capparelli, Whitaker-Menezes, et al., 2012; Maes, Rubio, Garg, & Agostinis, 2013).

In addition to modulating secretion in senescent fibroblasts, autophagy in cancer-associated fibroblasts (CAFs) may directly fuel cancer cell metabolism. Autophagic senescent CAFs release metabolites such as glutamine, ketone bodies, and glycolytic intermediates that may promote tumor growth and metastasis. These studies raise the possibility that autophagy in the tumor stroma is important for the continued growth of the tumor (Ko et al., 2011; Martinez-Outschoorn et al., 2010; Salem et al., 2012).

5.4. Autophagy inhibition in cancer therapy

The increased dependence of tumors on altered metabolism is an attractive therapeutic target. In addition to targeting metabolic enzymes, targeting autophagy may provide a similar benefit. However, such an approach is complicated by the multifaceted role of autophagy in tumor formation and progression (Cheong, Lu, Lindsten, & Thompson, 2012). Increased autophagy has been observed in tumor cells following numerous anticancer treatments and is proposed to represent a common adaptive stress response that enables tumor cells to survive these therapeutic insults (Fig. 2.4C). This has motivated significant interest in combining autophagy inhibition with other agents to synergistically eliminate cancer cells. Readers are referred to several reviews for additional information (Amaravadi et al., 2011; Eisenberg-Lerner & Kimchi, 2009; Høyer-hansen & Jäättelä, 2008).

Notably, certain targeted therapies that enhance autophagy *in vitro* may benefit from combined autophagy inhibition. Autophagy is upregulated in response to erlotinib in NSCLC cell lines and combined treatment with chloroquine, an antimalarial that inhibits autophagy, enhances erlotinib sensitivity (Li, Lam, Mak, Zheng, & Ho, 2013). Similarly, gastrointestinal stromal tumors exhibit enhanced autophagy in response to imatinib, which lessens the therapeutic benefit. Combined inhibition of autophagy with imatinib treatment increased the number of cells undergoing apoptosis, both *in vitro* and *in vivo*, and reduced the outgrowth of resistant cells (Gupta et al., 2010). Moreover, upon treatment with the VEGF-neutralizing antibody bevacizumab, increased autophagy due to hypoxia promotes tumor cell survival and resistance to this antiangiogenic therapy (Hu et al., 2012). In contrast, inhibition of erlotinib-induced autophagy in human NSCLC xenografts *in vivo* by inducible expression of a Beclin 1 tyrosine phosphomimetic mutant resulted in partial chemoresistance (Wei et al., 2013), suggesting that the effects of autophagy inhibition may vary depending upon the autophagy step targeted, *in vitro* versus *in vivo* studies, or due to other differences in tumor type or experimental systems.

In the previously mentioned examples, autophagy is targeted due to its induction in response to therapy, but autophagy inhibition can also synergize with therapies that do not normally promote autophagic flux. For example, combining autophagy inhibition with immunotherapy could increase efficacy. Hypoxia-induced autophagy prevents lung cancer cells from cytolytic T-cell mediated cell death, but inhibition of autophagy combined with immunotherapy may provide a powerful and tumor-specific therapy (Noman et al., 2011). Another synergistic approach involves targeting the proteasome pathway and autophagy in tumor cells that are prone to ER stress. Autophagy inhibitors in combination with proteasome inhibitors increase suppression of proliferation and induce apoptosis in hepatocellular carcinoma (Hui et al., 2012). Additional studies in multiple myeloma cells also show the same increased sensitivity to the combination of proteasome inhibitors and autophagy inhibitors *in vitro* (Kawaguchi et al., 2011).

Importantly, one should recognize that many studies of autophagy inhibition as anticancer therapy have employed the lysosomal inhibitor HCQ. Hence, an important caveat for these experiments is that the cytotoxic effects of HCQ and similar agents are likely to involve processes other than autophagy. To date, the precise contributions of autophagy inhibition toward the efficacy of these antimalarials remain uncertain. Moreover, compensatory pathways, such as CMA, may influence the efficacy of autophagy inhibition as a therapeutic approach. For example, autophagy inhibition in combination with the HDACi vorinostat in a sensitive T-cell lymphoma cell line results in decreased cell death, but the resulting vorinostat-resistant subclones become partially resensitized by the inhibition of CMA (Dupéré-Richer et al., 2013).

While it remains controversial whether autophagy can mediate cell death, several studies demonstrate that genetic knockdown of autophagy blocks tumor cell death induced by oncogenic RAS (Elgendy, Sheridan, Brumatti, & Martin, 2011) or by various chemotherapeutic agents (Janku, McConkey, Hong, & Kurzrock, 2011; Notte, Leclere, & Michiels, 2011). Furthermore, acute inhibition of autophagy can limit chemotherapy responses *in vivo* by preventing autophagy-dependent anticancer immune responses (Michaud et al., 2011). Thus, additional studies are needed to further clarify the contexts in which autophagy inhibition will be beneficial in the treatment of cancer, but as these studies have shown, autophagy inhibition as a clinical therapy will not be straightforward.

6. CONCLUSION

Autophagy and metabolism in cancer cells are inexorably linked. The cross regulation of these processes acts to buffer cancer cells from the environmental and internal stresses caused by excessive proliferation. As more targeted therapies are being designed and tested, unintended consequences on autophagy, both positive and negative, must be considered in order to predict and combat side effects and resistance mechanisms. Therefore, further understanding of how autophagy contributes to cancer cell metabolism will provide insight into how to better treat cancers.

ACKNOWLEDGMENTS

J. D. is supported by the NIH (CA126792), the DOD Breast Cancer Research Program (W81XWH-11-1-0130 and W81XWH-12-1-0505), and the Samuel Waxman Cancer Research Foundation. B. L. is supported by the NIH (CA109618) and CPRIT (RP120718-P1).

REFERENCES

Amaravadi, R. K., Lippincott-schwartz, J., Yin, X., Weiss, W. A., Takebe, N., Timmer, W., et al. (2011). Principles and current strategies for targeting autophagy for cancer treatment. *Clinical Cancer Research, 17*(4), 654–666.

Amaravadi, R., Yu, D., & Lum, J. (2007). Autophagy inhibition enhances therapy-induced apoptosis in a Myc-induced model of lymphoma. *The Journal of Clinical Investigation, 117*(2), 326–336. http://dx.doi.org/10.1172/JCI28833.

Avivar-Valderas, A., Salas, E., Bobrovnikova-Marjon, E., Diehl, J. A., Nagi, C., Debnath, J., et al. (2011). PERK integrates autophagy and oxidative stress responses to promote survival during extracellular matrix detachment. *Molecular and Cellular Biology, 31*(17), 3616–3629. http://dx.doi.org/10.1128/MCB.05164-11.

B'chir, W., Maurin, A.-C., Carraro, V., Averous, J., Jousse, C., Muranishi, Y., et al. (2013). The eIF2α/ATF4 pathway is essential for stress-induced autophagy gene expression. *Nucleic Acids Research, 41*(16), 7683–7699. http://dx.doi.org/10.1093/nar/gkt563.

Belaid, A., Cerezo, M., Chargui, A., Corcelle-Termeau, E., Pedeutour, F., Giuliano, S., et al. (2013). Autophagy plays a critical role in the degradation of active RHOA, the control of cell cytokinesis and genomic stability. *Cancer Research, 73*(14), 4311–4322. http://dx.doi.org/10.1158/0008-5472.CAN-12-4142.

Bellot, G., Garcia-Medina, R., Gounon, P., Chiche, J., Roux, D., Pouysségur, J., et al. (2009). Hypoxia-induced autophagy is mediated through hypoxia-inducible factor induction of BNIP3 and BNIP3L via their BH3 domains. *Molecular and Cellular Biology, 29*(10), 2570–2581. http://dx.doi.org/10.1128/MCB.00166-09.

Berhends, C., Sowa, M. E., Gygi, S. P., & Harper, J. W. (2010). Network organization of the human autophagy system. *Nature, 466*(7302), 68–76. http://dx.doi.org/10.1038/nature09204.

Bjørkøy, G., Lamark, T., Brech, A., Outzen, H., Perander, M., Overvatn, A., et al. (2005). p62/SQSTM1 forms protein aggregates degraded by autophagy and has a protective effect on huntingtin-induced cell death. *The Journal of Cell Biology*, *171*(4), 603–614. http://dx.doi.org/10.1083/jcb.200507002.

Boya, P., Reggiori, F., & Codogno, P. (2013). Emerging regulation and functions of autophagy. *Nature Cell Biology*, *15*(7), 713–720. http://dx.doi.org/10.1038/ncb2788.

Burgess, D. J. (2013). Metabolism: Glutamine connections. *Nature Reviews Cancer*, *13*(5), 293. http://dx.doi.org/10.1038/nrc3515.

Capparelli, C., Chiavarina, B., Whitaker-Menezes, D., Pestell, T. G., Pestell, R. G., Hulit, J., et al. (2012). CDK inhibitors (p16/p19/p21) induce senescence and autophagy in cancer-associated fibroblasts, "fueling" tumor growth via paracrine interactions, without an increase in neo-angiogenesis. *Cell Cycle*, *11*(19), 3599–3610.

Capparelli, C., Guido, C., Whitaker-menezes, D., Bonuccelli, G., Balliet, R., Timothy, G., et al. (2012). Autophagy and senescence in cancer-associated fibroblasts metabolically supports tumor growth and metastasis via glycolysis and ketone production. *Cell Cycle*, *11*(12), 2285–2302.

Capparelli, C., Whitaker-Menezes, D., Guido, C., Balliet, R., Pestell, T. G., Howell, A., et al. (2012). CTGF drives autophagy, glycolysis and senescence in cancer-associated fibroblasts via HIF1 activation, metabolically promoting tumor growth. *Cell Cycle*, *11*(12), 2272–2284.

Chan, E. Y. (2009). mTORC1 phosphorylates the ULK1-mAtg13-FIP200 autophagy regulatory complex. *Science Signaling*, *2*(84), pe51. http://dx.doi.org/10.1126/scisignal.284pe51.

Chen, N., Eritja, N., Lock, R., & Debnath, J. (2013). Autophagy restricts proliferation driven by oncogenic phosphatidylinositol 3-kinase in three-dimensional culture. *Oncogene*, *32*(20), 2543–2554. http://dx.doi.org/10.1038/onc.2012.277.

Cheong, H., Lindsten, T., & Thompson, C. B. (2012). Autophagy and ammonia. *Autophagy*, *8*(1), 122–123.

Cheong, H., Lu, C., Lindsten, T., & Thompson, C. B. (2012). Therapeutic targets in cancer cell metabolism and autophagy. *Nature Biotechnology*, *30*(7), 671–678. http://dx.doi.org/10.1038/nbt.2285.

Cuervo, A. M., Terleckyh, S. R., Dice, J. F., & Knecht, E. (1994). Selective binding and uptake of ribonuclease A and glyceraldehyde-3-phosphate dehydrogenase by isolated rat liver lysosomes. *The Journal of Biological Chemistry*, *269*(42), 26374–26380.

Dang, C. V. (1999). c-Myc target genes involved in cell growth, apoptosis, and metabolism. *Molecular and Cellular Biology*, *19*(1), 1–11.

DeBerardinis, R. J., Lum, J. J., Hatzivassiliou, G., & Thompson, C. B. (2008). The biology of cancer: Metabolic reprogramming fuels cell growth and proliferation. *Cell Metabolism*, 7(1), 11–20. http://dx.doi.org/10.1016/j.cmet.2007.10.002.

Debnath, J. (2011). The multifaceted roles of autophagy in tumors-implications for breast cancer. *Journal of Mammary Gland Biology and Neoplasia*, *16*(3), 173–187. http://dx.doi.org/10.1007/s10911-011-9223-3.

Degenhardt, K., Mathew, R., Beaudoin, B., Bray, K., Anderson, D., Chen, G., et al. (2006). Autophagy promotes tumor cell survival and restricts necrosis, inflammation, and tumorigenesis. *Cancer Cell*, *10*(1), 51–64. http://dx.doi.org/10.1016/j.ccr.2006.06.001.

DeNardo, D. G., Johansson, M., & Coussens, L. M. (2008). Immune cells as mediators of solid tumor metastasis. *Cancer Metastasis Reviews*, *27*, 11–18.

Dice, J. F., Chiang, H., Spencer, E. P., & Backer, J. M. (1986). Regulation of catabolism of microinjected ribonuclease A. *The Journal of Biological Chemistry*, *261*(15), 6853–6859.

Dupéré-Richer, D., Kinal, M., Ménasché, V., Nielsen, T. H., Del Rincon, S., Pettersson, F., et al. (2013). Vorinostat-induced autophagy switches from a death-promoting to a cytoprotective signal to drive acquired resistance. *Cell Death & Disease*, *4*, e486. http://dx.doi.org/10.1038/cddis.2012.210.

Duran, A., Linares, J. F., Galvez, A. S., Wikenheiser, K., Flores, J. M., Diaz-Meco, M. T., et al. (2008). The signaling adaptor p62 is an important NF-kappaB mediator in tumorigenesis. *Cancer Cell, 13*(4), 343–354. http://dx.doi.org/10.1016/j.ccr.2008.02.001.

Durán, R. V., Oppliger, W., Robitaille, A. M., Heiserich, L., Skendaj, R., Gottlieb, E., et al. (2012). Glutaminolysis activates Rag-mTORC1 signaling. *Molecular Cell, 47*(3), 349–358. http://dx.doi.org/10.1016/j.molcel.2012.05.043.

Eisenberg-Lerner, A., & Kimchi, A. (2007). DAP kinase regulates JNK signaling by binding and activating protein kinase D under oxidative stress. *Cell Death and Differentiation, 14*(11), 1908–1915. http://dx.doi.org/10.1038/sj.cdd.4402212.

Eisenberg-Lerner, A., & Kimchi, A. (2009). The paradox of autophagy and its implication in cancer etiology and therapy. *Apoptosis, 14*(4), 376–391. http://dx.doi.org/10.1007/s10495-008-0307-5.

Elgendy, M., Sheridan, C., Brumatti, G., & Martin, S. J. (2011). Oncogenic Ras-induced expression of Noxa and Beclin-1 promotes autophagic cell death and limits clonogenic survival. *Molecular Cell, 42*(1), 23–35. http://dx.doi.org/10.1016/j.molcel.2011.02.009.

Eng, C. H., Yu, K., Lucas, J., White, E., & Abraham, R. T. (2010). Ammonia derived from glutaminolysis is a diffusible regulator of autophagy. *Science Signaling, 3*(119), ra31. http://dx.doi.org/10.1126/scisignal.2000911.

Ermak, G., Sojitra, S., Yin, F., Cadenas, E., Cuervo, A. M., & Davies, K. J. (2012). Chronic expression of RCAN1-1L protein induces mitochondrial autophagy and metabolic shift from oxidative phosphorylation to glycolysis in neuronal cells. *The Journal of Biological Chemistry, 287*(17), 14088–14098. http://dx.doi.org/10.1074/jbc.M111.305342.

Fimia, G. M., Stoykova, A., Romagnoli, A., Giunta, L., Di Bartolomeo, S., Nardacci, R., et al. (2007). Ambra1 regulates autophagy and development of the nervous system. *Nature, 447*(7148), 1121–1125. http://dx.doi.org/10.1038/nature05925.

Gaglio, D., Metallo, C. M., Gameiro, P. A., Hiller, K., Danna, L. S., Balestrieri, C., et al. (2011). Oncogenic K-Ras decouples glucose and glutamine metabolism to support cancer cell growth. *Molecular Systems Biology*, 7(523), 523. http://dx.doi.org/10.1038/msb.2011.56.

Gozuacik, D., & Kimchi, A. (2004). Autophagy as a cell death and tumor suppressor mechanism. *Oncogene, 23*(16), 2891–2906. http://dx.doi.org/10.1038/sj.onc.1207521.

Guo, J. Y., Chen, H.-Y., Mathew, R., Fan, J., Strohecker, A. M., Karsli-Uzunbas, G., et al. (2011). Activated Ras requires autophagy to maintain oxidative metabolism and tumorigenesis. *Genes & Development, 25*(5), 460–470. http://dx.doi.org/10.1101/gad.2016311.

Guo, J. Y., Karsli-Uzunbas, G., Mathew, R., Aisner, S. C., Kamphorst, J. J., Strohecker, A. M., et al. (2013). Autophagy suppresses progression of K-ras-induced lung tumors to oncocytomas and maintains lipid homeostasis. *Genes & Development, 27*(13), 1447–1461. http://dx.doi.org/10.1101/gad.219642.113.

Gupta, A., Roy, S., Lazar, A. J. F., Wang, W.-L., McAuliffe, J. C., Reynoso, D., et al. (2010). Autophagy inhibition and antimalarials promote cell death in gastrointestinal stromal tumor (GIST). *Proceedings of the National Academy of Sciences of the United States of America, 107*(32), 14333–14338. http://dx.doi.org/10.1073/pnas.1000248107.

Han, J. M., Jeong, S. J., Park, M. C., Kim, G., Kwon, N. H., Kim, H. K., et al. (2012). Leucyl-tRNA synthetase is an intracellular leucine sensor for the mTORC1-signaling pathway. *Cell, 149*(2), 410–424. http://dx.doi.org/10.1016/j.cell.2012.02.044.

Haq, R., Shoag, J., Andreu-Perez, P., Yokoyama, S., Edelman, H., Rowe, G. C., et al. (2013). Oncogenic BRAF regulates oxidative metabolism via PGC1α and MITF. *Cancer Cell, 23*(3), 302–315. http://dx.doi.org/10.1016/j.ccr.2013.02.003.

Hara, T., Nakamura, K., Matsui, M., Yamamoto, A., Nakahara, Y., Suzuki-Migishima, R., et al. (2006). Suppression of basal autophagy in neural cells causes neurodegenerative disease in mice. *Nature, 441*(7095), 885–889. http://dx.doi.org/10.1038/nature04724.

Ho, J., de Moura, M. B., Lin, Y., Vincent, G., Thorne, S., Duncan, L. M., et al. (2012). Importance of glycolysis and oxidative phosphorylation in advanced melanoma. *Molecular Cancer*, *11*(1), 76. http://dx.doi.org/10.1186/1476-4598-11-76.

Høyer-hansen, M., & Jäättelä, M. (2008). Autophagy: An emerging target for cancer therapy. *Autophagy*, *4*(5), 574–580.

Hsu, P. P., & Sabatini, D. M. (2008). Cancer cell metabolism: Warburg and beyond. *Cell*, *134*(5), 703–707. http://dx.doi.org/10.1016/j.cell.2008.08.021.

Hu, Y.-L., DeLay, M., Jahangiri, A., Molinaro, A. M., Rose, S. D., Carbonell, W. S., et al. (2012). Hypoxia-induced autophagy promotes tumor cell survival and adaptation to antiangiogenic treatment in glioblastoma. *Cancer Research*, *72*(7), 1773–1783. http://dx.doi.org/10.1158/0008-5472.CAN-11-3831.

Hui, B., Shi, Y.-H., Ding, Z.-B., Zhou, J., Gu, C.-Y., Peng, Y.-F., et al. (2012). Proteasome inhibitor interacts synergistically with autophagy inhibitor to suppress proliferation and induce apoptosis in hepatocellular carcinoma. *Cancer*, *118*(22), 5560–5571. http://dx.doi.org/10.1002/cncr.27586.

Inami, Y., Waguri, S., Sakamoto, A., Kouno, T., Nakada, K., Hino, O., et al. (2011). Persistent activation of Nrf2 through p62 in hepatocellular carcinoma cells. *The Journal of Cell Biology*, *193*(2), 275–284. http://dx.doi.org/10.1083/jcb.201102031.

Ishikawa, K., Takenaga, K., Akimoto, M., Koshikawa, N., Yamaguchi, A., Imanishi, H., et al. (2008). ROS-generating mitochondrial DNA mutations can regulate tumor cell metastasis. *Science*, *320*(5876), 661–664. http://dx.doi.org/10.1126/science.1156906.

Israelsen, W. J., Dayton, T. L., Davidson, S. M., Fiske, B. P., Hosios, A. M., Bellinger, G., et al. (2013). PKM2 isoform-specific deletion reveals a differential requirement for pyruvate kinase in tumor cells. *Cell*, *155*(2), 397–409. http://dx.doi.org/10.1016/j.cell.2013.09.025.

Itakura, E., Kishi, C., Inoue, K., & Mizushima, N. (2008). Beclin 1 forms two distinct phosphatidylinositol 3-kinase complexes with mammalian Atg14 and UVRAG. *Molecular Biology of the Cell*, *19*(December), 5360–5372. http://dx.doi.org/10.1091/mbc.E08-01-0080.

Itakura, E., Kishi-Itakura, C., & Mizushima, N. (2012). The hairpin-type tail-anchored SNARE syntaxin 17 targets to autophagosomes for fusion with endosomes/lysosomes. *Cell*, *151*(6), 1256–1269. http://dx.doi.org/10.1016/j.cell.2012.11.001.

Janku, F., McConkey, D., Hong, D., & Kurzrock, R. (2011). Autophagy as a target for anticancer therapy. *Nature Reviews. Clinical Oncology*, *8*(9), 528–539.

Johansen, T., & Lamark, T. (2011). Selective autophagy mediated by autophagic adapter proteins. *Autophagy*, 7(3), 279–296. http://dx.doi.org/10.4161/auto.7.3.14487.

Jung, H. S., Chung, K. W., Won Kim, J., Kim, J., Komatsu, M., Tanaka, K., et al. (2008). Loss of autophagy diminishes pancreatic beta cell mass and function with resultant hyperglycemia. *Cell Metabolism*, *8*(4), 318–324. http://dx.doi.org/10.1016/j.cmet.2008.08.013.

Jung, C. H., Jun, C. B., Ro, S., Kim, Y., Otto, N. M., Cao, J., et al. (2009). ULK-Atg13-FIP200 complexes mediate mTOR signaling to the autophagy machinery. *Molecular Biology of the Cell*, *20*(April), 1992–2003. http://dx.doi.org/10.1091/mbc.E08-12-1249.

Jung, C. H., Ro, S.-H., Cao, J., Otto, N. M., & Kim, D.-H. (2010). mTOR regulation of autophagy. *FEBS Letters*, *584*(7), 1287–1295. http://dx.doi.org/10.1016/j.febslet.2010.01.017.

Kang, M. R., Kim, M. S., Oh, J. E., Kim, Y. R., Song, S. Y., Kim, S. S., et al. (2009). Frameshift mutations of autophagy-related genes ATG2B, ATG5, ATG9B and ATG12 in gastric and colorectal cancers with microsatellite instability. *The Journal of Pathology*, *217*, 702–706. http://dx.doi.org/10.1002/path.2509.

Kang, T.-W., Yevsa, T., Woller, N., Hoenicke, L., Wuestefeld, T., Dauch, D., et al. (2011). Senescence surveillance of pre-malignant hepatocytes limits liver cancer development. *Nature*, *479*(7374), 547–551. http://dx.doi.org/10.1038/nature10599.

Kang, R., Zeh, H. J., Lotze, M. T., & Tang, D. (2011). The Beclin 1 network regulates autophagy and apoptosis. *Cell Death and Differentiation*, *18*(4), 571–580. http://dx.doi.org/10.1038/cdd.2010.191.

Kawaguchi, T., Miyazawa, K., Moriya, S., Ohtomo, T., Che, X.-F., Naito, M., et al. (2011). Combined treatment with bortezomib plus bafilomycin A1 enhances the cytocidal effect and induces endoplasmic reticulum stress in U266 myeloma cells: Crosstalk among proteasome, autophagy-lysosome and ER stress. *International Journal of Oncology*, *38*(3), 643–654. http://dx.doi.org/10.3892/ijo.2010.882.

Kim, T.-H., Hur, E., Kang, S.-J., Kim, J.-A., Thapa, D., Lee, Y. M., et al. (2011). NRF2 blockade suppresses colon tumor angiogenesis by inhibiting hypoxia-induced activation of HIF-1α. *Cancer Research*, *71*(6), 2260–2275. http://dx.doi.org/10.1158/0008-5472.CAN-10-3007.

Kim, M. S., Jeong, E. G., Ahn, C. H., Kim, S. S., Lee, S. H., & Yoo, N. J. (2008). Frameshift mutation of UVRAG, an autophagy-related gene, in gastric carcinomas with microsatellite instability. *Human Pathology*, *39*(7), 1059–1063. http://dx.doi.org/10.1016/j.humpath.2007.11.013.

Kim, J., Kim, Y. C., Fang, C., Russell, R. C., Kim, J. H., Fan, W., et al. (2013). Differential regulation of distinct Vps34 complexes by AMPK in nutrient stress and autophagy. *Cell*, *152*(1–2), 290–303. http://dx.doi.org/10.1016/j.cell.2012.12.016.

Kim, J., Kundu, M., Viollet, B., & Guan, K.-L. (2011). AMPK and mTOR regulate autophagy through direct phosphorylation of Ulk1. *Nature Cell Biology*, *13*(2), 132–141. http://dx.doi.org/10.1038/ncb2152.

Kimmelman, A. C. (2011). The dynamic nature of autophagy in cancer. *Genes & Development*, *25*(19), 1999–2010. http://dx.doi.org/10.1101/gad.17558811.

Kirkin, V., Lamark, T., Sou, Y.-S., Bjørkøy, G., Nunn, J. L., Bruun, J.-A., et al. (2009). A role for NBR1 in autophagosomal degradation of ubiquitinated substrates. *Molecular Cell*, *33*(4), 505–516. http://dx.doi.org/10.1016/j.molcel.2009.01.020.

Klionsky, D. (2013). An overview of autophagy: Morphology, mechanism and regulation. *Antioxidants & Redox Signaling*, *20*(3), 460–473. http://dx.doi.org/10.1089/ars.2013.5371.

Klionsky, D. J., & Emr, S. (2000). Autophagy as a regulated pathway of cellular degradation. *Science*, *290*(5497), 1717–1721.

Ko, Y.-H., Lin, Z., Flomenberg, N., Pestell, R. G., Howell, A., Sotgia, F., et al. (2011). Glutamine fuels a vicious cycle of autophagy in the tumor stroma and oxidative mitochondrial metabolism in epithelial cancer cells. *Cancer Biology & Therapy*, *12*(12), 1085–1097. http://dx.doi.org/10.4161/cbt.12.2.18671.

Koga, H., Martinez-vicente, M., Macian, F., Verkhusha, V. V., & Cuervo, A. M. (2011). A photoconvertible fluorescent reporter to track chaperone-mediated autophagy. *Nature Communications*, *2*(386). http://dx.doi.org/10.1038/ncomms1393.

Komatsu, M., Kurokawa, H., Waguri, S., Taguchi, K., Kobayashi, A., Ichimura, Y., et al. (2010). The selective autophagy substrate p62 activates the stress responsive transcription factor Nrf2 through inactivation of Keap1. *Nature Cell Biology*, *12*(3), 213–223. http://dx.doi.org/10.1038/ncb2021.

Komatsu, M., Waguri, S., Chiba, T., Murata, S., Iwata, J., Tanida, I., et al. (2006). Loss of autophagy in the central nervous system causes neurodegeneration in mice. *Nature*, *441*(7095), 880–884. http://dx.doi.org/10.1038/nature04723.

Komatsu, M., Waguri, S., Ueno, T., Iwata, J., Murata, S., Tanida, I., et al. (2005). Impairment of starvation-induced and constitutive autophagy in Atg7-deficient mice. *The Journal of Cell Biology*, *169*(3), 425–434. http://dx.doi.org/10.1083/jcb.200412022.

Komiya, K., Uchida, T., Ueno, T., Koike, M., Abe, H., Hirose, T., et al. (2010). Free fatty acids stimulate autophagy in pancreatic β-cells via JNK pathway. *Biochemical and Biophysical Research Communications*, *401*(4), 561–567. http://dx.doi.org/10.1016/j.bbrc.2010.09.101.

Kon, M., Kiffin, R., Koga, H., Chapochnick, J., Macian, F., Varticovski, L., et al. (2011). Chaperone-mediated autophagy is required for tumor growth. *Science Translational Medicine, 3*(109), 109ra117. http://dx.doi.org/10.1126/scitranslmed.3003182.

Kroemer, G., Mariño, G., & Levine, B. (2010). Autophagy and the integrated stress response. *Molecular Cell, 40*(2), 280–293. http://dx.doi.org/10.1016/j.molcel.2010.09.023.

Las, G., Serada, S. B., Wikstrom, J. D., Twig, G., & Shirihai, O. S. (2011). Fatty acids suppress autophagic turnover in β-cells. *The Journal of Biological Chemistry, 286*(49), 42534–42544. http://dx.doi.org/10.1074/jbc.M111.242412.

Lee, J. W., Jeong, E. G., Soung, Y. H., Nam, S. W., Lee, J. Y., Yoo, N. J., et al. (2006). Decreased expression of tumour suppressor Bax-interacting factor-1 (Bif-1), a Bax activator, in gastric carcinomas. *Pathology, 38*(4), 312–315. http://dx.doi.org/10.1080/00313020600820880.

Levine, B., & Kroemer, G. (2008). Autophagy in the pathogenesis of disease. *Cell, 132*(1), 27–42. http://dx.doi.org/10.1016/j.cell.2007.12.018.

Li, M., Hou, Y., Wang, J., Chen, X., Shao, Z.-M., & Yin, X.-M. (2011). Kinetics comparisons of mammalian Atg4 homologues indicate selective preferences toward diverse Atg8 substrates. *The Journal of Biological Chemistry, 286*(9), 7327–7338. http://dx.doi.org/10.1074/jbc.M110.199059.

Li, Y.-Y., Lam, S.-K., Mak, J. C.-W., Zheng, C.-Y., & Ho, J. C.-M. (2013). Erlotinib-induced autophagy in epidermal growth factor receptor mutated non-small cell lung cancer. *Lung Cancer, 81*(3), 354–361. http://dx.doi.org/10.1016/j.lungcan.2013.05.012.

Liang, C., Feng, P., Ku, B., Dotan, I., Canaani, D., Oh, B.-H., et al. (2006). Autophagic and tumour suppressor activity of a novel Beclin1-binding protein UVRAG. *Nature Cell Biology, 8*(7), 688–699. http://dx.doi.org/10.1038/ncb1426.

Liang, C., Feng, P., Ku, B., & Oh, B. (2007). UVRAG: A new player in autophagy and tumor cell growth. *Autophagy, 3*(1), 69–71.

Liang, X. H., Jackson, S., Seaman, M., Brown, K., Kempkes, B., Hibshoosh, H., et al. (1999). Induction of autophagy and inhibition of tumorigenesis by beclin 1. *Nature, 402*(6762), 672–676. http://dx.doi.org/10.1038/45257.

Liang, C., Lee, J., Inn, K., Gack, M. U., Li, Q., Roberts, E. A., et al. (2008). Beclin1-binding UVRAG targets the class C Vps complex to coordinate autophagosome maturation and endocytic trafficking. *Nature Cell Biology, 10*(7), 776–787. http://dx.doi.org/10.1038/ncb1740.

Lin, T., Chen, Y., Kensicki, E., Li, A. Y., Kong, M., Li, Y., et al. (2012). Resetting glutamine-dependent metabolism and oxygen consumption. *Autophagy, 8*(10), 1477–1493.

Lock, R., & Debnath, J. (2011). Ras, autophagy and glycolysis. *Cell Cycle, 10*(10), 1516–1517. http://dx.doi.org/10.4161/cc.10.10.15434.

Lock, R., Roy, S., Kenific, C. M., Su, J. S., Salas, E., Ronen, S. M., et al. (2011). Autophagy facilitates glycolysis during Ras-mediated oncogenic transformation. *Molecular Biology of the Cell, 22*(2), 165–178. http://dx.doi.org/10.1091/mbc.E10-06-0500.

Lorin, S., Tol, M. J., Bauvy, C., Strijland, A., Poüs, C., Verhoeven, A. J., et al. (2013). Glutamate dehydrogenase contributes to leucine sensing in the regulation of autophagy. *Autophagy, 9*(6), 850–860. http://dx.doi.org/10.4161/auto.24083.

Lozy, F., & Karantza, V. (2012). Autophagy and cancer cell metabolism. *Seminars in Cell & Developmental Biology, 23*(4), 395–401. http://dx.doi.org/10.1016/j.semcdb.2012.01.005.

Lum, J. J., Bauer, D. E., Kong, M., Harris, M. H., Li, C., Lindsten, T., et al. (2005). Growth factor regulation of autophagy and cell survival in the absence of apoptosis. *Cell, 120*(2), 237–248. http://dx.doi.org/10.1016/j.cell.2004.11.046.

Lv, L., Li, D., Zhao, D., Lin, R., Chu, Y., Zhang, H., et al. (2011). Acetylation targets the M2 isoform of pyruvate kinase for degradation through chaperone-mediated autophagy and promotes tumor growth. *Molecular Cell, 42*(6), 719–730. http://dx.doi.org/10.1016/j.molcel.2011.04.025.

Maes, H., Rubio, N., Garg, A. D., & Agostinis, P. (2013). Autophagy: Shaping the tumor microenvironment and therapeutic response. *Trends in Molecular Medicine*, *19*, 428–446. http://dx.doi.org/10.1016/j.molmed.2013.04.005.

Marcilla-Etxenike, A., Martín, M. L., Noguera-Salvà, M. A., García-Verdugo, J. M., Soriano-Navarro, M., Dey, I., et al. (2012). 2-Hydroxyoleic acid induces ER stress and autophagy in various human glioma cell lines. *PLoS One*, 7(10), e48235. http://dx.doi.org/10.1371/journal.pone.0048235.

Mariño, G., Salvador-Montoliu, N., Fueyo, A., Knecht, E., Mizushima, N., & López-Otín, C. (2007). Tissue-specific autophagy alterations and increased tumorigenesis in mice deficient in Atg4C/autophagin-3. *The Journal of Biological Chemistry*, *282*(25), 18573–18583. http://dx.doi.org/10.1074/jbc.M701194200.

Martinez-Outschoorn, U. E., Casey, T., Lin, Z., Whitaker-Menezes, D., Chiavarina, B., Zhou, J., et al. (2010). Autophagy in cancer associated fibroblasts promotes tumor cell survival: Role of hypoxia, HIF1 induction and NFκB activation in the tumor stromal microenvironment. *Cell Cycle*, *9*(17), 3515–3533. http://dx.doi.org/10.4161/cc.9.17.12928.

Martino, L., Masini, M., Novelli, M., Beffy, P., Bugliani, M., Marselli, L., et al. (2012). Palmitate activates autophagy in INS-1E β-cells and in isolated rat and human pancreatic islets. *PLoS One*, 7(5), e36188. http://dx.doi.org/10.1371/journal.pone.0036188.

Massey, A. C., Kaushik, S., Sovak, G., Kiffin, R., & Cuervo, A. M. (2006). Consequences of the selective blockage of chaperone-mediated autophagy. *Proceedings of the National Academy of Sciences of the United States of America*, *103*(15), 5805–5810. http://dx.doi.org/10.1073/pnas.0507436103.

Mathew, R., Karp, C. M., Beaudoin, B., Vuong, N., Chen, G., Chen, H.-Y., et al. (2009). Autophagy suppresses tumorigenesis through elimination of p62. *Cell*, *137*(6), 1062–1075. http://dx.doi.org/10.1016/j.cell.2009.03.048.

Matsunaga, K., Saitoh, T., Tabata, K., Omori, H., Satoh, T., Kurotori, N., et al. (2009). Two Beclin 1-binding proteins, Atg14L and Rubicon, reciprocally regulate autophagy at different stages. *Nature Cell Biology*, *11*(4), 385–396. http://dx.doi.org/10.1038/ncb1846.

Mei, S., Ni, H., Manley, S., Bockus, A., Kassel, K. M., Luyendyk, J. P., et al. (2011). Differential roles of unsaturated and saturated fatty acids on autophagy and apoptosis in hepatocytes. *The Journal of Pharmacology and Experimental Therapeutics*, *339*(2), 487–498. http://dx.doi.org/10.1124/jpet.111.184341.

Michaud, M., Martins, I., Sukkurwala, A., Adjemian, S., Ma, Y., Pellegatti, P., et al. (2011). Autophagy-dependent anticancer immune responses induced by chemotherapeutic agents in mice. *Science*, *334*(6062), 1573–1577.

Mihaylova, M. M., & Shaw, R. J. (2011). The AMPK signalling pathway coordinates cell growth, autophagy and metabolism. *Nature Cell Biology*, *13*(9), 1016–1023. http://dx.doi.org/10.1038/ncb2329.

Miller, D. M., Thomas, S. D., Islam, A., Muench, D., & Sedoris, K. (2012). c-Myc and cancer metabolism. *Clinical Cancer Research*, *18*(20), 5546–5553. http://dx.doi.org/10.1158/1078-0432.CCR-12-0977.

Mizushima, N. (2007). Autophagy: Process and function. *Genes & Development*, *21*(22), 2861–2873. http://dx.doi.org/10.1101/gad.1599207.

Mizushima, N. (2010). The role of the Atg1/ULK1 complex in autophagy regulation. *Current Opinion in Cell Biology*, *22*(2), 132–139. http://dx.doi.org/10.1016/j.ceb.2009.12.004.

Mortensen, M., Soilleux, E. J., Djordjevic, G., Tripp, R., Lutteropp, M., Sadighi-Akha, E., et al. (2011). The autophagy protein Atg7 is essential for hematopoietic stem cell maintenance. *The Journal of Experimental Medicine*, *208*(3), 455–467. http://dx.doi.org/10.1084/jem.20101145.

Moruno-Manchón, J. F., Pérez-Jiménez, E., & Knecht, E. (2013). Glucose induces autophagy under starvation conditions by a p38 MAPK-dependent pathway. *The Biochemical Journal*, *449*(2), 497–506. http://dx.doi.org/10.1042/BJ20121122.

Murrow, L., & Debnath, J. (2013). Autophagy as a stress-response and quality-control mechanism: Implications for cell injury and human disease. *Annual Review of Pathology*, *8*, 105–137. http://dx.doi.org/10.1146/annurev-pathol-020712-163918.

Nakatogawa, H., Suzuki, K., Kamada, Y., & Ohsumi, Y. (2009). Dynamics and diversity in autophagy mechanisms: Lessons from yeast. *Nature Reviews Molecular Cell Biology*, *10*(7), 458–467. http://dx.doi.org/10.1038/nrm2708.

Narita, M., Young, A. R. J., Arakawa, S., Samarajiwa, S. A., Nakashima, T., Yoshida, S., et al. (2011). Spatial coupling of mTOR and autophagy augments secretory phenotypes. *Science*, *332*(6032), 966–970. http://dx.doi.org/10.1126/science.1205407.

Neufeld, T. (2010). TOR-dependent control of autophagy: Biting the hand that feeds. *Current Opinion in Cell Biology*, *22*(2), 157–168. http://dx.doi.org/10.1016/j.ceb.2009.11.005.

Noman, M. Z., Janji, B., Kaminska, B., Van Moer, K., Pierson, S., Przanowski, P., et al. (2011). Blocking hypoxia-induced autophagy in tumors restores cytotoxic T-cell activity and promotes regression. *Cancer Research*, *71*(18), 5976–5986. http://dx.doi.org/10.1158/0008-5472.CAN-11-1094.

Notte, A., Leclere, L., & Michiels, C. (2011). Autophagy as a mediator of chemotherapy-induced cell death in cancer. *Biochemical Pharmacology*, *82*(5), 427–434. http://dx.doi.org/10.1016/j.bcp.2011.06.015.

Ogata, M., Hino, S., Saito, A., Morikawa, K., Kondo, S., Kanemoto, S., et al. (2006). Autophagy is activated for cell survival after endoplasmic reticulum stress. *Molecular and Cellular Biology*, *26*(24), 9220–9231. http://dx.doi.org/10.1128/MCB.01453-06.

Ogier-Denis, E., Pattingre, S., El Benna, J., & Codogno, P. (2000). Erk1/2-dependent phosphorylation of Galpha-interacting protein stimulates its GTPase accelerating activity and autophagy in human colon cancer cells. *The Journal of Biological Chemistry*, *275*(50), 39090–39095. http://dx.doi.org/10.1074/jbc.M006198200.

Onodera, J., & Ohsumi, Y. (2005). Autophagy is required for maintenance of amino acid levels and protein synthesis under nitrogen starvation. *The Journal of Biological Chemistry*, *280*(36), 31582–31586. http://dx.doi.org/10.1074/jbc.M506736200.

Papandreou, I., Lim, A. L., Laderoute, K., & Denko, N. C. (2008). Hypoxia signals autophagy in tumor cells via AMPK activity, independent of HIF-1, BNIP3, and BNIP3L. *Cell Death and Differentiation*, *15*(10), 1572–1581. http://dx.doi.org/10.1038/cdd.2008.84.

Pattingre, S., Bauvy, C., & Codogno, P. (2003). Amino acids interfere with the ERK1/2-dependent control of macroautophagy by controlling the activation of Raf-1 in human colon cancer HT-29 cells. *The Journal of Biological Chemistry*, *278*(19), 16667–16674. http://dx.doi.org/10.1074/jbc.M210998200.

Pattingre, S., Tassa, A., Qu, X., Garuti, R., Liang, X. H., Mizushima, N., et al. (2005). Bcl-2 antiapoptotic proteins inhibit Beclin 1-dependent autophagy. *Cell*, *122*(6), 927–939. http://dx.doi.org/10.1016/j.cell.2005.07.002.

Qin, L., Wang, Z., Tao, L., & Wang, Y. (2010). ER stress negatively regulates AKT/TSC/mTOR pathway to enhance autophagy. *Autophagy*, *6*(2), 239–247. Retrieved from, http://www.ncbi.nlm.nih.gov/pubmed/20104019.

Qu, X., Yu, J., Bhagat, G., Furuya, N., Hibshoosh, H., Troxel, A., et al. (2003). Promotion of tumorigenesis by heterozygous disruption of the beclin 1 autophagy gene. *The Journal of Clinical Investigation*, *112*(12), 1809–1820. http://dx.doi.org/10.1172/JCI200320039.

Rabinowitz, J. D., & White, E. (2010). Autophagy and metabolism. *Science*, *330*(6009), 1344–1348. http://dx.doi.org/10.1126/science.1193497.

Salem, A. F., Whitaker-menezes, D., Lin, Z., Martinez-outschoorn, U. E., Tanowitz, H. B., Al-zoubi, M. S., et al. (2012). Two-compartment tumor metabolism: Autophagy in the tumor microenvironment and oxidative mitochondrial metabolism (OXPHOS) in cancer cells. *Cell Cycle*, *11*(13), 2545–2556. http://dx.doi.org/10.4161/cc.20920.

Sancak, Y., Bar-Peled, L., Zoncu, R., Markhard, A. L., Nada, S., & Sabatini, D. M. (2010). Ragulator-Rag complex targets mTORC1 to the lysosomal surface and is necessary for its activation by amino acids. *Cell*, *141*(2), 290–303. http://dx.doi.org/10.1016/j.cell.2010.02.024.

Sancak, Y., Peterson, T. R., Shaul, Y. D., Lindquist, R. A., Thoreen, C. C., Bar-Peled, L., et al. (2008). The Rag GTPases bind raptor and mediate amino acid signaling to mTORC1. *Science*, *320*(5882), 1496–1501. http://dx.doi.org/10.1126/science.1157535.

Sanchez, C. G., Penfornis, P., Oskowitz, A. Z., Boonjindasup, A. G., Cai, D. Z., Dhule, S. S., et al. (2011). Activation of autophagy in mesenchymal stem cells provides tumor stromal support. *Carcinogenesis*, *32*(7), 964–972. http://dx.doi.org/10.1093/carcin/bgr029.

Santos, C. R., & Schulze, A. (2012). Lipid metabolism in cancer. *The FEBS Journal*, *279*(15), 2610–2623. http://dx.doi.org/10.1111/j.1742-4658.2012.08644.x.

Scherz-Shouval, R., Shvets, E., Fass, E., Shorer, H., Gil, L., & Elazar, Z. (2007). Reactive oxygen species are essential for autophagy and specifically regulate the activity of Atg4. *The EMBO Journal*, *26*(7), 1749–1760. http://dx.doi.org/10.1038/sj.emboj.7601623.

Sheen, J.-H., Zoncu, R., Kim, D., & Sabatini, D. M. (2011). Defective regulation of autophagy upon leucine deprivation reveals a targetable liability of human melanoma cells in vitro and in vivo. *Cancer Cell*, *19*(5), 613–628. http://dx.doi.org/10.1016/j.ccr.2011.03.012.

Shpilka, T., Weidberg, H., Pietrokovski, S., & Elazar, Z. (2011). Atg8: An autophagy-related ubiquitin-like protein family. *Genome Biology*, *12*(7), 226. http://dx.doi.org/10.1186/gb-2011-12-7-226.

Singh, R., & Cuervo, A. M. (2012). Lipophagy: Connecting autophagy and lipid metabolism. *International Journal of Cell Biology*, *2012*, 282041. http://dx.doi.org/10.1155/2012/282041.

Singh, R., Kaushik, S., Wang, Y., Xiang, Y., Novak, I., Komatsu, M., et al. (2009). Autophagy regulates lipid metabolism. *Nature*, *458*(7242), 1131–1135. http://dx.doi.org/10.1038/nature07976.

Singh, A., Misra, V., Thimmulappa, R. K., Lee, H., Ames, S., Hoque, M. O., et al. (2006). Dysfunctional KEAP1-NRF2 interaction in non-small-cell lung cancer. *PLoS Medicine*, *3*(10), e420. http://dx.doi.org/10.1371/journal.pmed.0030420.

Son, J., Lyssiotis, C. A., Ying, H., Wang, X., Hua, S., Ligorio, M., et al. (2013). Glutamine supports pancreatic cancer growth through a KRAS-regulated metabolic pathway. *Nature*, *496*(7443), 101–105. http://dx.doi.org/10.1038/nature12040.

Strohecker, A. M., Guo, J. Y., Karsli-Uzunbas, G., Price, S. M., Chen, G. J., Mathew, R., et al. (2013). Autophagy sustains mitochondrial glutamine metabolism and growth of BRAFV600E-driven lung tumors. *Cancer Discovery*, *3*(11), 1272–1285. http://dx.doi.org/10.1158/2159-8290.CD-13-0397.

Takahashi, Y., Coppola, D., Matsushita, N., Cualing, H. D., Sun, M., Sato, Y., et al. (2007). Bif-1 interacts with Beclin 1 through UVRAG and regulates autophagy and tumorigenesis. *Nature Cell Biology*, *9*(10), 1142–1151. http://dx.doi.org/10.1038/ncb1634.

Takamura, A., Komatsu, M., Hara, T., Sakamoto, A., Kishi, C., Waguri, S., et al. (2011). Autophagy-deficient mice develop multiple liver tumors. *Genes & Development*, *25*(8), 795–800. http://dx.doi.org/10.1101/gad.2016211.

Tan, S. H., Shui, G., Zhou, J., Li, J. J., Bay, B.-H., Wenk, M. R., et al. (2012). Induction of autophagy by palmitic acid via protein kinase C-mediated signaling pathway independent of mTOR (mammalian target of rapamycin). *The Journal of Biological Chemistry*, *287*(18), 14364–14376. http://dx.doi.org/10.1074/jbc.M111.294157.

Terés, S., Lladó, V., Higuera, M., Barceló-Coblijn, G., Martin, M. L., Noguera-Salvà, M. A., et al. (2012). 2-Hydroxyoleate, a nontoxic membrane binding anticancer drug, induces glioma cell differentiation and autophagy. *Proceedings of the National Academy of Sciences of the United States of America*, *109*(22), 8489–8494. http://dx.doi.org/10.1073/pnas.1118349109.

Thurston, T. L. M., Ryzhakov, G., Bloor, S., von Muhlinen, N., & Randow, F. (2009). The TBK1 adaptor and autophagy receptor NDP52 restricts the proliferation of ubiquitin-coated bacteria. *Nature Immunology*, *10*(11), 1215–1221. http://dx.doi.org/10.1038/ni.1800.

Tomohiro, Y., & Klionsky, D. J. (2007). Endoplasmic reticulum stress: A new pathway to induce autophagy. *Autophagy*, *3*(2), 160–162.

Vakifahmetoglu-Norberg, H., Kim, M., Xia, H.-G., Iwanicki, M. P., Ofengeim, D., Coloff, J. L., et al. (2013). Chaperone-mediated autophagy degrades mutant p53. *Genes & Development*, *27*(15), 1718–1730. http://dx.doi.org/10.1101/gad.220897.113.

Vander Heiden, M. G., Cantley, L. C., & Thompson, C. B. (2009). Understanding the Warburg effect: The metabolic requirements of cell proliferation. *Science*, *324*(5930), 1029–1033. http://dx.doi.org/10.1126/science.1160809.

Vazquez, F., Lim, J.-H., Chim, H., Bhalla, K., Girnun, G., Pierce, K., et al. (2013). PGC1α expression defines a subset of human melanoma tumors with increased mitochondrial capacity and resistance to oxidative stress. *Cancer Cell*, *23*(3), 287–301. http://dx.doi.org/10.1016/j.ccr.2012.11.020.

Wang, Y., Singh, R., Massey, A. C., Kane, S. S., Kaushik, S., Grant, T., et al. (2008). Loss of macroautophagy promotes or prevents fibroblast apoptosis depending on the death stimulus. *The Journal of Biological Chemistry*, *283*(8), 4766–4777. http://dx.doi.org/10.1074/jbc.M706666200.

Wang, R. C., Wei, Y., An, Z., Zou, Z., Xiao, G., Bhagat, G., et al. (2012). Akt-mediated regulation of autophagy and tumorigenesis through Beclin 1 phosphorylation. *Science*, *338*(6109), 956–959. http://dx.doi.org/10.1126/science.1225967.

Wei, Y., Pattingre, S., Sinha, S., Bassik, M., & Levine, B. (2008). JNK1-mediated phosphorylation of Bcl-2 regulates starvation-induced autophagy. *Molecular Cell*, *30*(6), 678–688. http://dx.doi.org/10.1016/j.molcel.2008.06.001.

Wei, Y., Zou, Z., Becker, N., Anderson, M., Sumpter, R., Xiao, G., et al. (2013). EGFR-mediated beclin 1 phosphorylation in autophagy suppression, tumor progression, and tumor chemoresistance. *Cell*, *154*(6), 1–16. http://dx.doi.org/10.1016/j.cell.2013.08.015.

Weidberg, H., Shpilka, T., Shvets, E., Abada, A., Shimron, F., & Elazar, Z. (2011). LC3 and GATE-16 N termini mediate membrane fusion processes required for autophagosome biogenesis. *Developmental Cell*, *20*(4), 444–454. http://dx.doi.org/10.1016/j.devcel.2011.02.006.

Weidberg, H., Shpilka, T., Shvets, E., Shinder, V., & Elazar, Z. (2010). LC3 and GATE-16/GABARAP subfamilies are both essential yet act differently in autophagosome biogenesis. *The EMBO Journal*, *29*, 1792–1802. http://dx.doi.org/10.1038/emboj.2010.74.

Wellen, K. E., & Thompson, C. B. (2010). Cellular metabolic stress: Considering how cells respond to nutrient excess. *Molecular Cell*, *40*(2), 323–332. http://dx.doi.org/10.1016/j.molcel.2010.10.004.

Wild, P., Farhan, H., McEwan, D. G., Wagner, S., Rogov, V. V., Brady, N. R., et al. (2011). Phosphorylation of the autophagy receptor optineurin restricts Salmonella growth. *Science*, *333*(6039), 228–233. http://dx.doi.org/10.1126/science.1205405.

Williams, T., Forsberg, L. J., Viollet, B., & Brenman, J. E. (2009). Basal autophagy induction without AMP-activated protein kinase under low glucose conditions. *Autophagy*, *5*(8), 1155–1165. Retrieved from, http://www.ncbi.nlm.nih.gov/pubmed/19844161.

Wojtkowiak, J. W., Rothberg, J. M., Kumar, V., Schramm, K. J., Haller, E., Proemsey, J. B., et al. (2012). Chronic autophagy is a cellular adaptation to tumor acidic pH microenvironments. *Cancer Research*, *72*(16), 3938–3947. http://dx.doi.org/10.1158/0008-5472.CAN-11-3881.

Wu, J. J., Quijano, C., Chen, E., Liu, H., Cao, L., Fergusson, M. M., et al. (2009). Mitochondrial dysfunction and oxidative stress mediate the physiological impairment induced by the disruption of autophagy. *Aging*, *1*(4), 425–437.

Yang, Z., & Klionsky, D. (2010). Mammalian autophagy: Core molecular machinery and signaling regulation. *Current Opinion in Cell Biology*, *22*(2), 124–131. http://dx.doi.org/10.1016/j.ceb.2009.11.014.

Yang, S., Wang, X., Contino, G., Liesa, M., Sahin, E., Ying, H., et al. (2011). Pancreatic cancers require autophagy for tumor growth. *Genes & Development*, *25*, 717–729. http://dx.doi.org/10.1101/gad.2016111.

Yin, L. I., Kharbanda, S., & Kufe, D. (2009). MUC1 oncoprotein promotes autophagy in a survival response to glucose deprivation. *International Journal of Oncology*, *34*, 1691–1699. http://dx.doi.org/10.3892/ijo.

Young, A. R. J., Narita, M., Ferreira, M., Kirschner, K., Sadaie, M., Darot, J. F. J., et al. (2009). Autophagy mediates the mitotic senescence transition. *Genes & Development*, *23*, 798–803. http://dx.doi.org/10.1101/gad.519709.798.

Yue, Z., Jin, S., Yang, C., Levine, A. J., & Heintz, N. (2003). Beclin 1, an autophagy gene essential for early embryonic development, is a haploinsufficient tumor suppressor. *Proceedings of the National Academy of Sciences of the United States of America*, *100*(25), 15077–15082. http://dx.doi.org/10.1073/pnas.2436255100.

Zhong, Y., Wang, Q. J., Li, X., Yan, Y., Backer, J. M., Chait, B. T., et al. (2009). Distinct regulation of autophagic activity by Atg14L and Rubicon associated with Beclin 1-phosphatidylinositol-3-kinase complex. *Nature Cell Biology*, *11*(4), 468–476. http://dx.doi.org/10.1038/ncb1854.

Zoncu, R., Bar-Peled, L., Efeyan, A., Wang, S., Sancak, Y., & Sabatini, D. M. (2011). mTORC1 senses lysosomal amino acids through an inside-out mechanism that requires the vacuolar H(+)-ATPase. *Science*, *334*(6056), 678–683. http://dx.doi.org/10.1126/science.1207056.

Zoncu, R., Efeyan, A., & Sabatini, D. M. (2011). mTOR: From growth signal integration to cancer, diabetes and ageing. *Nature Reviews Molecular Cell Biology*, *12*(1), 21–35. http://dx.doi.org/10.1038/nrm3025.

CHAPTER THREE

Regulation of Cancer Metabolism by Oncogenes and Tumor Suppressors

Raffaella Iurlaro[1], Clara Lucía León-Annicchiarico[1], Cristina Muñoz-Pinedo[2]

Cell Death Regulation Group, Bellvitge Biomedical Research Institute (IDIBELL), Barcelona, Spain
[1]These authors contributed equally.
[2]Corresponding author: e-mail address: cmunoz@idibell.cat

Contents

Abstract

Cell proliferation requires the coordination of multiple signaling pathways as well as the provision of metabolic substrates. Nutrients are required to generate such building blocks and their form of utilization differs to significant extents between malignant tissues and their nontransformed counterparts. Thus, oncogenes and tumor suppressor genes regulate the proliferation of cancer cells also by controlling their metabolism. Here, we discuss the central anabolic functions of the signaling pathways emanating from mammalian target of rapamycin, MYC, and hypoxia-inducible factor-1. Moreover, we analyze how oncogenic proteins like phosphoinositide-3-kinase, AKT, and RAS, tumor suppressors such as phosphatase and tensin homolog, retinoblastoma, and p53, as well as other factors associated with the proliferation or survival of cancer cells, such as NF-κB, regulate cellular metabolism.

Methods in Enzymology, Volume 542
ISSN 0076-6879
http://dx.doi.org/10.1016/B978-0-12-416618-9.00003-0

ABBREVIATIONS

AMPK AMP-activated protein kinase
COX cytochrome c oxidase
GLS1 glutaminase 1
HIF-1 hypoxia-inducible factor 1
IκB inhibitor of κB proteins
LDH lactate dehydrogenase
LKB1 liver kinase B1
mTOR mammalian (or mechanistic) target of rapamycin
PDH pyruvate dehydrogenase
PDK1 pyruvate dehydrogenase kinase 1
PHD prolyl-4-hydroxylase domain protein
pRb retinoblastoma protein
PtdIns(3,4,5) P3 phosphatidylinositol-3,4,5-trisphosphate
PTEN phosphatase and tensin homologue
SCO2 synthesis of cytochrome c oxidase 2
SREBP sterol regulatory element-binding protein
TIGAR TP53 (tumor protein 53)-induced glycolysis and apoptosis regulator
TSC1/2 tuberous sclerosis 1/2
VHL von Hippel–Lindau

1. INTRODUCTION

Most oncogenes and tumor suppressor genes encode proteins that promote cellular proliferation or cell cycle arrest. In recent years, we are learning that proliferation is tightly coupled with metabolic changes. For this reason, cancer metabolism is an area of intense research, since the metabolism of cancer cells can be exploited for therapeutic purposes (Munoz-Pinedo, El Mjiyad, & Ricci, 2012). In accordance to the normal function of their encoded proteins, oncogenes or tumor suppressors regulate cellular metabolism (Vander Heiden, Cantley, & Thompson, 2009). This is an intrinsic part of their program to reduce or promote cell proliferation. Oncogenes promote glucose and amino acid uptake and metabolism in order to make new lipids, nucleotides, and proteins. Conversely, tumor suppressors upregulate mitochondrial respiration and Krebs (TCA) cycle (see review by Frezza and colleagues, Chapter 1 of this volume). We will discuss how several oncogenes and tumor suppressors regulate cellular metabolism.

2. HIF-1: REGULATOR OF HYPOXIC RESPONSES AND CANCER METABOLISM

Highly proliferating tumor cells are characterized by a hypoxic microenvironment due to the increased oxygen consumption, which stimulates metabolic reprogramming (Vaupel, Thews, & Hoeckel, 2001). The master regulator of cellular responses to low oxygen is hypoxia-inducible factor 1 (HIF-1), a transcription factor induced by hypoxic conditions and whose levels are increased in many human cancers even under normoxia (Semenza, 2010). Under normal oxygen conditions, HIF-1 is degraded by the proteasome after prolyl hydroxylation by prolyl-4-hydroxylase domain proteins (PHDs) and ubiquitination by the tumor suppressor von Hippel–Lindau (VHL) (Kaelin & Ratcliffe, 2008; Fig. 3.1). HIF-1 can also be constitutively activated by genetic alterations, such as the loss of function of VHL in renal cancer cells, or due to the accumulation of metabolites such as fumarate or succinate (Boulahbel, Duran, & Gottlieb, 2009). Cancer cells frequently undergo oxygen shortage which inhibits the prolyl hydroxylases and stabilizes HIF-1, which induces the expression of hundreds of genes involved in angiogenesis, metabolism, apoptosis, and proliferation.

The major metabolic effect of HIF-1 is to trigger the switch from mitochondrial oxidative phosphorylation (OXPHOS) to anaerobic glycolysis. HIF-1 induces the expression of glucose transporters (GLUT-1, GLUT-3) to enhance glucose uptake and it upregulates glycolytic enzymes and the lactate dehydrogenase A (LDHA) subunit to stimulate the conversion of pyruvate into lactate (Brahimi-Horn, Chiche, & Pouyssegur, 2007; Semenza, 2011; Fig. 3.1). Importantly, HIF-1 activates the pyruvate dehydrogenase kinase 1 (PDK1; Kim, Tchernyshyov, Semenza, & Dang, 2006; McFate et al., 2008), a negative regulator of pyruvate dehydrogenase (PDH). PDH converts pyruvate into acetyl-CoA to enter the Krebs cycle in the mitochondria (Fig. 3.1). The effect of inhibiting PDH is the inhibition of mitochondrial oxygen consumption and reduction of ROS production, and this promotes anaerobic glycolysis and thus the Warburg effect (Papandreou, Cairns, Fontana, Lim, & Denko, 2006).

HIF-1 also controls respiration by regulating expression and stability of the cytochrome oxidase subunits cytochrome c oxidase (COX)4-1 and COX4-2 (Fukuda et al., 2007). Additionally, HIF-1 upregulates the expression of the proteins BNIP3 and BNIP3L, which trigger mitochondrial

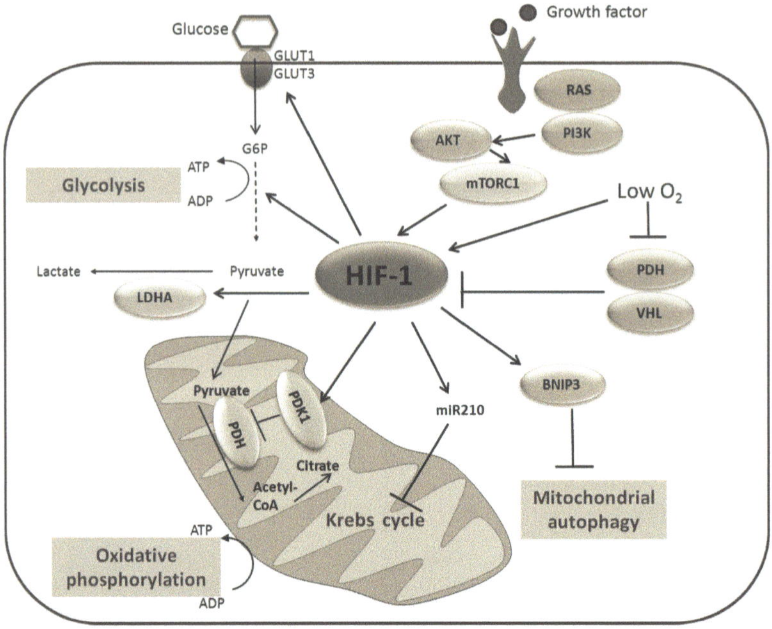

Figure 3.1 *Regulation of cancer metabolism by HIF-1.* HIF-1 switches metabolism from oxidative respiration to anaerobic glycolysis. Hypoxia induces HIF-1 by blocking its inhibitors prolyl-4-hydroxylase domain proteins (PHDs) and von Hippel–Lindau (VHL) protein that need O_2 to exert their functions. Once activated, HIF-1 upregulates the glucose transporters GLUT1 and GLUT3, thus enhancing glucose uptake. HIF-1 induces the expression of almost every enzyme of the glycolytic pathway and lactate dehydrogenase A (LDHA), thus resulting in lactate production. Importantly, HIF-1 induces the pyruvate dehydrogenase kinase 1 (PDK1) that phosphorylates pyruvate dehydrogenase (PDH) blocking the entry of pyruvate into the mitochondria. HIF-1 also induces the expression of miR210, inhibiting important enzymes of Krebs cycle, and upregulates the protein BNIP3 that promotes mitochondrial autophagy. (See the color plate.)

autophagy, another possible mechanism by which HIF-1 reduces oxidative metabolism (Zhang et al., 2008). HIF-1 can also activate the transcription of miR-210, a microRNA which blocks the expression or activity of some enzymes of the Krebs cycle and the Complex I of the electron transport chain (Chen, Li, Zhang, Huang, & Luthra, 2010; Favaro et al., 2010; Fig. 3.1).

3. THE PI3K–AKT–PTEN PATHWAY REGULATES METABOLISM

The PI3K–AKT pathway is one of the main prosurvival pathways activated in human cancers. The phosphatidylinositol 3-kinases (PI3Ks) are a family of proteins that phosphorylate phoshoinositides at the D-3 position of the inositol ring, and their functions are linked to different biological roles, like regulation of cell growth, organismal metabolism, cell proliferation, and vesicle trafficking (Cantley, 2002; Engelman, Luo, & Cantley, 2006).

The best known effector downstream of PI3K is AKT (also known as Protein Kinase B, PKB). Oncogenic mutations in PI3K increase the PI3K and AKT signaling, promoting factor-independent growth and increasing cell invasion and metastasis (Manning & Cantley, 2007). Activated AKT is also an important driver of oncogenic metabolism. It was recognized early that AKT activation drives the glycolytic metabolism of tumor cells (Fig. 3.2; Elstrom et al., 2004). Activation of AKT increases cellular glucose uptake by inducing the expression and membrane translocation of glucose transporters (Barthel et al., 1999; Kohn, Summers, Birnbaum, & Roth, 1996). AKT also

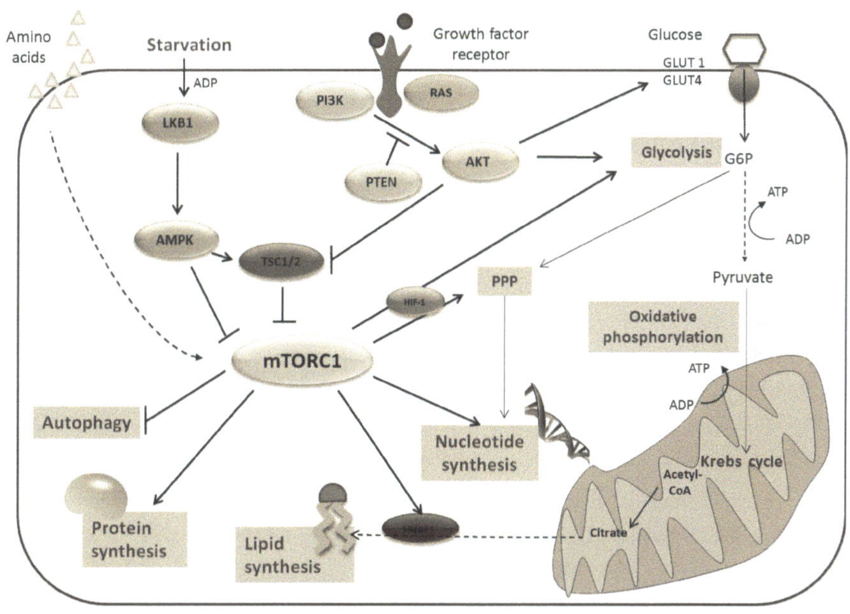

Figure 3.2 *Regulation of cancer metabolism by the PI3K–AKT–PTEN and LKB1–AMPK–mTORC1 pathways.* Growth factor receptors activate Ras and phosphatidylinositol 3-kinase (PI3K) leading to the activation of AKT. Once activated, AKT induces glycolysis by regulating glycolytic enzymes and glucose transporters. These effects are counteracted by the phosphatase and tensin homologue (PTEN). AKT can indirectly activate the mTORC1 pathway that promotes lipid, protein, and nucleotide synthesis, contributing to the building of bioblocks necessary for tumor proliferation. Under stress conditions, the AMP-activated protein kinase (AMPK) activation through the liver kinase B1 (LKB1), opposes glycolytic metabolism in part by inhibiting mTORC1. PPP, pentose phosphate pathway. (See the color plate.)

increases glycolysis by activating the enzyme phosphofructokinase-1 (PFK1) through phosphorylation of phosphofructokinase-2 (PFK2) (Deprez, Vertommen, Alessi, Hue, & Rider, 1997), which leads to allosteric activation of PFK1. In addition, AKT stimulates the mammalian (or mechanistic) target of rapamycin (mTOR) pathway, thus promoting many other metabolic branches as we will discuss below.

PI3K/AKT signaling pathway can be inhibited by the tumor suppressor gene phosphatase and tensin homologue (PTEN). PTEN dephosphorylates phosphatidylinositol-3,4,5-trisphosphate (PtdIns(3,4,5) P3), the second messenger generated by the activation of PI3K, and the main activator of AKT, thereby inhibiting the PI3K–AKT–mTOR pathway. The main functions of PTEN are the regulation of cell growth, metabolism, and survival, and thus it has an important tumor-suppressive ability (Carracedo & Pandolfi, 2008). Even a slight decrease of PTEN levels, or a fine change in *PTEN* gene expression, is sufficient to induce cancer susceptibility (Alimonti et al., 2010). Consistently, loss of PTEN promotes glycolysis (Tandon et al., 2011) and elevation of PTEN levels can reverse the cancer metabolic reprogramming from glycolysis to OXPHOS (Garcia-Cao et al., 2012). For example, transgenic mice carrying additional copies of *PTEN* (referred to as Super-PTEN mice), are less prone to cancer development. In this model, PTEN elevation resulted in a healthier metabolism, with systemic metabolic reprogramming; mice display increased oxygen consumption and energy expenditure, higher mitochondrial biogenesis increasing the mitochondrial ATP production, and an important reduction of body fat accumulation. Cells derived from these mice show reduced glucose and glutamine uptake, increased mitochondrial OXPHOS, and resistance to oncogenic transformation (Garcia-Cao et al., 2012). Conversely, in nontransformed thyrocytes of a PTEN-deficient mouse model, the constitutive PTEN deficiency caused a downregulation of Krebs cycle and OXPHOS, defective mitochondria and reduction of respiration with compensatory glycolysis. In this case, the metabolic switch to glycolysis is driven by PI3K-dependent AMP-activated protein kinase (AMPK) inactivation (Antico Arciuch, Russo, Kang, & Di Cristofano, 2013).

4. mTOR CONTROLS ANABOLISM AND IT IS INHIBITED BY AMPK UPON METABOLIC STRESS

mTOR is a serine/threonine kinase that is part of two distinct complexes, TORC1 and TORC2, which have different sensitivity to rapamycin. We will discuss the role of the rapamycin sensitive complex, mTORC1, which controls cell growth and metabolism in response to environmental signals (Wullschleger, Loewith, & Hall, 2006). The mTOR pathway is one of the most deregulated signaling pathways in human cancer, and growth-factor-independent activation of mTORC1 is observed in up to 80% of tumors, across nearly all lineages (Guertin & Sabatini, 2007; Menon & Manning, 2009). mTOR is also deregulated in metabolic disorders, such as obesity and type 2 diabetes. Mice with hyperactive mTORC1 signaling in the liver display metabolic abnormalities, including defects in glucose and lipid homeostasis, and subsequently develop hepatocellular carcinoma (Menon et al., 2012).

mTOR integrates diverse signals to regulate cell growth: growth factors, nutrients, oxygen, energy, and several forms of stress. mTOR, downstream of PI3K, responds to growth factors via the inactivation of tuberous sclerosis (TSC)1 and TSC2 by AKT; these proteins are negative regulators of mTORC1 (Manning & Cantley, 2007; Fig. 3.2). Nutrients, particularly amino acids, also regulate mTORC1 signaling, which controls protein translation. The molecular mechanism by which mTORC1 senses intracellular amino acids is not fully understood, but it requires the Rag GTPases (Kim, Goraksha-Hicks, Li, Neufeld, & Guan, 2008; Sancak et al., 2010).

mTOR regulates many anabolic pathways. Through regulation of HIF1 it activates glycolysis and the pentose phosphate pathway (PPP) (Figs. 3.1 and 3.2), and by activating the transcription factor sterol regulatory element-binding protein (SREBP)1, it also stimulates lipid synthesis (Düvel et al., 2010; Fig. 3.2). Nucleotide synthesis is also regulated by mTOR in two different manners: through regulation of the PPP and by activation of an enzyme of pyrimidine synthesis (Ben-Sahra, Howell, Asara, & Manning, 2013; Robitaille et al., 2013). Thus, cells with active mTOR are stimulated to proliferate by making all necessary building blocks.

mTOR is inhibited in conditions of nutritional stress by the AMPK. Tumors under metabolic stress adapt to these conditions by altering the liver kinase B1 (LKB1)–AMPK pathway (Sebbagh, Olschwang, Santoni, & Borg, 2011). As a result, the LKB1–AMPK pathway works as a metabolic checkpoint and inhibits cancer metabolic reprogramming (Jones et al., 2005; Kuhajda, 2008). AMPK is an ATP sensor that checks and regulates cellular energy homeostasis. AMPK is activated in response to nutrient deprivation or hypoxia, when ATP levels decline and the AMP and ADP levels increase (Fig. 3.2) (Hardie, 2011; Xiao et al., 2011). Under conditions of energy stress, LKB1 (serine–threonine kinase LKB1) acts as the main upstream kinase that activates AMPK (Shaw, Bardeesy, et al., 2004; Woods et al.,

2003). Once activated, AMPK can target a wide range of downstream metabolic pathways, especially the mTOR pathway. During energetic stress, AMPK can inhibit mTORC1 through two different mechanism; phosphorylating TSC2 (Corradetti, Inoki, Bardeesy, DePinho, & Guan, 2004; Inoki, Zhu, & Guan, 2003; Shaw, Kosmatka, et al., 2004) or by direct phosphorylation of Raptor, a component of mTORC1 (Scott, Norman, Hawley, Kontogiannis, & Hardie, 2002). LKB1-deficient cells and mutant mice for LKB1, or MEFs deficient for TSC2, show hyperactive mTORC1 signaling in response to energy stress (Shaw, Bardeesy, et al., 2004). Thus, AMPK alters important cellular responses, like cell growth, proliferation and autophagy (Shackelford et al., 2009). The lack of AMPK signaling increase tumorigenesis and enhances the glycolytic metabolism in cancer cells (Faubert et al., 2012). However, AMPK can also promote survival of tumor cells: LKB1 deficiency reduces the AMPK signaling in tumor cells (Godlewski et al., 2010; Shackelford & Shaw, 2009; Zheng et al., 2009), and deletion of LKB1 makes the cells more sensitive to nutrient deprivation (Shaw, Bardeesy, et al., 2004). Additionally, by inhibiting lipid synthesis and promoting lipid oxidation, AMPK contributes to maintenance of NADPH levels thus mitigating redox stress (Jeon, Chandel, & Hay, 2012).

5. c-MYC PROMOTES AEROBIC ANABOLISM

c-Myc has been reported to be the master regulator of metabolic processes involved in cell proliferation. Myc is deregulated in many human cancers in which it triggers tumorogenesis through the transcriptional modulation of many genes. In fact, it has been recently proposed that Myc is a "general" transcription factor, in the sense that high levels of c-Myc in tumor cells produce elevated levels of transcripts from the existing gene expression program of tumor cells (Lin et al., 2012). This includes genes involved in glucose metabolism, nucleotide, lipid, amino acid, and protein synthesis (Dang, 2013; Li & Simon, 2013). Once activated, c-Myc binds, with its cofactor Max, to the consensus sequences called "E-boxes" present in genes driven by all three RNA polymerases, resulting in ribosomal RNA synthesis and ribosome biogenesis, necessary to build the increasing cell mass (Grandori et al., 2005; van Riggelen, Yetil, & Felsher, 2010).

c-Myc also regulates mitochondrial biogenesis by inducing the expression of genes involved in mitochondrial structure and function, such as *TFAM* which encodes a protein involved in mitochondrial transcription and mitochondrial DNA replication (Li, 2005; Fig. 3.3). To trigger biomass

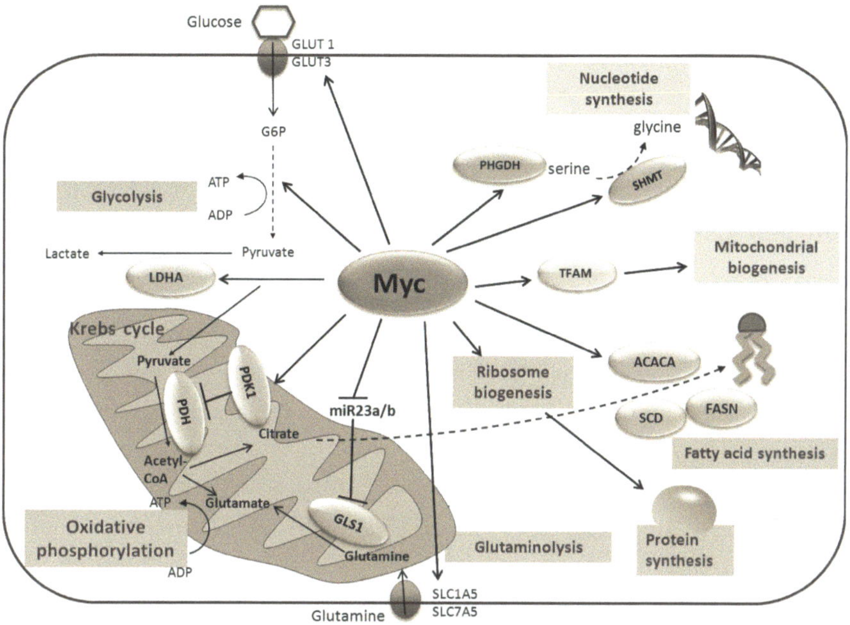

Figure 3.3 *Myc regulates cancer metabolism.* Myc promotes cancer cell metabolism at several levels. Myc upregulates the glucose transporters GLUT1 and GLUT3 increasing glucose uptake. It induces several glycolytic enzymes such as the lactate dehydrogenase A (LDHA) resulting in lactate production. Like HIF-1, Myc induces pyruvate dehydrogenase kinase 1 (PDK1) expression, which prevents pyruvate entry into the mitochondria. Myc also regulates glutaminolysis: it upregulates glutamine transporters SLC1A5 and SLC7A5 and induces glutaminase 1 (GLS1) expression. Myc also promotes biomass accumulation essential for proliferating tumor cells. It regulates ribosome biogenesis, mitochondrial biogenesis, and several enzymes involved in fatty acids synthesis such as acetyl-CoA carboxylase (ACACA), fatty acid synthetase (FASN), and stearoyl-CoA desaturase (SCD). Additionally, Myc regulates enzymes involved in nucleotide synthesis such as phosphoglycerate dehydrogenase (PHGDH) and serine hydroxymethyltransferase (SHMT). (See the color plate.)

accumulation necessary for cell proliferation, c-Myc induces the expression of almost every glycolytic gene, redirecting cells to glucose consumption for ATP but also for biomolecule production. c-Myc also stimulates the transcription of LDHA that is necessary for c-Myc mediated tumorigenesis in some models (Shim et al., 1997; Fig. 3.3).

Like HIF-1, c-Myc regulates other important glycolytic enzymes such as hexokinase 2 -that phosphorylates glucose to make glucose-6-phosphate- and PDK1 -which phosphorylates and inhibits PDH, blocking the entry of pyruvate into the mitochondria (Kim, Gao, Liu, Semenza, & Dang,

2007; Fig. 3.3). It has been shown by *in vivo* imaging techniques that in c-Myc-driven liver tumors pyruvate is converted preferentially to lactate (Hu et al., 2011). Interestingly, metabolic changes were detected prior to the appearance of tumors: in pretumor tissues, an accumulation of alanine due to increased expression of transaminases was observed.

c-Myc also controls glutamine metabolism, achieved through regulation of mitochondrial glutaminase 1 (GLS1) expression (Gao et al., 2009). Glutamine is converted to glutamate by GLS1, whose expression is increased in c-Myc-dependent tumors. Glutamate then enters the Krebs cycle to produce ATP or glutathione. There are evidences that GLS1 is regulated by c-Myc also at posttranscriptional level. c-Myc suppresses the expression of two miRNAs, miR-23a and miR-23b, which target GLS1 in its 3′UTR, resulting in increased glutaminase expression and glutamine metabolism. c-Myc also stimulates the transport of glutamine inside the cell by increasing the expression of the glutamine transporters SLC1A5 and SLC7A5 (Fig. 3.3).

It has been shown that c-Myc can regulate nucleotide biosynthesis by transcriptional regulation of several key enzymes, redirecting glycolysis to the synthesis of serine and glycine that are essential for nucleotide building (Mannava et al., 2008). Recently, Myc has also been associated to lipid synthesis as many enzymes of fatty acid biosynthesis are its direct targets and they contribute to the building of bioblocks needed in the c-Myc-driven proliferation program (Loven et al., 2012; Fig. 3.3). Thus, Myc has been shown to activate all pathways necessary to build new cells.

6. RAS STIMULATES GLYCOLYSIS AND THE PPP

The Ras family encompasses a number of small GTPases that transduce signals to induce proliferation, including the metabolic switch. Transfection of a constitutively activated form of Ras is sufficient to stimulate glycolysis and the PPP (Vizan et al., 2005). Ras proteins are activated downstream of growth factors or they are constitutively active in tumors, and they signal through MAP kinases and/or through PI3K. Some of the metabolic effects of Ras, thus, may be mediated through the PI3K/AKT/mTOR pathway, while other effects can be due to stimulation of Myc. H-Ras, for instance, upregulates Glut-1 mRNA through the PI3-kinase pathway. This effect is indirect, through the PI3K-mediated upregulation of HIF-1 (Chen, Pore, Behrooz, Ismail-Beigi, & Maity, 2001). Since Ras can indirectly regulate HIF-1, it can regulate metabolism in the same manner, and this is for instance the case in colon cancer cells with hyperactivated KRas, in which KRas inhibits mitochondrial metabolism through activation of HIF-1 (Chun et al., 2010).

Pancreatic tumors often carry activating KRAS mutations. In these cells, KRas regulates multiple metabolic pathways at the transcriptional level. It stimulates glucose uptake and it channels glucose intermediates into the hexosamine biosynthesis and PPPs. These effects are mediated by MAP kinases and Myc (Ying et al., 2012). Additionally, pancreatic ductal adenocarcinomas have recently been shown to depend on a nonclassical glutamine utilization pathway stimulated transcriptionally by Kras. Kras directs the metabolism of these cells in toward the use of glutamine as a source of pyruvate and NADPH to maintain the cellular redox balance (Son et al., 2013).

Ras is also a regulator of autophagy, a cellular process that can provide nutrients by self-digestion of intracellular components. This process is also responsible for clearance of damaged mitochondria. Ras-mediated transformation induces autophagy, which is required to maintain mitochondrial metabolic functions in Ras-driven tumors (Guo et al., 2011). In these tumors, knockdown of essential autophagy genes can promote the accumulation of abnormal mitochondria unable to metabolize lipids through fatty acid oxidation (White, 2013). Similarly, tumors driven by a Ras downstream effector, the oncogene BRAF, rely on autophagy to maintain healthy mitochondria and glutamine metabolism (Strohecker et al., 2013).

7. NF-kappaB REGULATES INFLAMMATION AND PROLIFERATION BUT ALSO METABOLISM

NF-κB is a transcription factor of the Rel-homology-domain family. Its subunit p65/RelA is the most important in transactivation of several target genes involved in immunity, inflammation, and proliferation. Its activity is tightly regulated by the inhibitors of κB proteins (IκBs) and the IκB kinase proteins (IKKs), and it results in the expression of growth factors, cytokines, and promotion of cell proliferation (Hayden & Ghosh, 2004). Although NF-κB is not considered a classical oncogene, its expression can be regulated by several oncogenes, suggesting a role of NF-κB in promotion of tumorogenesis (Basseres & Baldwin, 2006). It has been reported that oncogenic H-Ras activates NF-κB (Finco et al., 1997) inducing lung tumor progression *in vivo* in a p53-dependent (Meylan et al., 2009) or independent manner (Bassères, Ebbs, Levantini, & Baldwin, 2010). In cells with mutated p53, the activation of Ras induces a metabolic switch from oxidative mitochondrial phosphorylation to aerobic glycolysis that has been related to NF-κB activation (Kawauchi, Araki, Tobiume, & Tanaka, 2008). In this model, the loss of p53 activity resulted in transcriptional activation of NF-κB that was essential for the enhanced glucose consumption and lactate production. GLUT3 expression was directly regulated by NF-κB, accordingly with the observed increase of glucose uptake in those cells. Recently, it has been shown that NF-κB activation by the epidermal growth factor receptor (EGFR) in cancer cells induces the expression of pyruvate kinase M2 (PKM2), triggering lactate production and glucose uptake (Yang et al., 2012). However, NF-κB has also been shown to contribute to tumorogenesis by sustaining mitochondrial function. This effect was mediated through p53 and its target synthesis of cytochrome c oxidase 2 (SCO2), which increases OXPHOS (Mauro et al., 2011). Although NF-κB is not a typical oncogene, all these findings suggest an involvement of NF-κB in metabolic reprogramming and tumorigenesis. However, the manner by which NF-κB regulates cancer metabolism is still unclear and may be context dependent.

8. RETINOBLASTOMA: SUPPRESSING TUMOROGENESIS AND ANABOLISM

The retinoblastoma protein (pRb) is one of the tumor suppressors whose role in cancer metabolism has been most extensively studied (Nicolay & Dyson, 2013). The major function of pRb is the inhibition of cell cycle progression exerted through repression of the E2F1 transcription factor. This function is reverted by pRb phosphorylation by cyclin D-CDK4/6, which inactivates Rb and promotes E2F1-mediated transcription. Many signals can regulate pRb expression; among those, AMPK has been shown to phosphorylate directly pRb controlling the G1/S phase transition based on the energy status of the cell (Dasgupta & Milbrandt, 2009). Recently, pRb was shown to regulate starvation-induced stress response in *Caenorhabditis elegans* (Cui, Cohen, Teng, & Han, 2013) and similar results have been recently provided in a Drosophila model, suggesting an involvement of pRb in cancer metabolism (Nicolay et al., 2013). This study shows that flies with mutant RBF1 (Drosophila Rb homolog) are hypersensitive to fasting conditions and present deregulated glutamine and nucleotide metabolism. Also human cancers with inactivated pRb show an increase in glutamine uptake due to upregulation of expression of the glutamine transporter ASCT2, and an increase in glutamine utilization in the Krebs cycle resulting in glutathione accumulation (Reynolds et al., 2014). pRb and E2F1 can regulate in an opposite way the oxidative metabolism, modulating the expression of different genes at their promoters. pRb deletion in murine erythrocytes causes a block in differentiation and impairs mitochondrial biogenesis uncovering a positive role of pRb on mitochondrial activity (Sankaran, Orkin, & Walkley, 2008), while other studies show that E2F1 induces a switch from oxidative to glycolytic metabolism by repressing multiple genes involved in mitochondrial function (Blanchet et al., 2011). Some studies have described a role of pRb in lipid metabolism, showing that pRb deletion induces E2F-dependent expression of fatty acid biosynthesis enzymes and SREBP (Shamma et al., 2009). Additionally, pRb has been shown to play a role in nucleotide metabolism by inhibiting enzymes such as dihydrofolate reductase and thymidylate synthase (Angus et al., 2002). All these data indicate a connection of pRb in cell cycle progression and regulation of tumor metabolism.

9. p53 REGULATES MULTIPLE METABOLIC PATHWAYS

p53 function is lost in most human cancers (Soussi & Beroud, 2001). p53 exerts an important defense mechanism against tumor development (Vousden & Ryan, 2009). It is a transcription factor that regulates a large range of functions like DNA damage response, apoptosis, and senescence. Mutations in p53 found in tumors can produce a variety of biological effects, for example: lack of control in cell cycle, defective apoptosis, and inefficient DNA repair (Resnick & Inga, 2003). In p53 knockout mice, tumor development is rapid and spontaneous (Donehower et al., 1992). p53 also plays an important role in metabolic stress response (Vousden & Ryan, 2009). Cells lacking p53 and deprived of glucose cannot undergo cell cycle arrest, since p53 controls a metabolic checkpoint. This makes p53-defective cells more sensitive than nontransformed cells to metabolic stress, what has led to propose the use of antiglycolytic drugs for therapy of p53-deficient tumors (Jones et al., 2005). p53 also responds to lack of serine and allows *de novo* synthesized serine to be channeled to production of reduced glutathione to counter oxidative stress (Maddocks et al., 2013). For this reason, p53-deficient cells are more sensitive to serine depletion.

As part of the antitumor activity of p53, it promotes glucose OXPHOS and it inhibits glycolysis (Fig. 3.4). Disruption of TP53 in mice promotes a significant decrease in oxygen consumption that closely correlates with p53

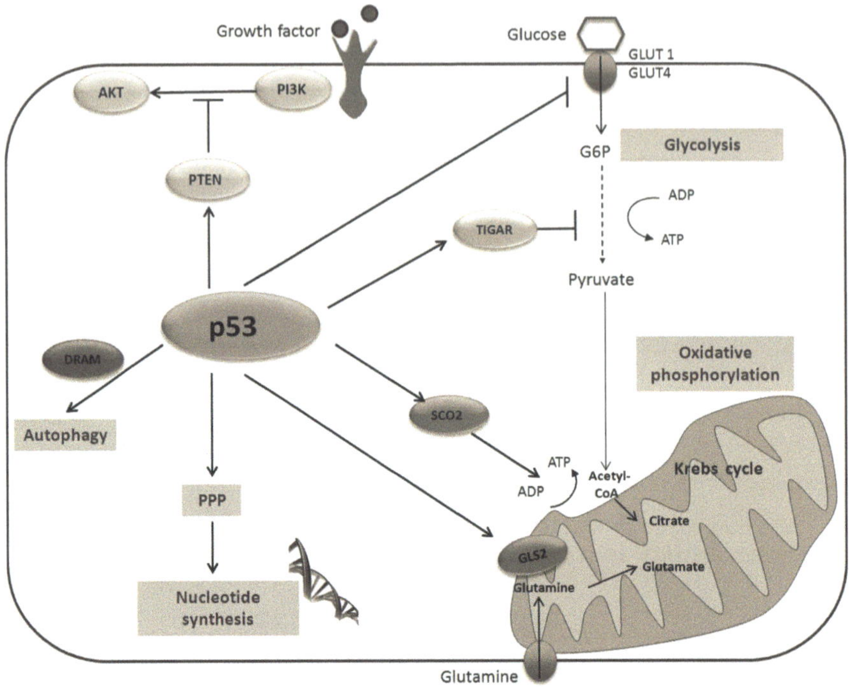

Figure 3.4 *p53 regulates multiple metabolic pathways.* p53 responds to metabolic stress and it can inhibit the tumorigenic metabolic switch by suppressing glycolysis and activating the phosphatase and tensin homologue (PTEN). p53 inhibits the transcription of GLUT1 and GLUT4 reducing glucose uptake and it upregulates the TP53 (tumor protein 53)-induced glycolysis and apoptosis regulator (TIGAR), which results in glycolysis inhibition. p53 increases the mitochondrial metabolism by activation of the synthesis of cytochrome c oxidase 2 (SCO2), thus promoting oxidative phosphorylation. p53 can also induce, contradictorily, prosurvival responses in cancer cells, for instance when it increases the flux through the pentose phosphate pathway (PPP) or glutamine utilization. P53 can regulate positively autophagy by increasing the expression of DRAM. (See the color plate.)

deficiency, as p53 increases OXPHOS through upregulation of the gene SCO2, whose product participates in the assembly of COX in the mitochondria (Matoba et al., 2006). p53 upregulates TP53-induced glycolysis and apoptosis regulator (TIGAR), an enzyme that decreases the levels of the glycolytic activator fructose-2,6-bisphosphate (Bensaad et al., 2006). It also inhibits glucose uptake by inhibiting the transcription of GLUT1 and GLUT4 (Schwartzenberg-Bar-Yoseph, Armoni, & Karnieli, 2004). p53 can also inhibit the glycolytic pathway indirectly by activating PTEN, thus inhibiting the PI3K pathway (Stambolic et al., 2001).

p53 is also involved in somewhat contradictory responses, since it has been associated with pathways that may support tumor growth and survival. For example, in some tumor cells it can increase the flux through the PPP, reducing oxidative stress and promoting anabolism, thus helping the growth of cancer cells (Vousden & Ryan, 2009). p53 is also able to contribute to glutaminolysis, an alternative fuel bioenergetic pathway, where glutamine is metabolized to produce α-ketoglutarate from glutamate in the Krebs cycle. This pathway is important in the process of oncogenic transformation: the enzyme which converts glutamine to glutamate, glutaminase 1 (GLS1/KGA) has been shown to help tumor development (Wang et al., 2010). p53 can play a role in the regulation of glutaminolysis by the activation of another isoform of glutaminase (GLS2/LGA), helping the cells produce ATP in periods of glucose deprivation (Hu et al., 2010; Suzuki et al., 2010). Both the activation of the PPP and glutaminolysis could have a function in reduction of oxidative stress.

Another function of p53 is related to autophagy. The control of p53 in autophagy is context specific, and it could work like a prodeath or cell survival mechanism. One of the ways by which p53 regulates autophagy is by upregulating damage regulated autophagy modulator (DRAM), a lysosomal protein that positively regulates autophagy (Crighton et al., 2006).

The family of transcription factors of p53 includes p63 and p73, both functional homologs with high sequential and structural similarity (Kaghad et al., 1997; Yang et al., 1998). These two members of the p53 family have functions that are markedly different from those of p53 (Allocati et al., 2012), but they also have many similarities and overlapping activity with p53, including the regulation of cellular metabolism (Berkers, Maddocks, Cheung, Mor, & Vousden, 2013). Tp63 and Tp73 genes are transcribed from two different promoters, and the final product can be either full length proteins that retain a full transactivation (TA) domain (TAp63 and TAp73) or N-terminally truncated isoforms (ΔNp63 and ΔNp73) (De Laurenzi & Melino, 2000). TAp63 can control fat and glucose metabolism, because is a positive regulator of the transcription of Sirt1, AMPKa2, and LKB1. TAp73 can promote cancer cell proliferation, controlling biosynthetic pathways and cellular antioxidant capacity through the regulation of glucose metabolism. TAp73 regulates the expression of glucose-6-phosphate dehydrogenase (G6PD), an enzyme involved in glucose metabolism through the PPP (Du et al., 2013). p73 can be negatively regulated by AMPKα by direct interaction without affecting p53, which represses the TAp73 transcription program (Lee, Lee, Sin, Kim, & Um, 2008).

And recently it was discovered that like p53, TAp73 is implicated in the maintenance of mitochondrial Complex IV (Rufini et al., 2012).

In summary, p53 opposes the PI3K pathway to inhibit anabolism, it promotes mitochondrial metabolism and it regulates oxidative stress. The metabolic roles of p53 may well be more important for its tumor suppressor abilities than its roles as a proapoptotic or prosenescent proteins, as recently revealed by a study employing a mutant that had lost these functions and still suppressed tumorogenesis (Li et al., 2012).

10. CONCLUSIONS

To date, a good number of oncogenes and tumor suppressors have been shown to play a role as regulators of metabolism. The vast literature is growing quickly, and we have only summarized here the roles of a few of these genes. However, many other proteins involved in cancer have been shown to play roles in metabolism, from the breast cancer associated receptor tyrosine kinase ErbB2 (Her2/neu) (Zhao et al., 2009) to the promyelocytic leukemia tumor suppressor (Carracedo et al., 2012) or many of the Bcl-2 family of antiapoptotic proteins (reviewed by Fulda and colleagues, Chapter 4 of this volume). Metabolic rewiring is such an important part of the cellular growth process that we will likely see this field expanding in the future.

ACKNOWLEDGMENTS

Studies in CMP's lab related to the topic of this review are supported by FIS grant PI13/00139. R. I. is supported by a fellowship of SUR of the ECO of the Government of Catalonia. We apologize to colleagues whose work could not be cited.

REFERENCES

Alimonti, A., Carracedo, A., Clohessy, J. G., Trotman, L. C., Nardella, C., & Egia, A. (2010). Subtle variations in Pten dose determine cancer susceptibility. *Nature Genetics*, *42*(5), 454–458.

Allocati, N., Di Ilio, C., & De Laurenzi, V. (2012). p63/p73 in the control of cell cycle and cell death. *Experimental Cell Research*, *318*(11), 1285–1290.

Angus, S. P., Wheeler, L. J., Ranmal, S. A., Zhang, X., Markey, M. P., & Mathews, C. K. (2002). Retinoblastoma tumor suppressor targets dNTP metabolism to regulate DNA replication. *Journal of Biological Chemistry*, *277*(46), 44376–44384. http://dx.doi.org/10.1074/jbc.M205911200.

Antico Arciuch, V. G., Russo, M. A., Kang, K. S., & Di Cristofano, A. (2013). Inhibition of AMPK and Krebs cycle gene expression drives metabolic remodeling of Pten-deficient preneoplastic thyroid cells. *Cancer Research*, *73*(17), 5459–5472.

Barthel, A., Okino, S. T., Liao, J., Nakatani, K., Li, J., & Whitlock, J. P., Jr. (1999). Regulation of GLUT1 gene transcription by the serine/threonine kinase Akt1. *The Journal of Biological Chemistry*, *274*(29), 20281–20286.

Basseres, D. S., & Baldwin, A. S. (2006). Nuclear factor-[kappa]B and inhibitor of [kappa]B kinase pathways in oncogenic initiation and progression. *Oncogene*, *25*(51), 6817.

Bassères, D. S., Ebbs, A., Levantini, E., & Baldwin, A. S. (2010). Requirement of the NF-kappaB subunit p65/RelA for K-Ras-induced lung tumorigenesis. *Cancer Research*, *70*(9), 3537–3546. http://dx.doi.org/10.1158/0008-5472.can-09-4290.

Bensaad, K., Tsuruta, A., Selak, M. A., Vidal, M. N., Nakano, K., & Bartrons, R. (2006). TIGAR, a p53-inducible regulator of glycolysis and apoptosis. *Cell*, *126*(1), 107–120.

Ben-Sahra, I., Howell, J. J., Asara, J. M., & Manning, B. D. (2013). Stimulation of de novo pyrimidine synthesis by growth signaling through mTOR and S6K1. *Science*, *339*(6125), 1323–1328.

Berkers, C. R., Maddocks, O. D., Cheung, E. C., Mor, I., & Vousden, K. H. (2013). Metabolic regulation by p53 family members. *Cell Metabolism*, *18*(5), 617–633.

Blanchet, E., Annicotte, J.-S., Lagarrigue, S., Aguilar, V., Clape, C., & Chavey, C. (2011). E2F transcription factor-1 regulates oxidative metabolism. *Nature Cell Biology*, *13*(9), 1146.

Boulahbel, H., Duran, R. V., & Gottlieb, E. (2009). Prolyl hydroxylases as regulators of cell metabolism. *Biochemical Society Transactions*, *37*(Pt 1), 291–294. http://dx.doi.org/10.1042/BST0370291, BST0370291 [pii].

Brahimi-Horn, M. C., Chiche, J., & Pouyssegur, J. (2007). Hypoxia signalling controls metabolic demand. *Current Opinion in Cell Biology*, *19*(2), 223.

Cantley, L. C. (2002). The phosphoinositide 3-kinase pathway. *Science*, *296*(5573), 1655–1657.

Carracedo, A., & Pandolfi, P. P. (2008). The PTEN–PI3K pathway: Of feedbacks and cross-talks. *Oncogene*, *27*(41), 5527–5541.

Carracedo, A., Weiss, D., Leliaert, A. K., Bhasin, M., de Boer, V. C., & Laurent, G. (2012). A metabolic prosurvival role for PML in breast cancer. *The Journal of Clinical Investigation*, *122*(9), 3088–3100.

Chen, Z., Li, Y., Zhang, H., Huang, P., & Luthra, R. (2010). Hypoxia-regulated microRNA-210 modulates mitochondrial function and decreases ISCU and COX10 expression. *Oncogene*, *29*(30), 4362.

Chen, C., Pore, N., Behrooz, A., Ismail-Beigi, F., & Maity, A. (2001). Regulation of glut1 mRNA by hypoxia-inducible factor-1. Interaction between H-ras and hypoxia. *The Journal of Biological Chemistry*, *276*, 9519.

Chun, S. Y., Johnson, C., Washburn, J. G., Cruz-Correa, M. R., Dang, D. T., & Dang, L. H. (2010). Oncogenic KRAS modulates mitochondrial metabolism in human colon cancer cells by inducing HIF-1alpha and HIF-2alpha target genes. *Molecular Cancer*, *9*, 293.

Corradetti, M. N., Inoki, K., Bardeesy, N., DePinho, R. A., & Guan, K.-L. (2004). Regulation of the TSC pathway by LKB1: Evidence of a molecular link between tuberous sclerosis complex and Peutz-Jeghers syndrome. *Genes & Development*, *18*(13), 1533–1538.

Crighton, D., Wilkinson, S., O'Prey, J., Syed, N., Smith, P., & Harrison, P. R. (2006). DRAM, a p53-induced modulator of autophagy, is critical for apoptosis. *Cell*, *126*(1), 121.

Cui, M., Cohen, M. L., Teng, C., & Han, M. (2013). The tumor suppressor Rb critically regulates starvation-induced stress response in C. elegans. *Current Biology*, *23*(11), 975.

Dang, C. V. (2013). MYC, metabolism, cell growth, and tumorigenesis. *Cold Spring Harbor Perspectives in Medicine*, *3*(8), a014217.

Dasgupta, B., & Milbrandt, J. (2009). AMP-activated protein kinase phosphorylates retinoblastoma protein to control mammalian brain development. *Developmental Cell*, *16*(2), 256.

De Laurenzi, V., & Melino, G. (2000). Evolution of functions within the p53/p63/p73 family. *Annals of the New York Academy of Sciences*, *926*, 90–100.

Deprez, J., Vertommen, D., Alessi, D. R., Hue, L., & Rider, M. H. (1997). Phosphorylation and activation of heart 6-phosphofructo-2-kinase by protein kinase B and other protein kinases of the insulin signaling cascades. *The Journal of Biological Chemistry*, *272*(28), 17269–17275.

Donehower, L. A., Harvey, M., Slagle, B. L., McArthur, M. J., Montgomery, C. A., Jr., & Butel, J. S. (1992). Mice deficient for p53 are developmentally normal but susceptible to spontaneous tumours. *Nature, 356*(6366), 215–221.

Du, W., Jiang, P., Mancuso, A., Stonestrom, A., Brewer, M. D., & Minn, A. J. (2013). TAp73 enhances the pentose phosphate pathway and supports cell proliferation. *Nature Cell Biology, 15*(8), 991–1000.

Düvel, K., Yecies, J. L., Menon, S., Raman, P., Lipovsky, A. I., & Souza, A. L. (2010). Activation of a metabolic gene regulatory network downstream of mTOR complex 1. *Molecular Cell, 39*(2), 171.

Elstrom, R. L., Bauer, D. E., Buzzai, M., Karnauskas, R., Harris, M. H., & Plas, D. R. (2004). Akt stimulates aerobic glycolysis in cancer cells. *Cancer Research, 64*(11), 3892–3899. http://dx.doi.org/10.1158/0008-5472.can-03-2904.

Engelman, J. A., Luo, J., & Cantley, L. C. (2006). The evolution of phosphatidylinositol 3-kinases as regulators of growth and metabolism. *Nature Reviews Genetics*, 7(8), 606–619.

Faubert, B., Boily, G., Izreig, S., Griss, T., Samborska, B., & Dong, Z. (2012). AMPK is a negative regulator of the Warburg effect and suppresses tumor growth in vivo. *Cell Metabolism, 17*(1), 113–124.

Favaro, E., Ramachandran, A., McCormick, R., Gee, H., Blancher, C., & Crosby, M. (2010). MicroRNA-210 regulates mitochondrial free radical response to hypoxia and krebs cycle in cancer cells by targeting iron sulfur cluster protein ISCU. *PLoS One, 5*(4), e10345. http://dx.doi.org/10.1371/journal.pone.0010345.

Finco, T. S., Westwick, J. K., Norris, J. L., Beg, A. A., Der, C. J., & Baldwin, A. S. (1997). Oncogenic Ha-Ras-induced signaling activates NF-kappaB transcriptional activity, which is required for cellular transformation. *Journal of Biological Chemistry, 272*(39), 24113–24116. http://dx.doi.org/10.1074/jbc.272.39.24113.

Fukuda, R., Zhang, H., Kim, J. W., Shimoda, L., Dang, C. V., & Semenza, G. L. (2007). HIF-1 regulates cytochrome oxidase subunits to optimize efficiency of respiration in hypoxic cells. *Cell, 129*(1), 111–122. http://dx.doi.org/10.1016/j.cell.2007.01.047, S0092-8674(07)00307-8 [pii].

Gao, P., Tchernyshyov, I., Chang, T.-C., Lee, Y.-S., Kita, K., & Ochi, T. (2009). c-Myc suppression of miR-23a/b enhances mitochondrial glutaminase expression and glutamine metabolism. *Nature, 458*(7239), 762.

Garcia-Cao, I., Song, M. S., Hobbs, R. M., Laurent, G., Giorgi, C., & de Boer, V. C. (2012). Systemic elevation of PTEN induces a tumor-suppressive metabolic state. *Cell, 149*(1), 49–62.

Godlewski, J., Nowicki, M. O., Bronisz, A., Nuovo, G., Palatini, J., & De Lay, M. (2010). MicroRNA-451 regulates LKB1/AMPK signaling and allows adaptation to metabolic stress in glioma cells. *Molecular Cell, 37*(5), 620.

Grandori, C., Gomez-Roman, N., Felton-Edkins, Z. A., Ngouenet, C., Galloway, D. A., & Eisenman, R. N. (2005). c-Myc binds to human ribosomal DNA and stimulates transcription of rRNA genes by RNA polymerase I. *Nature Cell Biology*, 7(3), 311.

Guertin, D. A., & Sabatini, D. M. (2007). Defining the role of mTOR in cancer. *Cancer Cell, 12*(1), 9.

Guo, J. Y., Chen, H. Y., Mathew, R., Fan, J., Strohecker, A. M., & Karsli-Uzunbas, G. (2011). Activated Ras requires autophagy to maintain oxidative metabolism and tumorigenesis. *Genes & Development, 25*(5), 460–470.

Hardie, D. G. (2011). Adenosine monophosphate-activated protein kinase: A central regulator of metabolism with roles in diabetes, cancer, and viral infection. *Cold Spring Harbor Symposia on Quantitative Biology, 76*, 155–164. http://dx.doi.org/10.1101/sqb.2011.76.010819.

Hayden, M. S., & Ghosh, S. (2004). Signaling to NF-KB. *Genes & Development, 18*(18), 2195–2224. http://dx.doi.org/10.1101/gad.1228704.

Hu, S., Balakrishnan, A., Bok, R. A., Anderton, B., Larson, P. E. Z., & Nelson, S. J. (2011). 13C-Pyruvate imaging reveals alterations in glycolysis that precede c-Myc-induced tumor formation and regression. *Cell Metabolism*, *14*(1), 131–142.

Hu, W., Zhang, C., Wu, R., Sun, Y., Levine, A., & Feng, Z. (2010). Glutaminase 2, a novel p53 target gene regulating energy metabolism and antioxidant function. *Proceedings of the National Academy of Sciences of the United States of America*, *107*(16), 7455–7460.

Inoki, K., Zhu, T., & Guan, K.-L. (2003). TSC2 mediates cellular energy response to control cell growth and survival. *Cell*, *115*(5), 577.

Jeon, S. M., Chandel, N. S., & Hay, N. (2012). AMPK regulates NADPH homeostasis to promote tumour cell survival during energy stress. *Nature*, *485*(7400), 661–665.

Jones, R. G., Plas, D. R., Kubek, S., Buzzai, M., Mu, J., & Xu, Y. (2005). AMP-activated protein kinase induces a p53-dependent metabolic checkpoint. *Molecular Cell*, *18*(3), 283.

Kaelin, W. G., & Ratcliffe, P. J. (2008). Oxygen sensing by metazoans: The central role of the HIF hydroxylase pathway. *Molecular Cell*, *30*(4), 393.

Kaghad, M., Bonnet, H., Yang, A., Creancier, L., Biscan, J. C., & Valent, A. (1997). Monoallelically expressed gene related to p53 at 1p36, a region frequently deleted in neuroblastoma and other human cancers. *Cell*, *90*(4), 809–819.

Kawauchi, K., Araki, K., Tobiume, K., & Tanaka, N. (2008). p53 regulates glucose metabolism through an IKK-NF-[kappa]B pathway and inhibits cell transformation. *Nature Cell Biology*, *10*(5), 611.

Kim, J. W., Gao, P., Liu, Y. C., Semenza, G. L., & Dang, C. V. (2007). Hypoxia-inducible factor 1 and dysregulated c-Myc cooperatively induce vascular endothelial growth factor and metabolic switches hexokinase 2 and pyruvate dehydrogenase kinase 1. *Molecular and Cellular Biology*, *27*, 7381.

Kim, E., Goraksha-Hicks, P., Li, L., Neufeld, T. P., & Guan, K. L. (2008). Regulation of TORC1 by Rag GTPases in nutrient response. *Nature Cell Biology*, *10*(8), 935–945.

Kim, J. W., Tchernyshyov, I., Semenza, G. L., & Dang, C. V. (2006). HIF-1-mediated expression of pyruvate dehydrogenase kinase: A metabolic switch required for cellular adaptation to hypoxia. *Cell Metabolism*, *3*(3), 177–185.

Kohn, A. D., Summers, S. A., Birnbaum, M. J., & Roth, R. A. (1996). Expression of a constitutively active Akt Ser/Thr kinase in 3T3-L1 adipocytes stimulates glucose uptake and glucose transporter 4 translocation. *The Journal of Biological Chemistry*, *271*(49), 31372–31378.

Kuhajda, F. P. (2008). AMP-activated protein kinase and human cancer: Cancer metabolism revisited. *International Journal of Obesity*, *32*(Suppl. 4), S36–S41.

Lee, Y. G., Lee, S. W., Sin, H. S., Kim, E. J., & Um, S. J. (2008). Kinase activity-independent suppression of p73α by AMP-activated kinase α (AMPKα). *Oncogene*, *28*(7), 1040–1052.

Li, F. (2005). Myc stimulates nuclearly encoded mitochondrial genes and mitochondrial biogenesis. *Molecular and Cellular Biology*, *25*, 6225.

Li, T., Kon, N., Jiang, L., Tan, M., Ludwig, T., & Zhao, Y. (2012). Tumor suppression in the absence of p53-mediated cell-cycle arrest, apoptosis, and senescence. *Cell*, *149*(6), 1269–1283.

Li, B., & Simon, M. C. (2013). Molecular Pathways: Targeting MYC-induced Metabolic Reprogramming and Oncogenic Stress in Cancer. *Clinical Cancer Research*, *19*(21), 5835–5841.

Lin, C. Y., Loven, J., Rahl, P. B., Paranal, R. M., Burge, C. B., & Bradner, J. E. (2012). Transcriptional amplification in tumor cells with elevated c-Myc. *Cell*, *151*(1), 56–67. http://dx.doi.org/10.1016/j.cell.2012.08.026, S0092-8674(12)01057-4 [pii].

Loven, J., Orlando, D. A., Sigova, A. A., Lin, C. Y., Rahl, P. B., & Burge, C. B. (2012). Revisiting global gene expression analysis. *Cell*, *151*(3), 476.

Maddocks, O. D., Berkers, C. R., Mason, S. M., Zheng, L., Blyth, K., & Gottlieb, E. (2013). Serine starvation induces stress and p53-dependent metabolic remodelling in cancer cells. *Nature*, *493*(7433), 542–546. http://dx.doi.org/10.1038/nature11743, nature11743 [pii].

Mannava, S., Grachtchouk, V., Wheeler, L. J., Im, M., Zhuang, D., & Slavina, E. G. (2008). Direct role of nucleotide metabolism in C-MYC-dependent proliferation of melanoma cells. *Cell Cycle*, 7(15), 2392.

Manning, B. D., & Cantley, L. C. (2007). AKT/PKB signaling: Navigating downstream. *Cell*, *129*(7), 1261–1274.

Matoba, S., Kang, J.-G., Patino, W. D., Wragg, A., Boehm, M., & Gavrilova, O. (2006). p53 regulates mitochondrial respiration. *Science*, *312*(5780), 1650–1653. http://dx.doi.org/10.1126/science.1126863.

Mauro, C., Leow, S. C., Anso, E., Rocha, S., Thotakura, A. K., & Tornatore, L. (2011). NF-kappaB controls energy homeostasis and metabolic adaptation by upregulating mitochondrial respiration. *Nature Cell Biology*, *13*(10), 1272–1279. http://dx.doi.org/10.1038/ncb2324, ncb2324 [pii].

McFate, T., Mohyeldin, A., Lu, H., Thakar, J., Henriques, J., Halim, N. D., & Verma, A. (2008). Pyruvate dehydrogenase complex activity controls metabolic and malignant phenotype in cancer cells. *The Journal of Biological Chemistry*, *283*(33), 22700–22708.

Menon, S., & Manning, B. D. (2009). Common corruption of the mTOR signaling network in human tumors. *Oncogene*, *27*(S2), S43.

Menon, S., Yecies, J. L., Zhang, H. H., Howell, J. J., Nicholatos, J., & Harputlugil, E. (2012). Chronic activation of mTOR complex 1 is sufficient to cause hepatocellular carcinoma in mice. *Science Signaling*, *5*(217), ra24.

Meylan, E., Dooley, A. L., Feldser, D. M., Shen, L., Turk, E., & Ouyang, C. (2009). Requirement for NF-[kgr]B signalling in a mouse model of lung adenocarcinoma. *Nature*, *462*(7269), 104.

Munoz-Pinedo, C., El Mjiyad, N., & Ricci, J. E. (2012). Cancer metabolism: Current perspectives and future directions. *Cell Death and Disease*, *3*, e248.

Nicolay, B. N., & Dyson, N. J. (2013). The multiple connections between pRB and cell metabolism. *Current Opinion in Cell Biology*, *25*(6), 735–740.

Nicolay, B. N., Gameiro, P. A., Tschöp, K., Korenjak, M., Heilmann, A. M., & Asara, J. M. (2013). Loss of RBF1 changes glutamine catabolism. *Genes & Development*, *27*(2), 182–196. http://dx.doi.org/10.1101/gad.206227.112.

Papandreou, I., Cairns, R. A., Fontana, L., Lim, A. L., & Denko, N. C. (2006). HIF-1 mediates adaptation to hypoxia by actively downregulating mitochondrial oxygen consumption. *Cell Metabolism*, *3*(3), 187–197.

Resnick, M. A., & Inga, A. (2003). Functional mutants of the sequence-specific transcription factor p53 and implications for master genes of diversity. *Proceedings of the National Academy of Sciences of the United States of America*, *100*(17), 9934–9939.

Reynolds, M. R., Lane, A. N., Robertson, B., Kemp, S., Liu, Y., Hill, B. G., et al. (2014). Control of glutamine metabolism by the tumor suppressor Rb. *Oncogene*, *33*(5), 556–566.

Robitaille, A. M., Christen, S., Shimobayashi, M., Cornu, M., Fava, L. L., & Moes, S. (2013). Quantitative phosphoproteomics reveal mTORC1 activates de novo pyrimidine synthesis. *Science*, *339*(6125), 1320–1323. http://dx.doi.org/10.1126/science.1228771, science.1228771, [pii].

Rufini, A., Niklison-Chirou, M. V., Inoue, S., Tomasini, R., Harris, I. S., & Marino, A. (2012). TAp73 depletion accelerates aging through metabolic dysregulation. *Genes & Development*, *26*(18), 2009–2014.

Sancak, Y., Bar-Peled, L., Zoncu, R., Markhard, A. L., Nada, S., & Sabatini, D. M. (2010). Ragulator-Rag complex targets mTORC1 to the lysosomal surface and is necessary for its activation by amino acids. *Cell*, *141*(2), 290–303.

Sankaran, V. G., Orkin, S. H., & Walkley, C. R. (2008). Rb intrinsically promotes erythropoiesis by coupling cell cycle exit with mitochondrial biogenesis. *Genes & Development*, *22*(4), 463–475. http://dx.doi.org/10.1101/gad.1627208.

Schwartzenberg-Bar-Yoseph, F., Armoni, M., & Karnieli, E. (2004). The tumor suppressor p53 down-regulates glucose transporters GLUT1 and GLUT4 gene expression. *Cancer Research, 64*(7), 2627–2633.

Scott, J. W., Norman, D. G., Hawley, S. A., Kontogiannis, L., & Hardie, D. G. (2002). Protein kinase substrate recognition studied using the recombinant catalytic domain of AMP-activated protein kinase and a model substrate. *Journal of Molecular Biology, 317*(2), 309–323.

Sebbagh, M., Olschwang, S., Santoni, M. J., & Borg, J. P. (2011). The LKB1 complex-AMPK pathway: The tree that hides the forest. *Familial Cancer, 10*(3), 415–424.

Semenza, G. L. (2010). Defining the role of hypoxia-inducible factor 1 in cancer biology and therapeutics. *Oncogene, 29*(5), 625–634.

Semenza, G. L. (2011). Regulation of metabolism by hypoxia-inducible factor 1. *Cold Spring Harbor Symposia on Quantitative Biology, 76*, 347–353. http://dx.doi.org/10.1101/sqb.2011.76.010678.

Shackelford, D. B., & Shaw, R. J. (2009). The LKB1-AMPK pathway: Metabolism and growth control in tumour suppression. *Nature Reviews. Cancer, 9*(8), 563.

Shackelford, D. B., Vasquez, D. S., Corbeil, J., Wu, S., Leblanc, M., & Wu, C. L. (2009). mTOR and HIF-1alpha-mediated tumor metabolism in an LKB1 mouse model of Peutz-Jeghers syndrome. *Proceedings of the National Academy of Sciences of the United States of America, 106*(27), 11137–11142.

Shamma, A., Takegami, Y., Miki, T., Kitajima, S., Noda, M., & Obara, T. (2009). Rb regulates DNA damage response and cellular senescence through E2F-dependent suppression of N-Ras isoprenylation. *Cancer Cell, 15*(4), 255.

Shaw, R. J., Bardeesy, N., Manning, B. D., Lopez, L., Kosmatka, M., & DePinho, R. A. (2004). The LKB1 tumor suppressor negatively regulates mTOR signaling. *Cancer Cell, 6*(1), 91–99.

Shaw, R. J., Kosmatka, M., Bardeesy, N., Hurley, R. L., Witters, L. A., & DePinho, R. A. (2004). The tumor suppressor LKB1 kinase directly activates AMP-activated kinase and regulates apoptosis in response to energy stress. *Proceedings of the National Academy of Sciences of the United States of America, 101*(10), 3329–3335.

Shim, H., Dolde, C., Lewis, B. C., Wu, C. S., Dang, G., & Jungmann, R. A. (1997). c-Myc transactivation of LDH-A: Implications for tumor metabolism and growth. *Proceedings of the National Academy of Sciences of the United States of America, 94*(13), 6658–6663.

Son, J., Lyssiotis, C. A., Ying, H., Wang, X., Hua, S., & Ligorio, M. (2013). Glutamine supports pancreatic cancer growth through a KRAS-regulated metabolic pathway. *Nature, 496*(7443), 101.

Soussi, T., & Beroud, C. (2001). Assessing TP53 status in human tumours to evaluate clinical outcome. *Nature Reviews. Cancer, 1*(3), 233–240.

Stambolic, V., MacPherson, D., Sas, D., Lin, Y., Snow, B., & Jang, Y. (2001). Regulation of PTEN transcription by p53. *Molecular Cell, 8*(2), 317–325.

Strohecker, A. M., Guo, J. Y., Karsli-Uzunbas, G., Price, S. M., Chen, G. J., Mathew, R., et al. (2013). Autophagy sustains mitochondrial glutamine metabolism and growth of BRAFV600E-driven lung tumors. *Cancer Discovery, 3*(11), 1272–1285.

Suzuki, S., Tanaka, T., Poyurovsky, M. V., Nagano, H., Mayama, T., & Ohkubo, S. (2010). Phosphate-activated glutaminase (GLS2), a p53-inducible regulator of glutamine metabolism and reactive oxygen species. *Proceedings of the National Academy of Sciences of the United States of America, 107*(16), 7461–7466.

Tandon, P., Gallo, C. A., Khatri, S., Barger, J. F., Yepiskoposyan, H., & Plas, D. R. (2011). Requirement for ribosomal protein S6 kinase 1 to mediate glycolysis and apoptosis resistance induced by Pten deficiency. *Proceedings of the National Academy of Sciences of the United States of America, 108*(6), 2361–2365.

Vander Heiden, M. G., Cantley, L. C., & Thompson, C. B. (2009). Understanding the Warburg effect: The metabolic requirements of cell proliferation. *Science*, *324*(5930), 1029–1033.

van Riggelen, J., Yetil, A., & Felsher, D. W. (2010). MYC as a regulator of ribosome biogenesis and protein synthesis. *Nature Reviews. Cancer*, *10*(4), 301.

Vaupel, P., Thews, O., & Hoeckel, M. (2001). Treatment resistance of solid tumors: Role of hypoxia and anemia. *Medical Oncology*, *18*(4), 243–259. http://dx.doi.org/10.1385/MO:18:4:243, MO:18:4:243 [pii].

Vizan, P., Boros, L. G., Figueras, A., Capella, G., Mangues, R., & Bassilian, S. (2005). K-ras codon-specific mutations produce distinctive metabolic phenotypes in NIH3T3 mice [corrected] fibroblasts. *Cancer Research*, *65*(13), 5512–5515.

Vousden, K. H., & Ryan, K. M. (2009). p53 and metabolism. *Nature Reviews. Cancer*, *9*(10), 691.

Wang, J. B., Erickson, J. W., Fuji, R., Ramachandran, S., Gao, P., & Dinavahi, R. (2010). Targeting mitochondrial glutaminase activity inhibits oncogenic transformation. *Cancer Cell*, *18*(3), 207–219.

White, E. (2013). Exploiting the bad eating habits of Ras-driven cancers. *Genes & Development*, *27*(19), 2065–2071. http://dx.doi.org/10.1101/gad.228122.113, 27/19/2065 [pii].

Woods, A., Johnstone, S. R., Dickerson, K., Leiper, F. C., Fryer, L. G., & Neumann, D. (2003). LKB1 is the upstream kinase in the AMP-activated protein kinase cascade. *Current Biology*, *13*(22), 2004–2008.

Wullschleger, S., Loewith, R., & Hall, M. N. (2006). TOR signaling in growth and metabolism. *Cell*, *124*(3), 471–484.

Xiao, B., Sanders, M. J., Underwood, E., Heath, R., Mayer, F. V., & Carmena, D. (2011). Structure of mammalian AMPK and its regulation by ADP. *Nature*, *472*(7342), 230–233.

Yang, A., Kaghad, M., Wang, Y., Gillett, E., Fleming, M. D., & Dötsch, V. (1998). p63, a p53 homologue at 3q27-29, encodes multiple products with transactivating, deathinducing, and dominant-negative activities. *Molecular Cell*, *2*(3), 305–316.

Yang, W., Xia, Y., Cao, Y., Zheng, Y., Bu, W., & Zhang, L. (2012). EGFR-induced and PKCε monoubiquitylation-dependent NF-kappaB activation upregulates PKM2 expression and promotes tumorigenesis. *Molecular Cell*, *48*(5), 771.

Ying, H., Kimmelman, A. C., Lyssiotis, C. A., Hua, S., Chu, G. C., & Fletcher-Sananikone,- E. (2012). Oncogenic Kras maintains pancreatic tumors through regulation of anabolic glucose metabolism. *Cell*, *149*(3), 656.

Zhang, H., Bosch-Marce, M., Shimoda, L. A., Tan, Y. S., Baek, J. H., & Wesley, J. B. (2008). Mitochondrial autophagy is an HIF-1-dependent adaptive metabolic response to hypoxia. *Journal of Biological Chemistry*, *283*(16), 10892–10903. http://dx.doi.org/10.1074/jbc.M800102200.

Zhao, Y. H., Zhou, M., Liu, H., Ding, Y., Khong, H. T., & Yu, D. (2009). Upregulation of lactate dehydrogenase A by ErbB2 through heat shock factor 1 promotes breast cancer cell glycolysis and growth. *Oncogene*, *28*(42), 3689–3701.

Zheng, B., Jeong, J. H., Asara, J. M., Yuan, Y. Y., Granter, S. R., & Chin, L. (2009). Oncogenic B-RAF negatively regulates the tumor suppressor LKB1 to promote melanoma cell proliferation. *Molecular Cell*, *33*(2), 237–247.

CHAPTER FOUR

Cross Talk Between Cell Death Regulation and Metabolism

Simone Fulda[1]
Institute for Experimental Cancer Research in Pediatrics, Goethe-University, Frankfurt, Germany
[1]Corresponding author: e-mail address: simone.fulda@kgu.de

Contents

Abstract

Accumulating evidence indicates that metabolism plays a critical role in the control of various cellular functions, including cell death. Thus, several mechanisms of programmed cell death have been shown to be controlled by metabolic cues. Since programmed cell death represents a fundamental process in various physiological and pathophysiological conditions, including oncogenesis, tumor progression, and resistance to therapy, the metabolic profile of cancer cells is expected to have a significant impact on all these phases of malignant transformation. Further insights into the signal transduction cascades that regulate different cell death pathways in response to metabolic fluctuations will likely result in the identification of potential targets for the development of novel therapeutic interventions. As the deregulation of cell metabolism as well as alterations in cell death pathways are involved in the pathogenesis of multiple human diseases other than cancer, this knowledge has a great translational potential in several areas of medicine.

1. INTRODUCTION

Programmed cell death is a fundamental process that plays an important role during numerous physiological as well as pathophysiological conditions (Lockshin & Zakeri, 2007). There are several types of programmed cell death in mammalian cells. Apoptosis and necrosis are among the best

Methods in Enzymology, Volume 542
ISSN 0076-6879
http://dx.doi.org/10.1016/B978-0-12-416618-9.00004-2

characterized forms of programmed cell death (Galluzzi et al., 2012). In addition, necroptosis has more recently been identified and has been described as RIP1- and/or RIP3-dependent form of regulated necrosis (Vandenabeele, Galluzzi, Vanden Berghe, & Kroemer, 2010). Compared to necrosis, the controlled sequence of molecular events is a typical feature of necroptosis (Vandenabeele et al., 2010). Apoptosis is defined by several morphological and biochemical parameters including internucleosomal fragmentation of DNA, membrane blebbing, formation of apoptotic bodies, and activation of intracellular enzymes termed caspases (Taylor, Cullen, & Martin, 2008). As these different forms of programmed cell death are critically involved in a variety of cellular regulatory mechanisms under both normal and pathological conditions, a better understanding of the molecular events that control these death modalities is critical in order to selectively interfere with these processes. There is mounting evidence over the last years, showing that cell metabolism has a profound impact on the control of programmed cell death and, more specifically, also on the type of cell death. This intersection between cell death signaling pathways and metabolic events is therefore an area of currently very high interest. This review in particular discusses the link between metabolism and necroptotic cell death.

2. SIGNALING TO PROGRAMMED CELL DEATH

Tumor necrosis factor (TNF)α-engaged programmed cell death pathways provide a prime example showing that the same stimulus can initiate different modes of programmed cell death in a context-dependent manner, including apoptosis and necroptosis. TNFα is a ligand of the death receptor superfamily that binds to its corresponding cell surface receptor TNF receptor 1 (TNFR1) on the plasma membrane. This initiates the intracellular signal transduction machinery that, under certain conditions, finally leads to apoptotic or necroptotic cell death. Upon receptor oligomerization, adapter molecules, for example, TNF receptor type 1-associated death domain protein (TRADD), receptor-interacting protein (RIP)1 as well as cellular inhibitor of apoptosis (cIAP)1, and cIAP2 proteins, are recruited to TNFR1, thereby building a multiprotein complex termed complex I. In order to stop the signaling machinery that is initiated upon binding of TNFα, the receptor is internalized and subject to changes in the composition of the multiprotein complex. TNFR1 complex II is then formed in the cytosol which is composed of Fas-Associated protein with Death Domain (FADD), RIP1, and caspase-8 as core elements to drive caspase-8 activation and apoptosis.

Alternatively, when caspase-8 activation is not achieved, RIP1 can interact with RIP3 to form the necrosome complex which then signals to necroptosis. This illustrates that, depending on the intracellular composition and availability of the signaling molecules and additional regulatory events, TNFα stimulation can signal to either apoptosis or necroptosis. TNFα-stimulated necroptosis can be considered as a prototype of necroptotic cell death that has extensively been studied to elucidate the signal transduction elements that are involved in mediating necroptosis (Chu, 2013). Furthermore, depending on regulatory steps, engagement of TNFα receptor can also mediate nuclear factor kappa B (NF-κB) activation, for example, under conditions under which cell death induction is blocked. Posttranslational modifications are considered to play an important role in the fine tuning of these different cellular fates. For example, RIP1 is a multifunctional protein which, depending on its posttranslational modifications, can be involved in apoptotic, necroptotic, and NF-κB signaling pathways. While non-degradative ubiquitination of RIP1 supports signaling to NF-κB activation, RIP1 in its deubiquitinated form functions as a mediator of cell death (Vandenabeele et al., 2010). Also, phosphorylation events play an important role in these regulatory steps. Both RIP1 and RIP3 have been shown to be phosphorylated and thereby activated in response to necroptotic stimuli (Vandenabeele et al., 2010).

Besides TNFα, a series of additional stimuli have been implicated in necroptotic cell death induction, including different chemical compounds, physical damage, ligands of the TNF receptor superfamily, pathogens, or physicochemical stress stimuli, including reactive oxygen species (ROS), ischemia–reperfusion injury, calcium overload, certain anticancer drugs, photodynamic therapy, or adenosine triphosphate (ATP) depletion (Vanlangenakker, Vanden Berghe, & Vandenabeele, 2012). This list of necroptotic stimuli points to a connection between metabolism and the necroptosis signal transduction machinery. Metabolic events can regulate the intracellular signal transduction cascade leading to necroptotic cell death, although little is yet known about the molecular details of this regulation.

3. BIOENERGETIC REGULATION OF PROGRAMMED CELL DEATH

Various intracellular signal transduction pathways and events are under the control of bioenergetic demands. Therefore, it is not surprising that also different forms of programmed cell death including apoptosis and necroptosis are tightly regulated by the availability of cellular metabolites and the overall bioenergetic state of the cell. In this respect, ATP represents a key molecule that is involved in the regulation of cell metabolism as well as multiple signal transduction pathways. This key position of ATP within the regulatory network of programmed cell death is illustrated by the fact that the decision between the initiation of apoptotic versus necrotic cell death has been shown to depend, at least in part, on the intracellular content of ATP (Eguchi, Shimizu, & Tsujimoto, 1997; Leist, Single, Castoldi, Kuhnle, & Nicotera, 1997). Interestingly, the presence or absence of ATP has been reported to provide a switch from apoptosis under conditions of high ATP levels toward necrosis when ATP content is depleted (Leist et al., 1997). This ATP-dependent molecular switch between apoptotic and necrotic cell death has been documented in response to different cell death stimuli including engagement of the CD95 receptor as well as treatment with staurosporine (Leist et al., 1997). In mechanistic terms, the crucial demand of ATP for the execution of apoptosis was demonstrated to be linked to the requirement of energy for nuclear DNA fragmentation (Leist et al., 1997). Accordingly, the exogenous supply of ATP was sufficient in this model of programmed cell death in T-cells to initiate the switch from necrotic toward apoptotic cell death (Leist et al., 1997). In addition to DNA fragmentation, a number of additional cellular events have been linked to the requirement of ATP including activation of caspase-9, activity of the ubiquitin proteasome machinery to degrade cellular proteins, the activity status of the translational apparatus, as well as the activity of enzymes that are involved in the repair of DNA damage such as poly(ADP-ribose) polymerase (PARP)1 (Saelens et al., 2005; Soldani & Scovassi, 2002; Sun et al., 2004; Zou, Li, Liu, & Wang, 1999). The role of PARP1 is not restricted to its ability to control DNA damage and repair processes but also extends to the modulation of metabolic events which are connected to the control of cell death. Under resting conditions, PARP1 is present in the nucleus where it functions as a sensor for DNA damage. In response to extensive DNA damage that cannot be repaired, PARP1 has been described to become overactivated (Zong, Ditsworth, Bauer, Wang, & Thompson, 2004). Since PARP1 activity depends on the consumption of ATP as well as nicotinamide adenine dinucleotide (NAD), this overactivation of PARP1 causes an acute event of bioenergetic failure, which in turn leads to induction of cell death (Zong et al., 2004). Apoptosis-inducing factor (AIF) is a key mediator in this model of cell death upon overactivation of PARP1, which is released from the intermembrane space of mitochondria and translocates into the nucleus (Yu et al., 2002). PARP1-triggered, AIF-mediated cell death has also been implicated to be instrumental for neuronal cell death upon oxygen–glucose depletion under conditions of cerebral ischemia (Culmsee et al., 2005). In the nucleus, AIF has been reported to cause a form of caspase-independent cell death which is characterized by large-scale DNA fragmentation (Boujrad, Gubkina, Robert, Krantic, & Susin, 2007). This metabolically linked PARP1-dependent cell death has been shown to occur under several different conditions, including depletion of oxygen or glucose, for example, during ischemic insults (Hong, Dawson, & Dawson, 2004).

Furthermore, bioenergetic regulation of programmed cell death has also been linked to the mitochondrial permeability transition pore complex (PTPC), a large multiprotein complex (Kroemer, Galluzzi, & Brenner, 2007). The molecular components of PTPC comprise adenine-nucleotide translocase (ANT) that resides within the inner mitochondrial membrane and regulates the exchange of ATP (Kroemer et al., 2007). Another component of PTPC is voltage-dependent anion channel (VDAC), which is part of the outer mitochondrial membrane (Giorgio et al., 2010). Pro- as well as antiapoptotic proteins of the Bcl-2 family are involved in the regulation of VDAC activity in addition to cell metabolites or ions (Shoshan-Barmatz & Golan, 2012). Yet another component of PTPC is cyclophilin D that is present in the mitochondrial matrix (Kroemer et al., 2007). Studies in cyclophilin D knockdown mice revealed the critical role of cyclophilin D in the control of necrotic cell death under ischemic conditions (Nakagawa et al., 2005; Schinzel et al., 2005).

In addition to PTPC, also cytoplasmic multiprotein complexes are involved in the regulation of cell metabolism and programmed cell death. For example, the necrosome complex containing RIP1 and RIP3 as core components has recently been linked to metabolic pathways (Schulze-Osthoff et al., 1992). Once activated, RIP3 has been shown to regulate the activity of metabolic enzymes involved in the control of sugar and protein systems. RIP3-regulated metabolic enzymes include glycogen phosphorylase (PYGL) that is responsible for the conversion of glycogen into glucose-1-phosphate (Schulze-Osthoff et al., 1992). The latter metabolite can fuel into glycolysis upon its modification into glucose-6-phospate, thereby fostering the mitochondrial metabolism, which in turn favors the production of ROS. Additional substrates of RIP3 during necroptotic signaling are glutamate–ammonia ligase (GLUL) and glutamate dehydrogenase-1 (GLUD-1) (Schulze-Osthoff et al., 1992). Both enzymes are involved in glutaminolysis, either in the cytosol or within the mitochondria, and an increased rate in glutaminolysis has been linked to the production of ROS. These events provide a link between RIP3-regulated glycogenolysis and glutaminolysis, ROS production, and necroptotic cell death.

Moreover, coenzyme A has recently been described to provide a link between active metabolism and cell death pathways. Coenzyme A was shown to promote calcium/calmodulin-dependent protein kinase II (CaMKII)-mediated phosphorylation of caspase-2, thereby inhibiting apoptosis (McCoy et al., 2013).

4. REDOX SIGNALING AND ROS: CRITICAL REGULATORS OF PROGRAMMED CELL DEATH

ROS species have for long been implicated in the control of programmed cell death (Ott, Gogvadze, Orrenius, & Zhivotovsky, 2007). In particular, during the execution of cell death, ROS have been implicated to be responsible for the damage of intracellular structures to such an extent that cells succumb to cell death. In addition to the destructive function of ROS, there is mounting evidence showing that redox processes may also function as signaling molecules that fine-tune signal transduction pathways in order to control cellular functions (Bindoli & Rigobello, 2013). While the role of ROS in the control of necrosis has been well established, far less is known about the implications of redox signaling in the regulation of necroptosis. It has been shown that the levels of ROS rapidly rise shortly after the engagement of the necroptotic signaling machinery (Vanden Berghe et al., 2010; Vanlangenakker et al., 2011). This points to an involvement of ROS as a mediator of necroptosis. In support of this notion, the addition of ROS scavengers has been described to attenuate necroptotic cell death (Laukens et al., 2011; Vanlangenakker et al., 2011). However, little is yet known about the intracellular localization where ROS originate during necroptosis. Since the respiratory chain is the site where oxygen is handled to generate energy, this site is also considered as the major site for the generation of ROS under various conditions including programmed cell death. More specifically, complexes I and III have been shown to represent the primary sites as mitochondrial ROS generators (Schulze-Osthoff et al., 1992). Also, mitochondrial ANT represents another possible site for mitochondrial ROS generation (Brenner, Subramaniam, Pertuiset, & Pervaiz, 2011). On theoretical grounds, ANT may provide a link to necroptotic signaling, since it has been reported to be under the control of RIP1, as RIP1 can negatively affect ANT activity (Temkin, Huang, Liu, Osada, & Pope, 2006). Besides mitochondria as a key generator of ROS, there are also extramitochondrial sites which have been implicated in the production of ROS during programmed cell death. The main extramitochondrial sources for ROS generation are the NAD(P)H oxidase (NOX) family of nicotinamide adenine dinucleotide phosphate (NADPH) oxidases (Bedard & Krause, 2007). The TNFR1 complex may provide a link between necroptosis signaling and ROS production. Accordingly, stimulation with TNFα has been shown to cause increased activity of NOX1 complex in an RIP1-dependent fashion (Kim, Morgan, Choksi, & Liu, 2007). To this end, the engagement of TNFR1 upon ligation of TNFα stimulates the association of NOX1 with RIP and additional signaling molecules which leads to production of ROS in a NOX1-dependent manner (Kim et al., 2007). However, the question as to whether or not ROS generation at the level of the TNFR1 represents a critical event in the signal transduction to necroptosis remains to be investigated in further detail in future studies.

5. DESTRUCTIVE ROS DURING THE EXECUTION OF CELL DEATH

As mentioned above, ROS have been implicated in the execution of necrosis since a long time. Following stimulation with TNFα, ROS were shown to be generated within the mitochondrial complex I of the mitochondrial respiratory chain (Schulze-Osthoff et al., 1992). These mitochondrial ROS were reported to be required for necrotic cell death upon treatment with TNFα (Schulze-Osthoff et al., 1992). The destructive function of ROS in this context was supported by data showing that structural damage to intracellular organelles, including the mitochondria and the endoplasmic reticulum, occurred in response to ROS generation (Festjens et al., 2006; Schulze-Osthoff et al., 1992). In addition to destructive ROS generated at the level of the mitochondria, the lysosomal compartment has also been implicated in cell death by ROS production. Lysosomes are permeable to hydrogen peroxide which, once inside the lysosomal space, can interact with ferrous iron in order to generate hydroxyl radical (Fenton reaction) (Vanlangenakker, Berghe, Krysko, Festjens, & Vandenabeele, 2008). This ROS is known to be highly reactive, thereby immediately causing damage to organelle membranes via lipid oxidation.

6. CONCLUSIONS

Programmed cell death pathways play a key regulatory role in various physiological processes and are disturbed in a variety of human diseases. Pathways to programmed cell death are tightly controlled by various regulatory mechanisms, including cell metabolism. There is an increasing list of metabolic events that can initiate different forms of programmed cell death. In addition, metabolic events can function as determinants of the way of how cells die. As more and more control points between programmed cell death networks and cell metabolism are identified, these connections offer the possibility to serve as regulatory switches. A better understanding of the molecular events that are involved in the regulation of programmed cell death by metabolic pathways will likely result in novel opportunities for therapeutic interventions.

ACKNOWLEDGMENTS

The expert secretarial assistance of C. Hugenberg is greatly appreciated. Work in the author's laboratory is supported by grants from the Deutsche Forschungsgemeinschaft, the Deutsche Krebshilfe, the Bundesministerium für Bildung und Forschung (01GM1104C), Else Kröner Fresenius-Stiftung, Wilhelm Sander-Stiftung, and the European Community.

REFERENCES

Bedard, K., & Krause, K. H. (2007). The NOX family of ROS-generating NADPH oxidases: Physiology and pathophysiology. *Physiological Reviews*, *87*, 245–313.
Bindoli, A., & Rigobello, M. P. (2013). Principles in redox signaling: From chemistry to functional significance. *Antioxidants & Redox Signaling*, *18*, 1557–1593.
Boujrad, H., Gubkina, O., Robert, N., Krantic, S., & Susin, S. A. (2007). AIF-mediated programmed necrosis: A highly regulated way to die. *Cell Cycle*, *6*, 2612–2619.
Brenner, C., Subramaniam, K., Pertuiset, C., & Pervaiz, S. (2011). Adenine nucleotide translocase family: Four isoforms for apoptosis modulation in cancer. *Oncogene*, *30*, 883–895.
Chu, W. M. (2013). Tumor necrosis factor. *Cancer Letters*, *328*, 222–225.
Culmsee, C., Zhu, C., Landshamer, S., Becattini, B., Wagner, E., Pellecchia, M., et al. (2005). Apoptosis-inducing factor triggered by poly(ADP-ribose) polymerase and Bid mediates neuronal cell death after oxygen-glucose deprivation and focal cerebral ischemia. *Journal of Neuroscience*, *25*, 10262–10272.
Eguchi, Y., Shimizu, S., & Tsujimoto, Y. (1997). Intracellular ATP levels determine cell death fate by apoptosis or necrosis. *Cancer Research*, *57*, 1835–1840.
Festjens, N., Kalai, M., Smet, J., Meeus, A., Van Coster, R., Saelens, X., et al. (2006). Butylated hydroxyanisole is more than a reactive oxygen species scavenger. *Cell Death and Differentiation*, *13*, 166–169.
Galluzzi, L., Vitale, I., Abrams, J. M., Alnemri, E. S., Baehrecke, E. H., Blagosklonny, M. V., et al. (2012). Molecular definitions of cell death subroutines: Recommendations of the

Nomenclature Committee on Cell Death 2012. *Cell Death and Differentiation*, *19*, 107–120.
Giorgio, V., Soriano, M. E., Basso, E., Bisetto, E., Lippe, G., Forte, M. A., et al. (2010). Cyclophilin D in mitochondrial pathophysiology. *Biochimica et Biophysica Acta*, *1797*, 1113–1118.
Hong, S. J., Dawson, T. M., & Dawson, V. L. (2004). Nuclear and mitochondrial conversations in cell death: PARP-1 and AIF signaling. *Trends in Pharmacological Sciences*, *25*, 259–264.
Kim, Y. S., Morgan, M. J., Choksi, S., & Liu, Z. G. (2007). TNF-induced activation of the Nox1 NADPH oxidase and its role in the induction of necrotic cell death. *Molecular Cell*, *26*, 675–687.
Kroemer, G., Galluzzi, L., & Brenner, C. (2007). Mitochondrial membrane permeabilization in cell death. *Physiological Reviews*, *87*, 99–163.
Laukens, B., Jennewein, C., Schenk, B., Vanlangenakker, N., Schier, A., Cristofanon, S., et al. (2011). Smac mimetic bypasses apoptosis resistance in FADD- or caspase-8-deficient cells by priming for tumor necrosis factor alpha-induced necroptosis. *Neoplasia*, *13*, 971–979.
Leist, M., Single, B., Castoldi, A. F., Kuhnle, S., & Nicotera, P. (1997). Intracellular adenosine triphosphate (ATP) concentration: A switch in the decision between apoptosis and necrosis. *Journal of Experimental Medicine*, *185*, 1481–1486.
Lockshin, R. A., & Zakeri, Z. (2007). Cell death in health and disease. *Journal of Cellular and Molecular Medicine*, *11*, 1214–1224.
McCoy, F., Darbandi, R., Lee, H. C., Bharatham, K., Moldoveanu, T., Royappa, G., et al. (2013). Metabolic activation of CaMKII by coenzyme A. *Molecular Cell*, *52*, 325–339.
Nakagawa, T., Shimizu, S., Watanabe, T., Yamaguchi, O., Otsu, K., Yamagata, H., et al. (2005). Cyclophilin D-dependent mitochondrial permeability transition regulates some necrotic but not apoptotic cell death. *Nature*, *434*, 652–658.
Ott, M., Gogvadze, V., Orrenius, S., & Zhivotovsky, B. (2007). Mitochondria, oxidative stress and cell death. *Apoptosis*, *12*, 913–922.
Saelens, X., Festjens, N., Parthoens, E., Vanoverberghe, I., Kalai, M., van Kuppeveld, F., et al. (2005). Protein synthesis persists during necrotic cell death. *Journal of Cell Biology*, *168*, 545–551.
Schinzel, A. C., Takeuchi, O., Huang, Z., Fisher, J. K., Zhou, Z., Rubens, J., et al. (2005). Cyclophilin D is a component of mitochondrial permeability transition and mediates neuronal cell death after focal cerebral ischemia. *Proceedings of the National Academy of Sciences of the United States of America*, *102*, 12005–12010.
Schulze-Osthoff, K., Bakker, A. C., Vanhaesebroeck, B., Beyaert, R., Jacob, W. A., & Fiers, W. (1992). Cytotoxic activity of tumor necrosis factor is mediated by early damage of mitochondrial functions. Evidence for the involvement of mitochondrial radical generation. *Journal of Biological Chemistry*, *267*, 5317–5323.
Shoshan-Barmatz, V., & Golan, M. (2012). Mitochondrial VDAC1: Function in cell life and death and a target for cancer therapy. *Current Medicinal Chemistry*, *19*, 714–735.
Soldani, C., & Scovassi, A. I. (2002). Poly(ADP-ribose) polymerase-1 cleavage during apoptosis: An update. *Apoptosis*, *7*, 321–328.
Sun, X. M., Butterworth, M., MacFarlane, M., Dubiel, W., Ciechanover, A., & Cohen, G. M. (2004). Caspase activation inhibits proteasome function during apoptosis. *Molecular Cell*, *14*, 81–93.
Taylor, R. C., Cullen, S. P., & Martin, S. J. (2008). Apoptosis: Controlled demolition at the cellular level. *Nature Reviews Molecular Cell Biology*, *9*, 231–241.
Temkin, V., Huang, Q., Liu, H., Osada, H., & Pope, R. M. (2006). Inhibition of ADP/ATP exchange in receptor-interacting protein-mediated necrosis. *Molecular and Cellular Biology*, *26*, 2215–2225.

Vandenabeele, P., Galluzzi, L., Vanden Berghe, T., & Kroemer, G. (2010). Molecular mechanisms of necroptosis: An ordered cellular explosion. *Nature Reviews Molecular Cell Biology, 11*, 700–714.

Vanden Berghe, T., Vanlangenakker, N., Parthoens, E., Deckers, W., Devos, M., Festjens, N., et al. (2010). Necroptosis, necrosis and secondary necrosis converge on similar cellular disintegration features. *Cell Death and Differentiation, 17*, 922–930.

Vanlangenakker, N., Berghe, T. V., Krysko, D. V., Festjens, N., & Vandenabeele, P. (2008). Molecular mechanisms and pathophysiology of necrotic cell death. *Current Molecular Medicine, 8*, 207–220.

Vanlangenakker, N., Vanden Berghe, T., Bogaert, P., Laukens, B., Zobel, K., Deshayes, K., et al. (2011). cIAP1 and TAK1 protect cells from TNF-induced necrosis by preventing RIP1/RIP3-dependent reactive oxygen species production. *Cell Death and Differentiation, 18*, 656–665.

Vanlangenakker, N., Vanden Berghe, T., & Vandenabeele, P. (2012). Many stimuli pull the necrotic trigger, an overview. *Cell Death and Differentiation, 19*, 75–86.

Yu, S. W., Wang, H., Poitras, M. F., Coombs, C., Bowers, W. J., Federoff, H. J., et al. (2002). Mediation of poly(ADP-ribose) polymerase-1-dependent cell death by apoptosis-inducing factor. *Science, 297*, 259–263.

Zong, W. X., Ditsworth, D., Bauer, D. E., Wang, Z. Q., & Thompson, C. B. (2004). Alkylating DNA damage stimulates a regulated form of necrotic cell death. *Genes & Development, 18*, 1272–1282.

Zou, H., Li, Y., Liu, X., & Wang, X. (1999). An APAF-1 cytochrome c multimeric complex is a functional apoptosome that activates procaspase-9. *Journal of Biological Chemistry, 274*, 11549–11556.

CHAPTER FIVE

Techniques to Monitor Glycolysis

Tara TeSlaa[*], **Michael A. Teitell**[*,†,‡,§,¶,||,#,1]
[*]Molecular Biology Institute, UCLA, Los Angeles, California, USA
[†]Department of Pathology and Laboratory Medicine, UCLA, Los Angeles, California, USA
[‡]Department of Bioengineering, UCLA, Los Angeles, California, USA
[§]Department of Pediatrics, UCLA, Los Angeles, California, USA
[¶]California NanoSystems Institute, UCLA, Los Angeles, California, USA
[||]Jonsson Comprehensive Cancer Center, UCLA, Los Angeles, California, USA
[#]Broad Center of Regenerative Medicine and Stem Cell Research, UCLA, Los Angeles, California, USA
[1]Corresponding author: e-mail address: mteitell@mednet.ucla.edu

Contents

Abstract

An increased flux through glycolysis supports the proliferation of cancer cells by providing additional energy in the form of ATP as well as glucose-derived metabolic intermediates for nucleotide, lipid, and protein biosynthesis. Thus, glycolysis and other metabolic pathways that control cell proliferation may represent valuable targets for therapeutic interventions and diagnostic procedures. In this context, the measurement of glucose uptake and lactate excretion by malignant cells may be useful to detect shifts in glucose catabolism, while determining the activity of rate-limiting glycolytic enzymes can provide insights into points of metabolic regulation. Moreover, metabolomic studies can be used to generate large, integrated datasets to track changes in carbon flux through glycolysis and its collateral anabolic pathways. As discussed here, these approaches can reveal and quantify the metabolic alterations that underlie malignant cell proliferation.

Methods in Enzymology, Volume 542
ISSN 0076-6879
http://dx.doi.org/10.1016/B978-0-12-416618-9.00005-4

1. INTRODUCTION

Glycolysis is the intracellular biochemical conversion of one molecule of glucose into two molecules of pyruvate with the concurrent generation of two molecules of ATP. Pyruvate is a metabolic intermediate with several potential fates including entrance into the tricarboxylic acid (TCA) cycle within mitochondria to produce NADH and $FADH_2$. These reducing agents subsequently donate electrons to the mitochondrial electron transport chain (ETC), which when fully coupled to the complex V ATP synthase of the mitochondrial inner membrane generates an additional 34 molecules of ATP per glucose. Alternatively, pyruvate can be converted into lactate in the cytosol by lactate dehydrogenase with concurrent regeneration of NAD^+ from NADH. Conversion of pyruvate to lactate blocks further ATP production, but the resultant increase in NAD^+ drives the first biochemical step in glycolysis (DeBerardinis, Lum, Hatzivassiliou, & Thompson, 2008). An increase in the flow of carbon metabolites through the glycolytic pathway, or glycolytic flux, can increase the rate of ATP production within cells despite being markedly less efficient at generating ATP compared to oxidative phosphorylation (Pfeiffer, Schuster, & Bonhoeffer, 2001).

In addition to generating ATP, glycolysis also supplies biosynthetic intermediates for cell growth and proliferation. For example, glucose-6-phosphate, the first cytosolic product of glucose metabolism, can shunt into the pentose phosphate pathway to drive NADPH generation from $NAPD^+$. NADPH reduces reactive oxygen species produced mainly by respiration to maintain cellular redox balance and to protect the genome from mutations. Carbon flux through the pentose phosphate pathway supplies metabolites for nucleotide biosynthesis that is required for DNA replication and RNA transcription. Another example is 3-phosphoglycerate, a glycolytic metabolite used to synthesize serine, glycine, and cysteine, which in turn supplies one carbon metabolism. Folate and methionine cycles, the components of one carbon metabolism, provide metabolites that support diverse cellular processes including methylation reactions, antioxidant defenses, lipid head group modifications, and nucleotide metabolism (Locasale, 2013).

Warburg (1956) first observed that proliferating tumor cells augment aerobic glycolysis, the conversion of glucose to lactate in the presence of oxygen, in contrast to nonmalignant cells that mainly respire when oxygen is available. This mitochondrial bypass, called the Warburg effect, occurs in rapidly proliferating cells including cancer cells, activated lymphocytes, and pluripotent stem cells. While the Warburg effect is energy inefficient, it is offset by an increased glycolytic flux to provide additional biosynthetic precursors to support rapid cancer cell proliferation (DeBerardinis et al., 2008). This energy compromise supports higher rates of nucleotide synthesis for DNA replication and RNA transcription, phospholipids for membrane production, and amino acids for protein translation to support increased cell division. The Warburg effect has been exploited for clinical diagnostic tests that use positron emission tomography (PET) scanning to identify increased cellular uptake of fluorinated glucose analogs such as ^{18}F-deoxyglucose.

Not all tumors, however, shift to glycolysis for energy production. Some diffuse large B cell lymphomas and glioblastomas remain dependent on oxidative phosphorylation for energy production (Caro et al., 2012; Marin-Valencia et al., 2012). Metabolic enzyme activity is heterogeneous between different tumors even within tumor classes, and glycolytic enzymes can be either increased or decreased in their expression (Hu et al., 2013). Glutamine and fatty acids can also be used by cancers as alternative sources of fuel to make ATP through oxidative phosphorylation (Le et al., 2012; Zaugg et al., 2011).

Although Warburg made his observations over 75 years ago, the detailed mechanisms and consequences of shifting metabolism toward glycolysis are only starting to be revealed. Pyruvate kinase isoform M2 (PKM2), an embryonic splice variant of the glycolytic enzyme pyruvate kinase (PK), is highly expressed in several types of cancer (Christofk, Vander Heiden, Harris, et al., 2008; Lim et al., 2012). PKM2 shows a decreased kinase activity that helps shunt glycolytic intermediates through biosynthetic pathways at the expense of respiration to CO_2 (Christofk, Vander Heiden, Harris, et al., 2008; Hitosugi et al., 2012). Phosphorylation of Tyr-105 of PKM2 causes the release of the allosteric activator of PKM2, 1,6-bisphosphate, which decreases its activity (Hitosugi et al., 2012). Another glycolytic enzyme, phosphoglycerate dehydrogenase, is amplified in human tumors and directs glycolytic carbon flux into serine biosynthesis instead of continued catabolism to pyruvate (Locasale et al., 2011; Possemato et al., 2011). An increased carbon flux through the serine biosynthesis pathway also supports glycine production, which is used for nucleotide biosynthesis and regulates cell proliferation (Jain et al., 2012).

2. MEASURING GLUCOSE UPTAKE AND LACTATE PRODUCTION

For cells in culture, glycolytic flux can be quantified by measuring glucose uptake and lactate excretion. Glucose uptake into the cell is through glucose transporters (Glut1–Glut4), whereas lactate excretion is through monocarboxylate transporters (MCT1–MCT4) at the cell membrane. A caveat to relatively straightforward measurements of extracellular fluid concentrations of glucose and lactate is that they do not provide information on other possible fates for glucose-derived carbons (Fig. 5.1). Extracellular measurements of glucose and lactate in cell culture media are relatively simple to perform, cell nondestructive, and provide a reasonable estimate of glycolytic flux (Table 5.1).

Glucose uptake into cells provides an intracellular estimate of glycolytic flux and can be quantified through detection of entrapped glucose analogs. This approach can quantify glucose uptake at the level of a single cell, but is cell destructive.

2.1. Extracellular glucose and lactate

Commercially available kits and instruments are available to quantify glucose and lactate levels within cell culture media. When using these methods, cell number must be considered as this can greatly affect results. Kit detection methods are usually colorimetric or fluorometric and are compatible with standard lab equipment such as spectrophotometers.

BioProfile Analyzers (Nova Biomedical) or Biochemistry Analyzers (YSI Life Sciences) can measure levels of both glucose and lactate in cell culture media. GlucCell (Cesco BioProducts) can measure only glucose levels in cell culture media. While each commercial method has a different detection protocol, the collection of culture media for analysis is the same. Generally for both extracellular glucose and lactate measurements, cells should be plated at equal densities with the culture media changed 24 h prior to collection. To account for differences in the rates of cell proliferation or death,

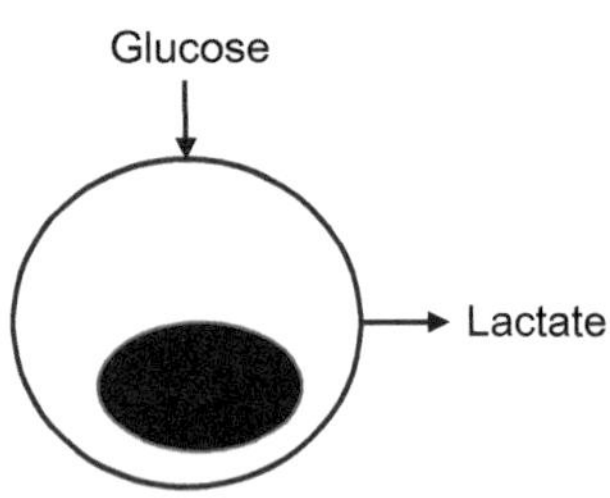

Figure 5.1 *Glucose uptake and lactate excretion approximate glycolytic flux.* Additional metabolic processes that siphon or add metabolic intermediates to the glycolytic pathway remain undefined, which may impact glycolytic flux measurements and complicate data interpretations.

Table 5.1 Methods to detect glucose uptake and lactate excretion

Method	Detectable metabolite		Equipment	Cell destructive?	Radioactive?	Advantages and disadvantages
	Glucose	Lactate				
Extracellular glucose or lactate kit	Y	Y	Spectrophotometer	N	N	Simple, standard equipment
Extracellular bioanalyzer	Y	Y	Bioanalyzer	N	N	Simple
ECAR measurement	N	Y	Seahorse XF analyzer	N	N	Coupled with oxygen consumption
[^{3}H]-2-DG or [^{14}C]-2-DG uptake	Y	N	Scintillation counter	Y	Y	Expensive tracers
^{18}FDG uptake	Y	N	Scintillation counter/ cyclotron	Y	Y	Short half-life
2-NBDG uptake	Y	N	Flow cytometer or fluorescent microscopy	Y	N	Single cell analysis

cells should be counted at the time of media collection for normalization of results to cell number.

2.1.1 Extracellular acidification rate by Seahorse XF analyzer

The Seahorse extracellular flux (XF) analyzer (Seahorse Bioscience) is a powerful tool for measuring glycolysis and oxidative phosphorylation (through oxygen consumption) simultaneously in the same cells. Glycolysis is determined through measurements of the extracellular acidification rate (ECAR) of the surrounding media, which is predominately from the excretion of lactic acid per unit time after its conversion from pyruvate (Wu et al., 2007). However, additional metabolic processes in cells, such as CO_2 production by the TCA cycle, can change the pH of the media complicating this analysis. Because ECAR is essentially a measurement of pH, a buffering agent, such as sodium bicarbonate, is not included in the assay medium. In addition, bicarbonate and media pH can play a role in regulating glycolysis, which can confound measurements of ECAR. To increase measurement accuracy, chemical inhibitors can be used to determine pH changes that are from lactate excretion versus other sources of media acidification (Fig. 5.2). The Seahorse XF analyzer requires only small cell numbers per assay condition and has injection ports for the addition of up to four inhibitors in each experiment. Available detectors use either a 24-well (XF24) or a 96-well (XF96) assay plate format.

To obtain reproducible data within the linear detection range of the XF analyzer, optimization of the cell number and inhibitor concentration is required. The XF24 analyzer has a two-step cell seeding protocol and generally requires 2×10^4–1×10^5 cells per well. Experimental cell density should be chosen based on three main factors: (1) an ECAR value of at least 20 mpH/min, which is the low end of a potential linear detection range, (2) retained morphology of the plated cells by visual inspection, and (3) a cell density that lies within the linear range of a graph of cell seeding density versus ECAR values (Fig. 5.3A). Metabolic inhibitors used in assays should be titrated to obtain the optimal concentration for maximal response for each cell type without toxicity, as assessed by ECAR and oxygen consumption rate (OCR) measurements (Fig. 5.3B). Loosely adherent and suspension cells can be attached to XF V7 cell culture assay plates by coating these plates with Cell-Tak (BD Bioscience). The example protocol below is for adherent cells used with the XF24 format.

To probe glycolysis, an assay called the glycolysis stress test is often used. In this assay, glucose, oligomycin, and 2-deoxyglucose (2-DG) are inserted through injection ports sequentially while measurements are being made.

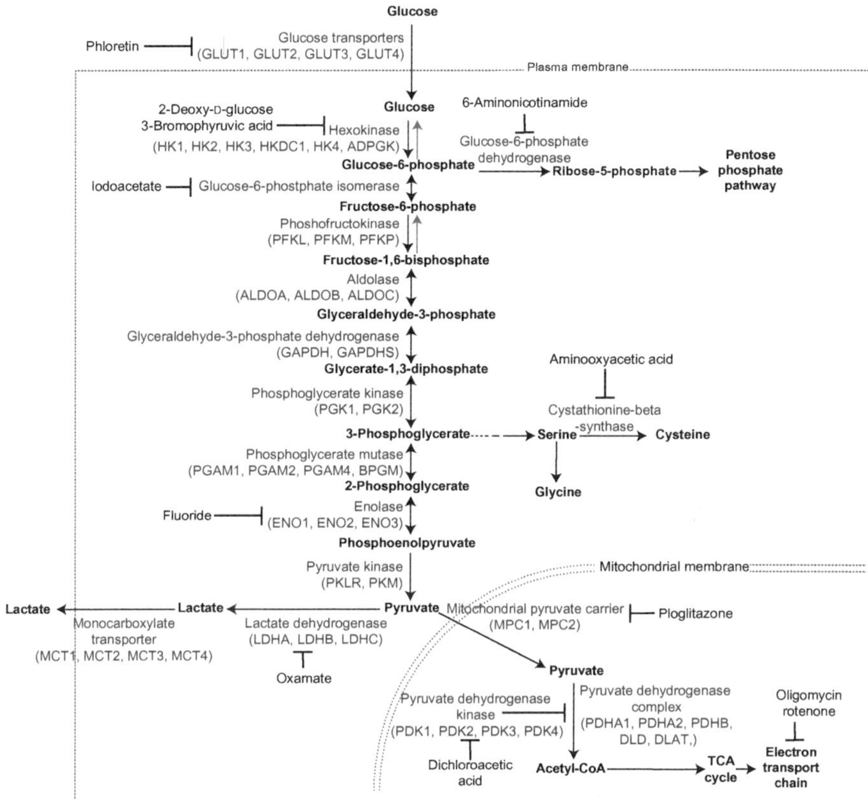

Figure 5.2 *Chemical inhibitors of glycolysis.* Inhibitors of glycolytic enzymes are used to quantify metabolite fluxes within the glycolytic pathway. Inhibition of glycolysis with 2-dexoxy-D-glucose (2-DG) is a common tool for measuring glycolysis (e.g., glycolysis stress test).

Glucose is supplied to feed glycolysis, and the difference between ECAR before and after addition of glucose is a measure of the glycolytic rate. Oligomycin inhibits ATP synthase in the ETC, which decreases the ATP/ADP ratio and drives glycolysis. The difference between ECAR before and after oligomycin addition is equal to the glycolytic reserve capacity of cells. 2-DG inhibits glycolysis and therefore provides a baseline ECAR measurement. ECAR after 2-DG addition accounts for the nonglycolytic ECAR of cells (Fig. 5.3C).

2.1.1.1 Protocol for ECAR measurement by Seahorse XF analyzer

Day 1: Seed cells and prepare a sensor cartridge

For adherent cells, wash cells with 1 × PBS, pH 7.4, and add 1 × trypsin until cells begin to visually detach. Add culture media with serum to

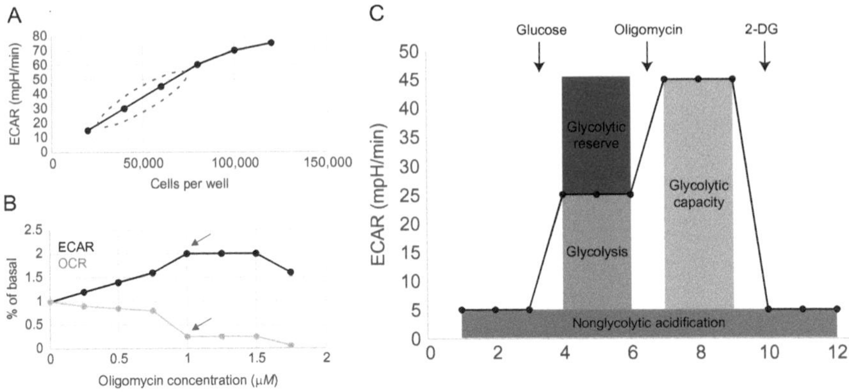

Figure 5.3 *ECAR measurements with a Seahorse XF analyzer.* (A) Cell number optimization for XF assays. Circle identifies region with acceptable cell densities. (B, C) Oligomycin optimization for glycolysis stress test. Shown are ECAR (B) and OCR (C) with an oligomycin drug titration. 1 μ*M* is the optimal oligomycin concentration in this case because it stimulates the maximum ECAR (arrow) and inhibits OCR (arrow). (C) Glycolysis, glycolytic capacity, glycolytic reserve, and nonglycolytic acidification can be measured with the glycolysis stress test.

deactivate trypsin and pipette up and down to create a uniform cell suspension. Count cells (e.g., hemocytometer) and resuspend the total so that the number of desired cells per well is within 100 μl (e.g., for 5×10^4 cells per well, suspend cells at a concentration of 5×10^5 cells per ml). Pipette 100 μl of cell suspension into each well of 24-well V7 plate (Seahorse Bioscience #100777-004). For accurate reproducible measurements, pipette up and down to make sure that cells are uniformly spread throughout the well. One well in each V7 plate row or column should lack cells to be used as a blank control. Put the seeded plate into a 37 °C, 5% CO_2 incubator to allow cells to adhere for 1–5 h. Add an additional 150 μl of media and allow cells to grow overnight in a 37 °C, 5% CO_2 incubator.

The XF24 and XF96 sensor cartridges (Seahorse Bioscience #100850-001) must be hydrated overnight for consistent results. Add XF calibrant, pH 7.4 (Seahorse Bioscience #100840-000) to each well of a Seahorse 24-well or 96-well plate. Put a sensor cartridge on top of the plate with sensors submerged in calibrant solution. Hydrate at 37 °C without CO_2 for up to 72 h. Turn on the Seahorse XF Analyzer to allow the instrument to warm to 37 °C before an assay.

Day 2: Prepare a cell plate for analysis

Warm XF DMEM media (Seahorse Bioscience # 102353-100) to 37 °C in a water bath, add L-glutamine to a final concentration of 2 m*M* and adjust the pH of the media to 7.4. Remove all but 50–100 μl of media from each well and add 1 ml of 37 °C XF DMEM media, pH 7.4. It is important during this process to always leave a small amount of media in each well so that cells do not dry out and the cell monolayer is not disrupted. Then remove all but 50 μl of media from each well again and add the desired volume of 37 °C XF DMEM media. Place a 24- or 96-well plate in a 37 °C incubator without CO_2 for 1 h before running an assay. CO_2 will alter the pH of the media and disrupt ECAR measurements.

Inhibitors should be adjusted to the proper concentration in the XF DMEM media. Since inhibitors will be added through injection ports in the cartridge, they should be at a concentration of 10× higher than the final dilution. Prepare 2 ml stocks of 100 m*M* glucose, 1 M 2-DG, and the optimized oligomycin concentration chosen by titration. Pipette 61.5 μl of glucose stock solution, 67.5 μl of oligomycin stock solution, and 75 μl of 2-DG stock solution into injection ports A, B, and C, respectively. Make sure to fill all A, B, and C injection ports including blank wells.

The XF24 and XF96 analyzers require the input of a protocol command sequence. Important parameters of this protocol are mix time, wait, and measure time. Cell lines with higher metabolic activity require shorter measurement times than cells with less metabolic activity. A good protocol to start with is a 3-min mix time, 3-min wait, and 3-min measure time. If OCR or ECAR values are too low, a protocol of 2-min mix time, 2-min wait, and 4- or 5-min measure time can improve the data quality.

After running the glycolysis stress test, normalize ECAR data to cell number or protein concentration and plot ECAR against time. Glycolysis, glycolytic capacity, and glycolytic reserve can be calculated with normalized ECAR values using the equations given below (Fig. 5.3C):

$$\text{Glycolysis} = \text{ECAR after addition of glucose} - \text{ECAR after 2} - \text{DG treatment}$$

$$\text{Glycolytic capacity} = \text{ECAR after oligomycin treatment} - \text{ECAR after 2} - \text{DG treatment}$$

$$\text{Glycolytic reserve} = \text{ECAR after oligomycin treatment} - \text{ECAR after addition of glucose}$$

$$\text{Nonglycolytic acidification} = \text{ECAR after 2} - \text{DG treatment}$$

2.2. Glucose analog uptake

To determine the glucose uptake rate by cells, a labeled isoform of glucose can be added to the cell culture media and then measured within cells after a given period of time. Two types of glucose analogs are typically used for these studies: radioactive glucose analogs, such as 2-deoxy-D-[1,2-^{3}H]-glucose, 2-deoxy-D-[1-^{14}C]-glucose, or 2-deoxy-2-(^{18}F)-fluoro-D-glucose (^{18}FDG), or fluorescent glucose analogs, such as 2-[*N*-(7-nitrobenz-2-oxa-1,3-diaxol-4-yl)amino]-2-deoxyglucose (2-NBDG). Measurements of radioactive glucose analog uptake require a scintillation counter, whereas 2-NBDG uptake is usually measured by flow cytometry or fluorescent microscopy. In contrast to extracellular approaches, these techniques require cell harvest and termination of the cell culture.

2.2.1 2-Deoxy-D-[1,2-^{3}H]-glucose or 2-deoxy-D-[1-^{14}C]-glucose

Glucose analog, 2-DG, is brought into cells by a glucose transporter and is phosphorylated within the cytosol by hexokinase (Jenkins, Furler, & Kraegen, 1986). Further metabolism of 2-DG is slow and phosphorylation by hexokinase traps 2-DG within a cell (Hansen, Gulve, & Holloszy, 1994). By using radioactive 2-deoxy-D-[1,2-^{3}H]-glucose or 2-deoxy-D-[1-^{14}C]-glucose, the accumulation of 2-DG can be measured to quantify glucose uptake.

2.2.1.1 Protocol for 2-deoxy-D-[1,2-^{3}H]-glucose or 2-deoxy-D-[1-^{14}C]-glucose uptake

Plate approximately 5×10^{4} cells per well in a 24-well plate. Allow time for the cells to attach and spread on the plate and switch to glucose-free or glucose low media. Add 1 μCi 2-deoxy-D-[1,2-^{3}H]-glucose (MP Biomedicals #0127088) or 2-deoxy-D-[1-^{14}C]-glucose (MP Biomedicals #0111012) and incubate for at least 20 min. Wash cells with 1× PBS, pH 7.4 (Life Technologies #10010-023), and lyse cells with 1% SDS lysis buffer (1% sodium dodecyl sulfate, 10 m*M* Tris, pH 7.5). Use a scintillation counter to quantify [^{3}H] or [^{14}C] and normalize readings to cell number.

2.2.2 2-Deoxy-2-(^{18}F)-fluoro-D-glucose

^{18}FDG PET scanning has been used in clinical cancer diagnostics since the 1990s, but use of ^{18}FDG for *in vitro* research protocols has been more limited, likely because of the short half-life of the radioactive fluorine (109.7 min) and limited access to a cyclotron for its manufacture for most research labs. Like other 2-DG analogs, ^{18}FDG is phosphorylated after transport into the cell and cannot be exported (Smith, 2000). ^{18}FDG accumulation occurs in many tumors, or other rapidly dividing cells, due to the Warburg effect, and increased analog trapping contrasts with surrounding normal tissues to provide signal discrimination for *in vivo* PET imaging. Use of ^{18}FDG for *in vitro* experiments requires consideration of factors that can influence metabolism including cell density and growth media (Mertens, Mees, Lambert, Van de Wiele, & Goethals, 2012). ^{18}FDG requires instrumentation to measured gamma radiation and increased safety procedures for working with radiation. Because the measurement ^{18}FDG uptake *in vitro* is uncommon and an alternate method involving fluorescent detection of glucose uptake is available, a detailed work protocol for ^{18}FDG uptake measurements is not provided here.

2.2.3 2-[N-(7-Nitrobenz-2-oxa-1,3-diaxol-4-yl)amino]-2-deoxyglucose

2-NBDG is taken into cells through glucose transporters and phosphorylated by hexokinase. However, 2-NBDG can be dephosphorylated, which can result in efflux from cells, but this analog still provides a good approximation of glucose uptake (O'Neil, Wu, & Mullani, 2005). Because 2-NBDG and glucose are both imported by glucose transporters, unlabeled glucose in the media will impact 2-NBDG uptake (O'Neil et al., 2005; Zou, Wang, & Shen, 2005). Therefore, the use of low-glucose media is optimal when tolerated by cells and available. 2-NBDG can be excited to fluoresce by 465 nm wavelength light and yields a 540-nm wavelength emission, which can be detected using a fluorescence detector channel or filter typically used for green fluorophores. A negative control sample should be included that is not incubated with 2-NBDG to set flow cytometer gating of negative and positive events and compensation when necessary (Fig. 5.4).

2.2.3.1 Protocol for 2-NBDG uptake

Approximately 5×10^4 cells should be plated per well. Cell confluence can affect metabolism and should be considered and adjusted appropriately. For each condition to be examined, count and plate cells at equal densities and allow time for cells, if they are adherent, to fully attach to the plate surface or substrate. Wash cells with $1 \times$ PBS, pH 7.4, followed by the addition of culture media (low glucose when possible) supplemented with 10–100 μM 2-NBDG (Life Technologies #N13195), and incubate for 1–12 h at 37 °C with 5% CO_2. A negative control should be incubated with the same culture media without adding 2-NBDG. The cell density, concentration of 2-NBDG, and time of incubation can be adjusted for optimal results.

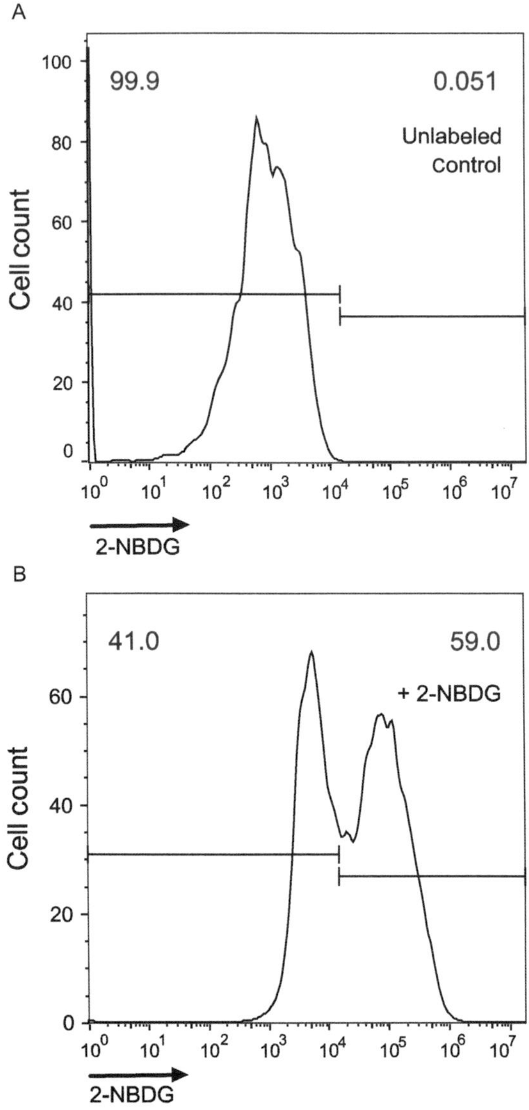

Figure 5.4 *Measurement of 2-NBDG uptake by flow cytometry.* (A) A negative control that has not been cultured with 2-NBDG. (B) A sample cultured with 2-NBDG.

For analysis by flow cytometry, add 1× trypsin (Life Technologies #25300054), harvest cells, wash twice with ice cold 1× PBS, pH 7.4, at 4 °C, and keep on ice. Keep the samples on ice to inhibit dephosphorylation and export of 2-NBDG from cells. Analyze the collected cells by flow cytometry within 30 min of harvest. A control sample lacking 2-NBDG should be used to set the flow cytometer compensation and gate parameters for 2-NBDG positive and negative events (Fig. 5.4).

3. MEASURING THE ACTIVITY OF RATE-LIMITING GLYCOLYTIC ENZYMES

While the activity of any single glycolytic enzyme is not a proxy for carbon flux through the entire pathway, specific enzymes limit the rate of glycolysis, and therefore, their activities control the maximum possible flux. Hexokinase, phosphofructokinase, and PK are the main rate-limiting enzymes in glycolysis.

Conversion from NAD^+ to NADH or $NADP^+$ to NADPH is often used in enzymatic assays because NADH and NADPH absorb light at 340 nm, while NAD^+ and $NADP^+$ lack absorbance at this wavelength (McComb, Bond, Burnett, Keech, & Bowers, 1976). By coupling the enzyme of interest with an NAD^+/NADH- or $NADP^+$/NADPH-dependent enzyme, the activity of the enzyme can be measured by the change in absorbance at 340 nm over time. Hexokinase, phosphofructokinase, and PK can be coupled with NAD^+/NADH- or $NADP^+$/NADPH-dependent enzymes by using enzymes that react with their products.

3.1. Hexokinase

Hexokinases phosphorylate the 6-hydroxyl of hexose using the γ-PO_4 of ATP as a donor group (Wilson, 1995). In humans, hexokinase 4 (HK4) is more specific to glucose than other hexoses and is referred to as glucokinase. An ADP-dependent glucokinase (ADPGK) has been recently discovered in mouse and human genomes, which transfers a phosphoryl group from ADP to glucose (Richter et al., 2012). Hexokinase 1 (HK1), hexokinase 2 (HK2), and hexokinase 3 (HK3) are allosterically inhibited by their product, glucose-6-phosphate, but HK4 is not sensitive to glucose-6-phosphate concentrations (Wilson, 1995). Hexokinase is the first key enzyme in glycolysis because it traps glucose within cells.

High expression of HK2 is associated with poor prognosis in hepatocellular carcinoma and brain metastasis (Palmieri et al., 2009; Peng, Lai, Pan, Hsiao, & Hsu, 2008). HK2 is required for tumor growth in multiple mouse models of cancer and is a potential target for therapeutics (Patra et al., 2013). Breast cancer cells regulate HK2 through miR-155, which increases expression of HK2 through multiple mechanisms (Jiang et al., 2012). To measure hexokinase activity, glucose is provided as a substrate, and glucose-6-phosphate dehydrogenase (G6PD) further metabolizes glucose-6-phosphate in a $NADP^+$ dependent manner. The absorbance of NADPH at 340 nm will increase as G6PD converts glucose-6-phoshate to fructose-6-phosphate.

3.1.1 Protocol for hexokinase activity assay

Wash cells with 1× PBS, pH 7.4, and incubate in lysis buffer (50 m*M* Tris–HCl, pH 7.5, 1 m*M* EDTA, 150 m*M* NaCl, 1% NP-40, 1 m*M* DTT, protease inhibitor cocktail) for 30 min on ice to lyse cells. Prepare reaction buffer (50 m*M* Tris–HCl, pH 7.5, 10 m*M* $MgCl_2$, 0.6 m*M* ATP (Sigma #A2383), 100 m*M* glucose (Sigma #G8270), 0.2 m*M* $NADP^+$ (Sigma #0505), and 0.1 units of glucose-6-phosphate dehydrogenase (Sigma #G4134)) per ml. Add 20 μg of fresh cell lysate to 1 ml of reaction buffer. A negative and positive control should be included without cell lysate and with 0.05 units of hexokinase (Sigma #H5000) added. Mix, incubate at 37 °C, and measure optical absorbance of the reaction at 340 nm with a spectrophotometer every 15 s for 10 min (Yi et al., 2012). Enzyme activity can be represented as the change in absorbance per minute, which should be calculated on a linear portion of the obtained curve.

3.2. Phosphofructokinase

Glucose-6-phosphate can enter the pentose phosphate pathway or continue through the glycolytic pathway. Phosphofructokinase 1 (PFK1) provides the first enzymatic step at which a glucose molecule becomes committed to glycolysis and therefore is subject to regulation (Nelson & Cox, 2008). PFK1 activity depends on the concentrations of AMP, ADP, and ATP with allosteric activation by AMP and ADP and allosteric inhibition by ATP. Additional allosteric inhibition is provided by citrate and activation by fructose 2,6-bisphosphate (Nelson & Cox, 2008).

PFK1 activity has been linked to the Warburg effect through the p53 target protein, TP-53-induced glycolysis and apoptosis regulator (TIGAR). TIGAR inhibits glycolysis and promotes the pentose phosphate pathway by reducing levels of fructose-2,6-bisphosphate, thereby reducing PFK1 activity (Bensaad et al., 2006). Glycosylation of the fructose-2,6-bisphosphate binding site on PFK1 also inhibits glycolytic flux which decreases cell proliferation, thereby reducing cancer aggressiveness (Yi et al., 2012). Measurement of phosphofructokinase activity can be achieved by providing other glycolytic enzymes, aldolase, triosephosphate isomerase, and glyceraldehyde 3-phosphate dehydrogenase, which are

NAD^+ dependent. An increase of NADH absorbance at 340 nm can be observed as fructose 1,6-bisphosphate is converted to 1,3-bisphosphoglycerate.

3.2.1 Protocol for phosphofructokinase activity assay

Wash cells with 1 × PBS, pH 7.4, and incubate in lysis buffer (50 m*M* Tris–HCl, pH 7.5, 1 m*M* EDTA, 150 m*M* NaCl, 1% NP-40, 1 m*M* DTT, protease inhibitor cocktail) for 30 min on ice to lyse cells. Prepare reaction buffer (50 m*M* Tris–HCl, pH 7.5, 5 m*M* $MgCl_2$, 5 m*M* ATP (Sigma #A2383), 0.2 m*M* NADH (Sigma #N4505), 100 m*M* KCl, 5 m*M* Na_2HPO_4, 5 m*M* $MgCl_2$, 0.01 AMP (Sigma #01930), 5 m*M* fructose-6-phosphate (Sigma #F3627), 5 units of triosephosphate isomerase (Sigma #T2507) per ml, 1 unit of aldolase (Sigma #A2714) per ml, and 10 units of glyceraldehyde-3-phosphate dehydrogenase (Sigma #G2267)) per ml. Add 20 μg of fresh cell lysate to 1 ml of reaction buffer. A negative and positive control should be included without cell lysate and with 0.05 units phosphofructokinase (Sigma #F0137) added. Measure optical absorbance of the reaction at 340 nm at room temperature every 15 s for 10 min with a spectrophotometer (Yi et al., 2012). Activity can be represented as the change in absorbance per minute, which should be calculated in a linear portion of the obtained curve.

3.3. Pyruvate kinase

PK catalyzes the final step in glycolysis by transferring a phosphoryl group from phosphoenolpyruvate (PEP) to ADP to form pyruvate and ATP. PK is active as a tetramer protein and is allosterically inhibited by ATP, acetyl-CoA, and long-chain fatty acids (Nelson & Cox, 2008). There are four PK isoforms including PKL, PKR, PKM1, and PKM2. PKL and PKR, the dominant PK isoforms in the liver and red blood cells, undergo allosteric activation by fructose-1,6-bisphosphate (Carbonell, Marco, Felíu, & Sols, 1973). Fructose-1,6-bisphosphate is also an allosteric activator of PKM2, which is a splice variant of the constitutively active PKM1 enzyme isoform (Dombrauckas, Santarsiero, & Mesecar, 2005). The product of PK, pyruvate, can be converted to lactate by lactate dehydrogenase, which is NADH dependent. The activity of PK can then be detected by a loss of absorbance at 340 nm.

3.3.1 Protocol for PK activity assay

Wash cells with 1 × PBS, pH 7.4, and incubate in lysis buffer (50 m*M* Tris–HCl, pH 7.5, 1 m*M* EDTA, 150 m*M* NaCl, 1% NP-40, 1 m*M* DTT, protease inhibitor cocktail) for 30 min on ice to lyse cells. Prepare reaction buffer (50 m*M* Tris–HCl, pH 7.5, 100 m*M* KCl, 5 m*M* $MgCl_2$, 0.6 m*M* ADP (Sigma #A2754), 0.5 m*M* PEP (Sigma #860077), 0.18 m*M* NAD^+ (Sigma #N1636), 10 μ*M* fructose-1,6-bisphosphate (Sigma #F6803), and 10 units of lactate dehydrogenase (Sigma #59747)) per ml. Add 10 μg of fresh cell lysate to 1 ml of reaction buffer. Measure optical absorbance of the reaction at 340 nm every 15 s for 10 min with a spectrophotometer (Christofk, Vander Heiden, Wu, Asara, & Cantley, 2008; Yi et al., 2012). A positive and a negative control should be run with no cell lysate and 0.05 units of PK (Sigma #P7768) added. Activity can be represented as the change in absorbance per minute, which should be calculated in a linear portion of the obtained curve.

4. METABOLITE MEASUREMENTS AND GLUCOSE TRACING

Unlabeled metabolites or ^{13}C-labeled metabolites can be quantified by mass spectrometry (MS) or nuclear magnetic resonance (NMR) spectrometry. Measurements of unlabeled metabolites provide steady-state data on the levels of all detectable metabolites within the cell. Using metabolite concentrations for extrapolation of flux information is difficult, however, because changes in the steady-state level of a metabolite can be due to increased or decreased flux through any of its associated metabolic pathways. Absolute quantification of an unlabeled metabolite is possible by adding a labeled standard for the metabolite of interest before MS analysis. Metabolite profiling was used to identify decreased serine biosynthesis as a result of PKM2 silencing (Chaneton et al., 2012).

The use of ^{13}C-labeled glucose provides glycolytic flux data by quantifying ^{13}C glucose-derived carbons in downstream metabolites. However, labeled substrates are costly and the label can be diluted by other carbon sources, making measurements challenging for some metabolites. However, detection of ^{13}C-glucose-derived carbons in glycolytic intermediates is robust; making it is a good method for analyzing glycolysis.

While chromatography combined with MS is becoming the most widely used approach for metabolomics, NMR spectroscopy also has advantages for detecting and quantifying metabolites. NMR spectroscopy is quantitative, fast, and reproducible but has a detection limit in the micromolar range (Pan & Raftery, 2007; Want, Cravatt, & Siuzdak, 2005). By contrast, MS is exceedingly sensitive in the nanomolar range, making it useful for complex biosamples. MS analysis, however, requires metabolite separation with gas chromatography (GC) or liquid chromatography (LC), which reduces the speed of data acquisition. GC/MS combines the separation of complex biosamples with the retained sensitivity of MS (Want et al., 2005). However, GC/MS has an additional derivatization step of sample preparation that is required for detecting nonvolatile or large, polar macromolecules (Dietmair, Timmins, Gray, Nielsen, & Krömer, 2010; Want et al., 2005). By contrast, LC/MS detects a broad range of metabolites with simple sample preparation. Following the separation of metabolites by LC, they must be ionized for analysis by MS. Liquid chromatography electrospray ionization mass spectrometry (LC/ESI-MS) combines an automated ionization method, electrospray ionization, with the simplicity of LC and the sensitivity of MS. Therefore, we provide metabolomics protocols that are suitable for use with LC/ESI-MS. Fourier transform ion cyclotron resonance MS (FT-ICR-MS), orbitrap MS, and multipass time of flight MS (multipass TOF-MS) have all be used for metabolomic analysis (Aharoni et al., 2002; Lei, Huhman, & Sumner, 2011). FT-ICR-MS has ultrahigh resolution and mass accuracy below 1 ppm, while orbitrap MS also has high resolving power and mass accuracy between 1 and 5 ppm (Lei et al., 2011).

4.1. Metabolite measurements

Metabolite extraction must be performed to obtain metabolites from cultured cells for LC/ESI-MS analysis. Metabolite extraction generally has two parts: quenching and extraction. Quenching first limits further metabolic activity and removes contaminants, which is then followed by a metabolite extraction step (Sellick, Hansen, Stephens, Goodacre, & Dickson, 2011). Metabolites with a high turnover rate, such as ATP, depend on complete and quick quenching methods for accurate detection. Chemical and physical properties of each metabolite require different methods of quenching and different extraction solutions making it challenging to find one method that works for all metabolites (Dietmair et al., 2010). Glucose-6-phosphate is particularly unstable and requires effective quenching for accurate measurements. Enzymatic assays to measure glucose-6-phosphate can be used to check on the success of quenching and extraction steps. For optimization of extraction, temperature and solvent can be adjusted in both the quenching and extraction step.

4.1.1 Protocol for metabolite extraction of adherent cells for LC/ESI-MS analysis

Approximately 1×10^6 cells are minimally required for detection of metabolites by LC/ESI-MS. Cells can be grown in six-well plates for direct extraction from each well. For normalization to cell number or protein concentration, an extra well of each condition should be plated in parallel for cell counting or protein quantification. Add fresh media to cells 24 h prior to metabolite extraction. Place the six-well plate on ice and wash cells with ice cold 1 × PBS, pH 7.4. Add 400 μl of ice cold HPLC grade methanol (Fisher #A454-4). To maximize quenching speed, methanol can be cooled on dry ice to achieve a temperature as low as −40 °C. Add an equal volume of Milli Q water and gently shake the plate to ensure mixing. Scrape cells off the plate and transfer to a 1.5-ml eppendorf tube. For separation of polar from nonpolar metabolites, add 400 μl of HPLC grade chloroform (Fisher C607-4). Briefly vortex each sample every 5 min for 15–30 min. Transfer the top layer, which is the aqueous phase containing polar molecules, to a new tube. The remaining bottom layer contains nonpolar metabolites, which can also be transferred into a fresh tube. Glycolytic intermediates are polar molecules and should be in the top layer. Dry the contents of both tubes with a vacuum concentrator. Samples can then be stored at −80 °C until analysis. Alternatively, samples can be suspended in 50 μl of initial LC mobile phase solution for immediate LC/ESI-MS analysis. MS measurements should be performed according to the methods described in the instrument documentation.

4.2. Metabolic flux analysis with stable glucose isotopes

^{13}C can replace any of the six carbons within a glucose molecule. [U-$^{13}C_6$]-glucose is ^{13}C labeled at all six carbons and is often used to trace the glucose contribution to the TCA cycle. [1,2-$^{13}C_2$]-glucose is the best tracer for glycolytic pathway metabolite measurements (Metallo, Walther, & Stephanopoulos, 2009). In addition, carbon flux through the pentose phosphate pathway is not well defined by [U-$^{13}C_6$]-glucose. Instead, [1,2-$^{13}C_2$]-glucose will lose one ^{13}C when it goes through the pentose phosphate pathway, which distinguishes metabolites that have gone through this pathway from those that have gone directly through glycolysis (Fig. 5.5).

Labeled glucose will compete for uptake into the cell with unlabeled glucose within the media. Therefore, glucose-free or low-glucose media is used for labeled glucose tracing flux experiments. Regular culture media should be replaced with [U-$^{13}C_6$]-glucose (Cambridge Isotope Labs #CLM-1396) or [1,2-$^{13}C_2$]-glucose (Cambridge Isotope Labs #CLM-504) supplemented

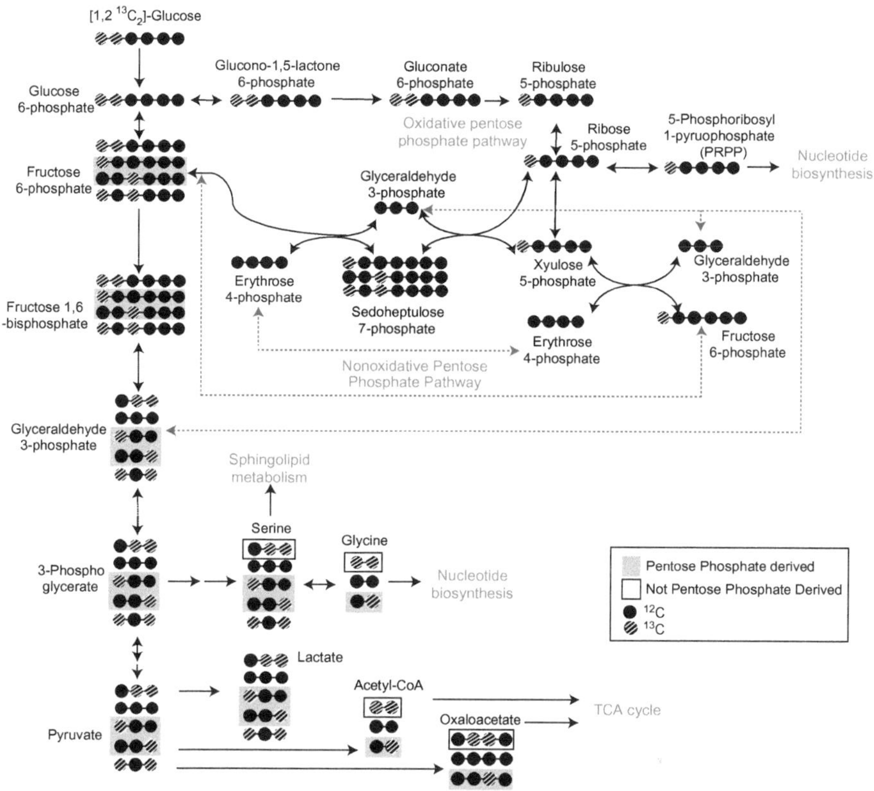

Figure 5.5 *Glycolytic pathway metabolite flux analysis with [1,2 $^{13}C_2$]-glucose.* Metabolites derived from the pentose phosphate pathway (gray boxes) can be distinguished from metabolites derived directly from glycolysis (black boxes). For simplicity, not all possible permutations of ^{12}C and ^{13}C are provided.

media 24 h prior to metabolite extraction. The same protocol can be used for extraction of metabolites for metabolic flux analysis as for measurement of unlabeled metabolites. Data analysis, however, is more complex and requires specialized software. Fiatflux, an open source software for use with isotopic tracer experiments, is an enabling tool for data analysis (Zamboni, Fischer, & Sauer, 2005). OpenFLUX and 13CFlux are also software packages available for use for with ^{13}C metabolic flux analysis (Quek, Wittmann, Nielsen, & Krömer, 2009; Weitzel et al., 2013).

4.3. Release of ^3H_2O from [5-^{3}H]-glucose

The release of ^3H_2O from [5-^{3}H]-glucose is a method to quantify glycolytic flux (Neely, Denton, England, & Randle, 1972). A single tritium at the C5

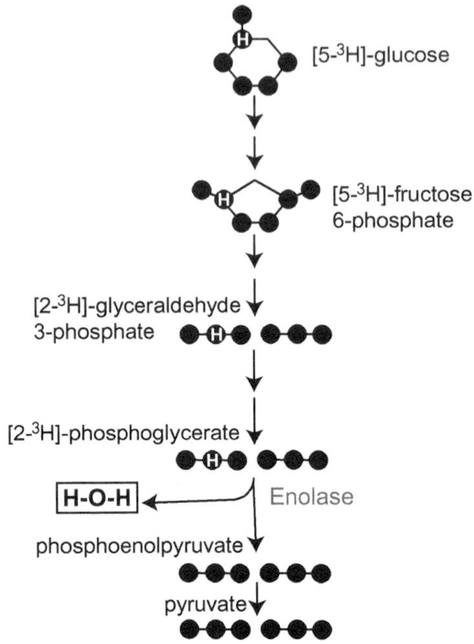

Figure 5.6 *Glycolytic flux measurements with [5-^{13}H]-glucose.* A single tritium present on 5C of glucose is released as water in the ninth step of glycolysis catalyzed by enolase.

of glucose is removed by a condensation reaction in the ninth step of glycolysis, which is catalyzed by enolase (Fig. 5.6). The resulting $^{3}H_2O$ diffuses freely into the cell culture media and is quantified by scintillation counting. The complete conversion of glucose to pyruvate is not measured by this reaction.

For detection of $^{3}H_2O$ using a liquid scintillation counter, the [5-^{3}H]-glucose must be removed from the cell supernatant. One method for separation is to allow $^{3}H_2O$ to diffuse into a cold H_2O solution while inhibiting the diffusion of [5-^{3}H]-glucose (Vander Heiden et al., 2011). An alternative method is to use ion chromatography to remove the glucose from the cell supernatant (Støttrup et al., 2010).

4.3.1 Protocol for measuring the release of $^{3}H_2O$ from [5-^{3}H]-glucose

Wash 1×10^6 cells with $1 \times$ PBS, pH 7.4, and resuspend in 1 ml Krebs buffer (126 m*M* NaCl, 2.5 m*M* KCl, 25 m*M* $NaHCO_3$, 1.2 m*M* NaH_2PO_4, 1.2 m*M* $MgCl_2$, and 2.5 m*M* $CaCl_2$). Incubate at 37 °C, 5% CO_2 for 30 min and then replace with Krebs buffer containing 10 μCi [5-^{3}H]-glucose (Perkin Elmer #NET531005MC) and 10 m*M* unlabeled glucose.

Table 5.2 Glucose tracers used to measure glycolysis and related metabolic pathways

Glucose tracer	Detection method	Detection precision Glycolysis	PPP	TCA cycle	Price
[U-$^{13}C_6$]-glucose	Mass spectrometry	Low	Low	High	$
[1,2-$^{13}C_2$]-glucose	Mass spectrometry	High	High	Ok	$$
[2-$^{13}C_1$]-glucose	Mass spectrometry	High	High	Ok	$$
[5-^{3}H]-glucose	Scintillation counter	High	No	No	$$$

PPP, pentose phosphate pathway.

Incubate at 37 °C, 5% CO_2 for 1 h. Spin down the sample and remove 500 μl of cell supernatant. Add 500 μl of 0.2 N HCl. Pack a 2-ml Poly-Prep chromatography column (Biorad #731-1550) with 2 ml of AG 1-X8 resin, 200–400 mesh, formate form (Biorad #140-1454). Wash the resin with 20 ml of 1 N NaOH and then with 10 ml of distilled water. Add 1 ml of supernatant solution to the column and collect the sample after glucose removal. Dissolve $^{3}H_2O$ into scintillation solution and quantify by beta-scintillation counting (Støttrup et al., 2010; Vander Heiden et al., 2010).

5. SUMMARY

Each of the discussed methods provides data on the glycolytic flux in mammalian cells, including cancer. Selecting which technique or combinations of techniques are appropriate for a given study depends on equipment availability, budget, and the type of information required. Extracellular glucose and lactate measurements and enzymatic activity assays can be performed with standard laboratory equipment but provide a limited amount of insight. Metabolomics approaches provide data on glycolysis as well as its offshoot pathways but require expensive equipment and laborious data analyses (Table 5.2). All of these methods can be used in combination to obtain a comprehensive assessment of glycolytic flux and downstream metabolic pathways in cancer cells.

ACKNOWLEDGMENTS

T. T. is supported by a Ruth L. Kirschstein National Research Service Award GM007185. M. A. T. is supported by CIRM grant RB1-01397, UC Discovery Awards Bio07-10663 and 178517, and NIH Grants GM073981, P01GM081621, CA156674, and CA90571.

REFERENCES

Aharoni, A., Ric de Vos, C. H., Verhoeven, H. A., Maliepaard, C. A., Kruppa, G., & Bino, R. (2002). Nontargeted metabolome analysis by use of Fourier transform ion cyclotron mass spectrometry. *Omics: A Journal of Integrative Biology*, *6*(3), 217–234.

Bensaad, K., Tsuruta, A., Selak, M. A., Vidal, M. N. C., Nakano, K., Bartrons, R., et al. (2006). TIGAR, a p53-inducible regulator of glycolysis and apoptosis. *Cell*, *126*(1), 107–120. http://dx.doi.org/10.1016/j.cell.2006.05.036.

Carbonell, J., Marco, R., Felíu, J. E., & Sols, A. (1973). Pyruvate kinase. *European Journal of Biochemistry*, *37*(1), 148–156.

Caro, P., Kishan, Amar U., Norberg, E., Stanley, I. A., Chapuy, B., Ficarro, Scott B., et al. (2012). Metabolic signatures uncover distinct targets in molecular subsets of diffuse large B cell lymphoma. *Cancer Cell*, *22*(4), 547–560. http://dx.doi.org/10.1016/j.ccr.2012.08.014.

Chaneton, B., Hillmann, P., Zheng, L., Martin, A. C. L., Maddocks, O. D. K., Chokkathukalam, A., et al. (2012). Serine is a natural ligand and allosteric activator of pyruvate kinase M2. *Nature*, *491*(7424), 458–462. http://dx.doi.org/10.1038/nature11540.

Christofk, H. R., Vander Heiden, M. G., Harris, M. H., Ramanathan, A., Gerszten, R. E., Wei, R., et al. (2008). The M2 splice isoform of pyruvate kinase is important for cancer metabolism and tumour growth. *Nature*, *452*(7184), 230–233. http://dx.doi.org/10.1038/nature06734.

Christofk, H. R., Vander Heiden, M. G., Wu, N., Asara, J. M., & Cantley, L. C. (2008). Pyruvate kinase M2 is a phosphotyrosine-binding protein. *Nature*, *452*(7184), 181–186. http://dx.doi.org/10.1038/nature06667.

DeBerardinis, R. J., Lum, J. J., Hatzivassiliou, G., & Thompson, C. B. (2008). The biology of cancer: Metabolic reprogramming fuels cell growth and proliferation. *Cell Metabolism*, 7(1), 11–20. http://dx.doi.org/10.1016/j.cmet.2007.10.002.

Dietmair, S., Timmins, N. E., Gray, P. P., Nielsen, L. K., & Krömer, J. O. (2010). Towards quantitative metabolomics of mammalian cells: Development of a metabolite extraction protocol. *Analytical Biochemistry*, *404*(2), 155–164. http://dx.doi.org/10.1016/j.ab.2010.04.031.

Dombrauckas, J. D., Santarsiero, B. D., & Mesecar, A. D. (2005). Structural basis for tumor pyruvate kinase M2 allosteric regulation and catalysis. *Biochemistry*, *44*(27), 9417–9429. http://dx.doi.org/10.1021/bi0474923.

Hansen, P. A., Gulve, E. A., & Holloszy, J. O. (1994). Suitability of 2-deoxyglucose for in vitro measurement of glucose transport activity in skeletal muscle. *Journal of Applied Physiology*, *76*(2), 979–985.

Hitosugi, T., Zhou, L., Elf, S., Fan, J., Kang, H.-B., Seo, Jae H., et al. (2012). Phosphoglycerate mutase 1 coordinates glycolysis and biosynthesis to promote tumor growth. *Cancer Cell*, *22*(5), 585–600. http://dx.doi.org/10.1016/j.ccr.2012.09.020.

Hu, J., Locasale, J. W., Bielas, J. H., O'Sullivan, J., Sheahan, K., Cantley, L. C., et al. (2013). Heterogeneity of tumor-induced gene expression changes in the human metabolic network. *Nature Biotechnology*, *31*(6), 522–529.

Jain, M., Nilsson, R., Sharma, S., Madhusudhan, N., Kitami, T., Souza, A. L., et al. (2012). Metabolite profiling identifies a key role for glycine in rapid cancer cell proliferation. *Science*, *336*(6084), 1040–1044. http://dx.doi.org/10.1126/science.1218595.

Jenkins, A. B., Furler, S. M., & Kraegen, E. W. (1986). 2-Deoxy-D-glucose metabolism in individual tissues of the rat in vivo. *International Journal of Biochemistry*, *18*(4), 311–318. http://dx.doi.org/10.1016/0020-711X(86)90036-4.

Jiang, S., Zhang, L.-F., Zhang, H.-W., Hu, S., Lu, M.-H., Liang, S., et al. (2012). A novel miR-155/miR-143 cascade controls glycolysis by regulating hexokinase 2 in breast cancer cells. *The EMBO Journal*, *31*(8), 1985–1998. http://dx.doi.org/10.1038/emboj.2012.45.

Le, A., Lane, Andrew N., Hamaker, M., Bose, S., Gouw, A., Barbi, J., et al. (2012). Glucose-independent glutamine metabolism via TCA cycling for proliferation and survival in B cells. *Cell Metabolism*, *15*(1), 110–121. http://dx.doi.org/10.1016/j.cmet.2011.12.009.

Lei, Z., Huhman, D. V., & Sumner, L. W. (2011). Mass spectrometry strategies in metabolomics. *Journal of Biological Chemistry*, *286*(29), 25435–25442.

Lim, J. Y., Yoon, S. O., Seol, S. Y., Hong, S. W., Kim, J. W., Choi, S. H., et al. (2012). Overexpression of the M2 isoform of pyruvate kinase is an adverse prognostic factor for signet ring cell gastric cancer. *World Journal of Gastroenterology: WJG*, *18*(30), 4037–4043.

Locasale, J. W. (2013). Serine, glycine and one-carbon units: Cancer metabolism in full circle. *Nature Reviews Cancer*, *13*(8), 572–583.

Locasale, J. W., Grassian, A. R., Melman, T., Lyssiotis, C. A., Mattaini, K. R., Bass, A. J., et al. (2011). Phosphoglycerate dehydrogenase diverts glycolytic flux and contributes to oncogenesis. *Nature Genetics*, *43*(9), 869–874. http://dx.doi.org/10.1038/ng.890.

Marin-Valencia, I., Yang, C., Mashimo, T., Cho, S., Baek, H., Yang, X.-L., et al. (2012). Analysis of tumor metabolism reveals mitochondrial glucose oxidation in genetically diverse human glioblastomas in the mouse brain in vivo. *Cell Metabolism*, *15*(6), 827–837. http://dx.doi.org/10.1016/j.cmet.2012.05.001.

McComb, R. B., Bond, L. W., Burnett, R. W., Keech, R. C., & Bowers, G. N. (1976). Determination of the molar absorptivity of NADH. *Clinical Chemistry*, *22*(2), 141–150.

Mertens, K., Mees, G., Lambert, B., Van de Wiele, C., & Goethals, I. (2012). In vitro 2-deoxy-2-[18F] fluoro-D-glucose uptake: Practical considerations. *Cancer Biotherapy and Radiopharmaceuticals*, *27*(3), 183–188.

Metallo, C. M., Walther, J. L., & Stephanopoulos, G. (2009). Evaluation of 13C isotopic tracers for metabolic flux analysis in mammalian cells. *Systems Biology for Biotechnological Innovation*, *144*(3), 167–174. http://dx.doi.org/10.1016/j.jbiotec.2009.07.010.

Neely, J. R., Denton, R. M., England, P. J., & Randle, P. J. (1972). The effects of increased heart work on the tricarboxylate cycle and its interactions with glycolysis in the perfused rat heart. *The Biochemical Journal*, *128*, 147–159.

Nelson, D. L., & Cox, M. M. (2008). *Lehninger principles of biochemistry*. New York: W.H. Freeman.

O'Neil, R. G., Wu, L., & Mullani, N. (2005). Uptake of a fluorescent deoxyglucose analog (2-NBDG) in tumor cells. *Molecular Imaging and Biology*, 7(6), 388–392.

Palmieri, D., Fitzgerald, D., Shreeve, S. M., Hua, E., Bronder, J. L., Weil, R. J., et al. (2009). Analyses of resected human brain metastases of breast cancer reveal the association between up-regulation of hexokinase 2 and poor prognosis. *Molecular Cancer Research*, 7(9), 1438–1445. http://dx.doi.org/10.1158/1541-7786.MCR-09-0234.

Pan, Z., & Raftery, D. (2007). Comparing and combining NMR spectroscopy and mass spectrometry in metabolomics. *Analytical and Bioanalytical Chemistry*, *387*(2), 525–527.

Patra, Krushna C., Wang, Q., Bhaskar, Prashanth T., Miller, L., Wang, Z., Wheaton, W., et al. (2013). Hexokinase 2 is required for tumor initiation and maintenance and its systemic deletion is therapeutic in mouse models of cancer. *Cancer Cell*, *24*(2), 213–228. http://dx.doi.org/10.1016/j.ccr.2013.06.014.

Peng, S.-Y., Lai, P.-L., Pan, H.-W., Hsiao, L.-P., & Hsu, H.-C. (2008). Aberrant expression of the glycolytic enzymes aldolase B and type II hexokinase in hepatocellular carcinoma are predictive markers for advanced stage, early recurrence and poor prognosis. *Oncology Reports*, *19*(4), 1045–1053.

Pfeiffer, T., Schuster, S., & Bonhoeffer, S. (2001). Cooperation and competition in the evolution of ATP-producing pathways. *Science*, *292*(5516), 504–507. http://dx.doi.org/10.1126/science.1058079.

Possemato, R., Marks, K. M., Shaul, Y. D., Pacold, M. E., Kim, D., Birsoy, K., et al. (2011). Functional genomics reveal that the serine synthesis pathway is essential in breast cancer. *Nature*, *476*(7360), 346–350. http://dx.doi.org/10.1038/nature10350.

Quek, L.-E., Wittmann, C., Nielsen, L., & Krömer, J. (2009). OpenFLUX: Efficient modelling software for 13C-based metabolic flux analysis. *Microbial Cell Factories*, *8*(1), 25.

Richter, S., Richter, J., Mehta, S., Gribble, A., Sutherland-Smith, A., Stowell, K., et al. (2012). Expression and role in glycolysis of human ADP-dependent glucokinase. *Molecular and Cellular Biochemistry*, *364*(1–2), 131–145. http://dx.doi.org/10.1007/s11010-011-1212-8.

Sellick, C. A., Hansen, R., Stephens, G. M., Goodacre, R., & Dickson, A. J. (2011). Metabolite extraction from suspension-cultured mammalian cells for global metabolite profiling. *Nature Protocols*, *6*(8), 1241–1249. http://dx.doi.org/10.1038/nprot.2011.366.

Smith, T. A. (2000). Mammalian hexokinases and their abnormal expression in cancer. *British Journal of Biomedical Science*, *57*(2), 170–178.

Støttrup, N. B., Løfgren, B., Birkler, R. D., Nielsen, J. M., Wang, L., Caldarone, C. A., et al. (2010). Inhibition of the malate–aspartate shuttle by pre-ischaemic aminooxyacetate loading of the heart induces cardioprotection. *Cardiovascular Research*, *88*(2), 257–266. http://dx.doi.org/10.1093/cvr/cvq205.

Vander Heiden, M. G., Christofk, H. R., Schuman, E., Subtelny, A. O., Sharfi, H., Harlow, E. E., et al. (2010). Identification of small molecule inhibitors of pyruvate kinase M2. *Biochemical Pharmacology*, *79*(8), 1118–1124. http://dx.doi.org/10.1016/j.bcp.2009.12.003.

Vander Heiden, M. G., Lunt, S. Y., Dayton, T. L., Fiske, B. P., Israelsen, W. J., Mattaini, K. R., et al. (2011). Metabolic pathway alterations that support cell proliferation. *Cold Spring Harbor Symposia on Quantitative Biology*, *76*, 325–334.

Want, E. J., Cravatt, B. F., & Siuzdak, G. (2005). The expanding role of mass spectrometry in metabolite profiling and characterization. *ChemBioChem*, *6*(11), 1941–1951. http://dx.doi.org/10.1002/cbic.200500151.

Warburg, O. (1956). On the origin of cancer cells. *Science*, *123*(3191), 309–314.

Weitzel, M., Nöh, K., Dalman, T., Niedenführ, S., Stute, B., & Wiechert, W. (2013). 13CFLUX2—High-performance software suite for 13C-metabolic flux analysis. *Bioinformatics*, *29*(1), 143–145.

Wilson, J. E. (1995). *Hexokinases. Reviews of physiology, biochemistry and pharmacology Vol. 126* (pp. 65–198). Berlin, Germany: Springer Berlin – Heidelberg.

Wu, M., Neilson, A., Swift, A. L., Moran, R., Tamagnine, J., Parslow, D., et al. (2007). Multiparameter metabolic analysis reveals a close link between attenuated mitochondrial bioenergetic function and enhanced glycolysis dependency in human tumor cells. *American Journal of Physiology Cell Physiology*, *292*(1), C125–C136.

Yi, W., Clark, P. M., Mason, D. E., Keenan, M. C., Hill, C., Goddard, W. A., et al. (2012). Phosphofructokinase 1 glycosylation regulates cell growth and metabolism. *Science*, *337*(6097), 975–980. http://dx.doi.org/10.1126/science.1222278.

Zamboni, N., Fischer, E., & Sauer, U. (2005). FiatFlux—A software for metabolic flux analysis from 13C-glucose experiments. *BMC Bioinformatics*, *6*(1), 209.

Zaugg, K., Yao, Y., Reilly, P. T., Kannan, K., Kiarash, R., Mason, J., et al. (2011). Carnitine palmitoyltransferase 1C promotes cell survival and tumor growth under conditions of metabolic stress. *Genes & Development*, *25*(10), 1041–1051.

Zou, C., Wang, Y., & Shen, Z. (2005). 2-NBDG as a fluorescent indicator for direct glucose uptake measurement. *Journal of Biochemical and Biophysical Methods*, *64*(3), 207–215.

CHAPTER SIX

Measurement of Enolase Activity in Cell Lysates

Keigo Fukano, Kazuhiro Kimura[1]
Department of Biomedical Sciences, Graduate School of Veterinary Medicine, Hokkaido University, Sapporo, Japan
[1]Corresponding author: e-mail address: k-kimura@vetmed.hokudai.ac.jp

Contents

Abstract

Enolase (EC 4.2.1.11) is a cytosolic metalloenzyme responsible for the conversion of 2-phosphoglycerate into phosphoenolpyruvate, the second to last step in glycolysis. In mammals, enolase is encoded by three homologous genes. These gene products not only possess distinct biochemical and immunological properties but also show different tissue distribution. Besides its glycolytic function, α-enolase plays a variety of roles in pathophysiological settings including oncogenesis, tumor progression, ischemia, and bacterial infection. The expression levels of α-enolase have been attributed diagnostic and prognostic value in a number of tumors. Furthermore, neuron-specific α-enolase is released into the cerebrospinal fluid as well as in the systemic circulation upon traumatic brain injury and ischemic episodes. Thus, the measurement of the enzymatic activity of enolase is relevant for diverse fields of investigation, including oncometabolism. Here, we described simple and rapid protocols to measure the activity of enolase in lysates from mammalian cells and tissues.

1. INTRODUCTION

Enolase, also known as phosphopyruvate hydratase, is a key glycolytic enzyme in the cytoplasm of prokaryotic and eukaryotic cells. It is a metalloenzyme that requires the metal ion such as Mg^{2+} and catalyzes the dehydration of 2-phosphoglycerate (2-PG) to phosphoenolpyruvate (PEP), in

Methods in Enzymology, Volume 542
ISSN 0076-6879
http://dx.doi.org/10.1016/B978-0-12-416618-9.00006-6

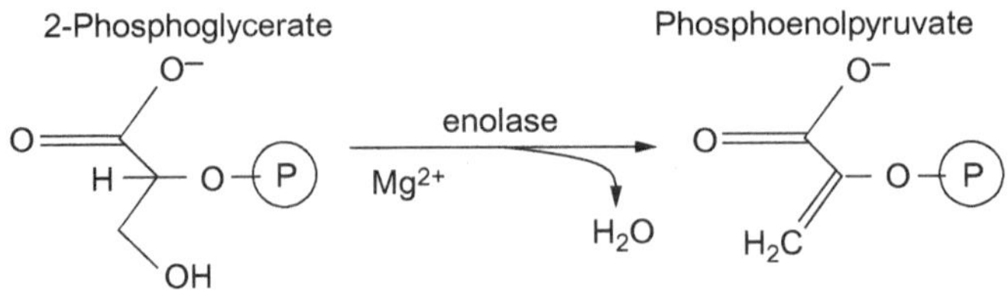

Figure 6.1 *Conversion of 2-phosphoglycerate to phosphoenolpyruvate by enolase.* Enolase, phosphopyruvate hydratase, catalyzes Mg^{2+}-dependent conversion of 2-phosphoglycerate to phosphoenolpyruvate in the glycolytic pathway.

the catabolic glycolytic pathway (Fig. 6.1). It can also catalyze the reverse reaction in the process of gluconeogenesis, depending on environmental concentrations of substrates.

In mammals, enolase is encoded by three homologous genes: *ENO1* which is ubiquitously expressed, *ENO2* which is expressed exclusively in neural tissues, and *ENO3* which is expressed in muscle tissues. Thus, the enzyme occurs as three isoforms: α-enolase coded by *ENO1* which is found in almost all tissues, β-enolase coded by *ENO3* which is predominantly found in muscle tissues, and γ-enolase coded by *ENO2* which is only found in neuron and neuroendocrine tissues (Marangos, Parma, & Goodwin, 1978; Merkulova et al., 1997; Pancholi, 2001). Three isoforms are characterized by distinct biochemical and immunological properties, in addition to different tissue distributions. Functional enzyme is a dimer made up of two homologous and heterologous isoforms, and only five types (αα, αβ, ββ, αγ, and γγ) of enolase isozyme can be found due to selective localization of β- and γ-enolases (Fletcher, Rider, & Taylor, 1976; Kato, Asai, Shimizu, Suzuki, & Ariyoshi, 1983; Royds, Parsons, Taylor, & Timperley, 1982). Interestingly, proportions of dimeric forms (αα, ββ, and αβ) of enolase in rat heart and skeletal muscle change during embryonic development. In both tissues, αα type is predominant in the fetus, although, as development progresses, αα type is replaced by αβ and ββ types in adult heart and by ββ type in adult striated muscle (Merkulova et al., 1997).

α-Enolase is considered to be a multifunctional protein, aside from its enzymatic function in the glycolytic pathway (Diaz-Ramos, Roig-Borrellas, Garcia-Melero, & Lopes-Alemany, 2012; Pancholi, 2001). Among its pleiotropic actions, α-enolase plays an important role in regulation of *c-myc* promoter activity in the form of an alternative translation product, *c-myc* promoter-binding protein 1 (MBP-1). It is localized in the nucleus and can bind to the *c-myc* P2 promoter and negatively regulates transcription of the proto-oncogene (Subramanian & Miller, 2001). In addition, α-enolase has been detected on the surface of hematopoietic cells such as monocytes, T cells and B cells, neuronal cells, and endothelial cells. Cell surface α-enolase serves as a receptor of plasminogen to enhance pericellular fibrinolytic activity and to lead extracellular matrix degradation (Godier & Hunt, 2012). α-Enolase has also been described as a neurotrophic factor, a heat-shock protein (HSP48), and a hypoxic stress protein. In the latter, it is suggested that upregulation of α-enolase contributes to hypoxia tolerance through nonglycolytic mechanisms (Aaronson, Graven, Tucci, McDonald, & Farber, 1995).

Multifunctional α-enolase plays variety roles on pathophysiological situations (Butterfield & Bader Lange, 2009; Capello, Ferri-Borgogno, Cappello, & Novelli, 2011; Diaz-Ramos et al., 2012; Godier & Hunt, 2012; Pancholi, 2001; Terrier et al., 2007). During tumor formation and expansion, tumor cells must increase glucose metabolism (Jin, DiPaola, Mathew, & White, 2007). Hypoxia is a common feature of solid tumors. Consistent with this, several reports have shown an upregulation of α-enolase in several types of cancer (Capello et al., 2011; Chang et al., 2006; Katayama et al., 2006; López-Pedrera et al., 2006), which may support anaerobic proliferation of tumor cells. In addition, α-enolase–plasminogen interaction on the cell surface is involved in promoting cell migration in tumor invasion and cancer metastasis (Capello et al., 2011; Godier & Hunt, 2012). Furthermore, the interaction mediates recruitment of monocytes to acutely inflamed tissue (Wygrecka et al., 2009). That is, it increases plasmin generation, promotes matrix degradation, and enhances monocyte migration. In rat heart ischemia–reperfusion model, α-enolase is induced in response to ischemic hypoxia and improves contractility of cardiomyocytes (Mizukami et al., 2004). The α-enolase–plasminogen interaction may also be involved in myogenesis and muscle regeneration (Diaz-Ramos et al., 2012).

Diagnostic and prognostic value of *ENO1* expression has been described in a number of tumors, as increased expression of α-enolase has been reported to correlate with progression of tumors (Capello et al., 2011; Diaz-Ramos et al., 2012). In addition, several posttranslational modifications such as acetylation, methylation, and phosphorylation have been found in α-enolase, and it is subjected to more acetylation, methylation, and phosphorylation in tumor cells than in normal tissues (Capello et al., 2011). Therefore, analysis of posttranslational modifications of α-enolase could also be of diagnostic and prognostic value in cancer, although it is uncertain whether such modifications affect the enzyme activity, localization, and stability. Furthermore, in many cancer patients, autoantibody against α-enolase has been detected. For example, in pancreatic ductal adenocarcinoma, α-enolase is upregulated and elicits a specific production of autoantibodies against isoforms phosphorylated on serine 419 (Tomaino et al., 2011). The presence of these autoantibodies is correlated with a significantly better clinical outcome in advanced patients treated with standard chemotherapy (Tomaino et al., 2011). Thus, levels of autoantibodies are also proposed as markers for diagnosis and prognosis of cancer.

Autoantibodies to α-enolase have been found in patients with chronic autoimmune diseases and inflammatory disorders such as rheumatoid arthritis, systemic sclerosis, and Hashimoto's encephalopathy (Diaz-Ramos et al., 2012; Terrier et al., 2007). In these diseases, the autoantibodies could induce endothelial injury through immune complex–complement activation, inhibit α-enolase–plasminogen interaction with perturbation of intravascular and pericellular fibrinolytic system, and induce apoptotic cell death. Autoantibodies to microbial α-enolase are seen in infectious disease and play a role in limiting microbial tissue invasion (Terrier et al., 2007). Thus, autoantibodies to α-enolase present in the sera of patients with autoimmune and infectious disease have potential diagnostic and prognostic value.

γ-Enolase is located in central and peripheral neurons and neuroendocrine cells and called neuron-specific enolase (NSE). This enzyme is released into the cerebrospinal fluid and blood when neural tissue is injured by traumatic brain injury and ischemic stroke (Ahmed, Wardlaw, & Whiteley, 2012; Meric, Gunduz, Turedi, Cakir, & Yandi, 2010). Tumors derived from neural or neuroendocrine tissue also release NSE into the blood, and therefore, NSE is a well-established tumor marker of small cell lung cancer and so on (Braga, Ferraro, Mozzi, Dolci, & Panteghini, 2013; Zhao & Luo, 2013). Therefore, measurement of enolase activity in biological fluids is of interest to investigators studying cancer cell metabolism.

We here describe simple and rapid *in vitro* protocols to measure enolase activity in native lysates from mammalian cells and tissues.

2. ASSAY OF ENOLASE ACTIVITY

2.1. Sample preparation

Enolase is localized in cytosol, nuclei, and plasma membrane, but majority of the enzyme is the most probably present in cytosol as a glycolytic enzyme. So preparation of cytosolic protein by any means may be able to use for total enolase activity assay. Two examples are shown as follows.

Animal tissues are removed and frozen immediately with liquid nitrogen and stored at −80 °C for several days. The tissues (0.1–0.3 g wet weight) are homogenized in 1 or 2 ml of 15 m*M* Tris–acetate (pH 6.5), 5 m*M* $MgSO_4$, and 1 m*M* EDTA. The homogenate is centrifuged at 100,000 × *g* for 1 h at 4 °C, and the supernatant is used for the assay of enolase activity (Kato, Ishiguro, Suzuki, Ito, & Semba, 1982).

Mammalian cells in culture are washed once with the buffer containing 250 m*M* sucrose, 20 m*M* HEPES (pH 7.5), 10 m*M* KCl, 1 m*M* $MgCl_2$, 1 m*M* EGTA, 1 m*M* EDTA, 1 m*M* dithiothreitol, and 1 m*M* phenylmethanesulfonylfluoride and homogenized in 1 ml of the buffer. The homogenate is then centrifuged at 15,000 × *g* for 15 min at 4 °C, and the resultant supernatant is recovered for enzymatic assay of enolase (Ishii et al., 2012).

RIPA buffer (25 m*M* Tris–HCl (pH 7.6), 150 m*M* NaCl, 1% NP-40, 1% sodium deoxycholate, 0.1% SDS) may be used, but the specific enolase activity is lower.

2.2. Assay

Enolase activity is measured by either direct or coupled methods. The direct method is to monitor PEP produced from 2-PG by enolase (Fig. 6.1; Ishii et al., 2012), while coupled method is to measure NADH decrease linked to conversion of PEP to lactate by a coupled enzyme system containing lactate dehydrogenase (LDH) and pyruvate kinase (PK) (Fig. 6.2; Hoorn, Flikweert, & Staal, 1974; Rider & Taylor, 1974).

2.2.1 Micro direct assay

a. Prepare a reaction buffer consisting of 20 m*M* imidazole HCl (pH 7.0), 400 m*M* KCl, and 1 m*M* magnesium acetate.
b. Put 50 μl of the reaction buffer into the tube.
c. Put 25 μl of the sample after appropriate dilution or dilution buffer into the tube.
d. Record absorbance of the solution at 240 nm at room temperature (25 °C).
e. Put 25 μl of 2-PG (1–2 m*M* as a final concentration) solution into the tube.
f. Record an increase in absorbance of the solution at 240 nm for 0–30 min (when time dependency is tested) after the addition of 2-PG at room temperature.
g. Calculate changes in the absorbance reflecting the conversion of 2-PG to PEP by enolase. Since molecular extinction coefficient for PEP is

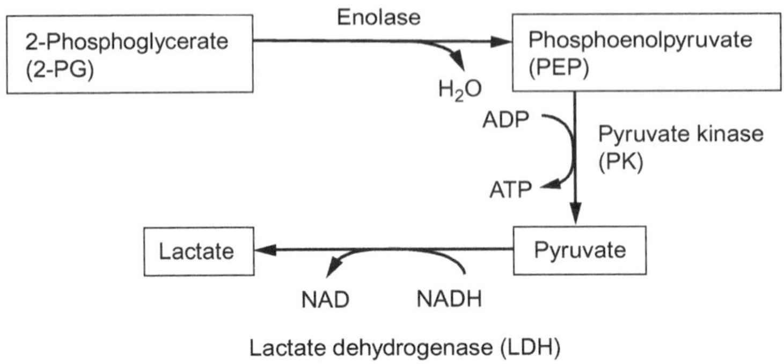

$$\text{Activity (U/mg)} = \frac{(\text{change in absorbance/min of sample} - \text{that of blank}) \times (\text{dilution factor})}{6.22 \times (\text{volume of sample used}) \times (\text{sample concentration in mg/ml})}$$

Figure 6.2 *Measurement of enolase activity by a coupled enzyme system of lactate dehydrogenase and pyruvate kinase.* Schema represents flow of the sequential conversion of substrates for enzymes involved, showing the last reaction is coupled with conversion of NADH to NAD. Thus, enolase activity can be monitored as a decrease in NADH absorbance at 340 nm and estimated by the equation.

$1.4 \times 10^3\ M^{-1}\ \text{cm}^{-1}$, the change of one absorbance unit at 240 nm means increase of PEP concentration in the reaction mixture from 0 to 714 μ*M* during the reaction time. Velocity of the reaction should be calculated from initial rates of absorbance increase.

h. Do not forget to make an appropriate blank/control tube.

Note

a. Due to limitation of amount of 2-PG in terms of cost, we recommend this micro assay system, although a spectrophotometer measurable in small volume (~1 μl) is necessary.

b. Mg^{2+} should be included in the reaction buffer. Some other divalent cations such as Zn^{2+} and Mn^{2+} work as a cofactor of enolase with lower efficacy (Baranowski & Wolna, 1975).

c. Inclusion of sodium fluoride (2 m*M*), phosphonoacetohydroxamate (2 μ*M*), or ENOblock (10 μ*M*) in the assay inhibits the enolase activity (Jung et al., 2013; Wedekind, Poyner, Reed, & Rayment, 1994). So you can confirm that specific changes in the absorbance are dependent on the enzyme.

d. Enolases from yeast and rabbit muscle as a positive control can be purchased from Oriental Yeast Co. (Tokyo, Japan) and Sigma-Aldrich (St. Louis, MO, USA), respectively. Alternatively mammalian blood which is diluted 1:1 with a hypotonic solution, sonicated, and centrifuged at 20,000 × *g* can be used.

e. We have experienced that inclusion of reduced glutathione (GSH, 5 m*M*) in the assay inhibits the enolase activity, due to acidification of the assay reagent by GSH. Be careful in pH in the reaction system if you include some additives.

2.2.2 Coupled assay

a. Freshly prepare a reaction buffer containing 100 m*M* triethanolamine (pH 7.4), 20 m*M* $MgCl_2$, 150 m*M* KCl, 0.6 m*M* β-NADH, 2 m*M* ADP, 20 units/ml PK, and 30 units/ml LDH.

b. Put 50 μl of the reaction buffer into the tube.

c. Put 25 μl of the sample after appropriate dilution or dilution buffer into the tube.

d. Record absorbance of the solution at 340 nm at room temperature (25 °C).

e. Put 25 μl of 2-PG (1–2 m*M* as a final concentration) solution into the tube.

f. Record absorbance of the solution at 340 nm for 5 min after the addition of 2-PG at room temperature.

g. Calculate changes in the absorbance reflecting the sequential conversion of 2-PG to PEP by enolase, PEP to pyruvate by PK, and pyruvate to lactate by LDH. Since molecular extinction coefficient for NADH is $6.22 \times 10^3\ M^{-1}\ \text{cm}^{-1}$, the change of one absorbance unit at 340 nm means increase of lactate concentration in the reaction mixture from 0–161 μ*M* during the reaction time.

h. Ordinary, enolase activity can be calculated by the equation shown in Fig. 6.2.

Note

a. Changes in NADH concentration could be monitored fluorometrically (NADH excitation 360 nm, emission 460 nm). This method is suitable for evaluating relative abundance of enolase activity and can be applied for high-throughput assay (Muller, Aquilanti, & DePhinho, 2012).

2.2.3 Isozyme-specific assay

Three isoforms of enolase have been identified: α-enolase, β-enolase, and γ-enolase. α-Enolase has been detected on most tissues, whereas β-enolase is expressed predominantly in muscle tissue, and γ-enolase is detected only in nervous tissue. These isoforms exist as both homodimers and heterodimers, and therefore, five types of dimeric enolase (isozymes) are found as αα, αβ, ββ, αγ, and γγ. Three isoforms can be distinguished immunologically, and five isozymes are separated by DEAE-cellulose chromatography (Rider & Taylor, 1974; Suzuki, Umeda, & Kato, 1980).

3. DEAE-CELLULOSE CHROMATOGRAPHY

a. Tissue extract from brain or muscle is dialyzed against 15 m*M* Tris–HCl (pH 7.9), 5 m*M* $MgCl_2$, and 0.1 m*M* EDTA.
b. The dialysate is applied on DEAE-cellulose column preequilibrated with the same buffer.
c. The αα type of enolase is recovered from flow-through (unbound) fraction.
d. The ββ, αβ, and αγ types of enolase are eluted with the same buffer containing 0.1 *M* NaCl after successful washing.
e. The γγ type of enolase is eluted with the same buffer containing 0.3 *M* NaCl.

REFERENCES

Aaronson, R. M., Graven, K. K., Tucci, M., McDonald, R. J., & Farber, H. W. (1995). Non-neuronal enolase is an endotherial hypoxic stress protein. *The Journal of Biological Chemistry*, *270*, 27752–27757.

Ahmed, O., Wardlaw, J., & Whiteley, W. N. (2012). Correlation of levels of neuronal and glial markers with radiological measures of infarct volume in ischaemic stroke: A systematic review. *Cerebrovascular Diseases*, *33*, 47–54.

Baranowski, T., & Wolna, F. (1975). Enolase from human muscle. *Methods in Enzymology*, *42*, 335–338.

Braga, F., Ferraro, S., Mozzi, R., Dolci, A., & Panteghini, M. (2013). Biological variation of neuroendocrine tumor markers chromogranin A and neuron-specific enoolase. *Clinical Biochemistry*, *46*, 148–151.

Butterfield, D. A., & Bader Lange, M. L. (2009). Multifunctional roles of enolase in Alzheimer's disease brain: Beyond altered glucose metabolism. *Journal of Neurochemistry*, *111*, 915–933.

Capello, M., Ferri-Borgogno, S., Cappello, P., & Novelli, F. (2011). α-Enolase: A promising therapeutic and diagnostic tumor target. *FEBS Journal*, *278*, 1064–1074.

Chang, G. C., Liu, K. J., Hsieh, C. L., Hu, T. S., Charoenfuprasert, S., Liu, H. K., et al. (2006). Identification of α-enolase as an autoantigen in lung cancer: Its overexpression is associated with clinical outcomes. *Clinical Cancer Research*, *12*, 5746–5754.

Diaz-Ramos, A., Roig-Borrellas, A., Garcia-Melero, A., & Lopes-Alemany, R. (2012). α-Enolase, a multifunctional protein: Its role on pathophysiological situations. *Journal of Biomedicine and Biotechnology*, *2012*, 156795. http://dx.doi.org/10.1155/2012/156795.

Fletcher, L., Rider, C. C., & Taylor, C. B. (1976). Enolase isoenzymes. III. Chromatographic and immunological characteristics of rat brain enolase. *Biochimica et Biophysica Acta*, *452*, 245–252.

Godier, A., & Hunt, B. J. (2012). Plasminogen receptors and their role in the pathogenesis of inflammatory, autoimmune and malignant disease. *Journal of Thrombosis and Haemostasis, 11*, 26–34.

Hoorn, R. K. J., Flikweert, J. P., & Staal, G. E. J. (1974). Purification and properties of enolase of human erythrocytes. *International Journal of Biochemistry, 55*, 845–852.

Ishii, T., Fukano, K., Shimada, K., Kamikawa, A., Okamatsu-Ogura, Y., Terao, A., et al. (2012). Proinsulin C-peptide activates α-enolase activity: Implications for C-peptide–cell membrane interaction. *Journal of Biochemistry, 152*, 53–62.

Jin, S., DiPaola, R. S., Mathew, R., & White, E. (2007). Metabolic catastrophe as a means to cancer cell death. *Journal of Cell Science, 120*, 379–383.

Jung, D.-W., Kim, W.-H., Park, S.-H., Lee, J., Kim, J., Su, D., et al. (2013). A unique small molecule inhibitor of enolase clarifies its role in fundamental biological processes. *ACS Chemical Biology, 8*, 1271–1282.

Katayama, M., Nakano, H., Ishiuchi, A., Wu, W., Oshima, R., Sakurai, J., et al. (2006). Protein pattern difference in the colon cancer cell lines examined by two-dimensional differential in-gel electrophoresis and mass spectrometry. *Surgery Today, 36*, 1085–1093.

Kato, K., Asai, R., Shimizu, A., Suzuki, F., & Ariyoshi, Y. (1983). Immunoassay of three enolase isozymes in human serum and in blood cells. *Clinica Chimica Acta, 127*, 353–363.

Kato, K., Ishiguro, Y., Suzuki, F., Ito, A., & Semba, R. (1982). Distribution of nervous system-specific forms of enolase in peripheral tissues. *Brain Research, 237*, 441–448.

López-Pedrera, C., Villalba, J. M., Siendones, E., Barbarroja, N., Gómez-Díaz, C., Rodríguez-Ariza, A., et al. (2006). Proteomic analysis of acute myeloid leukemia: Identification of potential early biomarkers and therapeutic targets. *Proteomics, 6*(Suppl. 1), S293–S299.

Marangos, P. J., Parma, A. M., & Goodwin, F. K. (1978). Functional properties of neuronal and glial isoenzymes of brain enolase. *Journal of Neurochemistry, 31*, 727–732.

Meric, E., Gunduz, A., Turedi, S., Cakir, E., & Yandi, M. (2010). The prognostic value of neuron-specific enolase in head trauma patients. *Journal of Emergency Medicine, 38*, 297–301.

Merkulova, T., Lucas, M., Jabet, C., Lamandé, N., Rouzeau, J. D., Gros, F., et al. (1997). Biochemical characterization of the mouse muscle-specific enolase: Developmental changes in electrophoretic variants and selective binding to other proteins. *Biochemical Journal, 323*, 791–800.

Mizukami, Y., Iwamatsu, A., Aki, T., Kimura, M., Nakamura, K., Nao, T., et al. (2004). ERK1/2 regulates intracellular ATP levels through α-enolase expression in cardiomyocytes exposed to ischemic hypoxia and reoxygenation. *The Journal of Biological Chemistry, 279*, 50120–50131.

Muller, F., Aquilanti, E., & DePhinho, R. (2012). In vitro enzymatic activity assay for enolase in mammalian cells in culture. *Protocol Exchange*. http://dx.doi.org/10.1038/protex.2012.040.

Pancholi, V. (2001). Multifunctional α-enolase: Its role in diseases. *Cellular and Molecular Life Sciences, 58*, 902–920.

Rider, C. C., & Taylor, C. B. (1974). Enolase isozymes in rat tissue: Electrophoretic, chromatographic, immunological and kinetic properties. *Biochimica et Biophysica Acta, 365*, 285–300.

Royds, J. A., Parsons, M. A., Taylor, C. B., & Timperley, W. R. (1982). Enolase isoenzyme distribution in the human brain and its tumours. *Journal of Pathology, 137*, 37–49.

Subramanian, A., & Miller, D. M. (2001). Structural analysis of α-enolase. *The Journal of Biological Chemistry, 275*, 5958–5965.

Suzuki, F., Umeda, Y., & Kato, K. (1980). Rat brain enolase isozymes: Purification of three forms of enolase. *Journal of Biochemistry, 87*, 1587–1594.

Terrier, B., Degand, N., Guilpain, P., Servettaz, A., Guillevin, L., & Mouthon, L. (2007). Alpha-enolase: A target of antibodies in infectious and autoimmune disease. *Autoimmunity Reviews, 6*, 176–182.

Tomaino, B., Cappello, P., Capello, M., Fredolini, C., Sperduti, I., Migliorini, P., et al. (2011). Circulating autoantibodies to phosphorylated alpha-enolase are a hallmark of pancreatic cancer. *Journal of Proteome Research, 1*, 105–112.

Wedekind, J. E., Poyner, R. R., Reed, G. H., & Rayment, I. (1994). Chelation of serine 39 to Mg^{2+} latches a gate at the active site of enolase: Structure of the bis(Mg^{2+}) complex of yeast enolase and the intermediate analogue phosphonoacetohydroxamate at 2.1-A resolution. *Biochemistry, 33*, 9333–9342.

Wygrecka, M., Marsh, L. M., Morty, R. E., Henneke, I., Guenther, A., Lohmeyer, J., et al. (2009). Enolase-1 promotes plasminogen-mediated recruitment of monocytes to the acutely inflamed lung. *Blood, 113*, 5588–5598.

Zhao, W.-X., & Luo, J. (2013). Serum neuron-specific enolase levels were associated with the prognosis of small cell lung cancer: Meta-analysis. *Tumor Biology, 34*, 3245–3248.

CHAPTER SEVEN

Extracellular Flux Analysis to Monitor Glycolytic Rates and Mitochondrial Oxygen Consumption

Martin Pelletier, Leah K. Billingham, Madhu Ramaswamy, Richard M. Siegel[1]

Autoimmunity Branch, Immunoregulation Section, National Institutes of Arthritis, Musculoskeletal and Skin Diseases (NIAMS), National Institutes of Health (NIH), Bethesda, Maryland, USA

[1]Corresponding author: e-mail address: siegelr@mail.nih.gov

Contents

Abstract

Evidence accumulating over the past decade has linked alterations in bioenergetic metabolism to the pathogenesis of several diseases, including inflammatory conditions and cancer. However, the mutual relationship between the effector functions and the metabolism of immune cells has begun to emerge only recently. Similar to malignant cells, both innate and adaptive immune cells undergo a metabolic reprogramming that is required for effector functions, *de facto* underlying the elicitation of a robust immune response. These changes allow immune cells not only to rapidly respond to pathogens

Methods in Enzymology, Volume 542
ISSN 0076-6879
http://dx.doi.org/10.1016/B978-0-12-416618-9.00007-8

or (pre)malignant cells but also to adapt to changing microenvironmental conditions. Targeting the metabolic alterations of malignant cells has been the subject of an intense wave of investigation, resulting in the identification of promising therapeutic strategies. Since the inflammatory milieu and the tumor microenvironment are similar, the metabolism of immune cells and its regulation has recently come under renewed interest as a target for immunotherapy. Here, we describe different tools and techniques to study the bioenergetic metabolism of cultured cells, using immune cells as a model. Our methodological approach relies on an extracellular flux analyzer, an instrument that enables the real-time measurement of the two central pathways used by living cells to generate adenosine triphosphate: glycolysis and oxidative phosphorylation. This instrument and similar technological innovations have transformed the study of cellular metabolism, unveiling its profound impact on various immunologic and oncological disorders.

1. INTRODUCTION

Adenosine triphosphate (ATP) is the primary source of cellular energy, and most of it is generated via glycolysis or oxidative phosphorylation (mitochondrial respiration). Cells break down glucose to pyruvate via glycolysis and then oxidize pyruvate via the tricarboxylic acid (TCA)/Krebs cycle and the mitochondrial electron transport chain (ETC). Fatty acids and amino acids can also be degraded to pyruvate, acetyl coenzyme A (acetyl CoA), and other intermediates to maintain ATP production (Vander Heiden, Cantley, & Thompson, 2009). Other metabolic pathways, such as the pentose phosphate pathway, mediate alternative metabolism of glucose for DNA or RNA synthesis in order to sustain cellular activation and proliferation (Chawla, Nguyen, & Goh, 2011). More than 80 years ago, the German biochemist Otto Heinrich Warburg was awarded the Nobel Prize after he discovered that cancer cells preferentially undergo glycolysis, even in the presence of oxygen (Warburg effect) (Soga, 2013). Immune cells, under some conditions, behave metabolically similar to cancer cells during an immune response. We will focus on immune cells as a model to study metabolic pathways using currently available techniques, all of which can be applied to the study of oncometabolism.

1.1. Energy metabolism and the regulation of immune cells

Energy metabolism ensures the proper function of the immune system, as immune cells require energy to maintain cellular viability and to perform specific immune functions. This includes energy consumption for motor functions (migration and phagocytosis), antigen processing and presentation, activation functions (signaling and proliferation), and effector functions (antibody synthesis, cytokine secretion, and cytotoxicity) (Buttgereit, Burmester, & Brand, 2000). Immune cells are highly adapted to the environments in which they circulate and home to during inflammatory responses. They often differ in their basal metabolic states and responses to activating stimuli. Adapted metabolic responses are pivotal to cellular survival and function. The failure of metabolism to match cellular needs may lead to impaired function and apoptosis, whereas excess metabolic function can lead to cell survival, heightened functionality, and hyperreactivity, leading to autoimmunity and inflammatory diseases (Jacobs et al., 2008; Rathmell, Elstrom, Cinalli, & Thompson, 2003).

1.2. Energy metabolism of myeloid cells

Myeloid cells such as polymorphonuclear neutrophils, monocytes, and macrophages rely on glycolysis to produce ATP in the hypoxic environment of inflamed and edematous tissue (Borregaard & Herlin, 1982; Lewis, Lee, Underwood, Harris, & Lewis, 1999; Sher, Wadee, & Joffe, 1983; Sumbayev, Nicholas, Streatfield, & Gibbs, 2009; Venge et al., 2003). The hypoxia-inducible factor-1 (HIF-1) is one of the principal mediators of metabolic adaptation to low concentrations of oxygen, and it emerged as an essential regulator of the survival and functions of myeloid cells in the inflammatory microenvironment (Cramer et al., 2003; Elks et al., 2011). Neutrophils are preadapted to hypoxic environments, having very few mitochondria (Maianski et al., 2004; Zucker-Franklin, 1968). Monocytes circulate in the high-oxygen environment of blood, which fuels oxidative phosphorylation, but upon migration to inflamed tissues, they must adapt immediately to the prevailing hypoxic conditions by switching to glycolysis (Gaber et al., 2011). During the differentiation process from monocytes to classical macrophages (or M1 macrophages), the localization of HIF-1 is shifted from the cytosol to the nucleus, leading to the expression of hypoxia-adjusted genes (Fangradt et al., 2012). HIF-1 was shown to upregulate the expression of glycolytic enzymes and downregulate the expression of mitochondrial ETC proteins, mediating a physiological glycolytic switch in response to hypoxia (Semenza, 2007). In contrast, the metabolism of alternatively activated macrophages (or M2 macrophages), which are involved in tissue repair and the regulation of inflammation and adaptive immunity, veers toward the use of fatty acid oxidation and mitochondrial respiration (Vats et al., 2006). In addition to hypoxia, inflammatory stimuli can induce HIF-1-dependent metabolic reprogramming toward glycolysis. Toll-like receptor agonists stimulate a metabolic transition to aerobic glycolysis in dendritic cells (DCs) and macrophages, a conversion essential for their maturation and function (Krawczyk et al., 2010; Tannahill et al., 2013). In DCs, nitric oxide is a key player in this switch by acting on components of the mitochondrial ETC to inhibit oxidative phosphorylation. The glycolytic switch allows for the maintenance of ATP levels and the survival of DCs when nitric oxide is produced during the inflammatory response (Everts et al., 2012). In phagocytes, the pentose phosphate pathway is important in generating NADPH for microbicidal functions regulated by NADPH oxidase (Pearce & Pearce, 2013).

1.3. Energy metabolism of lymphoid cells

Naive resting lymphocytes primarily use oxidative phosphorylation to generate ATP through catabolism of glucose, lipids, and amino acids. Without external cues, resting T cells internalize and degrade the glucose transporter GLUT-1, which results in the inefficient use of glucose needed to maintain T-cell viability (Maciver et al., 2008). The activation of T cells by TCR agonists and the binding of cytokines, hormones, growth factors, or other extrinsic signals result in a metabolic shift to glycolysis. This glycolytic switch occurs to support the high metabolic demand of clonal expansion (Jones & Thompson, 2007). The cytokines interleukin (IL)-4, IL-7, and CD28 costimulation increase the surface expression of glucose transporters and induce glycolysis (Jacobs et al., 2008; Jones & Thompson, 2007). Memory T cells have a very different metabolism than their activated counterparts and rely mostly on mitochondrial fatty acid oxidation for development and maintenance (van der Windt et al., 2012). B cells increase glycolysis through phosphatidylinositol 3-kinase activity following B-cell receptor (BCR) cross-linking, while the coengagement of the BCR and the FcγRIIB receptor inhibits glucose utilization (Doughty et al., 2006).

1.4. Altered metabolism and diseases

Alterations in energy metabolism are often associated with the pathophysiology of diseases such as obesity, diabetes, and cancer (Dang & Semenza, 1999; Li et al., 2012). The link between cancer and altered metabolism is well recognized (Cairns, Harris, & Mak, 2011; Soga, 2013), as cancer cells often utilize aerobic glycolysis, a phenomenon known as the Warburg effect, to meet the needs of rapidly dividing cells (Vander Heiden et al., 2009).

Alterations to fundamental chemical processes of macromolecules (carbohydrates, proteins, lipids, and nucleic acids) occur in cancer cells in the microenvironment of the tumor, where the availability of crucial nutrients such as glucose, glutamine, and oxygen is limited (Newsholme, Crabtree, & Ardawi, 1985; Vander Heiden et al., 2009). One such alteration is increased hypoxia, which leads to cells relying on reductive carboxylation of glutamine-derived α-ketoglutarate almost exclusively for lipid synthesis (Metallo et al., 2012). In fact, glutamine-dependent reductive carboxylation has been identified as the major source of acetyl CoA in ETC-deficient cancer cells (Mullen et al., 2012). Hypoxic and transformed cells also scavenge unsaturated fatty acids from lysophospholipids to support their growth (Kamphorst et al., 2013). In some cases, cancerous cells display other metabolic profiles, such as the reverse Warburg effect, where the malignant cells use oxidative stress to extract nutrients from surroundings cells (Martinez-Outschoorn et al., 2011), or metabolic symbiosis, in which cancer cells within the same tumor may simultaneously use glycolysis and oxidative phosphorylation (Ho et al., 2012).

Targeting cancer cell metabolism is the subject of intense research and has already generated a series of new potentially useful therapeutic agents (Butler, Zhao, Munoz-Pinedo, Lu, & Tan, 2013; Teicher, Linehan, & Helman, 2012; Zhang & Yang, 2013). Interestingly, many features of the cancer microenvironment, such as limited nutrient supply, hypoxia, and angiogenesis, are also prevalent in the inflammatory milieu (Konisti, Kiriakidis, & Paleolog, 2012; Purushothaman, Meerarani, & Moreno, 2007; Purushothaman et al., 2011; Taylor & Sivakumar, 2005). Evidence suggests that energy metabolism is crucial for the maintenance of chronic inflammation, in terms of both the energy supply and the control of the immune response through metabolic signals (Inoki, Kim, & Guan, 2012; Nutsch & Hsieh, 2011; Powell, Pollizzi, Heikamp, & Horton, 2012).

Modifications in mitochondrial metabolism have also been associated with inflammatory diseases. For example, mitochondrial reactive oxygen species (ROS) were found to be important in inducing proinflammatory cytokine production in a genetic autoinflammatory disease termed tumor necrosis factor (TNF) receptor-associated periodic syndrome (TRAPS), which is associated with mutations in the type 1 receptor of the TNF cytokine (TNFR1). Treatment with general ROS inhibitors, as well as specific mitochondrial ROS inhibitors, reduced cytokine production not only in cells harboring TNFR1 mutations but also in activated normal cells, indicating that the specific inhibition of ROS from the mitochondria represents a novel therapeutic strategy for the treatment of inflammation (Bulua et al., 2011). Interestingly, cells from patients with TRAPS or from mice engineered to express TRAPS-associated *Tnfrsf1a* mutations (C33Y heterozygote or T50M heterozygote) exhibit increased oxygen consumption and respiratory capacity, alterations that correlated with higher levels of mitochondrial ROS observed in these cells (Bulua et al., 2011).

2. TECHNIQUES TO EVALUATE IMMUNE CELL BIOENERGETICS

There are many different methods available for evaluating and understanding cellular bioenergetics, both *in vivo* and *in vitro*. Recent advances in technology and increases in commercially available kits have made many of these techniques more accessible and affordable. The radionuclide labeling of cells, such as is used in positron emission tomography, has long been used to track cellular movement and is commonly used in conjunction with computed tomography or magnetic resonance imaging (MRI) in oncology (Carty, Shortt, Shelly, Eustace, & O'Connell, 2010). More recently, MRI-based immune cell tracking using iron oxide-based and ^{19}F-based probes has been used to measure the partial pressure of molecular oxygen in tumor cells *in vivo* (Ahrens & Bulte, 2013; Kadayakkara, Janjic, Pusateri, Young, & Ahrens, 2010). *In vitro* assays allow for the further elucidation of cellular metabolism. The commonly used techniques will be discussed before an in-depth description of extracellular flux analysis is provided.

2.1. Individual metabolite assays

One way to analyze the metabolic state of cells and understand the metabolic features of cell types and diseases is through the measurement of distinct and specific metabolites and nutrients. Though tools for measuring such metabolic intermediates as pyridine nucleotides and amino acids have been available for decades (Kunin & Tannen, 1979; Lowry, Passonneau, Schulz, & Rock, 1961), the development of commercial kits has increased the ease and convenience of determining the abundance of nutrients and metabolic intermediates, such as ATP (Schafer et al., 2009; Yuneva, Zamboni, Oefner, Sachidanandam, & Lazebnik, 2007). Unfortunately, the determination of the levels and ratios of metabolites is only indicative of the cellular metabolic state at a certain point in time, and individual metabolite levels may yield less information than a global snapshot of metabolic pathways via metabolome analysis.

2.2. Metabolome profiling

Advancements in technologies such as NMR spectroscopy and mass spectrometry combined with the growing number of known metabolite databases have made it easier to precisely and accurately identify and quantify the amount of a given metabolite in a sample (Kell, 2004; Wishart et al., 2009). This approach allows for the quantification of a wide array of known metabolites (the metabolome) present in a biological sample, such as a cell or whole organism, at different time points. When combined with genomic approaches, this can be a powerful tool for understanding how cells regulate metabolic functions and can provide insight as to which metabolites are key in a given process. This approach has been used to identify succinate as an inflammatory signal important for the induction of IL-1β (Tannahill et al., 2013) and has suggested a role for unsaturated fatty acids in the mitigation of tissue damage in leprosy (Amaral et al., 2013). However, this powerful technique is relatively expensive and, as with individual metabolite assays, only represents a cell's metabolic state at a specific time point, rather than a metabolic flux.

2.3. Metabolic flux measurement and phenotypic microarrays

The use of radiolabeled metabolites, such as acetate and glucose, has long been used in tracer uptake-based techniques. By observing the accumulation of metabolites with autoradiography, rates of metabolism can be estimated (Ackermann & Lear, 1989; Buxton, Schwaiger, Nguyen, Phelps, & Schelbert, 1988; Lear & Ackermann, 1990). The accumulation of acetate metabolites, for instance, has shown that after uptake, acetate can be metabolized to acetyl CoA and enter the Krebs cycle (Lear & Ackermann, 1990). This method has been improved upon by recent technological developments, such as the metabolomic phenotyping array from Biolog (Phenotype MicroArrays™), that allow the *in vitro* testing of hundreds of metabolic pathways at once. These arrays expose cells to a singular carbon source (pyruvate, glutamine, glucose, etc.) and then use a redox dye to determine colorimetrically which carbon sources lead to NADH production, which is indicative of cellular metabolism. This can be used to determine either a cellular metabolic fingerprint or an alteration of cellular metabolism in disease. This technique has been used to demonstrate a role for androgen in amino acid metabolism in prostate cancer (Putluri et al., 2011) and reduced tryptophan metabolism in autism spectrum disorders (Boccuto et al., 2013). This technique can inform as to how different treatments or conditions result in a metabolic phenotype without the use of radioactive materials and proves to be a better metabolic flux indicator than metabolome and steady-state techniques. Although the utilization of carbon sources can be viewed in real time, the system does not yet allow for the observation of cellular reaction to perturbations as they happen.

2.4. Receptor expression and activity

Rather than directly measuring metabolites and metabolic by-products, the bioenergetics and metabolism of a cell can be evaluated via analyzing the expression and activity of nutrient transporters. Leukocytes greatly differ in their expression of nutrient receptors. Neutrophils and monocytes, for example, express higher levels of glucose receptors than other leukocytes and *in vitro* activation with phorbol 12-myristate 13-acetate or lipopolysaccharides increases the surface expression of glucose receptors across immune cells (Maratou et al., 2007; Tannahill et al., 2013). There are many tools available to correlate the expression of glucose or amino acid transporters to metabolic status, including Western blotting for protein expression, fluorescence-activated cell sorting, real-time quantitative PCR, and radiometric uptake assays using radiolabeled metabolites.

2.5. Extracellular flux analyzer

The rate of oxygen consumed by cells (oxygen consumption rate—OCR) can be used as an indicator of mitochondrial respiration (Fig. 7.1). Three methods are commonly used to measure OCR: electron paramagnetic resonance (EPR), the Clark oxygen electrode, and quenched-fluorescence oxygen-sensing assay (Diepart et al., 2010). EPR allows continuous monitoring of oxygen in cells and tissues when coupled with oxygen-sensitive probes (James, Jackson, Grinberg, & Swartz, 1995). Although relatively easy, this technique is rather expensive and does not allow the measurement of multiple samples simultaneously. The Clark oxygen electrode, invented by Leland Clark in 1959, allows the determination of OCR by polarography (Li & Graham, 2012). Although this reliable method is still used in many laboratories, it requires large amount of material and is more suitable for isolated mitochondria and permeabilized cells. The introduction more than 20 years ago of high-resolution respirometry, based on highly sensitive electrodes, minimization of oxygen diffusion, and advanced software, allowed the measurement of small changes in cellular respiration as well as subtle differences in respiratory effects of inhibitors in intact cells (Gnaiger,

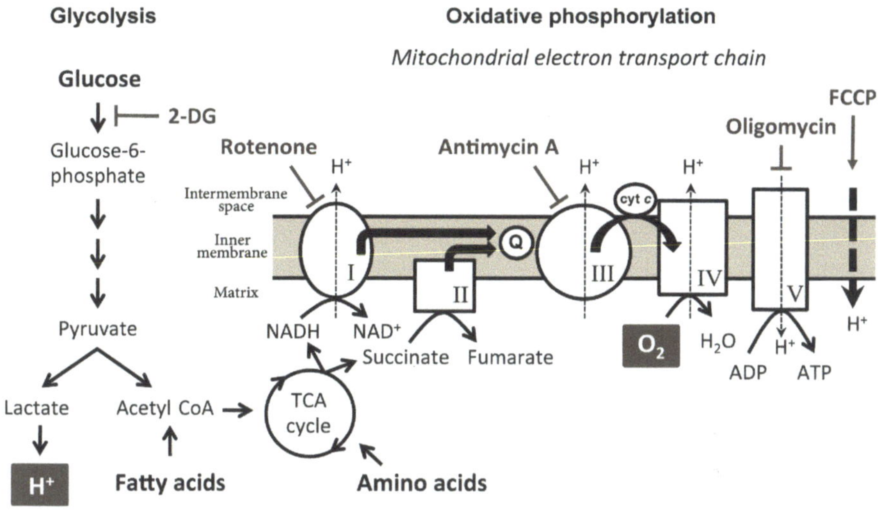

Figure 7.1 *Schematic representation of cellular bioenergetic pathways.* Cells generate energy in the form of ATP by a combination of two main metabolic pathways: glycolysis and oxidative phosphorylation. Glycolysis metabolizes glucose through a series of enzymatic reactions to pyruvate, which can be either reduced to lactate or oxidized via the TCA cycle and the mitochondrial ETC to generate ATP. Fatty acids and amino acids can also be broken down to pyruvate, acetyl CoA, and other intermediates for ATP production. The various compounds used to examine the bioenergetic profile of cells are shown at their respective sites of action. The glucose analog 2-DG cannot undergo conversion to glucose-6-phosphate, thereby competitively inhibiting the glycolytic ATP production. Rotenone and antimycin A block complex I and complex III of the mitochondrial ETC, respectively, while oligomycin inhibits complex V by blocking its proton channel. FCCP is an uncoupling agent that permeabilizes the inner mitochondrial membrane to protons, forcing the mitochondria to increase the flow of electrons (and thus oxygen consumption) to maintain the membrane potential. Block arrows represent the electron flow in the mitochondrial ETC. The two parameters measured with an extracellular flux analyzer, namely, the rate of oxygen (O_2) consumption and the measure of protons (H^+) released during lactate production, are depicted in boxes. ADP, adenosine diphosphate; ATP, adenosine triphosphate; Q, coenzyme Q; cyt *c*, cytochrome *c*; H^+, proton; TCA, tricarboxylic acid; ETC, electron transport chain; acetyl CoA, acetyl coenzyme A; 2-DG, 2-deoxyglucose; complex I, NADH dehydrogenase; complex II, succinate dehydrogenase; complex III, coenzyme Q: cytochrome-*c* oxidoreductase; complex IV, cytochrome-c oxidase; complex V, ATP synthase; FCCP, trifluorocarbonylcyanide phenylhydrazone.

Steinlechner-Maran, Mendez, Eberl, & Margreiter, 1995; Heerlein, Schulze, Hotz, Bartsch, & Mairbaurl, 2005; Hutter, Unterluggauer, Garedew, Jansen-Durr, & Gnaiger, 2006). Quenched-fluorescence oxygen-sensing methods are usually preferred for evaluating mitochondrial respiration in intact cells (Rolo, Palmeira, & Cortopassi, 2009). These assays are based on the ability of oxygen to quench the excited state of probes, such as the MitoXpress phosphorescent oxygen-sensitive probe developed by Luxcel. The depletion of oxygen by the cells increases the probe signal, reflecting changes in mitochondrial activity (Hynes et al., 2006).

ATP generation can also occur through the conversion of glucose to lactate independently of oxygen. The measure of protons produced indirectly via lactate released from cultured cells can be used as an indicator of glycolysis and is provided by measuring the acidification of the medium (extracellular acidification rate—ECAR) (Fig. 7.1). The XF Extracellular Flux Analyzer from Seahorse Bioscience is an instrument that measures bioenergetic parameters in real time with the advantage of providing information on both OCR and ECAR in living cells (Bulua et al., 2011; Chacko et al., 2013; Dranka et al., 2011; Nicholls et al., 2010). In addition, cells can be metabolically perturbed by successive addition of different compounds that shift the bioenergetic profile (Nicholls et al., 2010). The mitochondrial stress test is performed by the sequential addition of the ATP synthase inhibitor oligomycin, the mitochondrial uncoupler trifluorocarbonylcyanide phenylhydrazone (FCCP), and a combination of the complex I inhibitor rotenone and complex III inhibitor antimycin A (Fig. 7.1). This test reveals three key parameters of mitochondria: ATP turnover, proton leak, and maximal respiration (Fig. 7.2). The glycolytic stress test is performed by sequentially adding glucose, oligomycin (to block mitochondrial respiration and force the cells to rely on glycolysis for ATP production), and 2-deoxyglucose (2-DG; glucose analog and inhibitor of glycolytic ATP production) (Fig. 7.1) and provides information on glycolysis, the glycolytic capacity, and the glycolytic reserve (Fig. 7.2). We and other groups have used oxygen consumption and glycolytic rates to examine cellular metabolism and identify metabolic derangements that correlate with and may cause immunologic dysfunction (Bulua et al., 2011; Everts et al., 2012; Macrae et al., 2013; Munoz-Planillo et al., 2013; Sena et al., 2013; Wang et al., 2011).

It is important, however, to keep in mind that the extracellular flux analysis has some limitations. The interpretation of OCR may be influenced by factors within immune cell metabolism that may influence oxygen consumption such as the oxidative burst in phagocytes (Ferrari, Souto, Franca, & Honorio-Franca, 2011) or other oxygen-consuming enzymes (Ehrismann et al., 2007). It is however possible to measure the oxidative burst of activated immune cells using an extracellular flux analyzer by determining the increased oxygen consumption induced by an activator of the

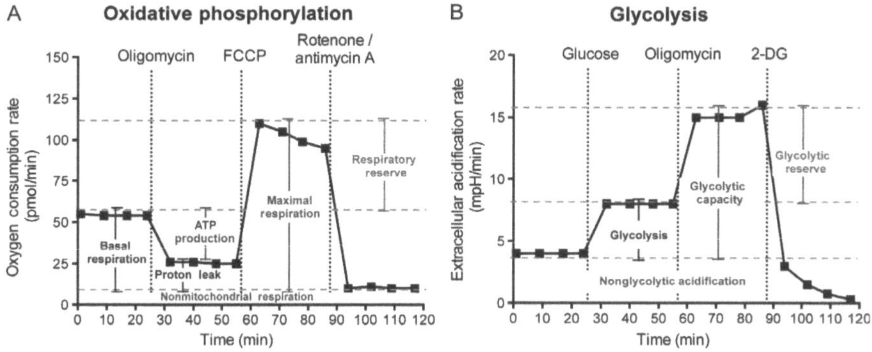

Figure 7.2 *Mitochondrial respiratory and glycolytic profiles of human peripheral blood mononuclear cells (PBMCs).* Human PBMCs were isolated from blood and seeded on a Cell-Tak-treated XF96 cell culture plate. Data were obtained with the XF Extracellular Flux Analyzer (Seahorse Bioscience). Error bars were eliminated for conceptual simplification. (A) Graph of the mitochondrial stress test detailing the four key parameters of mitochondrial function (basal respiration, ATP turnover, proton leak, and maximal respiration) with the sequential use of oligomycin (ATP synthase inhibitor), FCCP (mitochondrial uncoupler), and rotenone+antimycin A (complex I and III inhibitors). The difference between the maximal and the basal respirations constitutes the respiratory reserve, the capacity of a cell to generate ATP via oxidative phosphorylation in response to increased demand for energy. Substrate oxidation and cell surface oxygen consumption result in nonmitochondrial respiration. (B) Graph of the glycolytic stress test detailing the parameters of cellular glycolysis (glycolysis, glycolytic capacity, and glycolytic reserve) obtained with the sequential use of glucose, oligomycin (to block mitochondrial respiration and force the cells to rely on glycolysis for ATP production), and 2-deoxyglucose (2-DG, glucose analog and inhibitor of glycolytic ATP production). Glycolytic reserve amounts to the difference between the glycolytic capacity and glycolysis and is indicative of the cellular ability to increase the glycolytic rate upon increased energy demand. Acidification of carbon dioxide, the end product of the TCA cycle, which can be converted to bicarbonate, contributes to nonglycolytic acidification.

respiratory burst after the inhibition of the ETC (Chacko et al., 2013). The extracellular flux analyzer can also be used to perform metabolic analysis under hypoxic conditions (Benavides, Liang, Dodson, Darley-Usmar, & Zhang, 2013), and advanced hypoxic protocols and hypoxia rate calculator programs are available from Seahorse Bioscience (www.seahorsebio.com).

3. BIOENERGETIC PROFILING OF IMMUNE CELLS USING THE XF EXTRACELLULAR FLUX ANALYZER

The following protocol can easily be adapted for understanding the bioenergetics of cancer, from both primary tumor cells or lymphomas and cancer cell lines such as the immortalized human T-cell line Jurkat, the human monocytic cell line THP-1, or the pre-B lymphoblast cell line JM1.

Certain parameters should be taken into account before performing an assay using an XF Extracellular Flux Analyzer. Optimization assays, including cell seeding density titration and compound injection titration, should be performed for every cell type to determine optimal nontoxic experimental conditions for accurate analysis. In addition, measuring bioenergetic profiles requires immobilization in XF cell culture microplates of cells that grow in suspension and usually do not adhere. This can be achieved by coating microplates with Matrigel, polylysine, or Cell-Tak (Chacko et al., 2013; Zhang et al., 2012). It may be important to determine whether the immobilization material by itself leads to unexpected activation of the cells.

3.1. Required materials

Devices

- XF Extracellular Flux Analyzer (Seahorse Bioscience)
- Fluorescence microplate reader
- Incubator—set to 37 °C without CO_2
- Incubator—set to 37 °C with CO_2
- pH meter

Reagents and consumables

- FluxPak (cell plates, calibrant, and sensor cartridges) (Seahorse Bioscience)
- XF assay medium (Seahorse Bioscience)[1]
- Sodium pyruvate (Life Technologies)
- 45% (2.5 *M*) glucose solution (Corning cellgro)
- Cell-Tak™ Cell and Tissue Adhesive (BD Biosciences)
- Sodium bicarbonate ($NaHCO_3$)
- Dimethyl sulfoxide (DMSO)
- Oligomycin, FCCP, rotenone, antimycin A, and 2-DG (Sigma)
- CyQUANT® Cell Proliferation Assay Kit (Life Technologies)
- Sterile water

[1] XF assays require a nonbuffered medium to accurately measure proton production and glycolysis. The constituents of XF assay medium are based on the formulation of Dulbecco's modified Eagle's medium. Sodium pyruvate (1 m*M* final concentration) and glucose should be added before use. Other unbuffered media such as RPMI or Krebs–Henseleit buffer can also be used.

3.2. Protocol

This protocol is optimized for an XF96 Extracellular Flux Analyzer. A flowchart describing the essential steps is illustrated in Fig. 7.3.

3.2.1 Before the day of assay

1. Add 200 μL of Seahorse Bioscience calibrant pH 7.4 to each well of a Seahorse Bioscience 96-well utility plate.[2]
2. Place sensor cartridge on top of the utility plate and store at 37 °C without CO_2 overnight.[3]
3. Turn on instrument and start XF software to allow instrument to stabilize at 37 °C.[4]
4. Treat XF microplates for immobilization of nonadherent immune cells:
   a. Prepare a working solution of Cell-Tak in sterile H_2O.[5]
   b. Coat the XF96 cell culture plate with 10 μL of working solution of Cell-Tak.[6]
   c. Add 40 μL of $NaHCO_3$ 0.1 *M* (pH 8.0) per well to neutralize and promote adsorption of Cell-Tak to the plate.
   d. Incubate overnight in a normal incubator set at 37 °C.[7]

3.2.2 Day of assay

In order to examine the bioenergetic profile of a specific cell type, special care should be used to obtain a highly purified population devoid of contaminant cells. Immune cells isolated from blood or other organs by standard procedure (i.e., Ficoll-Paque™ or Lymphoprep™) should be further purified by magnetic selection (negative or positive selection) or cell sorting before assay.

[2] Volumes and cell numbers should be adjusted accordingly if an XF24 Extracellular Flux Analyzer is used.

[3] The hydrated cartridge can be kept at 37 °C without CO_2 for up to 72 h. However, it is highly recommended to wrap parafilm around the edges of the cartridge to prevent evaporation.

[4] The XF Extracellular Flux Analyzer can be turned on the day of the experiment but may need a few hours to stabilize at 37 °C prior to performing the assay.

[5] Cell-Tak is a formulation of nonimmunogenic polyphenolic proteins extracted from the marine mussel.

[6] Manufacturer's instructions recommend a starting density of 3.5 μg Cell-Tak/cm^2 of surface area, which represents approximately 0.45 μg of Cell-Tak per well in a XF96 cell culture plate. To prepare enough working solution for one XF96 cell plate, dilute 30 μL of Cell-Tak at ~1.5 mg/mL in 1 mL sterile H_2O. There is no need to add the adhesive solution to background wells.

[7] Alternatively, coating and adsorption of Cell-Tak to the cell culture plate can be performed for 30 min at 37 °C on the day of the assay.

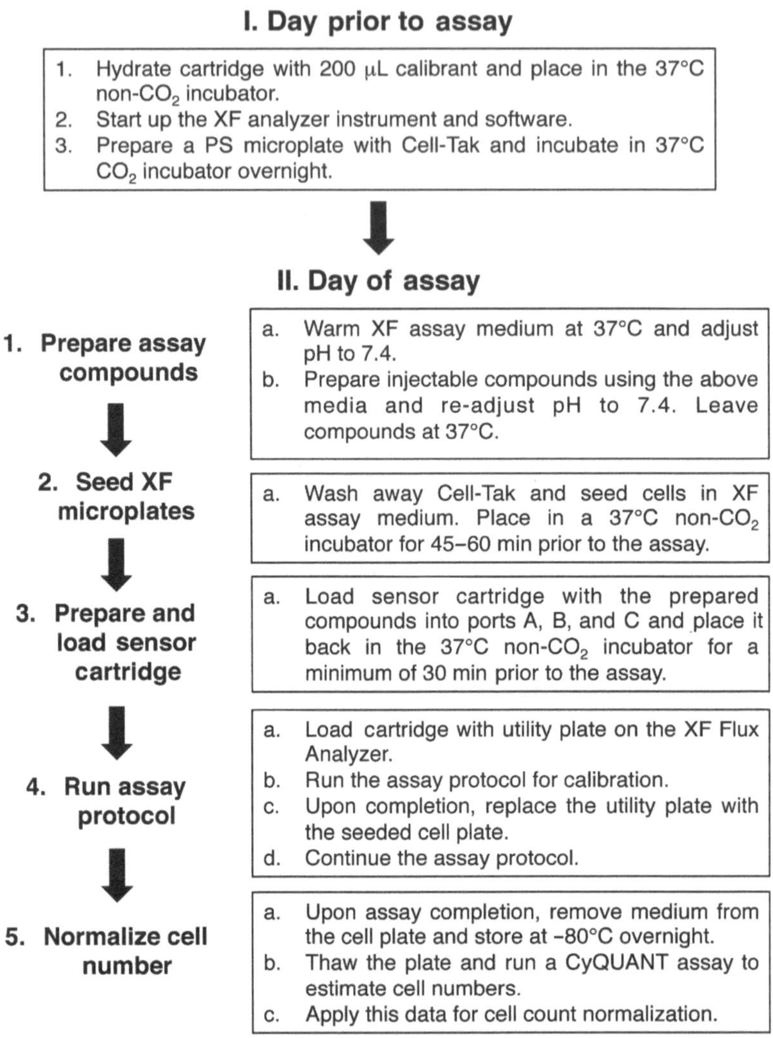

Figure 7.3 *Experimental flowchart.* Sequential steps to perform a mitochondrial stress test and/or a glycolytic stress test with immune cells on the XF96 Extracellular Flux Analyzer. See protocol and footnotes for details.

3.2.2.1 Preparation of compounds

Stock solutions of oligomycin, FCCP, rotenone, and antimycin A are typically made in DMSO, aliquoted, and stored at −20 °C until use. A titration of the optimal final concentration of each compound should be determined before the assay. Suggested concentrations used with success with human peripheral blood cells are mentioned later:

1. Warm the appropriate XF assay medium to 37 °C and adjust pH to 7.4.[8]
2. Prepare the compounds as follows:

   Mitochondrial stress test (prepared in XF assay medium containing glucose)

   40 μ*M* oligomycin (8 ×)
   13.5 μ*M* FCCP (9 ×)
   10 μ*M* rotenone + 1 μ*M* antimycin A (10 ×)

   The final concentrations in the XF96 cell culture plate are 5 μ*M* oligomycin, 1.5 μ*M* FCCP, and 1 μ*M* rotenone + 100 n*M* antimycin A.[9]

   Glycolytic stress test (prepared in XF assay medium without glucose)

   200 m*M* glucose (8 ×)
   45 μ*M* oligomycin (9 ×)
   1 *M* 2-DG (10 ×)

   The final concentrations in the XF96 cell culture plate are 25 m*M* glucose, 5 μ*M* oligomycin, and 100 m*M* 2-DG.[10]
3. Adjust pH to 7.4 and keep compounds at 37 °C prior until loading.

3.2.2.2 Seeding cells for assays on Cell-Tak-treated plates

Cell number titration is highly recommended before the assay. Maximum cell density for human or murine immune cells is usually achieved with ~300,000–400,000 cells/well:

1. Warm XF assay medium to 37 °C and adjust pH to 7.4.[11]
2. Pour off or aspirate the Cell-Tak solution from the wells of the XF96 cell culture plate and wash once with 200 μL of sterile H_2O to remove bicarbonate.[12]
3. Add XF assay medium in the appropriate background wells.[13]

[8] Ideally, the pH should be adjusted between 7.38 and 7.42.

[9] Oligomycin and FCCP at a final concentration of 1 and 0.5 μ*M*, respectively, were found to be optimal in the mitochondrial stress test of human $CD4^+$ and $CD8^+$ T cells.

[10] Final concentration of glucose in the assay can range from 5 to 25 m*M* to evaluate the glycolysis parameter. We usually use the same concentration of glucose as is used for cell culture and functional assays.

[11] For the mitochondrial stress test, XF assay medium should contain 5–25 m*M* glucose, as used for culture conditions. For the glycolytic stress test, the XF assay medium should not contain glucose. After adjustment of pH, medium should be filtered if sterility has been compromised.

[12] Plate can be air-dried and stored at 2–8 °C up to 2 weeks or with desiccant up to 4 weeks.

[13] At least four background wells are needed for the assay. Wells typically used as background are A1, A12, H1, and H12.

4. Seed purified immune cells on Cell-Tak-treated XF96 cell culture plate in 140 μL of XF assay medium.[14]
5. Centrifuge the plate at room temperature for 2 min at 2000 rpm (about 700 × *g*).
6. Transfer plate to 37 °C incubator without CO_2 and let sit for 45–60 min.[15]

3.2.2.3 Loading sensor cartridge

1. Load 20 μL of the compounds into the injector ports as follows:[16]
   - Mitochondrial stress test
     - Port A, oligomycin 8 ×
     - Port B, FCCP 9 ×
     - Port C, rotenone + antimycin A 10 ×
   - Glycolytic stress test
     - Port A, glucose 8 ×
     - Port B, oligomycin 9 ×
     - Port C, 2-DG 10 ×
2. Transfer plate to 37 °C incubator without CO_2 and let sit for 30 min.

3.2.2.4 Assay protocol

1. Load sensor cartridge and utility plate.
2. When calibration is completed, replace the utility plate with the pre-incubated cell plate.
3. Proceed following XF assay protocol.[17]
4. After the assay is complete, remove the medium from all the wells with a multichannel pipette.[18]
5. Seal the microplate with sealing tape or parafilm and store overnight at −80 °C.[19]

[14] If a fourth compound needs to be injected, cells should be seeded in 120 μL of assay medium. Alternatively, if only one or two compounds will be injected, cells should be seeded in 180 or 160 μL of assay medium, respectively. Cells should be loaded at least in triplicate. We typically avoid loading the entire triplicate on the outside wells due to the "edge effect," as results measured from the wells on the edge are often different from wells toward the center of the plate.

[15] Most cell types should adhere in 30 min.

[16] The recommended injection volume is 25 μL, but injections can range from 20 to 25 μL. All wells, including background or blank wells, need to have medium or compound loaded in the port to ensure proper injection in all wells.

[17] For immune cells, we typically use a mixing time of 4 min and a measuring time of 3 min.

[18] All liquid must be removed from the wells. However, it is not recommended to remove the liquid with a vacuum system to avoid aspirating cells.

[19] Frozen microplates can be stored at −80 °C for up to 4 weeks. The freezing step is important for efficient cell lysis.

3.2.2.5 Normalization of cell number

The normalization of cell numbers is essential when comparing different cell types in the same assay. Normalization can normally be achieved by using a colorimetric protein assay method such as the bicinchoninic acid, the Bradford, or the Lowry protein assays. However, due to the low numbers of cells used in XF96 cell culture plates, we favor a fluorometric assay that does not rely on the protein content or the metabolic status of the cell but is rather based on cell DNA content.

1. Thaw the microplate at room temperature.
2. Prepare the CyQUANT® reagent following manufacturer's instructions.[20]
3. Add 100 μL per well, including one background well as negative control.
4. Incubate 2–5 min at room temperature, protected from light.
5. Measure the sample fluorescence using a fluorescence microplate reader with filters appropriate for ~480 nm excitation and ~520 nm emission maxima (filter sets typically used for fluorescein).
6. Calculate the average fluorescence value of all the wells of one cell type that will be used as a reference (wild type, unstimulated, etc.).
7. Divide the fluorescence values of all the wells, including the wells with the cells of reference, with the calculated average fluorescence value to obtain the normalization values.[21]
8. Apply the normalization values to the data.

4. SUMMARY

In recent years, the use of the extracellular flux analyzer has become a prevalent means for understanding the biology and bioenergetics of immune cells in the context of vaccines, disease, and inflammation. The combination of the mitochondrial and glycolytic stress tests reveals whether cells engage both aerobic glycolysis and oxidative phosphorylation or preferentially use one of the two pathways. The extracellular flux analyzer can be used to evaluate the bioenergetic profiles of immune cell subsets purified from peripheral blood (Fig. 7.4). Blood monocytes efficiently use both oxidative phosphorylation and glycolysis. These cells display high basal respiration that almost doubles in the presence of FCCP. Monocyte glycolysis (measured by

[20] By using the recommended dilution of the CyQUANT® GR dye, the assay has a linear detection range from 50 to 50,000 cells. We use this assay with success with 300,000–400,000 cells by doubling the dye concentration.

[21] The values of the wells containing the cells of reference should be about 1.

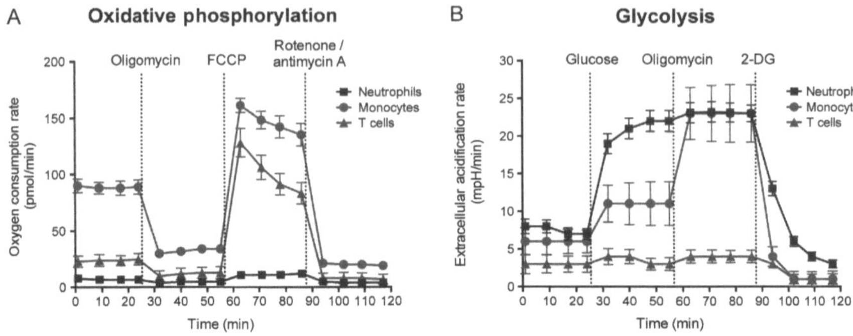

Figure 7.4 *Bioenergetic profiles of circulating immune cell subsets.* Human PBMCs and granulocytes were isolated from the blood of healthy donors by density gradient centrifugation using Ficoll-Paque™. Negative magnetic separation was used to further purify monocytes and CD4$^+$ T cells from the PBMC fraction by negative magnetic separation, while neutrophils were purified from the granulocyte fraction. Cells (400,000 cells/well) were seeded in warm XF assay medium on a Cell-Tak-treated XF96 cell culture plate, centrifuged, and incubated at 37 °C without CO_2 for 45 min. Data were acquired with a XF96 Extracellular Flux Analyzer. Cell number was normalized using CyQUANT reagent according to manufacturer's instructions. (A) Oxygen consumption rates were measured using sequential addition of oligomycin (5 μ*M*), FCCP (1.5 μ*M*), and rotenone +antimycin A (1 μ*M* and 100 n*M*, respectively). (B) Extracellular acidification rates were measured in real time using sequential addition of glucose (25 m*M*), oligomycin (5 μ*M*), and 2-deoxyglucose (2-DG; 100 m*M*). (See the color plate.)

ECAR) increases in the presence of glucose and even more in the presence of oligomycin, indicating that these cells possess both an important respiratory reserve and a glycolytic reserve. Resting T cells, in contrast, rely almost exclusively on oxidative phosphorylation. T-cell basal respiration is lower than monocytes, but their maximal respiration triples in the presence of FCCP. The poor glycolytic capability of resting T cells is indicated by their inability to increase lactate production in the presence of glucose, even after respiration is blocked by oligomycin. Conversely, neutrophils are highly glycolytic, with the highest glucose-stimulated lactate production, which is near maximal, given the minimal increase in lactate production when oligomycin is added. Neutrophils consume hardly any oxygen and, even in the presence of FCCP, commensurate with their low mitochondrial content.

In addition to using this method for understanding basal bioenergetics, the extracellular flux analyzer can be used to evaluate bioenergetic profiles following cellular activation. For example, the metabolic changes that occur throughout the process of T-cell activation are easily discernable by

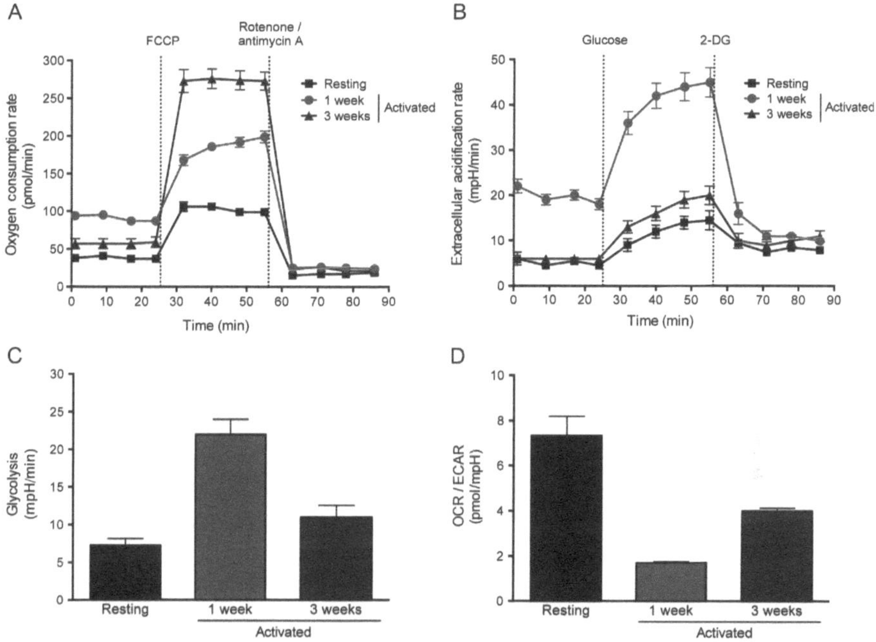

Figure 7.5 *Activation-induced glycolytic switch of CD8$^+$ T cells.* Total human CD8$^+$ T cells were purified using a negative selection kit from a normal donor apheresis sample (resting). The purified CD8$^+$ T cells were stimulated for 48 h with anti-CD3/CD28 and then expanded in media with IL-2 for either 5 days (1 week activated) or 3 weeks (3 week activated). (A) Oxygen consumption rates were determined at baseline or after FCCP addition following the mitochondrial stress test protocol. (B) Extracellular acidification rates were evaluated after sequential addition of glucose and 2-deoxyglucose (2-DG) following the glycolytic stress test protocol. (C) Glycolysis was determined by subtracting the extracellular acidification rates following the addition of glucose from non-glycolytic acidification from the data shown in (B). (D) OCR/ECAR ratios representing the relative contributions of basal respiration versus glycolysis are shown for each cell type. Derived from the data in (A) and (B). (See the color plate.)

examining their OCR and ECAR (Fig. 7.5). As stated earlier, resting CD8$^+$ T cells greatly rely on oxidative phosphorylation to produce ATP, while glycolysis is very inefficient (Fig. 7.5A and B). One week following TCR activation, CD8$^+$ T cells become highly glycolytic, as evidenced by increased lactate production (Fig. 7.5C), but still undergo a residual level of oxidative phosphorylation (as seen by the OCR/ECAR ratio in Fig. 7.5D). After 3 weeks of activation and culture in IL-2, maximal respiratory capacity remains high, whereas glycolytic capacity reverts to a state similar to a resting T cell (Fig. 7.5A–D) or CD8$^+$ T memory cells (van der Windt et al., 2012, 2013).

Besides obtaining mitochondrial and glycolytic stress test responses, the investigator can determine the bioenergetic profiles of cells exposed to other metabolic inhibitors or molecules of interest (Dranka et al., 2011). For example, the use of the CPT1 inhibitor etomoxir can indicate whether the spare respiratory capacity in cells is dependent on mitochondrial fatty acid oxidation (van der Windt et al., 2012). Mitochondrial or glycolytic profile of cells can also be studied in response to external cell stimulators, such as polyinosinic–polycytidylic acid (poly I:C) (Pantel et al., 2014), the antibiotics nigericin and gramicidin (Munoz-Planillo et al., 2013), ATP, or pathogens like *Chlamydia pneumonia* (Shimada et al., 2011). Finally, it is possible to directly interrogate individual components of the mitochondrial ETC in permeabilized cells using specific substrates and inhibitors of the ETC complexes (Salabei, Gibb, & Hill, 2014).

The study of cellular metabolism and its regulation has now become a major area of investigation, and targeting metabolic pathways to alter diseases has sparked wide interest in the fields of oncology and immunology. The use of extracellular flux analyzer is a powerful tool for investigating the cellular bioenergetics and reduces the need of using other more arduous techniques.

REFERENCES

Ackermann, R. F., & Lear, J. L. (1989). Glycolysis-induced discordance between glucose metabolic rates measured with radiolabeled fluorodeoxyglucose and glucose. *Journal of Cerebral Blood Flow and Metabolism, 9*, 774–785.

Ahrens, E. T., & Bulte, J. W. (2013). Tracking immune cells in vivo using magnetic resonance imaging. *Nature Reviews. Immunology, 13*, 755–763.

Amaral, J. J., Antunes, L. C., de Macedo, C. S., Mattos, K. A., Han, J., Pan, J., et al. (2013). Metabonomics reveals drastic changes in anti-inflammatory/pro-resolving polyunsaturated fatty acids-derived lipid mediators in leprosy disease. *PLoS Neglected Tropical Diseases, 7*, e2381.

Benavides, G. A., Liang, Q., Dodson, M., Darley-Usmar, V., & Zhang, J. (2013). Inhibition of autophagy and glycolysis by nitric oxide during hypoxia-reoxygenation impairs cellular bioenergetics and promotes cell death in primary neurons. *Free Radical Biology & Medicine, 65*, 1215–1228.

Boccuto, L., Chen, C. F., Pittman, A. R., Skinner, C. D., McCartney, H. J., Jones, K., et al. (2013). Decreased tryptophan metabolism in patients with autism spectrum disorders. *Molecular Autism, 4*, 16.

Borregaard, N., & Herlin, T. (1982). Energy metabolism of human neutrophils during phagocytosis. *Journal of Clinical Investigation, 70*, 550–557.

Bulua, A. C., Simon, A., Maddipati, R., Pelletier, M., Park, H., Kim, K. Y., et al. (2011). Mitochondrial reactive oxygen species promote production of proinflammatory cytokines and are elevated in TNFR1-associated periodic syndrome (TRAPS). *Journal of Experimental Medicine, 208*, 519–533.

Butler, E. B., Zhao, Y., Munoz-Pinedo, C., Lu, J., & Tan, M. (2013). Stalling the engine of resistance: Targeting cancer metabolism to overcome therapeutic resistance. *Cancer Research*, *73*, 2709–2717.

Buttgereit, F., Burmester, G. R., & Brand, M. D. (2000). Bioenergetics of immune functions: Fundamental and therapeutic aspects. *Immunology Today*, *21*, 192–199.

Buxton, D. B., Schwaiger, M., Nguyen, A., Phelps, M. E., & Schelbert, H. R. (1988). Radiolabeled acetate as a tracer of myocardial tricarboxylic acid cycle flux. *Circulation Research*, *63*, 628–634.

Cairns, R. A., Harris, I. S., & Mak, T. W. (2011). Regulation of cancer cell metabolism. *Nature Reviews. Cancer*, *11*, 85–95.

Carty, F., Shortt, C. P., Shelly, M. J., Eustace, S. J., & O'Connell, M. J. (2010). Whole-body imaging modalities in oncology. *Seminars in Musculoskeletal Radiology*, *14*, 68–85.

Chacko, B. K., Kramer, P. A., Ravi, S., Johnson, M. S., Hardy, R. W., Ballinger, S. W., et al. (2013). Methods for defining distinct bioenergetic profiles in platelets, lymphocytes, monocytes, and neutrophils, and the oxidative burst from human blood. *Laboratory Investigation*, *93*, 690–700.

Chawla, A., Nguyen, K. D., & Goh, Y. P. (2011). Macrophage-mediated inflammation in metabolic disease. *Nature Reviews. Immunology*, *11*, 738–749.

Cramer, T., Yamanishi, Y., Clausen, B. E., Forster, I., Pawlinski, R., Mackman, N., et al. (2003). HIF-1alpha is essential for myeloid cell-mediated inflammation. *Cell*, *112*, 645–657.

Dang, C. V., & Semenza, G. L. (1999). Oncogenic alterations of metabolism. *Trends in Biochemical Sciences*, *24*, 68–72.

Diepart, C., Verrax, J., Calderon, P. B., Feron, O., Jordan, B. F., & Gallez, B. (2010). Comparison of methods for measuring oxygen consumption in tumor cells *in vitro*. *Analytical Biochemistry*, *396*, 250–256.

Doughty, C. A., Bleiman, B. F., Wagner, D. J., Dufort, F. J., Mataraza, J. M., Roberts, M. F., et al. (2006). Antigen receptor-mediated changes in glucose metabolism in B lymphocytes: Role of phosphatidylinositol 3-kinase signaling in the glycolytic control of growth. *Blood*, *107*, 4458–4465.

Dranka, B. P., Benavides, G. A., Diers, A. R., Giordano, S., Zelickson, B. R., Reily, C., et al. (2011). Assessing bioenergetic function in response to oxidative stress by metabolic profiling. *Free Radical Biology & Medicine*, *51*, 1621–1635.

Ehrismann, D., Flashman, E., Genn, D. N., Mathioudakis, N., Hewitson, K. S., Ratcliffe, P. J., et al. (2007). Studies on the activity of the hypoxia-inducible-factor hydroxylases using an oxygen consumption assay. *Biochemical Journal*, *401*, 227–234.

Elks, P. M., van Eeden, F. J., Dixon, G., Wang, X., Reyes-Aldasoro, C. C., Ingham, P. W., et al. (2011). Activation of hypoxia-inducible factor-1alpha (Hif-1alpha) delays inflammation resolution by reducing neutrophil apoptosis and reverse migration in a zebrafish inflammation model. *Blood*, *118*, 712–722.

Everts, B., Amiel, E., van der Windt, G. J., Freitas, T. C., Chott, R., Yarasheski, K. E., et al. (2012). Commitment to glycolysis sustains survival of NO-producing inflammatory dendritic cells. *Blood*, *120*, 1422–1431.

Fangradt, M., Hahne, M., Gaber, T., Strehl, C., Rauch, R., Hoff, P., et al. (2012). Human monocytes and macrophages differ in their mechanisms of adaptation to hypoxia. *Arthritis Research & Therapy*, *14*, R181.

Ferrari, C. K., Souto, P. C., Franca, E. L., & Honorio-Franca, A. C. (2011). Oxidative and nitrosative stress on phagocytes' function: From effective defense to immunity evasion mechanisms. *Archivum immunologiae et therapiae experimentalis*, *59*, 441–448.

Gaber, T., Schellmann, S., Erekul, K. B., Fangradt, M., Tykwinska, K., Hahne, M., et al. (2011). Macrophage migration inhibitory factor counterregulates dexamethasone-mediated suppression of hypoxia-inducible factor-1 alpha function and differentially

influences human CD4+ T cell proliferation under hypoxia. *Journal of Immunology, 186*, 764–774.
Gnaiger, E., Steinlechner-Maran, R., Mendez, G., Eberl, T., & Margreiter, R. (1995). Control of mitochondrial and cellular respiration by oxygen. *Journal of Bioenergetics and Biomembranes, 27*, 583–596.
Heerlein, K., Schulze, A., Hotz, L., Bartsch, P., & Mairbaurl, H. (2005). Hypoxia decreases cellular ATP demand and inhibits mitochondrial respiration of a549 cells. *American Journal of Respiratory Cell and Molecular Biology, 32*, 44–51.
Ho, J., de Moura, M. B., Lin, Y., Vincent, G., Thorne, S., Duncan, L. M., et al. (2012). Importance of glycolysis and oxidative phosphorylation in advanced melanoma. *Molecular Cancer, 11*, 76.
Hutter, E., Unterluggauer, H., Garedew, A., Jansen-Durr, P., & Gnaiger, E. (2006). High-resolution respirometry—A modern tool in aging research. *Experimental Gerontology, 41*, 103–109.
Hynes, J., Marroquin, L. D., Ogurtsov, V. I., Christiansen, K. N., Stevens, G. J., Papkovsky, D. B., et al. (2006). Investigation of drug-induced mitochondrial toxicity using fluorescence-based oxygen-sensitive probes. *Toxicological Sciences, 92*, 186–200.
Inoki, K., Kim, J., & Guan, K. L. (2012). AMPK and mTOR in cellular energy homeostasis and drug targets. *Annual Review of Pharmacology and Toxicology, 52*, 381–400.
Jacobs, S. R., Herman, C. E., Maciver, N. J., Wofford, J. A., Wieman, H. L., Hammen, J. J., et al. (2008). Glucose uptake is limiting in T cell activation and requires CD28-mediated Akt-dependent and independent pathways. *Journal of Immunology, 180*, 4476–4486.
James, P. E., Jackson, S. K., Grinberg, O. Y., & Swartz, H. M. (1995). The effects of endotoxin on oxygen consumption of various cell types *in vitro*: An EPR oximetry study. *Free Radical Biology & Medicine, 18*, 641–647.
Jones, R. G., & Thompson, C. B. (2007). Revving the engine: Signal transduction fuels T cell activation. *Immunity, 27*, 173–178.
Kadayakkara, D. K., Janjic, J. M., Pusateri, L. K., Young, W. B., & Ahrens, E. T. (2010). In vivo observation of intracellular oximetry in perfluorocarbon-labeled glioma cells and chemotherapeutic response in the CNS using fluorine-19 MRI. *Magnetic Resonance in Medicine, 64*, 1252–1259.
Kamphorst, J. J., Cross, J. R., Fan, J., de Stanchina, E., Mathew, R., White, E. P., et al. (2013). Hypoxic and Ras-transformed cells support growth by scavenging unsaturated fatty acids from lysophospholipids. *Proceedings of the National Academy of Sciences of the United States of America, 110*, 8882–8887.
Kell, D. B. (2004). Metabolomics and systems biology: Making sense of the soup. *Current Opinion in Microbiology, 7*, 296–307.
Konisti, S., Kiriakidis, S., & Paleolog, E. M. (2012). Hypoxia—A key regulator of angiogenesis and inflammation in rheumatoid arthritis. *Nature Reviews. Rheumatology, 8*, 153–162.
Krawczyk, C. M., Holowka, T., Sun, J., Blagih, J., Amiel, E., DeBerardinis, R. J., et al. (2010). Toll-like receptor-induced changes in glycolytic metabolism regulate dendritic cell activation. *Blood, 115*, 4742–4749.
Kunin, A. S., & Tannen, R. L. (1979). Regulation of glutamate metabolism by renal cortical mitochondria. *American Journal of Physiology, 237*, F55–F62.
Lear, J. L., & Ackermann, R. F. (1990). Evaluation of radiolabeled acetate and fluoroacetate as potential tracers of cerebral oxidative metabolism. *Metabolic Brain Disease, 5*, 45–56.
Lewis, J. S., Lee, J. A., Underwood, J. C., Harris, A. L., & Lewis, C. E. (1999). Macrophage responses to hypoxia: Relevance to disease mechanisms. *Journal of Leukocyte Biology, 66*, 889–900.
Li, Z., & Graham, B. H. (2012). Measurement of mitochondrial oxygen consumption using a Clark electrode. *Methods in Molecular Biology, 837*, 63–72.

Li, K. J., Wu, C. H., Hsieh, S. C., Lu, M. C., Tsai, C. Y., & Yu, C. L. (2012). Deranged bioenergetics and defective redox capacity in T lymphocytes and neutrophils are related to cellular dysfunction and increased oxidative stress in patients with active systemic lupus erythematosus. *Clinical & Developmental Immunology, 2012*, 548516.

Lowry, O. H., Passonneau, J. V., Schulz, D. W., & Rock, M. K. (1961). The measurement of pyridine nucleotides by enzymatic cycling. *Journal of Biological Chemistry, 236*, 2746–2755.

Maciver, N. J., Jacobs, S. R., Wieman, H. L., Wofford, J. A., Coloff, J. L., & Rathmell, J. C. (2008). Glucose metabolism in lymphocytes is a regulated process with significant effects on immune cell function and survival. *Journal of Leukocyte Biology, 84*, 949–957.

Macrae, K., Stretton, C., Lipina, C., Blachnio-Zabielska, A., Baranowski, M., Gorski, J., et al. (2013). Defining the role of DAG, mitochondrial function, and lipid deposition in palmitate-induced proinflammatory signaling and its counter-modulation by palmitoleate. *Journal of Lipid Research, 54*, 2366–2378.

Maianski, N. A., Geissler, J., Srinivasula, S. M., Alnemri, E. S., Roos, D., & Kuijpers, T. W. (2004). Functional characterization of mitochondria in neutrophils: A role restricted to apoptosis. *Cell Death and Differentiation, 11*, 143–153.

Maratou, E., Dimitriadis, G., Kollias, A., Boutati, E., Lambadiari, V., Mitrou, P., et al. (2007). Glucose transporter expression on the plasma membrane of resting and activated white blood cells. *European Journal of Clinical Investigation, 37*, 282–290.

Martinez-Outschoorn, U. E., Pavlides, S., Howell, A., Pestell, R. G., Tanowitz, H. B., Sotgia, F., et al. (2011). Stromal-epithelial metabolic coupling in cancer: Integrating autophagy and metabolism in the tumor microenvironment. *International Journal of Biochemistry & Cell Biology, 43*, 1045–1051.

Metallo, C. M., Gameiro, P. A., Bell, E. L., Mattaini, K. R., Yang, J., Hiller, K., et al. (2012). Reductive glutamine metabolism by IDH1 mediates lipogenesis under hypoxia. *Nature, 481*, 380–384.

Mullen, A. R., Wheaton, W. W., Jin, E. S., Chen, P. H., Sullivan, L. B., Cheng, T., et al. (2012). Reductive carboxylation supports growth in tumour cells with defective mitochondria. *Nature, 481*, 385–388.

Munoz-Planillo, R., Kuffa, P., Martinez-Colon, G., Smith, B. L., Rajendiran, T. M., & Nunez, G. (2013). K(+) efflux is the common trigger of NLRP3 inflammasome activation by bacterial toxins and particulate matter. *Immunity, 38*, 1142–1153.

Newsholme, E. A., Crabtree, B., & Ardawi, M. S. (1985). The role of high rates of glycolysis and glutamine utilization in rapidly dividing cells. *Bioscience Reports, 5*, 393–400.

Nicholls, D. G., Darley-Usmar, V. M., Wu, M., Jensen, P. B., Rogers, G. W., & Ferrick, D. A. (2010). Bioenergetic profile experiment using C2C12 myoblast cells. *Journal of Visualized Experiments, 2511*.

Nutsch, K., & Hsieh, C. (2011). When T cells run out of breath: The HIF-1alpha story. *Cell, 146*, 673–674.

Pantel, A., Teixeira, A., Haddad, E., Wood, E. G., Steinman, R. M., & Longhi, M. P. (2014). Direct type I IFN but not MDA5/TLR3 activation of dendritic cells is required for maturation and metabolic shift to glycolysis after poly IC stimulation. *PLoS Biology, 12*, e1001759.

Pearce, E. L., & Pearce, E. J. (2013). Metabolic pathways in immune cell activation and quiescence. *Immunity, 38*, 633–643.

Powell, J. D., Pollizzi, K. N., Heikamp, E. B., & Horton, M. R. (2012). Regulation of immune responses by mTOR. *Annual Review of Immunology, 30*, 39–68.

Purushothaman, K. R., Meerarani, P., & Moreno, P. R. (2007). Inflammation and neovascularization in diabetic atherosclerosis. *Indian Journal of Experimental Biology, 45*, 93–102.

Purushothaman, K. R., Purushothaman, M., Muntner, P., Lento, P. A., O'Connor, W. N., Sharma, S. K., et al. (2011). Inflammation, neovascularization and intra-plaque hemorrhage are associated with increased reparative collagen content: Implication for plaque progression in diabetic atherosclerosis. *Vascular Medicine*, *16*, 103–108.

Putluri, N., Shojaie, A., Vasu, V. T., Nalluri, S., Vareed, S. K., Putluri, V., et al. (2011). Metabolomic profiling reveals a role for androgen in activating amino acid metabolism and methylation in prostate cancer cells. *PLoS One*, *6*, e21417.

Rathmell, J. C., Elstrom, R. L., Cinalli, R. M., & Thompson, C. B. (2003). Activated Akt promotes increased resting T cell size, CD28-independent T cell growth, and development of autoimmunity and lymphoma. *European Journal of Immunology*, *33*, 2223–2232.

Rolo, A. P., Palmeira, C. M., & Cortopassi, G. A. (2009). Biosensor plates detect mitochondrial physiological regulators and mutations *in vivo*. *Analytical Biochemistry*, *385*, 176–178.

Salabei, J. K., Gibb, A. A., & Hill, B. G. (2014). Comprehensive measurement of respiratory activity in permeabilized cells using extracellular flux analysis. *Nature Protocols*, *9*, 421–438.

Schafer, Z. T., Grassian, A. R., Song, L., Jiang, Z., Gerhart-Hines, Z., Irie, H. Y., et al. (2009). Antioxidant and oncogene rescue of metabolic defects caused by loss of matrix attachment. *Nature*, *461*, 109–113.

Semenza, G. L. (2007). Hypoxia-inducible factor 1 (HIF-1) pathway. *Science's STKE: Signal Transduction Knowledge Environment*, *2007*(407), cm8.

Sena, L. A., Li, S., Jairaman, A., Prakriya, M., Ezponda, M., Hildeman, D. A., et al. (2013). Mitochondria are required for antigen-specific T cell activation through ROS signaling. *Immunity*, *38*, 225–236.

Sher, R., Wadee, A., & Joffe, M. (1983). The enhancement of eosinophil function by lymphocyte supernatants. *Clinical and Experimental Immunology*, *51*, 525–534.

Shimada, K., Crother, T. R., Karlin, J., Chen, S., Chiba, N., Ramanujan, V. K., et al. (2011). Caspase-1 dependent IL-1beta secretion is critical for host defense in a mouse model of *Chlamydia pneumoniae* lung infection. *PLoS One*, *6*, e21477.

Soga, T. (2013). Cancer metabolism: Key players in metabolic reprogramming. *Cancer Science*, *104*, 275–281.

Sumbayev, V. V., Nicholas, S. A., Streatfield, C. L., & Gibbs, B. F. (2009). Involvement of hypoxia-inducible factor-1 HiF(1alpha) in IgE-mediated primary human basophil responses. *European Journal of Immunology*, *39*, 3511–3519.

Tannahill, G. M., Curtis, A. M., Adamik, J., Palsson-McDermott, E. M., McGettrick, A. F., Goel, G., et al. (2013). Succinate is an inflammatory signal that induces IL-1β through HIF-1α. *Nature*, *496*, 238–242.

Taylor, P. C., & Sivakumar, B. (2005). Hypoxia and angiogenesis in rheumatoid arthritis. *Current Opinion in Rheumatology*, *17*, 293–298.

Teicher, B. A., Linehan, W. M., & Helman, L. J. (2012). Targeting cancer metabolism. *Clinical Cancer Research*, *18*, 5537–5545.

Vander Heiden, M. G., Cantley, L. C., & Thompson, C. B. (2009). Understanding the Warburg effect: The metabolic requirements of cell proliferation. *Science*, *324*, 1029–1033.

van der Windt, G. J., Everts, B., Chang, C. H., Curtis, J. D., Freitas, T. C., Amiel, E., et al. (2012). Mitochondrial respiratory capacity is a critical regulator of CD8(+) T cell memory development. *Immunity*, *36*, 68–78.

van der Windt, G. J., O'Sullivan, D., Everts, B., Huang, S. C., Buck, M. D., Curtis, J. D., et al. (2013). CD8 memory T cells have a bioenergetic advantage that underlies their rapid recall ability. *Proceedings of the National Academy of Sciences of the United States of America*, *110*, 14336–14341.

Vats, D., Mukundan, L., Odegaard, J. I., Zhang, L., Smith, K. L., Morel, C. R., et al. (2006). Oxidative metabolism and PGC-1beta attenuate macrophage-mediated inflammation. *Cell Metabolism*, *4*, 13–24.
Venge, P., Moberg, L., Bjornsson, E., Bergstrom, M., Langstrom, B., & Hakansson, L. (2003). Mechanisms of basal and cytokine-induced uptake of glucose in normal human eosinophils: Relation to apoptosis. *Respiratory Medicine*, *97*, 1109–1119.
Wang, R., Dillon, C. P., Shi, L. Z., Milasta, S., Carter, R., Finkelstein, D., et al. (2011). The transcription factor Myc controls metabolic reprogramming upon T lymphocyte activation. *Immunity*, *35*, 871–882.
Wishart, D. S., Knox, C., Guo, A. C., Eisner, R., Young, N., Gautam, B., et al. (2009). HMDB: A knowledgebase for the human metabolome. *Nucleic Acids Research*, *37*, D603–D610.
Yuneva, M., Zamboni, N., Oefner, P., Sachidanandam, R., & Lazebnik, Y. (2007). Deficiency in glutamine but not glucose induces MYC-dependent apoptosis in human cells. *Journal of Cell Biology*, *178*, 93–105.
Zhang, J., Nuebel, E., Wisidagama, D. R., Setoguchi, K., Hong, J. S., Van Horn, C. M., et al. (2012). Measuring energy metabolism in cultured cells, including human pluripotent stem cells and differentiated cells. *Nature Protocols*, *7*, 1068–1085.
Zhang, Y., & Yang, J. M. (2013). Altered energy metabolism in cancer: A unique opportunity for therapeutic intervention. *Cancer Biology & Therapy*, *14*, 81–89.
Zucker-Franklin, D. (1968). Electron microscopic studies of human granulocytes: Structural variations related to function. *Seminars in Hematology*, *5*, 109–133.

CHAPTER EIGHT

Conventional Techniques to Monitor Mitochondrial Oxygen Consumption

Hélène Simonnet[*,1], Arnaud Vigneron[*,1], Jacques Pouysségur[†,‡,2]

[*]Centre de Recherche en Cancérologie de Lyon UMR INSERM U1052/CNRS 5286 Centre Léon Bérard, Lyon, France

[†]Institute of Research on Cancer and Aging (IRCAN) CNRS-Inserm-UNS UMR 7284, U 1081 Centre A. Lacassagne, Nice, France

[‡]Centre Scientifique de Monaco (CSM), Monaco

[1]Equal contribution of authors.

[2]Corresponding author: e-mail address: pouysseg@unice.fr

Contents

Abstract

Following several key discoveries on hypoxia-inducible factors, we have observed an explosion of studies investigating how the hypoxic microenvironment provokes bioenergetic alterations. This is particularly relevant for cancer cells, as they are often exposed to hypoxic conditions in the course of tumor progression. Thus, interest in the measurement of oxygen consumption at the tissue, cell, or mitochondrion level has been revived. Here, we describe the basic principles of cellular respiration and survey some of the conventional methods for measuring O_2 consumption in intact or permeabilized cells.

Methods in Enzymology, Volume 542
ISSN 0076-6879
http://dx.doi.org/10.1016/B978-0-12-416618-9.00008-X

1. INTRODUCTION

Monitoring oxygen (O_2) consumption is an efficient way to follow metabolic activity of cells, tissues, or whole organisms. Beyond metabolic analysis, O_2 consumption is now also used to analyze cytotoxicity in drug screening development as an early marker of cell death. As a reflect of mitochondrial activity, respiration studies are essential to understand the physiology of several diseases such as metabolic syndromes and also, as described by Otto Warburg almost one century ago, cancers and tumor progression.

Investigating cancer cell respiration addresses several questions: Does the cancer cell rely on respiration for its ATP production or for maintenance of mitochondrial anabolic activity? Does it receive a great amount of oxygen *in situ*? If respiration is slow, is this a mitochondrial dysfunction and at which site of the electron transport chain (ETC), or a particular oxygenation status of the tumor *in situ*? Do oncogenes/oncosuppressors and cancer drugs impact the normal respiratory function? (see review by Godinot, de Laplanche, Hervouet, & Simonnet, 2007).

2. PRINCIPLES OF CELL RESPIRATION MEASUREMENTS

2.1. Before choosing a method and a model to measure O_2 uptake, the different laws ruling this uptake should be kept in mind

Normal living cells in higher organisms take up dissolved oxygen (O_2) at a high rate, but many physical and chemical constraints exist between atmospheric oxygen and complex IV of the mitochondrial respiratory ETC.

In an aqueous solution in contact with air, the partial pressure (pO_2) equilibrates with that of air (21% O_2, i.e., a pO_2 of 160 mmHg at sea level). Stirring the solution is a means to greatly accelerate equilibration. Oxygen diffuses easily across biological membranes, but its flux (J_{diff}, in mol s^{-1}) from one point to another across the exchange surface (S) is dependent on the distance (Δx) and the concentration gradient (ΔC), meaning that $J_{\text{diff}} = SD\Delta C\Delta x^{-1}$ (where D represents the diffusion coefficient of oxygen).

After less than 1 mm, oxygen diffusion rate is generally not sufficient for the oxygen demand of the cell in an unstirred aqueous medium.

Whereas maximal in the in arterial blood, the lowest O_2 concentration is found deep in the tissues and inside cells at the site of its conversion to H_2O (complex IV, in the inner mitochondrial membrane).

The diffusion coefficient depends on solubility, which differs between water and solutions as oxygen is less soluble in the presence of salts and proteins. In blood, oxygen is mostly transported by hemoglobin (in 100 ml arterial blood: 0.3 ml dissolved O_2 and 19.7 ml hemoglobin-bound O_2). Erythrocytes in capillaries are generally separated from the cell surface by 10–200 μm, and their high O_2 content provides favorable diffusion gradients into cells. Such is not the case in the surroundings of an electrode or an oxygen-sensitive probe in hemoglobin-free *ex vivo* systems. For this reason, measurements of O_2 consumption can be made either on cell or mitochondria suspensions in stirred aqueous solutions or in cultured attached cells with an O_2 probe in the vicinity of the cells.

Lastly, whereas O_2 is almost freely permeant across biological membranes, nutrients have to be transported to the mitochondria and metabolized before reaching the ETC (Fig. 8.1). These parameters regulating the function of the respiratory chain either from its entry or from its end-point are thus essential to consider in choosing the correct model for the question asked. One could see how oxygen profiles have been explored in mice by noninvasive measurement (Torres Filho, Leunig, Yuan, Intaglietta, & Jain, 1994) or in human using pO_2 histography (Vaupel, Höckel, & Mayer, 2007).

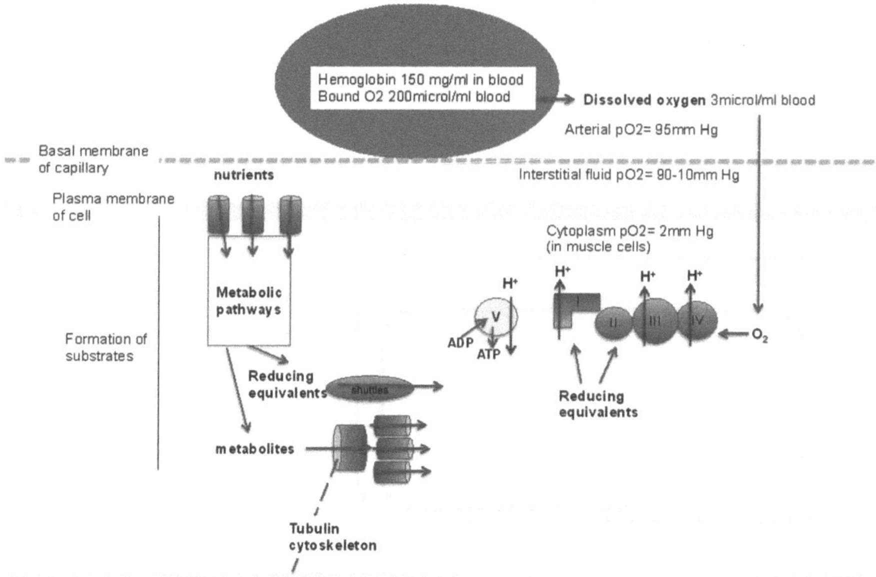

Figure 8.1 Oxygen delivery from hemoglobin to interstitial fluid, cytoplasm, and mitochondria.

2.2. How to choose a model?

Model 1. The determination of *in situ* concentration of oxygen will be covered by chapter 10 of this volume.

Model 2. Entire nonpermeabilized cells: In this model, O_2 consumption corresponds to normal mitochondrial activity integrated with other metabolic pathways acting in the cell (glycolysis, peroxysomal β-oxydation, anabolism, etc.). In these conditions, metabolite exchanges between cytoplasm and mitochondria are essential to the control of respiratory chain. Fatty acid entry into mitochondria through the carnitine shuttle is one of several substrates strongly regulating the O_2 consumption by the mitochondria. Another very important exchange is the inverted transport of ATP and ADP through the Ant exchanger/creatine kinase system. This system allows to consume rapidly the ATP coming from ATP synthase to maintain high level of ADP in mitochondria and to transport efficiently the energetic power of ATP through the form of creatine phosphate, a more stable and diffusible compound.

The structural shape and networks of mitochondria and the connections made with others organelles are also maintained in entire cells. These connections, especially with endoplasmic reticulum, are now recognized to be an integral part of mitochondrial activity. Mitochondrial Ca^{2+} homeostasis is controlled at this level, and it regulates dehydrogenase enzymes necessary for the Krebs cycle activity and reduced cofactor production.

Alternatively, the role of the mitochondrial network can also be tested in this condition. Mitochondria can fuse in the cell, a mechanism important for their homeostasis allowing them to exchange metabolite and enzymes but also to control their inner membrane polarity. Mitochondrial inner membrane polarity can be passed from one mitochondrion to another rescuing defective mitochondria or establishing different profiles of mitochondrial activity. For example, mitochondria with high polarity due to substrate overload can "pass" their polarity to fasted mitochondria to increase overall ATP production.

With entire cell models, intrinsic activity of respiratory chain complexes is difficult to test, since their activity is integrated with a myriad of different parameters. However, this system is ideal for observing the effects of a modulation of one metabolic pathway connected directly or indirectly to the mitochondrial activity. Extracellular medium alteration, genetic manipulation or cell permeable inhibitors can be used to dissect these connections. The contribution of other O_2-consuming enzymes can be measured through complete inhibition of the respiratory chain and subtracted from other analyses (for review see Pesta & Gnaiger, 2012).

Model 3. Permeabilized cells: here the substrates are supplied by the experimenter to the mitochondria in their natural structural microenvironment including interactions with the cytoskeleton and endoplasmic reticulum. Advantages of permeabilized cell models are the opportunity to feed the mitochondria with a specific substrate to analyze the mitochondrial response to this substrate. By this means, electron entry in the respiratory chain can be controlled more precisely. An intact microstructural environment controls substrate entry, especially at the outer membrane level. Voltage-dependent anion channel (VDAC) is the main transporter in this membrane, and its conductance is strongly repressed by microtubule interaction. Beyond these parameters, O_2 consumption reflects mitochondrial function with respect to different respiratory complexes, the maximal capacity of O_2 consumption in standard concentrations of substrates and its control (coupling) by the complex V/ATP synthase.

Model 4. Isolated mitochondria: From a technical point of view, a mitochondrial suspension is highly concentrated in ETC, and therefore, O_2 consumption with this model is higher than with cells. This is especially interesting if the respiratory capacity is low, such as in many cancer cells. However, if the cells are weakened by processes such as senescence, mitophagy, or death, cell fractionation could be difficult to achieve and/or may introduce bias. The protocol is the same as that of permeabilized cells. In the aspect of its interpretation, this model reflects genuine mitochondrial respiration and excludes extramitochondrial O_2 consumption. However, the control of VDAC opening by microtubules, or intramitochondrial concentration of calcium regulation which can be altered in transformed or malignant cells, is lost in this model (for review Saks et al., 2012). Similarly, mitochondria fragment during the isolation procedure, preventing any analysis of mitochondrial network effects on respiration.

2.3. Summary of available devices

i. Polarographic respirometry allows measurement of cell or mitochondria suspensions in a closed chamber.

Hansatech and Oroboros companies propose integrated systems to measure oxygen consumption by polarography with a Clark's electrode.

Briefly, oxygen diffuses through a Teflon membrane, which is permeable to uncharged gases but not to water. A platinum/silver/KCl coupled electrode reduces oxygen and oxidizes silver, giving rise to a current. Though the current is proportional to oxygen concentration, consumption of oxygen by the platinum cathode is maintained at a low level by polarization of the system at a low level. These two devices operate in a closed chamber of 2 ml, with a stirring magnet to homogenize oxygen and biological material. The volume can be adjusted to the amount of cells or mitochondria to analyze. The stirred medium warrants free access of the cells/mitochondria to O_2. Therefore, it is the capacity of respiration that is determined. Whereas mitochondria function is easily measured with these systems, whole cell or tissue analysis requires cell dissociation and suspension. Cell detachment from the extracellular matrix is a strong stress signal that greatly disturbs metabolism through several mechanisms, commonly resulting in a form of cell death called anoïkis. These effects take place in a few minutes, potentially affecting cellular O_2 consumption during the time of the experiment. Growth of cells on matrix beads made of collagen allows these problems to be circumvented by maintaining cell adhesion.

ii. Optical sensor probes to determine O_2 consumption in a defined medium in multiwell plates. Seahorse Biosciences provide an integrated system to measure oxygen consumption in attached cultured cells or in isolated mitochondria with an optical probe per individual well on the plate. The probes are brought close to the cells to investigate respiration in a very small closed chamber. This system is described in Chapter 7 of this volume. Similarly, plates coated with engineered porphyrine can be used to monitor oxygen consumption. These molecules present a near infrared fluorescence that is quenched in the presence of oxygen. They are mainly used for high-throughput cytotoxicity measurement since O_2 consumption is one of the most sensitive parameters for cell toxicity.

Here, we will describe a typical protocol to investigate cell respiration in both devices.

3. CONVENTIONAL MEASUREMENT OF OXYGEN CONSUMPTION

3.1. General features of oxymetry recordings

ETC function is mainly regulated at two levels. The first one is constituted by the quantity and the quality of substrate able to give electrons to the transport chain. The second one is the block imposed on the electron transfer by the proton gradient (membrane potential) between the inter membrane space and the mitochondrial matrix. Since electron transfer through the ETC is energetically coupled to proton extraction from the matrix, an increase of the gradient is thermodynamically unfavorable to ETC activity. By reducing this gradient, ATP synthase activity is a strong activator of the ETC.

Different mitochondrial states of respiration have been defined by Chance and Williams in their former experiments with mitochondria suspensions fed with external substrates and ADP (Chance & Williams, 1955). The term "state" was next extended to resembling situations in other models.

Recordings with mitochondria suspensions show that in isotonic salt buffer containing phosphate, there is almost no O_2 decay without substrates or ADP (state 1). In state 2, a basal decay is obtained after addition of substrate(s). In Fig. 8.2, addition of succinate at time zero is followed by a small decay of O_2 concentration. Further addition of ADP activates ATP synthase which partially or totally dissipates the proton gradient depending on the concentration of ADP in the matrix: this activated state is termed state 3. When exogenous and endogenous ADP is exhausted, the proton gradient increases to its maximum and respiration slows down to a resting coupled state termed *state 4*.

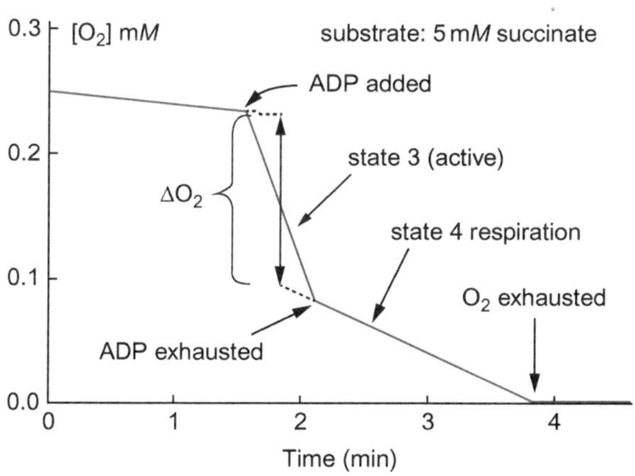

Figure 8.2 Oxygraphic O_2 consumption recording from Clark's electrode with a suspension of isolated mitochondria. Note the different states following sequential substrate additions as defined by Chance and Williams (1955).

Maximal states of respiration can also be achieved by using uncoupling agents such as proton ionophores, which dissipate the proton gradient. This maximally activated state is sometimes called state 3. Respiration in the presence of a maximal proton gradient can be achieved with inhibitors of the ATP synthase complex at the level of its proton channel: this is sometimes called state 4.

3.2. O_2 consumption in nonpermeabilized, intact cells

1. Measurement medium: culture medium without serum, 50 ml. Depending of the addressed question, substrate concentration in the medium can be modulated. Glucose (1–20 m*M*), glutamine (0.5–2 m*M*), pyruvate (0.5–2 m*M*), and palmitate (100–600 μ*M*) are some of the different substrates specifically added to the medium to test cell respiration.
2. Preparation of cell suspensions: Detach cells from their support by the appropriate technique (usually trypsination) and wash the suspension by centrifugation at 3000 rpm in measurement medium and resuspend in 0.5 ml of measurement medium. Keep the cells on ice and use them in the following two hours. Determine protein concentration in a 10-μl sample. Alternatively, amount of material to use can be normalized through cell number or DNA quantification. When cells are grown on matrix beads, beads can be harvested through centrifugation (1000 rpm) and resuspended directly in measurement medium.
3. Equilibration: Pipette 2 ml of measurement medium into the oxygraph chamber, introduce a clean magnetic stir bar, and close the stopper carefully to avoid any air bubbles. Equilibrate the electrode during 10–30 min at 30 °C and determine baseline.
4. Measurement: Pipette out the medium. Add 1.8 ml of fresh medium and 0.2 ml of cell suspension containing 0.3–3.0 mg protein and measure the slope during 1–10 min.

5a. Add 1 m*M* KCN, an inhibitor of complex IV, through the stopper to specifically inhibit mitochondrial respiration. Measure the slope during 5 min.

5b. Alternatively the following membrane permeant inhibitors can be added at this stage. Rotenone (0.1 μ*M*) inhibits complex I, and the subsequent decrease of the slope represents the contribution of the pathways feeding this entry to the cell respiration.

5c. Alternatively, the coupling of the ETC and ATP synthase can be estimated by addition of oligomycin (1 μ*M*), an inhibitor of ATP synthase at the level of its proton channel. The remaining slope of oxygen flux represents (1) the proton leak at maximal mitochondrial membrane potential (leak respiration); (2) the uncoupled respiration, including electrons binding to O_2 to form reactive oxygen species; (3) oxygen flux in other oxidases or peroxidases; and (4) cation cycling (Ca^{2+}, K^+).
In this state, disruption of the mitochondrial gradient of protons will be achieved by adding the proton ionophore FCCP (0.2–1 μ*M*). This will cause the slope to increase to the maximal flux of O_2, representing the respiration capacity of the cells.

6. Wash thoroughly the chamber with fresh medium and ethanol (95%) to remove hydrophobic mitochondrial inhibitor.

7. Repeat the assay with appropriate control cells. For example, cell line X, cell line X transfected with empty vector, and cell line X transfected with vector containing the gene of interest (oncoprotein or effector protein).

8. If the respiration is too slow or too fast, adjust the number of cells.

3.3. O_2 consumption in permeabilized cells or mitochondria

1. Medium: potassium-rich medium (KCl 120 m*M*; KH_2PO_4 5 m*M*; Tris–HCl 10 m*M*, pH 7.4; $MgCl_2$ 3 m*M*; EDTA 1 m*M*).

2. Trypsination, washing, and protein determination as done for intact cells protocol.

2a. Alternatively prepare freshly isolated mitochondria and determine protein content. Isolated mitochondria have to be kept on ice at all time and are stable for 2 h.

3. Equilibration. Pipette 2 ml of measurement medium into the oxygraph chamber, introduce a clean magnetic stir bar, and close the stopper carefully to avoid any air bubbles. Equilibrate electrode during 10–30 min at 30 °C and determine baseline.

4. Measurement. Pipette out the medium. Add 1.8 ml of fresh medium and 0.2 ml of cell or mitochondria suspension containing 0.3–3.0 mg protein. Let temperature equilibrate during 2 min and then add digitonin (0.1 mg/mg protein) through the stopper to permeabilize the plasma membrane. (It is better to determine the optimal concentration of digitonin appropriate for your cells, typically 0.1–3 mg digitonin/mg protein.) Do not use digitonin with isolated mitochondria.

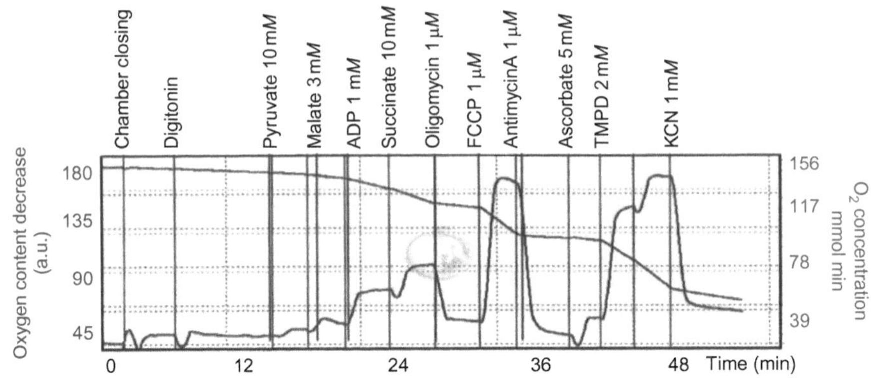

Figure 8.3 Recordings of O_2 flux in permeabilized cancer cells, using an Oroboros oxygraph. Lower curve represents oxygen consumed per unit of time; upper curve represents oxygen concentration against time. *Hervouets (2006), with permission.*

5. After 2 min, add the complex I substrates pyruvate (10 m*M*) and malate (3 m*M*) and then ADP (1 m*M*) which is required for complex V (ATPase) function. The flux is increased (Fig. 8.3) and the coupled respiration (complex I to complex IV coupled to ATP synthesis) will be recorded.
6. After the time required to obtain linear O_2 consumption (typically 1–10 min), add 10 m*M* succinate and let develop a linear slope: an additional increase can be recorded (Fig. 8.3), representing the capacity of respiration due to complex II.
7. In order to know to which extent respiration through the ETC is coupled to the phosphorylation of ADP by complex V, the latter can be inhibited at the level of its proton channel: add 1 μ*M* oligomycin. The drop in oxygen consumption represents coupled respiration. As soon as a linear slope can be stabilized, the proton gradient can be dissipated by addition of the uncoupling proton ionophore FCCP (1 μ*M*). The slope will increase to the maximal speed of oxygen consumption since the ETC is no longer restrained by proton accumulation in the intermembrane space.
8. In this uncoupled state, add 1 μ*M* antimycin A to inhibit complex III and therefore assess its activity. The slopes generally drop to basal level.
9. Even in this state, complex IV/cytochrome oxidase can be assessed if electron donors for cytochrome *c* are added to the medium. Add 2 m*M* ascorbate and 0.5 m*M* TMPD (*N*,*N*,*N*′,*N*′-tetramethyl-*p*-phenylenediamine dihydrochloride) to activate complex IV. Ascorbate

has to be added before TMPD to maintain its reduced state. As soon as a linear slope is obtained, add 1 m*M* KCN to assess complex IV activity. These different enzyme activities should be only used to compare tissue or cell preparations under the same protocol.

REFERENCES

Chance, B., & Williams, G. R. (1955). Respiratory enzymes in oxidative phosphorylation. III. The steady state. *Journal of Biological Chemistry, 217*, 409–427.

Godinot, C., de Laplanche, E., Hervouet, E., & Simonnet, H. (2007). Review. Actuality of Warburg's views in our understanding of renal cancer metabolism. *Journal of Bioenergetics and Biomembranes, 39*, 235–241.

Hervouet, Eric (2006). *Mécanismes impliqués dans la dérégulation de la biogenèse mitochondriale de tumeurs rénales: Carcinomes à cellules claires et oncocytomes,* PhD Thesis, Lyon.

Pesta, D., & Gnaiger, E. (2012). High-resolution respirometry: OXPHOS protocols for human cells and permeabilized fibers from small biopsies of human muscle. *Methods in Molecular Biology, 810*, 25–58.

Saks, V., Kuznetsov, A. V., Gonzalez-Granillo, M., Tepp, K., Timohhina, N., Karu-Varikmaa, M., et al. (2012). Intracellular energetic units regulate metabolism in cardiac cells. *Journal of Molecular and Cellular Cardiology, 52*, 419–436.

Torres Filho, I. P., Leunig, M., Yuan, F., Intaglietta, M., & Jain, R. K. (1994). Noninvasive measurement of microvascular and interstitial oxygen profiles in a human tumor in SCID mice. *Proceedings of the National Academy of Sciences of the United States of America, 91*, 2081–2085.

Vaupel, P., Höckel, M., & Mayer, A. (2007). Review. Detection and characterization of tumor hypoxia using pO2 histography. *Antioxidants & Redox Signaling, 9*, 1221–1235.

CHAPTER NINE

Use of Safranin for the Assessment of Mitochondrial Membrane Potential by High-Resolution Respirometry and Fluorometry

Gerhard Krumschnabel[*], **Andrea Eigentler**[†], **Mario Fasching**[*], **Erich Gnaiger**[*,†,1]
[*]OROBOROS INSTRUMENTS, Innsbruck, Austria
[†]Daniel Swarovski Research Laboratory, Department of Visceral, Transplant and Thoracic Surgery, Medical University of Innsbruck, Innsbruck, Austria
[1]Corresponding author: e-mail address: erich.gnaiger@oroboros.at

Contents

Abstract

The mitochondrial transmembrane potential ($\Delta\psi_{mt}$ or mtMP) is directly influenced by oxidative phosphorylation (OXPHOS). The exact nature of the interactions between respiration (flux) and mtMP (force) under various physiological and pathological conditions remains unclear, partially due to methodological limitations. Here, we describe a combination of high-resolution respirometry and fluorometry based on the OROBOROS Oxygraph-2k and the widely applied mtMP indicator safranin. The analysis of OXPHOS in mouse brain homogenates revealed that, at commonly applied concentrations, safranin inhibits Complex I-driven OXPHOS capacity, primarily targeting the phosphorylation system, but has no effects on LEAK respiration. Conversely, Complex II-driven OXPHOS capacity was inhibited by <20% by safranin concentrations normally used

Methods in Enzymology, Volume 542
ISSN 0076-6879
http://dx.doi.org/10.1016/B978-0-12-416618-9.00009-1

for mtMP monitoring. The mtMP was higher in the LEAK state without adenylates than at identical LEAK respiration after ADP stimulation and Complex V inhibition with oligomycin. The maximal electron transfer system (ETS) capacity was reached in uncoupler titrations before the mtMP fully collapsed, whereas respiration was inhibited at increasing uncoupler concentrations, resulting in the progressive reduction of mtMP. In a pharmacologically induced state of Complex II dysfunction, mtMP was rather insensitive to the inhibition of OXPHOS to 50% of its normal capacity, but robustly responded to inhibitors when respiration was limited by substrate depletion. The optimal concentration of uncoupler supporting maximal ETS capacity varied as a function of pharmacological intervention. Taken together, the combined measurement of respiration and mtMP greatly enhances the informative potential of OXPHOS studies. The respirometric validation of inhibitory and uncoupling effects is mandatory for any fluorophore employed to assess mtMP in any respiratory state, tissue type, and pathophysiological condition. The methodological issues analyzed herein are relevant for the study of mitochondrial respiration in a wide variety of setting, including cancer cell metabolism.

ABBREVIATIONS

Ama antimycin A (2.5 μ*M*), CIII inhibitor

CI Complex I, respiratory state with CI-linked substrates

CI+II substrate state with a cocktail of CI- and CII-linked substrates, supporting convergent electron transfer into the Q-junction

CII Complex II, respiratory state with CII-linked substrates

D ADP, adenosine diphosphate

Δ*j* flux control factor; Δj_{E-P} is the apparent excess ETS capacity over *P*

ΔpH pH difference across the inner mt-membrane. In the literature on bioenergetics, the term "proton gradient" is frequently used. The confusion of a gradient (a slope or change per unit distance) with a difference (across a membrane with undefined thickness) has important theoretical consequences, including the inappropriate comparison with Ohm's law that relates to a potential gradient rather than a potential difference (Gnaiger, 1989).

ETS, *E* electron transfer system, ETS capacity

FCCP, F carbonylcyanide-*p*-trifluoromethoxyphenyl hydrazone, protonophore dissolved in ethanol

G glutamate (10 m*M*)

LEAK, *L* resting state of respiration compensating for the proton leak, proton slip, and cation cycling

M malate (2 or 0.5 m*M*)

Mna malonic acid (5 m*M*), CII inhibitor

mt mitochondria, mitochondrial

mtMP mitochondrial membrane potential, $\Delta\psi_{mt}$

O2k OROBOROS Oxygraph-2k

OXPHOS, *P* oxidative phosphorylation, OXPHOS capacity, state of coupled respiration at saturating (ADP)

P pyruvate (5 m*M*)

Rot rotenone (0.3 or 0.5 μ*M*), CI inhibitor

ROX residual oxygen consumption, measured after inhibition of respiratory complexes or after substrate deprivation
S succinate (10 m*M*)
Saf safranin
Thom tissue homogenate

1. INTRODUCTION

Aerobic generation of metabolizable energy in the form of ATP is based on the coupling between electron flow through the electron transfer system (ETS) and synthesis of ATP through the F_0F_1-ATP synthase. The mechanistic link between these processes of oxidative phosphorylation (OXPHOS), as understood on the basis of Peter Mitchell's chemiosmotic theory (Mitchell, 1961), is the coupling of electron transfer to proton translocation. Energy [$J = V \cdot C$] is captured as electrochemical proton potential [V] across the inner mitochondrial (mt) membrane, composed of a chemical potential difference related to ΔpH (see list of abbreviations) and an electric potential difference, $\Delta\psi_{mt}$ or mtMP. This composite driving force is utilized to fuel the ATP synthase reaction. The overall consequences of these OXPHOS events are (i) generation of ATP, (ii) compensation of proton and electron leaks, (iii) consumption of oxygen, J_{O_2}, when O_2 is reduced to water in the course of respiration, and (vi) balanced generation (by the ETS) and utilization (by the phosphorylation system) of the chemiosmotic potential difference across the inner mt-membrane. Therefore, a strong link exists between the force of the mtMP and the flux of oxygen consumption. In fact, postulating a high degree of interdependence, mtMP is often used as an indirect readout of mitochondrial metabolic state and activity (Scaduto & Grotyohann, 1999; Solaini, Sgarbi, Lenaz, & Baracca, 2007). However, while the existence of such a link is well established and generally accepted, the interaction between mtMP and oxygen consumption is under debate and far from being fully elucidated. Complicating the matter, both parameters are not exclusively linked with each other but may additionally be governed by a number of independent variables. Since mtMP is an intermediate of electron transfer and ATP synthesis, mtMP depends on and exerts control over these processes (Nicholls, 2004). Taken together, determination of the interaction and interdependence of mtMP and mitochondrial respiration remains a challenging subject of research and application in mitochondrial medicine. Further progress can be made by the simultaneous measurement of both parameters in a multisensor system, allowing an in-depth analysis of functional aspects of mitochondrial energetics.

As experimental model, we used a tissue homogenate (Thom) prepared from mouse brain. Compared to other sample preparations such as tissue slices, effective concentrations of metabolites and various control variables affecting mitochondrial function can be defined more precisely, and the problem of tissue heterogeneity is minimized. Compared to isolated mitochondrial preparations, Thom requires smaller amounts of tissue.

In this chapter, we describe a protocol for application of the widely used fluorescence indicator safranin for estimation of mtMP in combination with high-resolution respirometry (HRR), an experimental approach of OXPHOS analysis for the precise and reproducible assessment of mitochondrial energetic aspects (Gnaiger, 2012). We evaluate some of the limitations of the use of safranin as mtMP indicator, an aspect that can be appreciated only in combination with respiratory measurements, but which would be fully overlooked when the mtMP indicator is applied with fluorometry as a sole tool for evaluation of the mitochondrial energetic status.

2. PREPARATION OF MITOCHONDRIAL THOM

2.1. Animals and tissue preparation

Male C57Bl/6N mice from Charles River Laboratories, Germany, 12–15 weeks old, were kept at 23 °C with 12/12 light/dark cycle with free access to standard laboratory rodent chow and water. After cervical dislocation, the whole brain was dissected from the skull and transferred into a Falcon tube containing ice-cold respiration medium to cover the entire tissue sample. Mitochondrial respiration medium, MiR06Cr (compare Gnaiger et al., 2000), was used throughout the preparation steps and during the subsequent experiments. MiR06Cr contains 110 m*M* sucrose, 20 m*M* HEPES, 10 m*M* KH_2PO_4, 20 m*M* taurine, 60 m*M* K-lactobionate, 3 m*M* $MgCl_2$, 0.5 m*M* EGTA, 1 g l^{-1} BSA, 20 m*M* creatine, 280 U ml^{-1} catalase, pH adjusted to 7.1 with KOH at 30 °C. The osmolarity is 330 mOsm and the ionic strength is 95 m*M* (Fasching, Eigentler, Fontana-Ayoub, & Gnaiger, 2013).

2.2. Homogenate preparation

The brain was placed on a filter paper to remove excess liquid. The wet weight (W_W, 340–430 mg) was determined with a microbalance (Mettler Toledo XS105DU), and the tissue was transferred into a small glass beaker

(10 ml) containing 2.5 ml of ice-cold, fresh MiR06Cr. The tissue was minced into small pieces using a pair of sharp forceps and transferred with the medium into a glass-to-teflon homogenizer (WiseStir® HS-30E; witeg Labortechnik GmbH, Germany). Homogenization was completed with eight full strokes (up/down cycles) by setting the speed of the homogenizer gear to 1000 rpm. The crude homogenate was centrifuged at a low speed of 1300 × *g* for 3 min at 4 °C (Hettich centrifuge, Rotina 380 R). The supernatant (first spin homogenate) was kept on ice until further use.

3. HRR AND DETERMINATION OF mtMP BY SAFRANIN

3.1. The Oxygraph-2k for HRR

The Oxygraph-2k (O2k; OROBOROS INSTRUMENTS, Innsbruck, Austria; Gnaiger, 2008) is a two-chamber multisensor instrument designed for HRR and simultaneous measurement of parameters for comprehensive OXPHOS analysis. In this study, we describe an application with the O2k-Fluorescence LED2-Module (Fasching, Sumbalova, & Gnaiger, 2013; Hickey et al., 2012). The 2.0-ml incubation chambers were kept at constant temperature by Peltier control (±0.001 °C). The medium with sample was stirred using white PVDF-coated stirrers (750 rpm). Changes of oxygen concentration over time are monitored with polarographic oxygen sensors; the negative time derivative is automatically corrected for instrumental background and is displayed in real time as oxygen flux, simultaneously with the fluorescence signals for both chambers. At the beginning of each experiment, the O2k chambers were filled with 2.5 ml of MiR06Cr and equilibrated with air at 37 °C (9.72 μM O_2/kPa). Various standard substrate–uncoupler–inhibitor titration protocols were applied for OXPHOS analysis (Gnaiger, 2012; Pesta & Gnaiger, 2012). Three O2k instruments (six chambers) were operated in parallel.

3.2. A standard safranin experiment

Safranin is a lipophilic cationic dye which accumulates in mitochondria according to the inside negative potential in energized mitochondria. In the course of this process, safranin undergoes a significant change in absorption and self-quenching of its fluorescence (Åkerman & Wikström, 1976; Perevoshchikova, Sorochkina, Zorov, & Antonenko, 2009). Within certain limits of mt-density and concentration of the dye, these spectral changes are linearly related to mtMP (Chinopoulos, Zhang, Thomas, Ten, & Starkov,

2011; Figueira, Melo, Vercesi, & Castilho, 2012; Perevoshchikova et al., 2009; Zanotti & Azzone, 1980). Measurement of alterations in fluorescence is much more sensitive than photometric absorbance measurements. Therefore, fluorometry with safranin is widely used. The results obtained on mtMP with safranin apply for any fluorescence cuvette or photometric device that might be used for this purpose.

The Fluorescence-Sensor Blue of the O2k-Fluorescence LED2 module (OROBOROS INSTRUMENTS, Austria) is equipped with a filter set for safranin (excitation at 495 mm and emission at 587 nm). The sensors are inserted through the front windows of the glass chambers and connected to the fluorescence control unit. Following air equilibration, the chambers were closed with black PEEK stoppers and black cover lids to block external light and confine the chamber volume to exactly 2.0 ml. 60–75 μl of the first spin homogenate were injected into the 2-ml O2k chamber, corresponding to a final CS activity (measured at 30 °C) of 1.9 ± 0.3 nmol s^{-1} ml^{-1} (mean ± SD for all experiments). For calibration, a 200 μ*M* stock solution of safranin dissolved in distilled water was titrated in five steps into the O2k chamber, up to a final safranin concentration of 2 μ*M*. A linear increase in the fluorescence signal was detected, reflecting the concentration of safranin in the chamber (Fig. 9.1). In the absence of respiratory inhibitors, a mtMP builds up on the basis of endogenous substrates, a corresponding amount of the dye accumulates in the mt matrix space (decline of fluorescence signal after initial maximal signals), and a fraction is nonspecifically bound to the mitochondria, such that the free safranin concentration is lower than the total safranin concentration added to the chamber.

Pyruvate, glutamate, and malate (PGM) were added as substrates of dehydrogenases generating NADH and transferring electrons to Complex I (CI). The fluorescence signal showed a sharp decrease, reflecting an increase of mtMP. This was followed by a progressive increase of fluorescence indicating a slow decrease of mtMP. At this point, ADP was added at high concentration (0.25 m*M*) to activate OXPHOS (State 3; Chance & Williams, 1955). This was associated with a decrease of mtMP (increase of fluorescence) due to the fact that protons are shuttled through the ATP synthase to drive ATP production. Next, the protonophore FCCP was titrated in 0.5 μ*M* steps to induce a progressive decrease of mtMP by uncoupling. Finally, 0.3 μ*M* rotenone (Rot; the previously published concentration is 0.5 μ*M*; Pesta & Gnaiger, 2012), 5 m*M* malonic acid (Mna), and 2.5 μ*M* antimycin A (Ama) were added to inhibit CI, CII, and CIII, respectively, leading to no further alterations of mtMP. A slight decay of the fluorescence signal over time could be observed (Fig. 9.1).

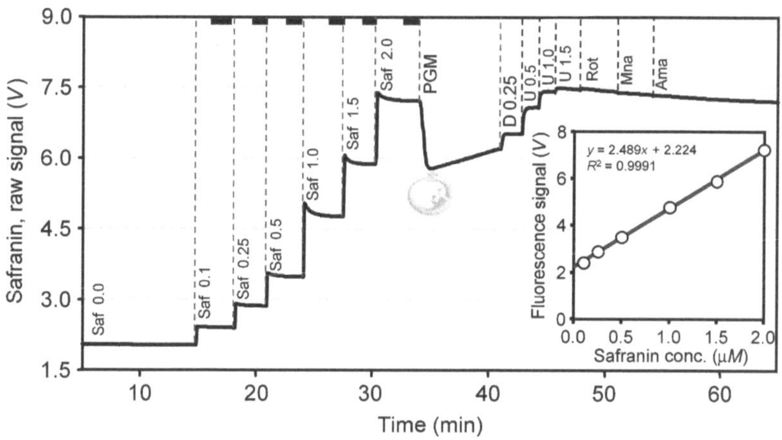

Figure 9.1 *Determination of changes of mitochondrial membrane potential with safranin.* Mouse brain homogenate (first spin) was incubated in the 2-ml chamber of the Oroboros Oxygraph-2k (O2k) with attached O2k-Fluorescence LED2-Module and Fluorescence-Sensor Blue with the filter set for safranin (excitation at 495 mm and emission at 587 nm). Safranin was titrated from 0.1 to 2 μ*M* final concentration (Saf 0.1–Saf 2.0). Horizontal bars indicate the marked sections selected for calibration by linear regression using the DatLab software (inset). Complex I-linked substrates pyruvate, glutamate, and malate (PGM; 5, 10, and 2 m*M*, respectively) were added (LEAK state without adenylates), followed by addition of 0.25 m*M* ADP (D0.25). Uncoupler titrations (0.5–1.5 μ*M* FCCP) diminished the membrane potential gradually up to a maximum fluorescence signal, without further increase after addition of inhibitors of Complex I (rotenone; Rot, 0.3 μ*M*), Complex II (malonate; Mna, 5 m*M*), and Complex III (antimycin A; Ama, 2.5 μ*M*). Exp. 2013-02-05_D-04.

In the absence of additional information, the interpretation of these data would apparently be that the mtMP shows a rather minor change upon transition from the LEAK state to the ADP-activated state, whereas there was a comparatively large decrease of mtMP upon uncoupling mitochondria. The cause for the progressive decrease of mtMP in the LEAK state (after addition of substrate in the absence of adenylates) is not immediately obvious but might be due to an increasing proton leak over time.

4. LIMITATIONS OF SAFRANIN AS mtMP INDICATOR

Soon after the introduction of safranin as mtMP indicator (Åkerman & Wikström, 1976), concerns were raised regarding possible side effects of the dye. A negative impact was noted of safranin above 40 μ*M* on ADP- and FCCP-stimulated respiration (Zanotti & Azzone, 1980) and at 10 μ*M* on mitochondrial calcium cycling (Harris & Baum, 1980; Valle, Pereira-da-Silva,

& Vercesi, 1986). Thus, researchers using safranin in fluorescence measurements at 5 μ*M* must have felt on the safe side. It is important to note that the concentration of safranin cannot be arbitrarily lowered to circumvent any side effects: first, at very low concentration, the signal-to-noise ratio becomes too low to obtain reliable data. Second, at very low concentrations, safranin does not adequately report mtMP. As shown by Perevoshchikova et al. (2009), the fraction of safranin nonspecifically bound amounts to 33% at 1 μ*M* safranin and to 5% at 10 μ*M* in deenergized mitochondria. The fraction of safranin accumulated in energized mitochondria is 73% and 86% at 1 and 10 μ*M* safranin, respectively. Further, at low concentrations of the dye, an increase of mtMP may be detectable as an increase of the signal due to enhanced fluorescence of safranin molecules in the hydrophobic milieu of the membrane, opposing the generally observed negative correlation of fluorescence output and mtMP due to signal quenching upon accumulation in the inner mt-membrane and matrix at higher potential.

Safranin is nonetheless widely applied in mitochondrial research. It is important, therefore, to analyze in more detail the inhibitory side effects of safranin and evaluate adequate experimental conditions for its application as a reporter of mtMP.

5. RESPIROMETRY WITH HOMOGENATE

We analyzed the complex mechanisms that induced the pattern of mtMP observed in Fig. 9.1 by determining oxygen consumption under identical conditions. In a respiratory control experiment in the absence of safranin (carrier; Fig. 9.2A), a very low oxygen flux was observed with endogenous substrates, corresponding to the classical State 1 (Chance & Williams, 1955). Addition of PGM as substrates supporting Complex I-linked respiration induced a sharp increase to the steady state of LEAK respiration (CI_L). 0.25 m*M* ADP exerted a strong stimulation of respiration, but maximum OXPHOS capacity was reached only at a higher ADP concentration of 2 m*M* (D2.0; after reoxygenation). The optimum uncoupler concentration for maximizing oxygen flux was 1.0 μ*M* FCCP. Since this condition is considered to reflect the maximum capacity for electron transfer thought the ETS, this is referred to as the ETS state (Gnaiger, 2009, 2012). Residual oxygen consumption (ROX) after inhibiting Complexes CI, CII, and CIII with rotenone, malonate, and Ama was practically zero.

Samples from the same homogenate were examined in parallel experiments with 2 μ*M* safranin applied before addition of substrate (Fig. 9.2B). LEAK respiration was identical to controls, but after addition of ADP, only

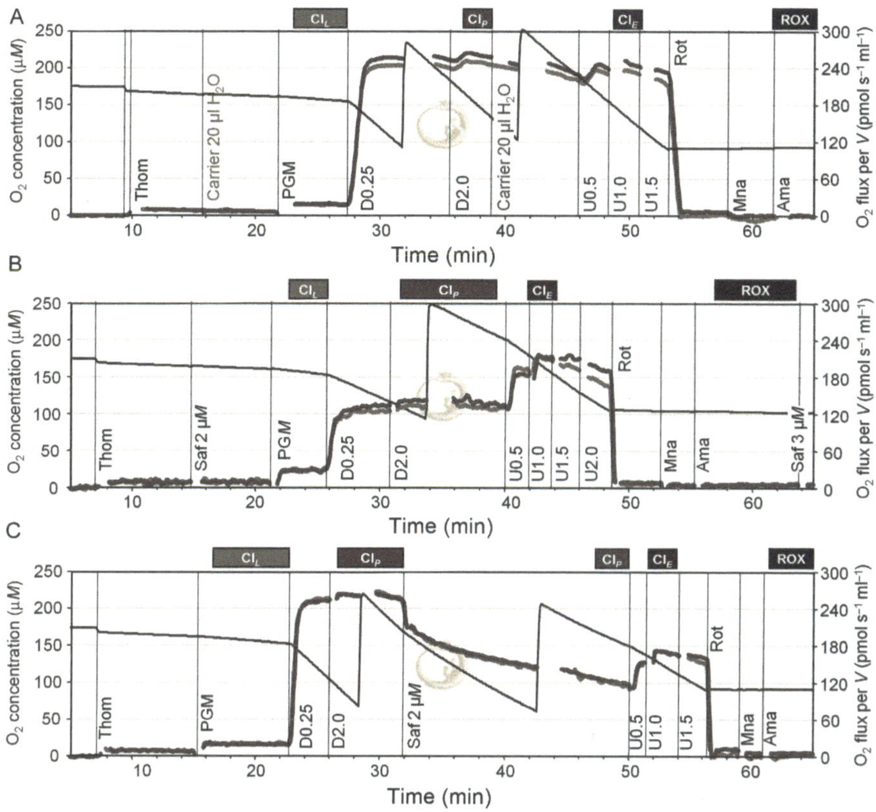

Figure 9.2 *High-resolution respirometry with mouse brain homogenate and substrate–uncoupler–inhibitor titration (SUIT) protocol with concentrations as described in Fig. 9.1.* Oxygen concentration (blue lines; left axes [μ*M*]) is shown for one chamber, whereas plots of oxygen flux per chamber volume (red and green lines, right axes [pmol s^{-1} ml^{-1}]) are superimposed for the two chambers of the O2k operated simultaneously. The plots of oxygen flux are interrupted by disturbances due to titrations or reoxygenations (3 μl of a 200 m*M* stock solution of H_2O_2). (A) Control experiment without safranin, with titration of carrier (20 μl H_2O) at different time points for the two chambers (red or green). (B) Addition of 2 μ*M* safranin into both chambers before substrate titration. (C) Addition of 2 μ*M* safranin into both chambers after inducing Complex I-linked OXPHOS with PGM and 2 m*M* ADP. Thom denotes addition of tissue homogenate (first spin), and bars above each panel denote CI-linked LEAK state (CI_L), OXPHOS state (CI_P), ETS state (CI_E), and residual oxygen consumption (ROX). Exp. 2013-02-13_KL-02, 2013-02-13_AB-02, 2013-02-13_CD-02. (See the color plate.)

about half the OXPHOS capacity was obtained. Whereas FCCP exerted a minor stimulatory effect in the controls (Fig. 9.2B), the apparent ETS excess capacity, expressed as $E-P$ flux control factor, $\Delta j_{E-P}=(E-P)/E$, was as high as 0.35 in the presence of safranin, compared to nearly zero in the controls (compare P/E ratio; Gnaiger, 2009). Since ETS capacity was significantly reduced by safranin, it obviously exerts an inhibitory effect on electron transfer, but the specific inhibition of the phosphorylation system (ATP synthase, ANT, and inorganic phosphate transporter) was much more pronounced. When safranin was added in the CI-linked OXPHOS state, a strong time dependence was observed of the inhibitory effect (Fig. 9.2C). Safranin exerted a dose-dependent inhibitory effect, resulting in about 70% inhibition of CI-linked respiration at 4 μM, a concentration close to the widely used concentration of 5 μM (Fig. 9.3A). Since sequential titrations were performed, a time effect on progressive inhibition of OXPHOS (Fig. 9.2C) is superimposed on the concentration effect.

Results with 2 μM safranin added either before (Fig. 9.2B) or after respiratory substrates and ADP (Fig. 9.2C) are summarized in Fig. 9.3B and C. CI-linked OXPHOS capacity was inhibited by approximately 50% relative to controls, whereas ETS capacity was diminished to a lesser extent by 15–35%. Therefore, the apparent excess ETS capacity, Δj_{E-P}, increased to 0.36 ± 0.03 and 0.31 ± 0.01 as compared to 0.00 ± 0.02 in controls. Δj_{E-P} is an expression of the limitation of OXPHOS capacity imposed by the phosphorylation system, the increase of which indicates that safranin acts primarily as an inhibitor of the phosphorylation system and exerts a relatively smaller inhibitory effect on ETS capacity.

6. OPTIMIZATION OF CONDITIONS FOR USE OF SAFRANIN

The apparent excess ETS capacity of zero observed in the mouse brain homogenate corresponds to the theoretical lower limit of Δj_{E-P}. It should not be regarded as a measure of the quality of a mitochondrial preparation, as it varies as an intrinsic characteristic of a cell or tissue type examined in a specific substrate state (Gnaiger, 2009), amounting to, for example, 0.15 and 0.04 in muscle tissue from nontrained and trained human individuals, respectively (Pesta et al., 2011), and is as high as 0.52 in human heart measured in the presence of a CI+II substrate combination (Lemieux, Semsroth, Antretter, Höfer, & Gnaiger, 2011). Three coupling sites (CI, CIII, and CIV) are involved in respiration with CI-linked substrates, whereas only two coupling sites are active in the pathway of electron transfer from succinate to oxygen. Hence, for a given oxygen flux in the OXPHOS state, the phosphorylation system is less active (the ATP/O_2 ratio is lower) and thus limits OXPHOS to a lesser extent with CII compared to CI-linked substrates. In addition, CII-linked OXPHOS capacity in mouse heart

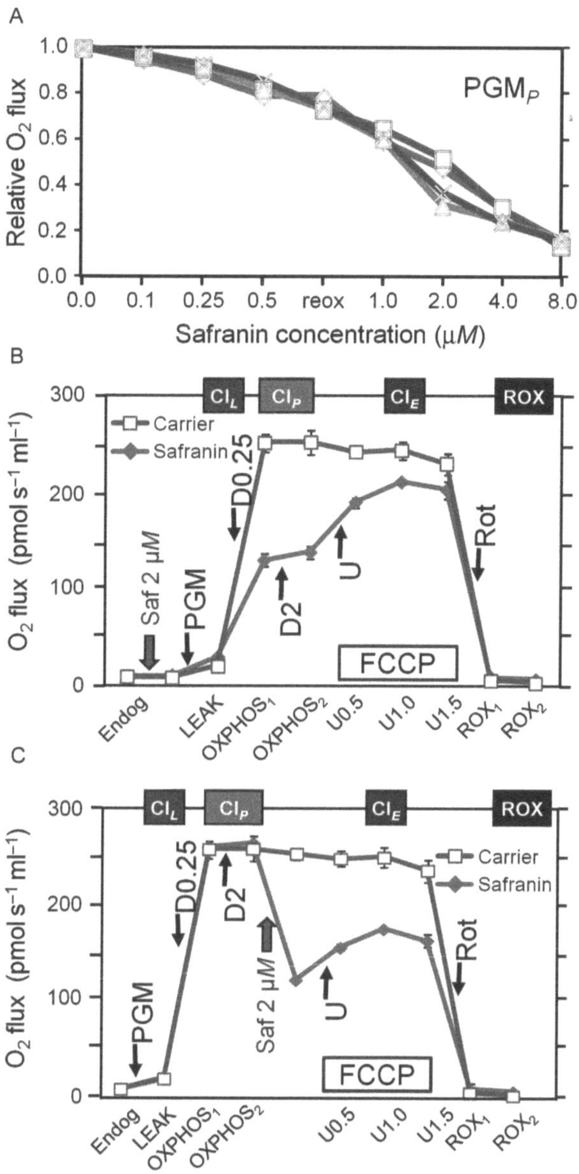

Figure 9.3 *Inhibition of CI-linked respiration by safranin (PGM, pyruvate + glutamate + malate) in mouse brain homogenate.* (A) Oxygen consumption rate as a function of safranin concentration (four assays), normalized to maximum OXPHOS (PGM_P). (B, C) Addition of safranin (Saf 2 μ*M*) before or after addition of substrates and ADP. Endog, respiration supported by endogenous substrates; LEAK, respiration supported by PGM without adenylates; $OXPHOS_1$ and $OXPHOS_2$, *P* at 0.25 and 2 m*M* ADP; U0.5, U1.0, U1.5, uncoupled respiration at 0.5, 1, and 1.5 μ*M* FCCP, without effect on carrier controls, but stimulation in the presence of safranin, indicating that the phosphorylation system was more strongly inhibited than ETS; ROX, residual oxygen consumption after inhibition by rotenone (Rot; ROX_1) and malonic acid and antimycin A (ROX_2). Data are means ± SD of four measurements on samples from two independent preparations. (See the color plate.)

mitochondria is lower than CI-linked OXPHOS capacity (Fasching, Sumbalova, et al., 2013). Any inhibition by safranin of the phosphorylation system and of respiratory complexes downstream of CI and CII, therefore, will exert a much smaller effect on CII-linked OXPHOS. Consequently, we tested respiration with succinate as a potentially superior substrate state for the application of safranin. An experimental example with safranin and carrier titration is shown in Fig. 9.4A and B.

After injection of first spin homogenate to the O2k-chamber, 0.3 μ*M* Rot was added followed by 10 m*M* succinate, establishing the CII-linked LEAK state. The OXPHOS state was induced by ADP at saturating concentration (D2.5; Fig. 9.4A and B). Safranin was titrated from 0.5 μ*M* to a final concentration of 8 μ*M*. FCCP was titrated to maximize respiration in the ETS state, and Mna + Ama were added to evaluate ROX. Figure 9.4C summarizes the results normalized to initial OXPHOS capacity. As expected, safranin exerted a much smaller inhibitory effect on OXPHOS in the CII compared to CI substrate state. In the concentration range between 0.25 and 2 μ*M*, safranin exerted no significant or only a small effect on OXPHOS and ETS capacity (Fig. 9.4D). In combination with fluorometry, however, safranin is typically added at the beginning of an experiment. Under these conditions, 2 μ*M* safranin showed an apparent inhibitory effect on respiration supported by succinate. After correction for ROX (Mna + Ama), safranin inhibition factors were not significant in the LEAK state (L_N and L_{Omy}), but were 0.18 and 0.16 for OXPHOS and ETS (Fig. 9.5). After inhibition of ATP synthase by oligomycin (Omy), LEAK respiration was identical to LEAK measured in the absence of adenylates, but uncoupler titration did not fully restore ETS to the level of OXPHOS capacity, which is frequently observed after oligomycin inhibition (Fig. 9.5).

In summary, these experiments indicate that safranin does not disturb mitochondrial respiration at concentrations up to 2 μ*M* in homogenate preparations incubated with Rot and succinate to an extent which excludes incubation with safranin. An application of combined fluorometry and HRR is presented in Fig. 9.6. After establishment of a baseline fluorescence signal in the presence of first spin homogenate, a near-complete release of the dye from the mitochondria was induced by addition of rotenone (which reduced ROX based on endogenous substrates), signifying mitochondrial depolarization. Subsequent addition of succinate caused an increase of respiration to succinate-supported LEAK and a concurrent repolarization of a maximum mtMP. Injecting ADP then induced a threefold increase of respiration to OXPHOS as well as a major depolarization of mtMP. Inhibition

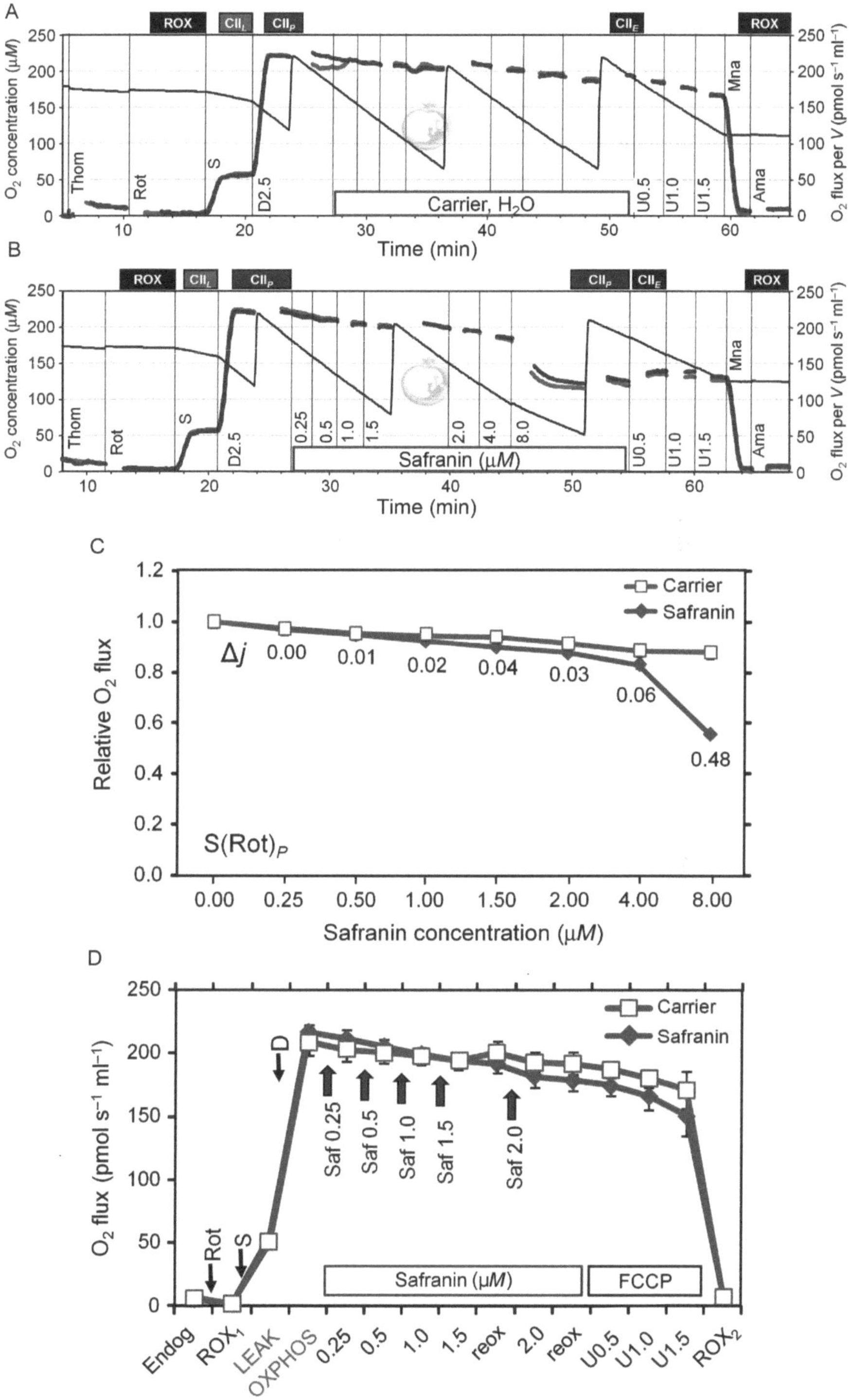

Figure 9.4 See figure caption on next page.

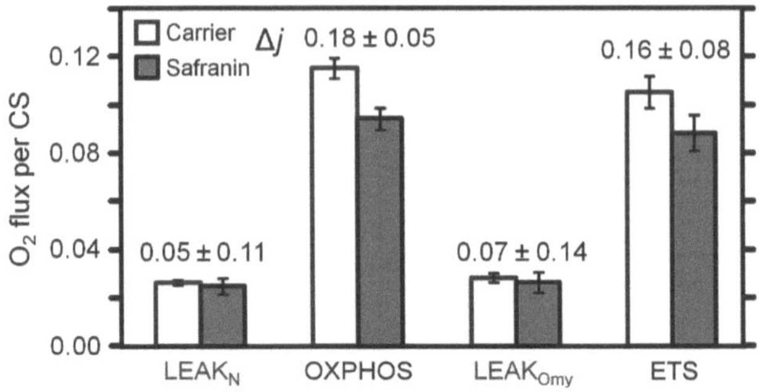

Figure 9.5 *Inhibition of Complex II-linked respiration by safranin (2 μM) in mouse brain homogenate.* Safranin or carrier was titrated before addition of homogenate to the incubation medium. Respiration is normalized for citrate synthase activity, CS, and corrected for residual oxygen consumption, ROX (Mna+Ama), in the presence of succinate ($LEAK_N$), ADP (OXPHOS), oligomycin ($LEAK_{Omy}$), and FCCP at 0.75 μ*M* (ETS). The inhibitory effect of 2 μ*M* safranin is expressed as the flux control factor $\Delta j = (\text{Con} - \text{Saf})/\text{Con}$. Data are means ± SD of eight measurements on samples from three independent preparations.

of ATP synthase by Omy caused a return of respiration to the LEAK state, L_{Omy}, with nearly identical oxygen flux compared to the LEAK state in the absence of external adenylates, L_{N}. This provides a test that both L_{N} and L_{Omy} are valid states reflecting LEAK respiration. In contrast, mtMP at L_{Omy} returned to a value close to but clearly above the previous L_{N} state. This suggests that with regard to mtMP, these LEAK states are not metabolically equivalent, an observation that is not readily explicable on the basis of our current knowledge. Finally, uncoupler titrations showed that ETS capacity was obtained when mtMP was reduced slightly below the potential observed

Figure 9.4—Cont'd *Inhibition of Complex II-linked respiration by safranin in mouse brain homogenate.* Safranin or carrier was added in the OXPHOS state, CII_P. (A, B) Oxygen concentration (blue lines, one chamber) and plots of oxygen flux superimposed for two chambers, with titration of carrier or safranin from 0.25 to 8 μ*M*. Thom, first spin tissue homogenate; Rot, rotenone; S, succinate; D2.5, 2.5 m*M* ADP. Bars above each panel denote CII-linked LEAK state (CII_L), OXPHOS state (CII_P), ETS state (CII_E), and residual oxygen consumption (ROX). (C) Respiration rate normalized to the OXPHOS state; the inhibitory effect is shown as $\Delta j = (\text{Carrier–Saf})/\text{Carrier}$, becoming significant at 4 μ*M* safranin. Data are means ± SD of 10 measurements on samples from four independent preparations. (D) Respiratory flux and safranin titration from 0.25 to 2.0 μ*M*; ROX_1, LEAK, and OXPHOS denote oxygen consumption after addition of rotenone, succinate, and 2.5 m*M* ADP, respectively. Data are means ± SD of three measurements on samples from two independent preparations. Exp. 2013-06-25_MN-02 (A) and 2013-06-25_OP-02 (B). (See the color plate.)

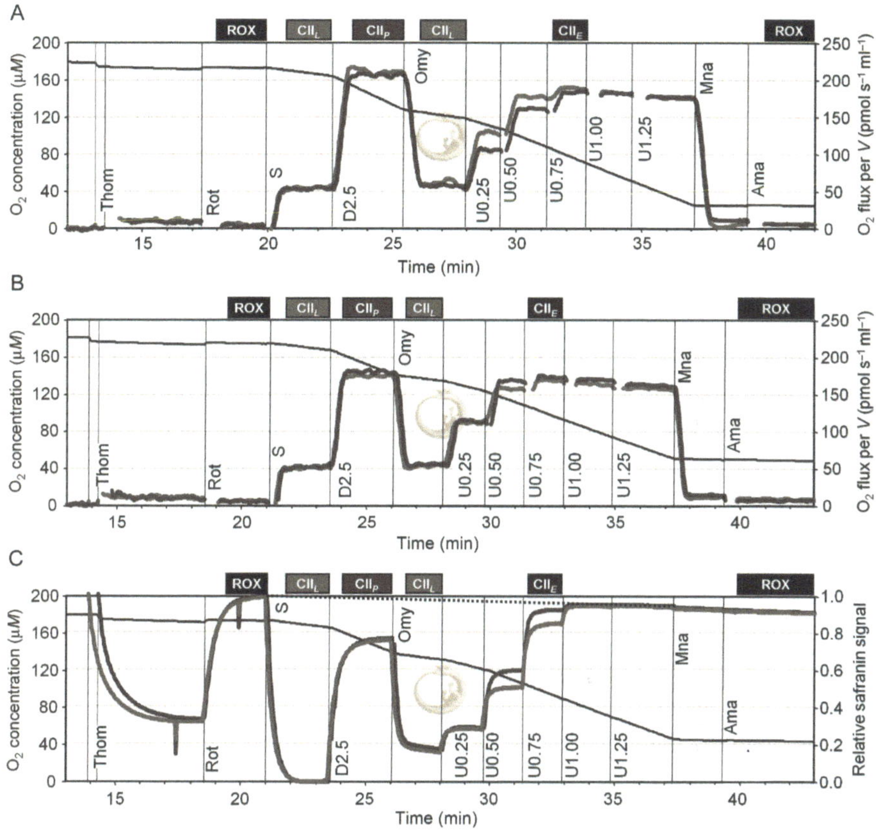

Figure 9.6 *Combined determination of oxygen consumption and mitochondrial membrane potential in the O2k high-resolution respirometer equipped with the O2k-Fluorescence LED2 module.* Oxygen concentration (blue plots, one chamber) and superimposed plots of oxygen flux or fluorescence (two chambers). Safranin was titrated before addition of first spin mouse brain homogenate (Thom). (A) Respiration without safranin. (B, C) Respiration and fluorescence changes with 2 μ*M* safranin. Rot, rotenone inhibiting endogenous respiration to the level of residual oxygen consumption at minimum mtMP (ROX; maximum fluorescence signal); S, succinate-stimulating LEAK respiration and maximizing mtMP in the absence of adenylates (CII_L); D2.5, 2.5 m*M* ADP-activating OXPHOS and reducing mtMP (CII_P); Omy, oligomycin inducing a LEAK state by inhibition of ATP synthase (CII_L); U0.25–U1.25, uncoupler titration from 0.25 to 1.25 μ*M* FCCP; optimum FCCP concentration for ETS capacity at 0.75 μ*M* (CII_E); Mna and Ama, malonic acid and antimycin A inhibit CII and CIII and reduce respiration to the level of residual oxygen consumption (ROX). The fluorescence signal shows an apparent drift in the uncoupled and inhibited state, which is extrapolated to the initial ROX state (dotted line). Exp. 2013-06-25_MN-04, 2013-06-25_OP-04. (See the color plate.)

in the OXPHOS state and was further depolarized when FCCP was increased to inhibitory concentrations. Addition of Mna and Ama reduced respiration to ROX but had no further impact on mtMP.

Taken together, these findings show that under appropriately selected conditions, safranin can be used as mtMP indicator. Validation of inhibition by combination with HRR offers a novel quantitative approach to the study of mtMP.

7. AN APPLICATION EXAMPLE

A set of experiments is presented to illustrate potential applications of combined HRR and fluorometry, simulating a pathological state where OXPHOS capacity is seriously compromised, for example, in brain tissue from a mouse Alzheimer model (Rhein et al., 2009). To this end, we used Mna at approximately 50% inhibition of CII-linked OXPHOS and determined how this would affect mtMP in relation to respirometric parameters.

The experiment depicted in Fig. 9.7 indicates that 0.2 m*M* Mna, as per experimental design, inhibited OXPHOS and ETS by approximately 50%. Whereas respiration on endogenous substrates and with succinate in the absence of ADP (CII_L) was not measurably decreased by this Mna concentration, a significant depolarization was observed of the mtMP upon addition of 0.2 m*M* Mna. If the fluorescence signal is normalized between its maximum in the CII_L state and the initial ROX state when respiratory fluxes are identical in the presence and absence of the inhibitor (Fig. 9.7), then the corresponding effect on the relative safranin signal is eliminated by the normalization procedure (Fig. 9.7B). The scope of inhibition of substrate-limited respiration is small, thus rotenone reduced respiration slightly to the level of ROX, whereas mtMP declined substantially. At identical LEAK respiration, the maximum mtMP in state CII_L was less in the partially Mna-inhibited state (Fig. 9.7C), using the fluorescence signal calibrated before addition of homogenate (compare Fig. 9.1). The strong inhibition of OXPHOS by 0.2 m*M* Mna exerted a minor effect on mtMP in the ADP-stimulated state (CII_P; Fig. 9.7C). However, the scope of depressing mtMP in the OXPHOS state is diminished, whereas the scope of inhibition of active respiration (*P*) is high. The sensitivity of mtMP or respiration to a defect, therefore, has to be interpreted relative to the actual range available for a response in any physiological state. Importantly, the lower range of mtMP can be interrogated with safranin at high resolution, whereas the

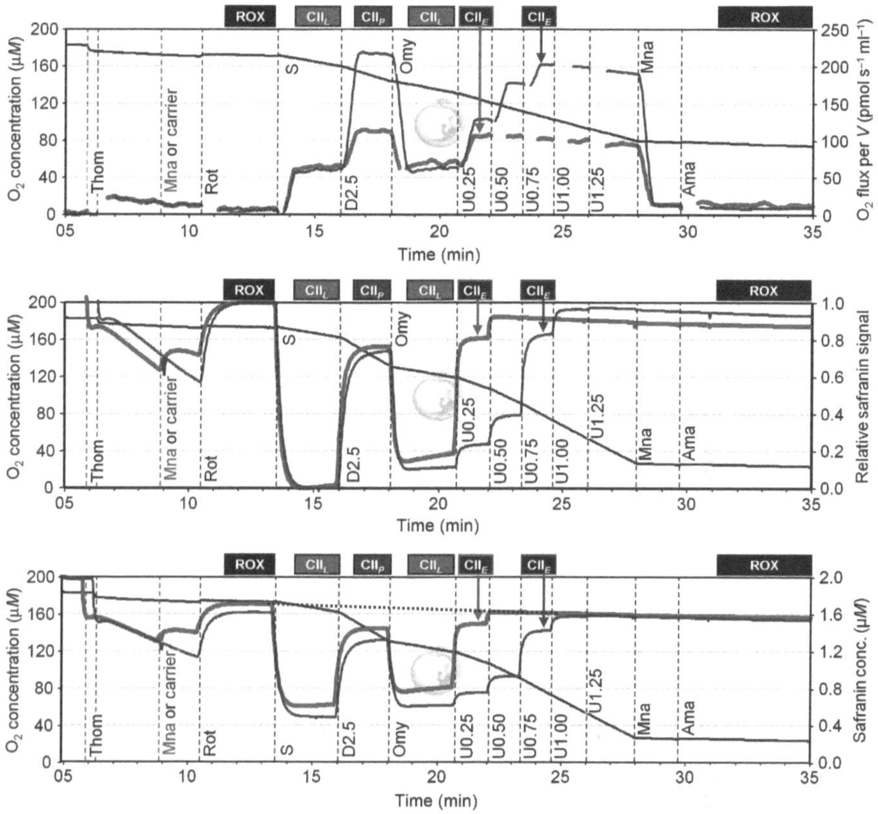

Figure 9.7 *Combined HRR and O2k-fluorometry in a model for a pathological condition compared to controls.* Oxygen concentration (blue plots, one chamber) and superimposed plots of oxygen flux or fluorescence (two chambers). Safranin was titrated before addition of first spin mouse brain homogenate (Thom). Carrier (H_2O) or 0.2 m*M* Mna (inhibiting CII-linked OXPHOS by 50%) was added before starting the SUIT protocol (see Fig. 9.6 for details). (A) Superimposed traces of oxygen concentration and fluxes of controls and Mna-treated cells. (B) Normalized fluorescence signals in controls and Mna-treated cells in the range between minimum and maximum values observed in the ROX and CII_L state. (C) Safranin concentration calibrated before addition of tissue homogenate. The fluorescence signal shows an apparent drift in the uncoupled and inhibited state, which is extrapolated to the initial ROX state (dotted line). Exp. 2013-08-09 MN-05, 2013-08-09 OP-04. (See the color plate.)

potentiometric measurement of mtMP with TPP^+ or $TPMP^+$ is intrinsically insensitive to changes of mtMP below 80–100 mV (Brand, 1995).

Inhibition of succinate dehydrogenase caused a dramatic increase of the sensitivity of respiration and mtMP to FCCP, which induced maximum ETS capacity at 0.25 μ*M* in Mna-treated samples versus 0.75 μ*M* in controls.

This illustrates the need for uncoupler titrations to evaluate optimum concentrations in each specific pathophysiological or pharmacological state. In addition, these experiments show that respiration and mtMP respond differently with high sensitivity to a mitochondrial defect and are therefore complementary diagnostic parameters in comprehensive OXPHOS analysis.

8. CONCLUSIONS

A respirometric validation of safranin revealed that this fluorophore cannot be used as an indicator of mtMP at high concentrations often applied previously, since it exerts a dose-dependent inhibitory effect on OXPHOS capacity which is particularly apparent when CI-linked respiration is examined. This also means that conclusions derived from studies using safranin only may be quite misleading, since such side effects of the dye remain undetectable if it is used without quality control by respirometry. As a way to circumvent this problem, we established conditions where the inhibitory effect of safranin on mitochondrial respiration was minimized, that is, in the concurrent presence of the CI inhibitor rotenone and the CII-linked substrate succinate, at a safranin concentration of 2 μM. In this setting, it is possible to conduct meaningful experiments that would, for example, allow the evaluation of pathological conditions in terms of their impact on mitochondrial functional states. However, as a cautionary final remark, it should be stated that for the use of safranin in other experimental models, may they be homogenates derived from other tissues or species or other types of mitochondrial preparations, we recommend re-evaluation of the exact experimental conditions.

REFERENCES

Åkerman, K. E., & Wikström, M. K. (1976). Safranine as a probe of the mitochondrial membrane potential. *FEBS Letters*, *68*, 191–197.

Brand, M. D. (1995). Measurement of mitochondrial protonmotive force. In G. C. Brown, & C. E. Cooper (Eds.), *Bioenergetics—A practical approach* (pp. 39–62). Oxford: Oxford University Press.

Chance, B., & Williams, G. R. (1955). Respiratory enzymes in oxidative phosphorylation: III. The steady state. *The Journal of Biological Chemistry*, *217*, 409–427.

Chinopoulos, C., Zhang, S. F., Thomas, B., Ten, V., & Starkov, A. A. (2011). Isolation and functional assessment of mitochondria from small amounts of mouse brain tissue. *Methods in Molecular Biology*, *793*, 311–324.

Fasching, M., Eigentler, A., Fontana-Ayoub, M., & Gnaiger, E. (2013). Mitochondrial respiration medium—MiR06. *Mitochondrial Physiology Network*, *14.13*, 1–4.

Fasching, M., Sumbalova, Z., & Gnaiger, E. (2013). O2k-fluorometry: HRR and H_2O_2 production in mouse brain mitochondria. *Mitochondrial Physiology Network*, *17.17*, 1–4.

Figueira, T. R., Melo, D. R., Vercesi, A. E., & Castilho, R. F. (2012). Safranine as a fluorescent probe for the evaluation of mitochondrial membrane potential in isolated organelles and permeabilized cells. *Methods in Molecular Biology, 810*, 103–117.

Gnaiger, E. (1989). Mitochondrial respiratory control: Energetics, kinetics and efficiency. In W. Wieser, & E. Gnaiger (Eds.), *Energy transformations in cells and organisms* (pp. 6–17). Stuttgart: Thieme.

Gnaiger, E. (2008). Polarographic oxygen sensors, the oxygraph and high-resolution respirometry to assess mitochondrial function. In J. A. Dykens & Y. Will (Eds.), *Mitochondrial dysfunction in drug-induced toxicity* (pp. 327–352). John Wiley.

Gnaiger, E. (2009). Capacity of oxidative phosphorylation in human skeletal muscle. New perspectives of mitochondrial physiology. *International Journal of Biochemistry and Cell Biology, 41*, 1837–1845.

Gnaiger, E. (2012). *Mitochondrial pathways and respiratory control.* An introduction to OXPHOS analysis. Mitochondrial physiology network. *Vol. 17.18* (pp. 64) (3rd ed.). Innsbruck: OROBOROS MiPNet Publications.

Gnaiger, E., Kuznetsov, A. V., Schneeberger, S., Seiler, R., Brandacher, G., Steurer, W., et al. (2000). Mitochondria in the cold. In G. Heldmaier, & M. Klingenspor (Eds.), *Life in the cold* (pp. 431–442). Heidelberg, Berlin, New York: Springer.

Harris, E. J., & Baum, H. (1980). Uptake of safranine by cardiac mitochondria. Competition with calcium ions and dependence on anions. *Biochemistry Journal, 192*, 551–557.

Hickey, A. J., Renshaw, G. M., Speers-Roesch, B., Richards, J. G., Wang, Y., Farrell, A. P., et al. (2012). A radical approach to beating hypoxia: Depressed free radical release from heart fibres of the hypoxia-tolerant epaulette shark (*Hemiscyllum ocellatum*). *Journal of Comparative Physiology B, 182*, 91–100.

Lemieux, H., Semsroth, S., Antretter, H., Höfer, D., & Gnaiger, E. (2011). Mitochondrial respiratory control and early defects of oxidative phosphorylation in the failing human heart. *International Journal of Biochemistry and Cell Biology, 43*, 1729–1738.

Mitchell, P. (1961). Coupling of phosphorylation to electron and hydrogen transfer by a chemi-osmotic type of mechanism. *Nature, 191*, 144–148.

Nicholls, D. G. (2004). Mitochondrial membrane potential and aging. *Aging Cell, 3*, 35–40.

Perevoshchikova, I. V., Sorochkina, A. I., Zorov, D. B., & Antonenko, Y. N. (2009). Safranine O as a fluorescent probe for mitochondrial membrane potential studied on the single particle level and in suspension. *Biochemistry (Mosc), 74*, 663–671.

Pesta, D., & Gnaiger, E. (2012). High-resolution respirometry. OXPHOS protocols for human cells and permeabilized fibres from small biopisies of human muscle. *Methods in Molecular Biology, 810*, 25–58.

Pesta, D., Hoppel, F., Macek, C., Messner, H., Faulhaber, M., Kobel, C., et al. (2011). Similar qualitative and quantitative changes of mitochondrial respiration following strength and endurance training in normoxia and hypoxia in sedentary humans. *American Journal of Physiology. Regulatory, Integrative and Comparative Physiology, 301*, R1078–R1087.

Rhein, V., Song, X., Wiesner, A., Ittner, L. M., Baysang, G., Meier, F., et al. (2009). Amyloid-beta and tau synergistically impair the oxidative phosphorylation system in triple transgenic Alzheimer's disease mice. *Proceedings of the National Academy of Sciences of the United States of America, 106*, 20057–20062.

Scaduto, R. C., Jr., & Grotyohann, L. W. (1999). Measurement of mitochondrial membrane potential using fluorescent rhodamine derivatives. *Biophysical Journal, 76*, 469–477.

Solaini, G., Sgarbi, G., Lenaz, G., & Baracca, A. (2007). Evaluating mitochondrial membrane potential in cells. *Bioscience Reports, 27*, 11–21.

Valle, V. G., Pereira-da-Silva, L., & Vercesi, A. E. (1986). Undesirable feature of safranine as a probe for mitochondrial membrane potential. *Biochemical and Biophysical Research Communications, 135*, 189–195.

Zanotti, A., & Azzone, G. F. (1980). Safranine as membrane potential probe in rat liver mitochondria. *Archives of Biochemistry and Biophysics, 201*, 255–265.

CHAPTER TEN

Kinetic Analysis of Local Oxygenation and Respiratory Responses of Mammalian Cells Using Intracellular Oxygen-Sensitive Probes and Time-Resolved Fluorometry

Alexander V. Zhdanov[1], Ruslan I. Dmitriev[1], James Hynes, Dmitri B. Papkovsky[2]

Biochemistry Department, University College Cork, Cork, Ireland

[1]These authors contributed equally to this work.

[2]Corresponding author: e-mail address: d.papkovsky@ucc.ie

Contents

Abstract

The cell-penetrating O_2-sensing probes based on phosphorescent Pt-porphyrins provide a means to accurately measure O_2 within cultured cells in a contact-less fashion. Via this approach, *in situ* oxygenation and respiratory responses can be monitored in real time on large number of cellular samples seeded in standard microtiter plates, by means of a commercial time-resolved fluorescence reader. Using neuronal PC12 cells as a model, here we describe the general strategy behind the use of phosphorescent Pt-porphyrins to measure O_2 within cultured cells as well as detailed protocols to implement this experimental approach, including cell preparation and probe loading, instrument and plate setup, system calibration, and data processing. Two main protocols are presented: (1) a basic setup and optimization for the measurement of intracellular O_2 under ambient O_2 concentration; (2) the calibration of the system required to perform measurements under low O_2 concentrations (hypoxia). A case study on the effects of bafilomycin A1 on PC12 cell function is included, illustrating how these methods can be combined with conventional bioenergetic markers to delineate how chemical and/or physical stimuli affect the metabolism of normal and malignant cells.

Methods in Enzymology, Volume 542
ISSN 0076-6879
http://dx.doi.org/10.1016/B978-0-12-416618-9.00010-8

ABBREVIATIONS

AntA antimycin A
Baf bafilomycin A1
$_c$Ca^{2+} cytosolic calcium
DiSBAC$_2$(3) bis-(1,3-diethylthiobarbituric acid) trimethine oxonol probe
DMSO dimethyl sulfoxide
dPC12 differentiated PC12 cells
FCCP carbonyl cyanide 4-(trifluorometoxy)phenylhydrazone
iO$_2$ intracellular O$_2$
$_m$Ca^{2+} mitochondrial calcium
NGF nerve growth factor
TMRM tetramethyl rhodamine methyl ester probe
TR-F time-resolved fluorescence
Val valinomycin
$\Delta\Psi_m$ mitochondrial membrane potential
$\Delta\Psi_p$ plasma membrane potential
τ probe phosphorescence lifetime

1. INTRODUCTION

Molecular oxygen (O_2) is one of the key substrates of the cell and oxidative phosphorylation machinery which, along with the other pathways, is used by the cell to generate energy in the form of ATP (Lenaz & Genova, 2009). Ambient air contains 20.86% of O_2 which corresponds to partial pressure of 158 mmHg or dissolved O_2 concentration of about 200 μ*M* (at normal pressure, air saturation, 37 °C). However, the oxygenation of cells and tissues in complex organisms such as mammals is much lower: ~97.5 mmHg in arteries and 1–40 mmHg in the brain (Jezek, Plecitá-Hlavatá, Smolková, & Rossignol, 2010). Deviations from "physiological normoxia" are implicated in many pathological conditions, including ischemia/stroke, cancer development, and neurological and metabolic disorders (Dave et al., 2001; Hertz, 2008; Tennant, Duran, & Gottlieb, 2010), and lead to energy stress, adaptive responses to hypoxia (Semenza, 2010), and, ultimately, cell survival or death.

Various *in vitro* and *ex vivo* models are used to study the effects of O_2 on cellular function (Francis & Wei, 2010; Hertz, 2008; Jezek et al., 2010). The importance of maintaining test cells under well-defined/controlled oxygenation conditions has been appreciated by researchers resulting in a wide use of hypoxic chambers. However, the chambers provide only macroscopic control of O_2 and do not inform on O_2 levels within cells and respiring specimens. A number of the extracellular O_2-sensing techniques allowing for the measurement of O_2 consumption rate are useful (Brand & Nicholls, 2011; Diepart et al., 2010; Ferrick, Neilson, & Beeson, 2008; Gerencser et al., 2009; Hynes et al., 2006; Nadanaciva et al., 2012; Ouattara et al., 2012; Robiolio, Rumsey, & Wilson, 1989; Vanderkooi, Maniara, Green, & Wilson, 1987; Wu, Luk, Lin, & Yuan, 2010; Wu et al., 2007; Zhdanov, Favre, et al., 2011), but they do not inform on local O_2 levels in cells (intracellular O_2 (iO_2)) and operate with sealed samples and changing O_2 levels. On the other hand, it has been shown that under controlled atmospheric pO_2 in a hypoxic chamber, the actual oxygenation of the cells and whole sample depends on many factors and may vary profoundly from the atmospheric level down to almost complete anoxia (Zhdanov, Dmitriev, & Papkovsky, 2011). Within actively respiring cell cultures, the pericellular (at cell surface) and iO_2 concentrations can be different from those measured by means of extracellular soluble probes in the bulk of the medium (Dmitriev, Zhdanov, Jasionek, & Papkovsky, 2012; Mik et al., 2006; Takahashi, Endoh, & Doi, 1999).

Therefore, monitoring of O_2 levels directly within respiring cells is important as this can provide more accurate information on their microenvironment, metabolic activity, and functional responses (Ferrick et al., 2008; Robiolio et al., 1989; Zhdanov, Hynes, Dmitriev, & Papkovsky, 2012). For example, local O_2 availability determines the activity of many cellular enzymes utilizing O_2 in oxidation and hydroxylation reactions. Having different affinities to O_2, these enzymes turn on and off at certain threshold O_2 levels and trigger a number of key metabolic pathways and adaptive responses. Thus, decreases in O_2 below 40 mmHg (60 μM in solution) inhibit prolyl and asparaginyl hydroxylases involved in the degradation of HIF (hypoxia-inducible factor) α subunits and subsequently activate HIF-dependent signaling pathways which direct the cells toward metabolic rearrangements and allow them to survive hypoxia and energy stress (Jiang, Semenza, Bauer, & Marti, 1996). By monitoring noninvasively iO_2 levels in different samples and experiments, one can study and interpret more reliably HIF signaling and changes in cellular function (Zhdanov, Dmitriev, & Papkovsky, 2012). Conversely, inadequate control of iO_2 may lead to misinterpretation of the observed biological effects and variability of results obtained with different samples and conditions, especially in long-term experiments in which cell activity may change over time and affect iO_2.

The cell-penetrating phosphorescent probes developed in recent years (Dmitriev & Papkovsky, 2012) have enabled direct, quantitative, contactless monitoring of iO_2. Compared to first-generation probes which had low cell-penetrating efficiency and brightness, laborious preparation, and use (Xu, Aylott, Kopelman, Miller, & Philbert, 2001), the new supramolecular (Dmitriev, Ropiak, Ponomarev, Yashunsky, & Papkovsky, 2011; Dmitriev, Ropiak, et al., 2010; Dmitriev, Zhdanov, Ponomarev, Yashunski, & Papkovsky, 2010; Koren, Dmitriev, Borisov, Papkovsky, & Klimant, 2012; Murase, Yoshihara, & Tobita, 2012; Neugebauer et al., 2008) and nanoparticle-based iO_2 probes (Fercher, Borisov, Zhdanov, Klimant, & Papkovsky, 2011; Kondrashina et al., 2012; Koo Lee et al., 2010; Napp et al., 2011) combine high brightness, broad cell specificity, convenient photophysical properties, and ease of use. Their fast (6–24 h) and efficient intracellular accumulation is achieved through passive nonspecific endocytosis (Dmitriev et al., 2011; Fercher et al., 2011) or direct translocation (Koren et al., 2012). As a result, iO_2-sensing assays can be performed by relatively simple means on a microscale, with different cell types and large number of samples. They have proved their utility in a number of complex physiological studies with cell lines and *ex vivo* models, including neuronal and cancer cells (Wotzlaw et al., 2011; Zhdanov, Dmitriev, Golubeva, Gavrilova, & Papkovsky, 2013; Zhdanov, Dmitriev, et al., 2012; Zhdanov, Ward, Prehn, & Papkovsky, 2008), and their use and application areas are growing (Papkovsky & Dmitriev, 2013).

Here, we describe standard protocols for the measurement of iO_2 with such probes, adapted for the analysis of multiple samples containing cells prepared on standard microplates (typically 96-well plates, but other substrates can also be used) and measured on commercial time-resolved fluorescence (TR-F) plate readers. This is aimed to help new users, who had no or very little previous experience with this technique, to set up such assays in their labs. A comprehensive troubleshooting guide for these assays and one representative case study are provided. The latter illustrates how the iO_2 method can be integrated in biological experiments and combined with measurement of other biomarkers and parameters, such as membrane potentials and intracellular Ca^{2+}, to gain a deeper insight into the processes taking place in the cell under normal and pathological conditions.

2. EXPERIMENTAL DESIGN AND PROTOCOLS

2.1. General strategy

Cell-penetrating phosphorescent O_2-sensing probes provide facile loading of mammalian cells by their simple addition to the medium. After incubation, the probe accumulates and is retained inside the cells without significant leakage (Dmitriev et al., 2011; Dmitriev, Ropiak, et al., 2010; Fercher et al., 2011). After removal of extracellular fraction by washing, the probe can inform in a nonchemical fashion on the *local* O_2, according to the known relationship between probe phosphorescent signal and O_2 concentration, described by the Stern–Volmer equation (Carraway, Demas, DeGraff, & Bacon, 1991; Stern & Volmer, 1919):

$$I_0/I = \tau_0/\tau = 1 + K_{s-v*}[O_2] = 1 + k_{q*}\tau_{0*}[O_2] \tag{10.1}$$

where I_0, τ_0 and I, τ are phosphorescence intensities and lifetimes of the probe at zero and given $[O_2]$, respectively, K_{s-v} is the Stern–Volmer quenching constant, and k_q—the bimolecular quenching rate constant. Increased O_2 concentration decreases probe phosphorescence intensity and lifetime in a reversible manner so that iO_2 levels and their changes can be monitored continuously and in real time. Phosphorescence lifetime (τ) is a preferred monitoring parameter as it is independent of probe concentration, photobleaching, and sample optical properties, thus providing more stable calibration and more robust iO_2 quantification.

Here, we describe the use of three different types of Pt-porphyrin-based iO_2 probes: (i) the MitoXpress®-Intra NanO2 probe based on polymeric nanoparticles impregnated with Pt-tetrakis-(pentafluorophenyl)porphine dye which possess high brightness, photo and chemical stability, and suitable for O_2 imaging (Fercher et al., 2011); (ii) Pt-coproporphyrin (PtCP) conjugated with cell-penetrating peptides (e.g., PEPP0, PtCPTE-CFR_9) which also demonstrated intracellular localization and cell specificity (Dmitriev et al., 2011; Dmitriev, Ropiak, et al., 2010; Dmitriev, Zhdanov, et al., 2010); and (iii) first-generation macromolecular, cell-impermeable probe MitoXpress®-Xtra delivered into cells by means of transfection reagents such as Endo-Porter (Zhdanov, Dmitriev, et al., 2011; Zhdanov, Ward, et al., 2010). The probes have been shown to work well in different cell types and media, over the whole physiological O_2 range 0–200 μM, with minimal (photo)toxic effects on cells upon prolonged exposure and optical measurements (Dmitriev et al., 2011; Dmitriev, Ropiak, et al., 2010;

Fercher et al., 2011; Hynes, Floyd, Soini, O'Connor, & Papkovsky, 2003). These probes, excitable at around 340–410 or 520–550 nm and emitting at 640–660 nm, are compatible with many PMT-based TR-F spectrometers and plate readers on which they show high signal-to-blank ratios and accurate lifetime-based detection and quantification of iO_2 by rapid lifetime determination (RLD) method (Sharman, Periasamy, Ashworth, & Demas, 1999). According to the RLD, TR-F intensity signals F_1, F_2 at two different delay times t_1, t_2 are measured and τ is calculated as follows:

$$\tau = (t_2 - t_1)/\ln(F_1/F_2) \tag{10.2}$$

This approach allows for the measurement of averaged τ and O_2 concentration across the cell layer at the bottom of microplate well (or other tissue culture vessel), with the assumption that O_2 distribution within the sample is uniform (confirmed by modeling and iO_2 imaging experiments (Fercher et al., 2011; Sharman et al., 1999; Zhdanov, Ogurtsov, Taylor, & Papkovsky, 2010)). On the other hand, the iO_2 assay performed with static respiring samples exposed to ambient air is influenced by a number of parameters (physical and biological variables), which require careful consideration and standardization. For common cell lines, growth medium, culturing, and probe loading conditions are usually defined *a priori*. However, for new cell types and models, basic optimization of probe loading and assay conditions should be considered with the main parameters being: (1) the density/number of cells, (2) the level of atmospheric O_2, and (3) the volume/height of the medium.

The density and respiratory activity of the cell layer determines O_2 flux across the sample and the resulting steady-state O_2 gradient (Zhdanov, Ogurtsov, et al., 2010). Working cell densities (seeding concentrations) have to be optimized considering the goals and conditions of the experiments. Reducing the cell density brings the iO_2 closer to atmospheric levels, while experiments under hypoxia may require lower cell numbers to avoid anoxia in the cell layer (Zhdanov, Ogurtsov, et al., 2010). On the other hand, at low cell numbers, probe intensity signals decrease proportionally and may become insufficient for accurate measurements. When monitoring respiratory responses to stimulation, relatively high cell densities are required to achieve partial deoxygenation in resting cells. However, excessive cell density may lead to anoxic conditions in the cell layer, whereas at low density the cells remain fully oxygenated. In these two conditions, respiratory responses become undetectable since probe signal tends to stay at its maximal or minimal level despite changes in cell respiration.

Atmospheric pO_2 has strong influence on iO_2 levels (Zhdanov, Ogurtsov, et al., 2010). Reduced pO_2 decreases relative oxygenation of the cell layer. Therefore, optimal pO_2 level is dependent on the biological model, measurement setup, and analytical task pursued. pO_2 in the measurement chamber can also be used to set the desired level of iO_2 in test sample.

Experimental medium usually occupies several millimeters or ¼–¾ of the total height of the microplate wells or culturing flask. Larger volumes and higher viscosity result in a greater barrier, lengthening diffusion times and thereby reducing O_2 in the cell layer. Increased barrier properties of medium can therefore be used to amplify the magnitude of response to cell stimulation. In a smaller volume, responses to changing conditions develop faster (mediated by O_2 diffusion through medium), but evaporation of the medium and meniscus effects becomes a major problem.

Other critical factors include sample temperature, pH control, and plate handling. Cell respiration and probe response show strong temperature dependence; therefore, accurate O_2 measurements necessitate effective temperature control across the test plate. This is achieved by appropriate instrument selection, use of only the center of a test plate, and filling edge wells with water/medium to improve heat exchange and humidification. pH also strongly affects cell metabolic activity and hence a proper buffering system, such as HEPES, pH 7.2, should be applied to maintain optimal pH, especially when conducting long-term experiments in a CO_2-free environment. Finally, although O_2 gradients within the wells are relatively stable under static conditions, sample disturbance can cause large alterations in O_2 levels requiring a considerable time to reequilibrate. Sudden agitation or temperature fluctuation must therefore be avoided. Effector addition during the monitoring should be made with minimal disturbance of the sample. Careful consideration of all the above factors will ensure reliable and reproducible measurements of iO_2, smooth iO_2 profiles, and sensitive detection of cellular responses.

Prior to starting the iO_2 experiments with a new biological model or measurement system, it is important to perform a set of initial tests which are detailed in the two protocols. Protocol 1 describes basic testing and optimization of cell loading with probe and iO_2 assay. Protocol 1 is conducted on a Checker Plate shown in Fig. 10.1A, in which cells are seeded at increasing cell density, loaded with probe, and measured on the TR-F reader in 200 μl at physiological temperature (37 °C) and ambient pO_2 (21%). To determine if the iO_2 assay is working properly, treatment of the

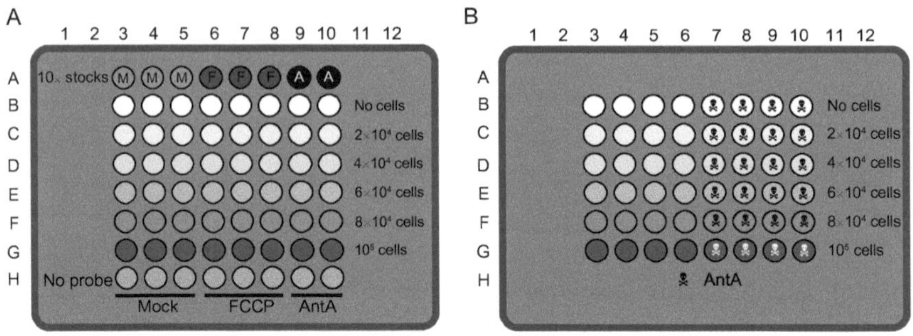

Figure 10.1 *Recommended plate layouts for the iO_2 assays.* (A) Checker Plate for Protocol 1. F, A, and M stand for FCCP (10 μ*M*), AntA (100 μ*M*), and DMSO mock (1%), respectively. (B) Calibration Plate for Protocol 2. Cells are seeded in central wells in standard media (0.2 ml/well); side wells are filled with medium to maintain humidity and uniform temperature across the plate.

cells with mitochondrial uncoupler carbonyl cyanide 4-(trifluorometoxy) phenylhydrazone (FCCP) or inhibitor antimycin A (AntA) is also included. To produce statistically plausible data, measurements are performed in three independent experiments in three to five replicates (may vary). These experiments allow the user to establish the iO_2 assay and select optimal conditions for further biological experiments under ambient conditions (21% O_2).

Protocol 2 describes system calibration and iO_2 measurements under hypoxic macroenvironment using Calibration Plate shown in Fig. 10.1B, which contains respiring and nonrespiring cells. The cells are plated at different densities, loaded with probe, treated with AntA to inhibit respiration (only wells marked as nonrespiring), and measured at several known pO_2 levels. At each pO_2 probe phosphorescence, lifetime is determined under steady-state conditions, and then pO_2 is changed to a different level. Gas equilibration is usually achieved in 30–90 min depending on the measurement setup. Since the use of hypoxia chambers is problematic below ~1% pO_2, zero O_2 point is produced by sample deoxygenation with glucose/glucose oxidase (Dmitriev, Ropiak, et al., 2010; Dmitriev, Zhdanov, et al., 2010; Fercher et al., 2011). The resulting set of five to six data points is used to determine analytical function which relates τ and iO_2 concentration and accounts for variations between instruments, cells, and measurement conditions. The results produced in Protocol 2 are used to calculate iO_2 in respiring samples and also select optimal conditions for experiments under hypoxia. Note that for NanO2 probe O_2 calibration remains stable in different cell types, while for other probes, it may vary slightly (Dmitriev et al., 2011; Dmitriev, Ropiak, et al., 2010; Dmitriev, Zhdanov, et al., 2010). For rough estimation of iO_2 concentration, one can use lifetime calibrations provided below or published in original papers.

If the main aim of your studies is to monitor relative changes in respiration or oxygenation (e.g., treated vs. untreated cells), probe calibration is not necessary and Protocol 2 can be skipped.

2.2. Materials

2.2.1 Principal equipment

Multilabel reader with TR-F capability, such as Victor 2, X4 (PerkinElmer, USA) or FLUOstar Omega (BMG, Germany), should have a red-sensitive photodetector, 380 ± 30 nm excitation and 650 ± 30 nm emission filters, and temperature control of microplate. For Protocol 2, the reader should be able to fit and operate inside a hypoxia chamber unless it has on-board O_2 control (e.g., FLUOstar Omega). Hypoxia chamber (Coy Laboratory Products, Grass Lake, MI, USA)—for probe calibration. Microbiological Safety Cabinet, table-top class II, with HEPA filter, CO_2 incubator.

2.2.2 Cell lines

PC12 rat pheochromocytoma, HCT116 human colon cancer, and mouse embryonic fibroblasts (MEF) cell lines from ATCC (LGC standards).

2.2.3 Reagent setup

Prepare stocks of FCCP (4 m*M*), AntA (10 m*M*), valinomycin (Val) (1 m*M*), and bafilomycin A1 (Baf) (500 μ*M*) in dimethyl sulfoxide (DMSO), aliquot them in 0.01–0.02 ml, and store at −18 °C for no longer than 6 months.

Dissolve phenol red-free DMEM in deionized water, add 10 m*M* glucose, 1 m*M* sodium pyruvate, 2 m*M* L-glutamine, 20 m*M* HEPES, pH 7.2, and 100 ng/ml NGF, and sterilize under the laminar flow hood, using 0.22 μm filter to produce Respiration Medium. Depending on cells types and experiments, medium composition may vary. Media with phenol red should be avoided as it reduces probe signals due to inner filter effect.

2.2.4 Protocol 1: Optimization of cell loading and iO_2 measurements

This protocol is performed under ambient pO_2 using PC12 pheochromocytoma cells as a model. However, it can also be carried out in a hypoxia chamber at reduced pO_2. Timing: 2–3 days.

1. Plate the stock of PC12 cells in 75-cm^2 flasks at 10^5–10^6 cells/ml in 30 ml of RPMI 1640 medium supplemented with 2 m*M* L-glutamine, 10% HS, 5% FBS, and 100 U/ml penicillin and 100 mg/ml streptomycin (P/S). Grow the cells in a humidified incubator (5% CO_2, 37 °C) for 24 h.
2. Transfer the suspension of cells from the flask to a 50-ml Falcon tube, spin it at 200 × *g* for 5 min at room temperature, and then gently aspirate the supernatant with a vacuum pump.
3. Using Pasteur pipette, resuspend the pellet of cells in 20 ml of prewarmed phosphate buffered saline (PBS). Spin down the cells again and aspirate the supernatant.
4. Resuspend the cells in 1 ml of 0.25% trypsin-1 mM EDTA solution, incubate for 1–2 min, and stop trypsinization by adding 10 ml of RPMI 1640 supplemented with 1% HS, L-glutamine, P/S.
5. Using a 20-ml plastic syringe, pass the suspension of cells through the 22½ G needle 8–10 times to break cell aggregates. Count the cells and dilute them to 2.5×10^5 cells/ml with RPMI 1640 supplemented with 1% HS, L-glutamine, P/S containing 100 ng/ml NGF.
6. Seed the cells on collagen IV-coated Checker Plate (Sarstedt) at 0–10^5 cells/well, according to the plate map in Fig. 10.1A. Adjust the volume of the medium to 200 μl.
7. Incubate the Checker Plate in humidified incubator at 5% CO_2, 37 °C for 24–48 h.

The protocol is cell specific. For differentiated cells (dPC12), step 7 includes culturing the cells for up to 5 days, changing medium every 48 h. Other cells are prepared using culturing procedures standard for these cells, depending on their cell type, size, and metabolic activity.

8. Prior to the experiment, reconstitute the vial of 1 × NanO2 probe in 0.1 ml of sterile water to produce 1 mg/ml solution; store at 4 °C for no longer than two weeks.
9. Dilute probe stock 1:100 in DMEM to produce working solution (10 μg/ml).
10. Take the plate with cells from the incubator and replace the medium in assay wells with 100 μl of probe working solution.
11. Incubate the plate at 37 °C in CO_2 incubator for 6–16 h.

For PEPP0 or MitoXpress™ probe recommended working dilutions are 10 and 1.2 μ*M*, respectively. Cell loading with MitoXpress™ probe is performed in the presence of 10 μ*M* of Endo-Porter (28 h for dPC12 cells).

Table 10.1 Instrument measurement settings for the iO_2 assays

Settings/instrument	Victor 2 (PE)	FLUOstar Omega (BMG)
Excitation filter	D340	TR-EXL (380 nm)
Emission filter	D642	655 BP (655 nm)
Measurement mode	TR-F	TR-F
Delay time 1 (F_1) (μs)	30	30
Delay time 2 (F_2) (μs)	70	70
Counting window (μs)	100	30
Counting cycle (μs)	1000	1000
Plate map	User defined	User defined
Assay type	Kinetic	Kinetic
Temperature (°C)	37	37
Plate scanning interval	User defined	User defined
Well reading time (s)	2–3	2–3
Number of repeats	20–99	20–250

Plate scanning interval can vary. For long-term monitoring, reading interval for each well can be 5–20 min, while for rapid respiratory responses to cell stimulation, it should be set at 1–3 min. To preserve O_2 gradients in the wells, plate movement should not be too fast (reading time should be >2 s/well). Typically, 24–48 wells are monitored on one plate.

12. Set up the TR-F reader for the measurements using the main settings given in Table 10.1 and plate map in Fig. 10.1A.
13. Allow the instrument to warm up for at least 20 min.
14. Take Checker Plate with cells from the incubator, wash assay wells twice with Respiration Medium, and add 200 μl of Respiration Medium (final volume).

Washing should be gentle and fast to minimize cell detachment and exposure to nonoptimal conditions such as low CO_2 and temperature. Manual aspiration is preferred; vacuum pump should only be used with strongly adhering cells and at very slow aspiration rate.

15. Dilute the stocks of mitochondrial effectors in Respiration Medium to produce 10 μ*M* FCCP, 100 μ*M* of AntA, and 1% DMSO mock (10 × solutions, 0.5–1.0 ml of each). Transfer them to the wells of Checker Plate as shown in Fig. 10.1A (raw A, 200 μl/well). Add 200 μl of Respiration Medium to the wells in raw H (blanks) and in columns 2 and

11 (for humidity and heat control). Concentrations of effectors must be optimized for each cell line.

16. Insert the plate in the reader, incubate for 10 min at 37 °C, and initiate kinetic measurements to monitor basal iO_2 levels for the resting cells for 10 measurement cycles.
17. Using instrument software, stop the measurement and withdraw the plate. Quickly and gently add 20 μl of 10 × effector solution according to the map in Fig. 10.1A; resume measurement under the same settings for 60–120 min.

Drug stocks must be prewarmed before adding them to the cells. This will minimize nonspecific temperature-dependent spikes in probe signal. Always consider the effect of solvent and perform control additions using solvent without drug.

18. When the experiment is completed, discard the plate and switch off the instrument.
19. Process measured data as follows:
    - **a.** Calculate τ values for each reading: $\tau = (t_2 - t_1)/\ln(F_1/F_2) = 40/\ln(F_1/F_2)$.
    - **b.** Plot and analyze F_1 profiles for each sample, to ensure consistency of optical signals. Analyze S/N ratio (probe/no probe), discard clear outliers, if any.
    - **c.** From the flat part of the profiles (step 16), determine average F_1 signals. Plot their relationship with cell numbers. Other variables on the plate can be probe concentration, loading time, and cell type.
    - **d.** Plot τ time profiles for different samples. Analyze their shape, consistency and variability (mean values, SD), basal levels, and effects of cell stimulation on τ.
20. Determine optimal cell number and measurement conditions for iO_2-sensing experiments. If cell loading appears to be low, probe concentration or loading time can be increased. Alternatively change to another O_2 probe.

2.2.4.1 Anticipated results

Figure 10.2A–C shows that cell loading and resulting average F_1 values depend on medium composition and loading time, the number of cells, and probe concentration (Dmitriev et al., 2011; Dmitriev, Zhdanov, et al., 2010; Fercher et al., 2011). Table 10.2 shows loading efficiency for the iO_2 probes with respect to different cell lines and primary cells tested so far in authors' laboratory. Left panel in Fig. 10.2C shows F_1 profiles;

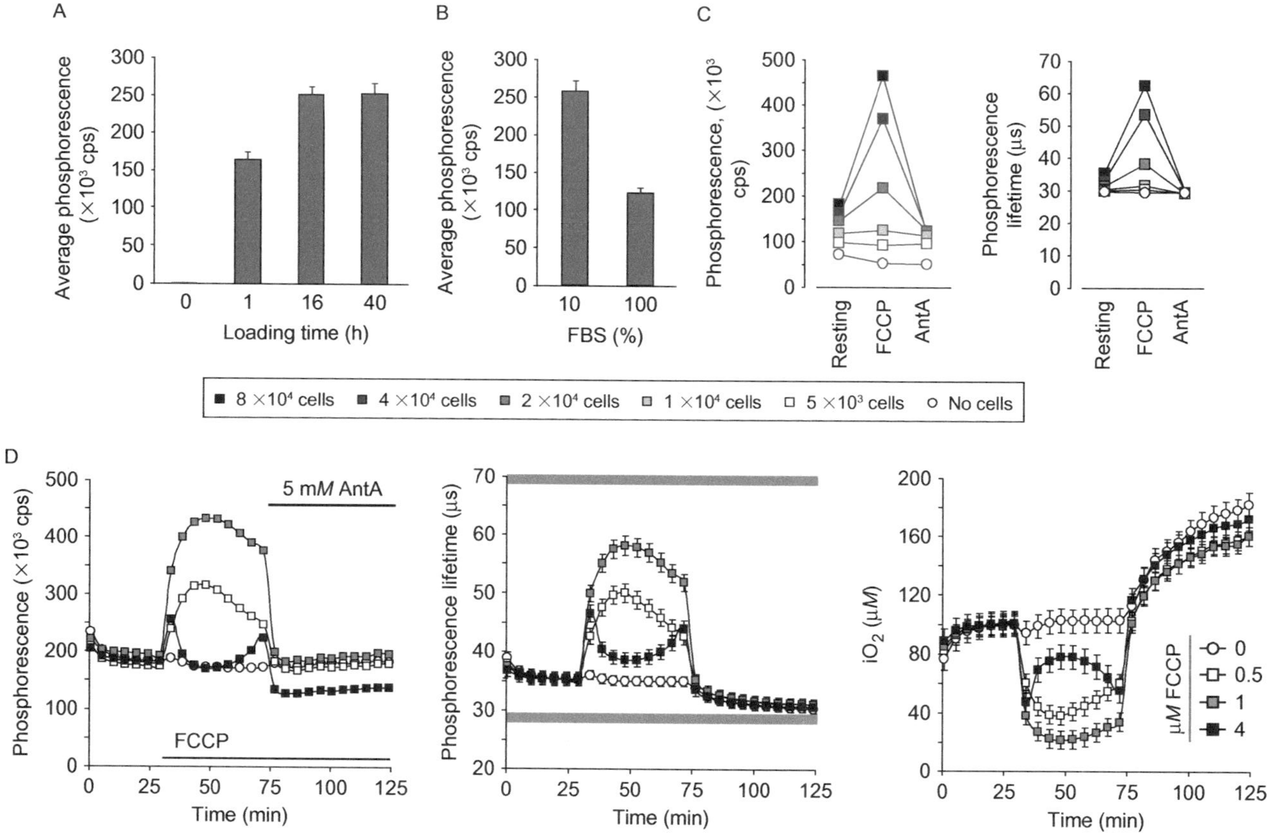

Figure 10.2 *Example results obtained in Protocol 1.* The effects of cell loading time (A) and serum concentration (B, 16 h loading) on TR-F intensity signals (F_1) from human colon cancer cells (HCT116) loaded with 10 μ*M* NanO2 probe in McCoy 5A medium at 37 °C. (C) Effects of cell seeding number on TR-F-intensity (left) and corresponding phosphorescence lifetime (right) in HCT116 cells at rest, upon uncoupling (at the peak of the response to 1 μ*M* FCCP) and upon AntA treatment (10 μ*M*). Cells were loaded with NanO2 probe (10 μg/ml, 7 h). (D) Phosphorescence intensity (left), corresponding lifetime (middle), and iO_2 concentration (right) profiles for the samples containing PC12 cells (seeded at 5×10^5 cells per well and differentiated) loaded with NanO2 probe (10 μg/ml, 20 h, 37 °C, RPMI medium). The period 0–30 min corresponds to resting cells, FCCP was added at 30 min (concentrations shown on right panel, μ*M*), then AntA at 75 min. Gray lines on the lifetime plot correspond to aerated (bottom) and fully deoxygenated (top) cells.

Table 10.2 Loading efficiency[a] of different cell lines with the various iO_2 probes

Cell line (medium)	NanO2 probe	PEPP0, 1, 3 (PtCP-CPP probes)	MitoXpress™ with Endo-Porter delivery
PC12 (1)	Medium	Medium	Medium
dPC12 (1,2)	Medium	High	Medium/high
SH-SY5Y (2)	High	High	Low
dSH-SY5Y (2)	Low	Low	Low
C2C12 (2)	n.t.	n.t.	Low/medium
dC2C12 (2)	n.t.	n.t.	Medium
MEF (2)	High	High	n.t.
HCT116 (3)	High	High	n.t.
HepG2 (2)	High	Medium	Low/medium
MCF-7 (2)	High	n.t.	n.t.
UOK262 (2)	High	n.t.	n.t.
MDA-MB-231	High	n.t.	n.t.
HeLa (2)	Medium/ high	High	Low
B-lymphocytes (1)	n.t.	n.t.	Low/medium
Rat cerebellar granule neurons, 7-day old (4)	n.t.	n.t.	Medium
Rat cortical neurons, E18 (4)	Low	low/medium	n.t.
SK-N-AS neurospheres (2)	High	n.t.	n.t.

[a]Assessed via TR-F signals from a confluent layer of loaded cells measured on a Victor® reader: low, <10,000 cps; medium, 30,000 cps; high, >80,000 cps. n.t., not tested; 1, 2, 3, and 4, RPMI, DMEM, McCoy's 5A, and DMEM F12/Ham, respectively; prefix d, differentiated cells.

F_1 exhibits significant inter- and intra-assay variability and dependence on cell seeding number and other parameters, and therefore, hard to use for O_2 quantification. Probe lifetime (τ), calculated using the RLD method (right panel), allows accurate and reproducible monitoring of cell responses to the treatment, as demonstrated using an activator (FCCP) and inhibitor (AntA) of respiration. Noteworthy, inhibition of respiration with AntA brings τ to the same value, regardless of cell number.

Figure 10.2D demonstrates sets of raw and processed data for one iO_2 measurement experiment with differentiated PC12 cells and their stimulation with metabolic effectors. Left panel shows F_1 profiles. The central panel shows calculated τ, which, in turn, is converted into O_2 concentrations (right panel), thus allowing monitoring of even small changes in cell oxygenation. The two gray lines below and above τ profiles correspond to air-saturated and fully deoxygenated conditions, respectively. The right panel shows the profiles of iO_2 concentration calculated from τ by applying the calibration function (see Section 2.2.5).

From the profiles in Fig. 10.2D, one can see that after initial temperature equilibration (~30 min), stable baseline is produced which corresponds to basal respiration and oxygenation of cell monolayer. The addition of FCCP at ~25 min results in a rapid increase in τ or reduction in iO_2. The response to FCCP is dose dependent: it progressively increases over the range 0–1.0 μ*M* causing deep and sustained deoxygenation of the cells due to their increased respiration. At higher concentrations, the response to FCCP decreases due to its toxic effects on the cells. DMSO addition (mock) only causes a small transient spike related to temperature change during the addition. On the other hand, blocking cell respiration by the addition of AntA (at ~75 min) increased oxygenation of the cells bringing iO_2 to air-saturated levels. In all these samples and experiments, the cells remained viable and maintained their ATP levels (not shown).

2.2.5 Protocol 2: Probe calibration and monitoring of cell oxygenation under hypoxia

This protocol is performed at reduced pO_2. Timing: 1 day (excluding preparation of the cells).

1. Plate cells according to the plate map in Fig. 10.1B. Prepare and load them with the iO_2 probe as described in Section 2.2.4 (steps 1–11).
2. Place the TR-F reader in the hypoxia chamber. Set up the chamber at 19.0% pO_2 and prepare the reader for the measurements as described in Section 2.2.4 (steps 12–13).
3. Take Calibration Plate from the incubator, wash the cells with Respiration Medium as described in Section 2.2.4 (step 14). Add 200 μl/well of Respiration Medium for measurement.
4. Prepare 100 μ*M* stock of AntA and add 20 μl according to the plate map provided in Fig. 10.1B, to produce nonrespiring cells. Leave respiring cells in the remaining wells. Add 200 μl of Respiration Medium to the surrounding wells.
5. Cover the plate with lid, transfer to the hypoxia chamber, and insert in the plate reader compartment. For this experiment which lasts several

hours, plate lid and a large volume (not less than 200 μl) of medium are necessary to minimize evaporation of liquid.

6. Initiate the measurements. Monitor the plate for 40–90 min taking readings every 2–10 min (see Fig. 10.2D).
7. After gas equilibration on the plate has been achieved (stable readings), pause the measurements and reset the chamber to 15% pO_2.
8. Repeat steps 6 and 7 moving gradually to 10.0%, 5.0%, 3.0%, and 1.0% pO_2.
9. Prepare fresh stock of glucose oxidase (500 μg/ml) and D-(+)-glucose (1 M). When the 1% pO_2 measurement is completed, add 30 μl of this stock to the wells with nonrespiring cells to achieve complete deoxygenation and then measure for a further 20–30 min to determine τ at 0% O_2.
10. When the measurement is completed, extract the plate and switch off the instrument.
11. Process experimental data as described in Section 2.2.4 (steps 19a and b).
12. Determine τ values for the nonrespiring and respiring cells which correspond to steady state at each pO_2 level (flat regions of τ profiles).
13. Calculate dissolved O_2 concentration for each pO_2 setting (% of O_2):

$$[O_2], \mu M = k * P_{\text{atm}} / P_{\text{st}} * pO_2,$$

where P_{atm}—ambient atmospheric pressure (mmHg), $P_{\text{st}} = 760$ mmHg, $k = 10.02\ \mu M/\%\ O_2$ at 37 °C.

14. Plot $[O_2]$ against corresponding τ values for the nonrespiring cells. Determine the mathematical function that describes best these experimental points. Examples of calibration data and analytical equation are shown in Fig. 10.2D.
15. Apply calibration equation to calculate $[O_2]$ in respiring cells (r) at different atmospheric pO_2 and calculate their relative deoxygenation with respect to nonrespiring cells (nr): $\% = [O_2]_r / [O_2]_{nr} {}^* 100$.
16. Apply calibration equation to convert lifetime profiles generated in Protocol 1 (step 19d) into profiles of iO_2 concentration in respiring cells. See example in Fig. 10.2C and D.

2.2.5.1 Anticipated results

Typical τ profiles of the iO_2 probe at different levels of atmospheric pO_2 in the hypoxia chamber (graded hypoxia), produced by nonrespiring (treated with 10 μM AntA) MEF cells, are shown in Fig. 10.3A. Figure 10.3B shows

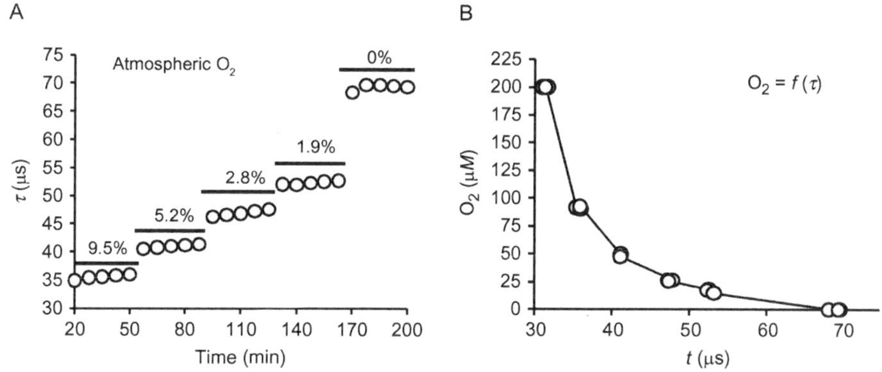

Figure 10.3 *Example results produced in Protocol 2.* (A) Phosphorescence lifetime profiles for nonrespiring (treated with 10 µ*M* AntA) MEF cells seeded at 3×10^5 cells per well, measured with the NanO2 probe (10 µg/ml, 16 h loading) at different levels of atmospheric pO_2 (indicated with bars), 37 °C. (B) The relationship between τ and iO_2 concentration for the nonrespiring cells in (A), from which the following calibration function was produced by fitting: $[iO_2]$, µ*M* = 18,576 * exp($-\tau$/6.8794).

O_2 calibration produced using steady-state τ values for nonrespiring cells at different pO_2 (shown in Fig. 10.3A). One can see that for respiring cells τ values are higher than for nonrespiring due to local O_2 gradients at cell monolayer, and they depend on cell density. Higher cell numbers (e.g., 30 vs. 6k) result in deeper deoxygenation (not shown). These data can also be used to optimize cell density in biological experiments under hypoxia or to maintain the required oxygenation level of respiring cells. The latter can be calculated from τ values using O_2 calibration function provided in Fig. 10.3 (Dmitriev et al., 2012).

2.2.6 Biological case study: Analysis of the effects of Baf A1 on the iO_2 and mitochondrial function of PC12 cells. Timing: 5–6 days

2.2.6.1 Experimental design and protocol

This biological case study illustrates how the iO_2 assays established in Protocols 1 and 2 can be applied to study the effects of drugs on cell metabolism, particularly to elaborate the mode of action of Baf (well-known V-ATPase inhibitor (Huss & Wieczorek, 2009)) on PC12 cells. In this study, a set of iO_2 assays were conducted using several different treatments with Baf, classical metabolic drugs, and important controls. In parallel, live cell imaging experiments were carried out looking at the other metabolic markers: tetramethyl rhodamine methyl ester probe (TMRM) for mitochondrial membrane potential ($\Delta\Psi_m$), bis-(1,3-diethylthiobarbituric acid) trimethine oxonol probe ($DiSBAC_2(3)$) for plasma membrane potential ($\Delta\Psi_p$), Fluo-4 for cytosolic calcium ($_cCa^{2+}$), and mitoCase12 (Case12-mito, Evrogen) for mitochondrial calcium ($_mCa^{2+}$). When compared, the results revealed marked and sustained activation of respiration of PC12 cells by Baf through mild mitochondrial uncoupling, thus demonstrating the considerable additional insight into cellular function and metabolism that can be gained from iO_2 analysis. Technical details of this study and its specific results are described below.

At nonquenching concentration of TMRM (20 n*M*), a decrease in probe fluorescence reports on $\Delta\Psi_m$ depolarization and vice versa. Since TMRM probe is sensitive to changes in both $\Delta\Psi_m$ and $\Delta\Psi_p$, $DiSBAC_2(3)$ probe which is only sensitive to $\Delta\Psi_p$ changes (Ward et al., 2007) should be monitored in parallel. Fluorescence of the Ca^{2+} probes is sensitive to changes in cytosolic (Fluo-4) and mitochondrial (mitoCase12) Ca^{2+} levels.

To measure iO_2 and respiratory responses to stimulation, dPC12 cells were seeded, grown and loaded with iO_2 probe as described in Section 2.2.4 (steps 1–11). To magnify the O_2 gradients and respiratory responses (Zhdanov, Ogurtsov, et al., 2010), nonstained PC12 cells were added (1.5×10^5), which settled at the bottom of the wells in 30 min. Then, 20 μl of 10 × stock solutions of Baf (0.4, 1.0, and 2.5 μ*M*) and Val (2.5 μ*M*) in Respiration Medium were added to the cells to produce final concentrations of Baf (0, 0.04, 0.1, and 0.25 μ*M*) and Val (0.25 μ*M*) in 200 μl of Respiration Medium. The plate was incubated for 30 min at 37 °C, then inserted in the TR-F reader, and measured for 30 min (10–20 readings at 21% pO_2) to determine basal levels of iO_2. After this, the plate was withdrawn, 20 μl of FCCP or DMSO stock (10 μ*M* and 0.25% in Respiration Medium, respectively) was gently applied to the wells with cells, and monitoring was resumed for further 60–120 min (Figs. 10.4 and 10.5). When the measurement was completed, the plate was extracted and experimental data were processed as described in Section 2.2.4 (step 19), and then τ profiles were converted into iO_2 profiles as described in Section 2.2.5 (step 16). For each treatment, TMRM intensities were monitored as described below. A similar experiment was carried out with the addition of 0.25 μ*M* Val to the cells pretreated with 0.25 μ*M* Baf for 30 min (Fig. 10.6D).

For live cell imaging, PC12 cell at were seeded at 2.5×10^4 cells per cm^2 and differentiated for 4–5 days with NGF on glass bottom mini dishes (MatTek) coated with a mixture of collagen IV (0.007%) and poly-D-lysine (0.003%). Transfection with a plasmid-encoding GFP-based mitoCase12 protein sensor (Zhdanov, Ward, et al., 2010) was carried out using

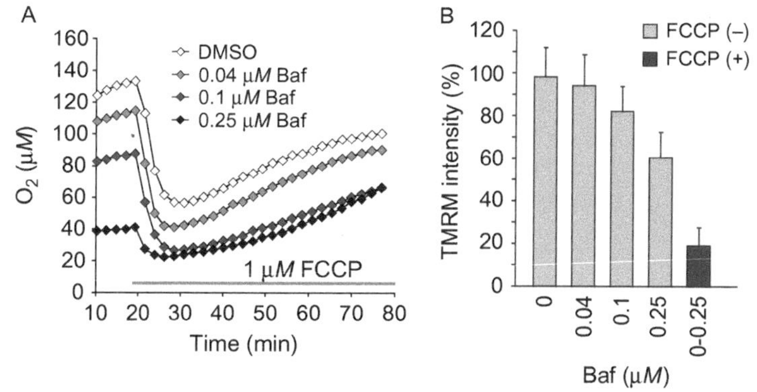

Figure 10.4 *The effect of Baf on dPC12 cells.* (A) Profiles of iO_2 show sustained, dose-dependent activation of respiration by Baf. The addition of FCCP causes further increase in respiration and deoxygenation of the cells. (B) TMRM staining shows only partial depolarization of the $\Delta\Psi_m$ by Baf and strong depolarization 10 min after the addition of FCCP(+).

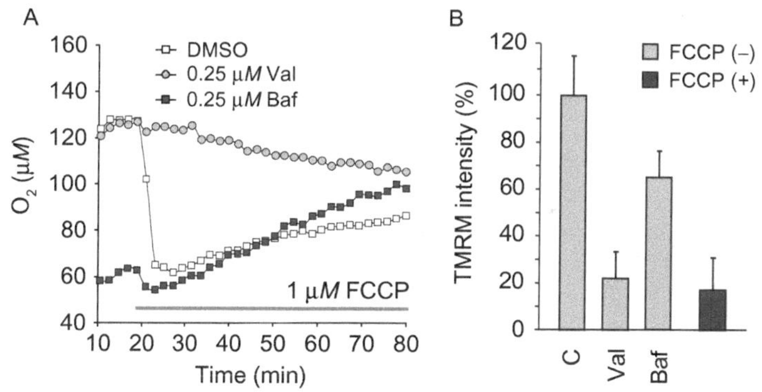

Figure 10.5 *Pretreatment of PC12 cells with Val and Baf for 30 min reveals drastically different effects.* (A) iO_2 levels and responses to FCCP for treated and untreated cells. (B) TMRM staining shows large depolarization of the $\Delta\Psi_m$ by Val and moderate by Baf. Subsequent treatment with FCCP reduces the $\Delta\Psi_m$ further (10 min after the addition—FCCP(+)).

Lipofectamine 2000 (Life Technologies, CA) and Opti-MEM I medium, as per the manufacturer procedure. Loading with fluorescent sensors was performed in Opti-MEM I medium with Ca^{2+} concentration adjusted to 2 mM as follows: 20 nM TMRM, 1 µM DiSBAC$_2$(3), and 2.5 Fluo-4 AM for 30 min. Deesterification of Fluo-4 AM was performed for 15 min prior to each experiment. During the measurements, TMRM and DiSBAC$_2$(3) were maintained in solution at 20 nM and 0.25 µM, respectively. Subsequent fluorescence imaging of the Fluo-4 and mitoCase12 probes was performed on an Olympus FV1000 confocal laser scanning microscope with controlled CO_2, humidity, and temperature, using excitation at 488 nm (5%, 2.5% of maximal laser power) and emission collected at 500–540 nm. The TMRM and $DiSBAC_2(3)$ probes were excited at 543 nm (1.5% and 1%, respectively) collecting emission at 555–600 nm. The probes were measured individually and also in pairs (TMRM and Fluo-4, Fluo-4 and $DiSBAC_2(3)$, TMRM, and mitoCase12), when their characteristics allowed simultaneous loading and measurement. When mitoCase12 was expressed at very high levels, it reduced the TMRM staining; therefore, only the cells with unaffected TMRM signal were analyzed. Acquisition of each spectral signal was done in sequential laser mode with emission gates adjusted to avoid overlaps. In all the experiments with drug addition (see Section 2.2.6.2), the differential interference contrast and fluorescence images were collected kinetically in two planes using 0.5 μm step, 20 s intervals, and a 60 × oil immersion objective. The resulting *z*-stacked images were analyzed using FV1000 Viewer software (Olympus), Excel, Adobe Photoshop, and Illustrator.

2.2.6.2 Results

It has been recently shown that Baf functions as a K^+ ionophore and causes mild uncoupling of the mitochondria in PC12 and other cells, leading to a sustained increase in respiration and a gradual decrease in iO_2 (Huss & Wieczorek, 2009; Zhdanov, Dmitriev, et al., 2011). The uncoupling by Baf is accompanied by a partial decrease in ${}_{m}Ca^{2+}$, flickering depolarization of the $\Delta\Psi_m$, and an increase in ${}_{c}Ca^{2+}$. Another K^+ ionophore Val and FCCP are also known to elevate ${}_{c}Ca^{2+}$, but these drugs strongly reduce ${}_{m}Ca^{2+}$ and $\Delta\Psi_m$.

To elaborate how changes in respiration and iO_2 correlate with the other bioenergetic parameters, we monitored PC12 cells upon treatment with Baf, Val, and FCCP using the iO_2 assay described in Section 2.2.4 and by live cell confocal fluorescence microscopy (experimental details can be found in Zhdanov, Dmitriev, et al., 2011). Cells treated with Baf for 30 min displayed a dose-dependent decrease in iO_2 (Fig. 10.4A) with the effect being observed at ~40 n*M* and above. In contrast, $\Delta\Psi_m$ (visualized with TMRM) began to decrease significantly only at >150 n*M* Baf (Fig. 10.4B), with iO_2 dropping below 60 μ*M*. Subsequent addition of 1 μ*M* FCCP rapidly depolarized $\Delta\Psi_m$ irrespective of Baf concentration and decreased iO_2 in all samples. Since cells partially retained the $\Delta\Psi_m$ even at 0.25 μ*M* Baf (Fig. 10.4B), the mitochondria were able to elevate respiration upon FCCP addition, although deoxygenating effect of FCCP was decreased.

The action of Val on PC12 cells was significantly different to that of Baf. The profile of iO_2 shows a rapid spike of deoxygenation followed by gradual reoxygenation returning to the level of resting cells after 30 min (Fig. 10.6D), by which time $\Delta\Psi_m$ was strongly depolarized (Fig. 10.6C). Unlike Baf, iO_2 in Val(+) cells was not affected much by FCCP (Fig. 10.5A). Notably, treatment with Val or Baf (<0.5 μ*M*) for 30–60 min did not affect the morphology and ATP levels in PC12 cells (not shown). Considering strong dissimilarity in iO_2 profiles for the two drugs, we analyzed the effects of Val on PC12 cells pretreated with Baf. Incubation with Baf for 30 min did not affect the $\Delta\Psi_p$, monitored with DiS-BAC$_2$(3), slightly decreased $_mCa^{2+}$ and $\Delta\Psi_m$, and strongly elevated $_cCa^{2+}$, as shown with Fluo-4 probe (Fig. 10.6A–C). The addition of Val to these cells did not influence $\Delta\Psi_p$ and $_cCa^{2+}$, whereas $_mCa^{2+}$ decreased further and

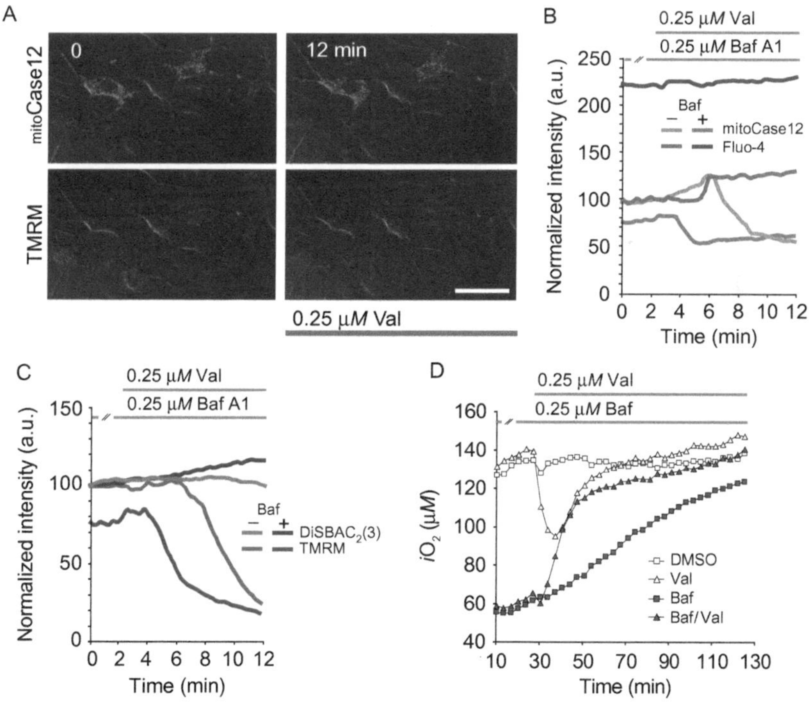

Figure 10.6 *Effects of Val on bioenergetic parameters of dPC12 cells pretreated with Baf (or mock) for 30 min.* (A) Confocal images demonstrate a decrease in $_mCa^{2+}$ (mitoCase12 probe) and strong depolarization of the $\Delta\Psi_m$ (TMRM probe) upon the addition of Val. (B–D) Profiles of cytosolic (Fluo-4 probe) and mitochondrial Ca^{2+} (B), $\Delta\Psi_m$ and $\Delta\Psi_p$ (DiSBAC$_2$(3) probe) (C), and iO_2 (D) for the cells treated with Baf and/or Val. (See the color plate.)

$\Delta\Psi_m$ fully depolarized in ~10 min. The addition of Val to control cells caused similar changes in $_mCa^{2+}$ and $\Delta\Psi_m$, while the level of $_cCa^{2+}$ increased. However, the most interesting effect was observed on iO_2 (Fig. 10.6D). Unlike the untreated control, cells pretreated with Baf underwent rapid reoxygenation upon the addition of Val, indicating that their respiration was inhibited or damaged. Further analysis revealed that in control cells, the early phase of $\Delta\Psi_m$ profile upon Val addition coincided with a transient increase in $_mCa^{2+}$ (Fig. 10.6B and C). In cells pretreated with Baf such an increase associated with elevated respiration was not observed.

Taken together, these data indicate that mitochondrial uncouplers FCCP and Val, while affecting a number of bioenergetic parameters, have very different effects on cell respiration, as can be devised from the iO_2 profiles. The multifunctional drug Baf also shows a unique metabolic pattern, and the iO_2-sensing technique can aid its detailed analysis and mechanistic description (Zhdanov, Dmitriev, et al., 2011, 2012).

ACKNOWLEDGMENTS

This work was supported by Science Foundation of Ireland (SFI), Grants 07/IN.1/B1804 and 12/RC/2276. We thank Dr. A. Fercher for his help with the optimization of measurement conditions of NanO2 probe on microplate reader.

REFERENCES

Brand, M. D., & Nicholls, D. G. (2011). Assessing mitochondrial dysfunction in cells. *Biochemical Journal*, *435*, 297–312.

Carraway, E. R., Demas, J. N., DeGraff, B. A., & Bacon, J. R. (1991). Photophysics and photochemistry of oxygen sensors based on luminescent transition-metal complexes. *Analytical Chemistry*, *63*, 337–342.

Dave, K. R., Saul, I., Busto, R., Ginsberg, M. D., Sick, T. J., & Perez-Pinzon, M. A. (2001). Ischemic preconditioning preserves mitochondrial function after global cerebral ischemia in rat hippocampus. *J of Cerebral Blood Flow and Metabolism*, *21*, 1401–1410.

Diepart, C., Verrax, J., Calderon, P. B., Feron, O., Jordan, B. F., & Gallez, B. (2010). Comparison of methods for measuring oxygen consumption in tumor cells in vitro. *Analytical Biochemistry*, *396*, 250–256.

Dmitriev, R. I., & Papkovsky, D. B. (2012). Optical probes and techniques for O(2) measurement in live cells and tissue. *Cellular and Molecular Life Sciences: CMLS*, *69*, 2025–2039.

Dmitriev, R. I., Ropiak, H. M., Ponomarev, G. V., Yashunsky, D. V., & Papkovsky, D. B. (2011). Cell-penetrating conjugates of coproporphyrins with oligoarginine peptides: Rational design and application for sensing intracellular O(2). *Bioconjugate Chemistry*, *22*, 2507–2518.

Dmitriev, R. I., Ropiak, H. M., Yashunsky, D. V., Ponomarev, G. V., Zhdanov, A. V., & Papkovsky, D. B. (2010). Bactenecin 7 peptide fragment as a tool for intracellular delivery of a phosphorescent oxygen sensor. *FEBS Journal*, *277*, 4651–4661.

Dmitriev, R. I., Zhdanov, A. V., Jasionek, G., & Papkovsky, D. B. (2012). Assessment of cellular oxygen gradients with a panel of phosphorescent oxygen-sensitive probes. *Analytical Chemistry, 84*, 2930–2938.

Dmitriev, R. I., Zhdanov, A. V., Ponomarev, G. V., Yashunski, D. V., & Papkovsky, D. B. (2010). Intracellular oxygen-sensitive phosphorescent probes based on cell-penetrating peptides. *Analytical Biochemistry, 398*, 24–33.

Fercher, A., Borisov, S. M., Zhdanov, A. V., Klimant, I., & Papkovsky, D. B. (2011). Intracellular O2 sensing probe based on cell-penetrating phosphorescent nanoparticles. *ACS Nano, 5*, 5499–5508.

Ferrick, D. A., Neilson, A., & Beeson, C. (2008). Advances in measuring cellular bioenergetics using extracellular flux. *Drug Discovery Today, 13*, 268–274.

Francis, K. R., & Wei, L. (2010). Human embryonic stem cell neural differentiation and enhanced cell survival promoted by hypoxic preconditioning. *Cell Death and Disease, 1*, e22.

Gerencser, A. A., Neilson, A., Choi, S. W., Edman, U., Yadava, N., Oh, R. J., et al. (2009). Quantitative microplate-based respirometry with correction for oxygen diffusion. *Analytical Chemistry, 81*, 6868–6878.

Hertz, L. (2008). Bioenergetics of cerebral ischemia: A cellular perspective. *Neuropharmacology, 55*, 289–309.

Huss, M., & Wieczorek, H. (2009). Inhibitors of V-ATPases: Old and new players. *Journal of Experimental Biology, 212*, 341–346.

Hynes, J., Floyd, S., Soini, A. E., O'Connor, R., & Papkovsky, D. B. (2003). Fluorescence-based cell viability screening assays using water-soluble oxygen probes. *Journal of Biomolecular Screening, 8*, 264–272.

Hynes, J., Marroquin, L. D., Ogurtsov, V. I., Christiansen, K. N., Stevens, G. J., Papkovsky, D. B., et al. (2006). Investigation of drug-induced mitochondrial toxicity using fluorescence-based oxygen-sensitive probes. *Toxicological Sciences, 92*, 186–200.

Jezek, P., Plecitá-Hlavatá, L., Smolková, K., & Rossignol, R. (2010). Distinctions and similarities of cell bioenergetics and the role of mitochondria in hypoxia, cancer, and embryonic development. *The International Journal of Biochemistry & Cell Biology, 42*, 604–622.

Jiang, B. H., Semenza, G. L., Bauer, C., & Marti, H. H. (1996). Hypoxia-inducible factor 1 levels vary exponentially over a physiologically relevant range of O2 tension. *American Journal of Physiology, 271*, C1172–C1180.

Kondrashina, A. V., Dmitriev, R. I., Borisov, S. M., Klimant, I., O'Brien, I., Nolan, Y. M., et al. (2012). A phosphorescent nanoparticle-based probe for sensing and imaging of (intra)cellular oxygen in multiple detection modalities. *Advanced Functional Materials, 22*, 4931–4939.

Koo Lee, Y.-E., Ulbrich, E. E., Kim, G., Hah, H., Strollo, C., Fan, W., et al. (2010). Near infrared luminescent oxygen nanosensors with nanoparticle matrix tailored sensitivity. *Analytical Chemistry, 82*, 8446–8455.

Koren, K., Dmitriev, R. I., Borisov, S. M., Papkovsky, D. B., & Klimant, I. (2012). Complexes of irIII-octaethylporphyrin with peptides as probes for sensing cellular O2. *ChemBioChem, 13*, 1184–1190.

Lenaz, G., & Genova, M. L. (2009). Structural and functional organization of the mitochondrial respiratory chain: A dynamic super-assembly. *International Journal of Biochemistry and Cell Biology, 41*, 1750–1772.

Mik, E. G., Stap, J., Sinaasappel, M., Beek, J. F., Aten, J. A., van Leeuwen, T. G., et al. (2006). Mitochondrial PO2 measured by delayed fluorescence of endogenous protoporphyrin IX. *Nature Methods, 3*, 939–945.

Murase, T., Yoshihara, T., & Tobita, S. (2012). Mitochondria-specific oxygen probe based on iridium complexes bearing triphenylphosphonium cation. *Chemistry Letters, 41*, 262–263.

Nadanaciva, S., Rana, P., Beeson, G. C., Chen, D., Ferrick, D. A., Beeson, C. C., et al. (2012). Assessment of drug-induced mitochondrial dysfunction via altered cellular respiration and acidification measured in a 96-well platform. *Journal of Bioenergetics and Biomembranes, 44*, 421–437.

Napp, J., Behnke, T., Fischer, L., Würth, C., Wottawa, M., Katschinski, D. M., et al. (2011). Targeted luminescent near-infrared polymer-nanoprobes for in vivo imaging of tumor hypoxia. *Analytical Chemistry, 83*, 9039–9046.

Neugebauer, U., Pellegrin, Y., Devocelle, M., Forster, R. J., Signac, W., Moran, N., et al. (2008). Ruthenium polypyridyl peptide conjugates: Membrane permeable probes for cellular imaging. *Chemical Communications (Cambridge)*, 5307–5309. http://dx.doi.org/10.1039/B810403D.

Ouattara, D. A., Prot, J.-M., Bunescu, A., Dumas, M.-E., Elena-Herrmann, B., Leclerc, E., et al. (2012). Metabolomics-on-a-chip and metabolic flux analysis for label-free modeling of the internal metabolism of HepG2/C3A cells. *Molecular BioSystems, 8*, 1908–1920.

Papkovsky, D. B., & Dmitriev, R. I. (2013). Biological detection by optical oxygen sensing. *Chemical Society Reviews, 42*, 8700–8732.

Robiolio, M., Rumsey, W. L., & Wilson, D. F. (1989). Oxygen diffusion and mitochondrial respiration in neuroblastoma cells. *American Journal of Physiology: Cell Physiology, 256*, C1207–C1213.

Semenza, G. L. (2010). HIF-1: Upstream and downstream of cancer metabolism. *Current Opinion in Genetics and Development, 20*, 51–56.

Sharman, K. K., Periasamy, A., Ashworth, H., & Demas, J. N. (1999). Error analysis of the rapid lifetime determination method for double-exponential decays and new windowing schemes. *Analytical Chemistry, 71*, 947–952.

Stern, O., & Volmer, M. (1919). The fading time of fluorescence. *Physikalishe Zeitschrift, 20*, 183–188.

Takahashi, E., Endoh, H., & Doi, K. (1999). Intracellular gradients of O2 supply to mitochondria in actively respiring single cardiomyocyte of rats. *American Journal of Physiology, 276*, H718–H724.

Tennant, D. A., Duran, R. V., & Gottlieb, E. (2010). Targeting metabolic transformation for cancer therapy. *Nature Reviews Cancer, 10*, 267–277.

Vanderkooi, J. M., Maniara, G., Green, T. J., & Wilson, D. F. (1987). An optical method for measurement of dioxygen concentration based upon quenching of phosphorescence. *Journal of Biological Chemistry, 262*, 5476–5482.

Ward, M. W., Huber, H. J., Weisová, P., Düssmann, H., Nicholls, D. G., & Prehn, J. H. (2007). Mitochondrial and plasma membrane potential of cultured cerebellar neurons during glutamate-induced necrosis, apoptosis, and tolerance. *The Journal of Neuroscience, 27*, 8238–8249.

Wotzlaw, C., Bernardini, A., Berchner-Pfannschmidt, U., Papkovsky, D., Acker, H., & Fandrey, J. (2011). Multifocal animated imaging of changes in cellular oxygen and calcium concentrations and membrane potential within the intact adult mouse carotid body ex vivo. *American Journal of Physiology: Cell Physiology, 301*, C266–C271.

Wu, C.-C., Luk, H.-N., Lin, Y.-T. T., & Yuan, C.-Y. (2010). A Clark-type oxygen chip for in situ estimation of the respiratory activity of adhering cells. *Talanta, 81*, 228–234.

Wu, M., Neilson, A., Swift, A. L., Moran, R., Tamagnine, J., Parslow, D., et al. (2007). Multiparameter metabolic analysis reveals a close link between attenuated mitochondrial bioenergetic function and enhanced glycolysis dependency in human tumor cells. *American Journal of Physiology: Cell Physiology, 292*, C125–C136.

Xu, H., Aylott, J. W., Kopelman, R., Miller, T. J., & Philbert, M. A. (2001). A real-time ratiometric method for the determination of molecular oxygen inside living cells using sol–gel-based spherical optical nanosensors with applications to rat C6 glioma. *Analytical Chemistry*, *73*, 4124–4133.

Zhdanov, A. V., Dmitriev, R. I., Golubeva, A. V., Gavrilova, S. A., & Papkovsky, D. B. (2013). Chronic hypoxia leads to a glycolytic phenotype and suppressed HIF-2 signaling in PC12 cells. *Biochimica et Biophysica Acta (BBA): General Subjects*, *1830*, 3553–3569.

Zhdanov, A. V., Dmitriev, R. I., & Papkovsky, D. B. (2011). Bafilomycin A1 activates respiration of neuronal cells via uncoupling associated with flickering depolarization of mitochondria. *Cellular and Molecular Life Sciences: CMLS*, *68*, 903–917.

Zhdanov, A. V., Dmitriev, R. I., & Papkovsky, D. B. (2012). Bafilomycin A1 activates HIF-dependent signalling in human colon cancer cells via mitochondrial uncoupling. *Bioscience Reports*, *32*, 587–595.

Zhdanov, A. V., Favre, C., O'Flaherty, L., Adam, J., O'Connor, R., Pollard, P. J., et al. (2011). Comparative bioenergetic assessment of transformed cells using a cell energy budget platform. *Integrative Biology (Cambridge)*, *3*, 1135–1142.

Zhdanov, A. V., Hynes, J., Dmitriev, R. I., & Papkovsky, D. B. (2012). O2 analysis on a fluorescence spectrometer or plate reader. In *Phosphorescent oxygen-sensitive probes* (pp. 29–69). Basel: Springer.

Zhdanov, A. V., Ogurtsov, V. I., Taylor, C. T., & Papkovsky, D. B. (2010). Monitoring of cell oxygenation and responses to metabolic stimulation by intracellular oxygen sensing technique. *Integrative Biology (Cambridge)*, *2*, 443–451.

Zhdanov, A. V., Ward, M. W., Prehn, J. H. M., & Papkovsky, D. B. (2008). Dynamics of intracellular oxygen in PC12 cells upon stimulation of neurotransmission. *Journal of Biological Chemistry*, *283*, 5650–5661.

Zhdanov, A. V., Ward, M. W., Taylor, C. T., Souslova, E. A., Chudakov, D. M., Prehn, J. H., et al. (2010). Extracellular calcium depletion transiently elevates oxygen consumption in neurosecretory PC12 cells through activation of mitochondrial Na(+)/Ca(2+) exchange. *Biochimica et Biophysica Acta*, *1797*, 1627–1637.

CHAPTER ELEVEN

Cell-Based Measurements of Mitochondrial Function in Human Subjects

Ju-Gyeong Kang, Ping-yuan Wang, Paul M. Hwang[1]
Center for Molecular Medicine, National Heart, Lung, and Blood Institute, National Institutes of Health, Bethesda, Maryland, USA
[1]Corresponding author: e-mail address: hwangp@mail.nih.gov

Contents

Abstract

There is growing evidence in the basic science field that aberrant metabolism plays an important role in tumorigenesis. Therefore, it is imperative to perform investigations in human subjects to determine (1) whether the metabolic observations made in model systems are applicable to humans; and (2) if indeed applicable, whether the metabolic alterations are clinically significant for cancer development. As an initial step, here we describe methods for measuring the mitochondrial metabolism of blood lymphocytes and skeletal muscle myoblasts that can be obtained from human subjects.

Methods in Enzymology, Volume 542
ISSN 0076-6879
http://dx.doi.org/10.1016/B978-0-12-416618-9.00011-X

1. INTRODUCTION

One proposed reason that cancer cells reprogram their metabolic activities is for survival and proliferation under adverse cellular environments. Although some types of cancer cells exhibit the "Warburg effect" (Warburg, 1956), growing evidence indicate that this increase in aerobic glycolysis is not intrinsically associated with cancer but, rather, that it may provide adaptive or growth advantages under specific conditions such as hypoxia (Zu & Guppy, 2004). Many cancer cells actually exhibit a concurrent increase in mitochondrial respiration and glycolysis as detailed in a comprehensive review of the Warburg effect (Koppenol, Bounds, & Dang, 2011). For example, increased oxygen consumption is associated with higher tumorigenic potential, and a subset of diffuse large B cell lymphoma has recently been shown to display a metabolic signature of increased oxidative phosphorylation and fatty acid oxidation (Caro et al., 2012; Ramanathan, Wang, & Schreiber, 2005). In our own recent translational study, we have observed that both lymphocytes and myoblasts of patients with Li–Fraumeni syndrome, an autosomal dominant premature cancer syndrome caused by germline mutations of *TP53*, exhibit increased oxidative metabolic activity (Wang et al., 2013). Therefore, further studies measuring mitochondrial function in tissue samples obtained from human subjects with different conditions may provide insights into the contribution of mitochondrial function in cancer pathogenesis that may have clinical implications.

2. BLOOD LYMPHOCYTE PROTOCOL

Human lymphocytes are prepared using a protocol as previously described with some modifications (Jackson, Devadas, Kwon, Pinto, & Williams, 2004). Peripheral blood mononuclear cells (PBMCs) are collected by density centrifugation using citrate-containing vacutainer cell preparation tubes (CPTs). The PBMCs are then activated on immobilized anti-CD3 and anti-CD28 antibodies for 3 days. Anti-CD3 antibody specifically activates the T cells to proliferate via the T-cell receptor complex. The stimulated T cells are subsequently cultured in RPMI media supplemented with IL-2 for 1–3 weeks and then attached to Seahorse XF-96 plates, which are pre-coated with Cell-TAK for metabolic analysis.

2.1. Materials

1. Citrate-containing vacutainer CPTs (Becton Dickinson, cat# 362761)
2. Human recombinant IL-2 (Boehringer Mannheim, cat# 799068)
3. Anti-human CD3e antibody (clone OKT3) (eBioscience, cat# 16-0037-85)
4. Anti-human CD28 antibody (clone CD28.2) (eBioscience, cat# 16-0289-85)
5. 10% BSA solution (MACS, cat# 130-091-376)
6. Phosphate-buffered saline (PBS), pH 7.4
7. 24-Well tissue culture plate
8. 50-ml sterile conical tube
9. Trypan blue
10. Dimethylsulfoxide (DMSO)
11. RPMI 1640 with GlutaMAX (Invitrogen, cat# 61870-036)
12. Fetal bovine serum (FBS) (certified, US origin, Invitrogen, cat# 16000-044)
13. Penicillin–streptomycin (P/S, 10,000 U/ml and 10,000 μg/ml, respectively)
14. β-Mercaptoethanol (BME, 55 m*M*) (Invitrogen cat # 21985)
15. Complete RPMI medium (supplemented with 10% FBS, 1% P/S, and 20 μ*M* BME)
16. XF96 4-port FluxPak (comprised of sensor cartridges, utility plates, cell culture microplates, and a calibrant solution) (Seahorse Bioscience cat# 102310)
17. XF assay medium (Seahorse Bioscience, cat# 102352)
18. BD Cell-Tak (BD Biosciences, cat# 354240)

2.2. Isolation of PBMCs

Note: Steps 1–3 are performed at room temperature, and the rest of the steps are at 4 °C or on ice with prechilled solutions.

1. If T-cell culturing as described in Section 2.3 is anticipated, 24 well plates need to be coated with anti-CD3e antibody the day before PBMC isolation. Prepare by pipetting 300 μl of an anti-CD3e antibody solution (5 μg/ml in sterile PBS solution) into each well of a 24-well tissue culture plate and incubate at 4 °C overnight (alternatively, 37 °C for 2 h may be an option).
2. Draw blood directly into citrate-containing vacutainer CPTs (~8 ml per tube).

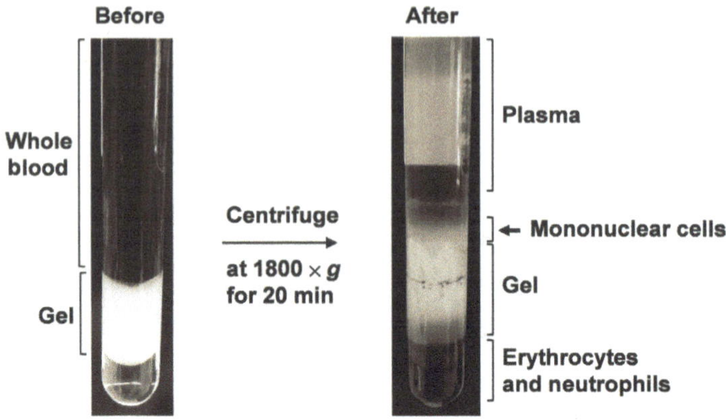

Figure 11.1 *Separation of lymphocytes from whole blood.* The appearance of a CPT filled with blood before and after centrifugation is shown. Density centrifugation allows the separation of peripheral mononuclear cells from other blood cell types and plasma. The mononuclear cell layer is gently transferred with a pipette for further processing and culturing. (See the color plate.)

3. Mix by gentle inversion and then centrifuge the tubes at $1800 \times g$ at room temperature for 20 min.
4. In a tissue culture hood, transfer the mononuclear cell layer (2–3 ml volume per CPT) containing lymphocytes and monocytes into a 50-ml sterile conical tube (Fig. 11.1).
5. Add at least three volumes of 0.5% BSA in PBS to the mononuclear cell layer to bring up the total volume to ~15 ml and then centrifuge at $300 \times g$ at 4 °C for 10 min.
6. Discard the supernatant, wash the cell pellet by resuspending in 10 ml of 0.5% BSA in PBS, and centrifuge at $300 \times g$ at 4 °C for 10 min.
7. Resuspend the cells again in 10 ml of 0.5% BSA in PBS and count with Trypan blue. The expected yield is $\sim 1 \times 10^6$ mononuclear cells/ml of blood.

2.3. T-cell culture

Note: The isolated peripheral blood mononuclear cells can be cultured at this point to expand T cells or they can be resuspended in FBS supplemented with 7% DMSO at a concentration of 2×10^6 cells/ml and stored in liquid nitrogen for future experiments. Human T cells can proliferate in culture for up to 3–4 weeks.

1. The PBMCs are centrifuged again at 300 × *g* at 4 °C for 5 min and then resuspended at 2×10^6 cells/ml in complete RPMI media containing 1 μg/ml anti-CD28 antibody.
2. Remove the previously incubated anti-CD3e antibody solution from each well of the 24-well tissue culture plate (see Section 2.2) and rinse once with PBS.
3. Dispense 1 ml of the PBMC suspension into each antibody-coated well to costimulate the T cells for 3 days.
4. Harvest the stimulated T cells by centrifuging at 300 × *g*, wash once with complete RPMI media, and resuspend at 1×10^6 cells/ml in the medium supplemented with IL-2 (10 U/ml). T cells can then be expanded for 1–3 weeks under this condition.

2.4. Measuring respiration in T cells using the Seahorse XF96 analyzer

Note: Unless otherwise specified, all 37 °C incubations should be performed in room air without CO_2 supplementation. Mitochondrial respiration can be measured using a Seahorse Analyzer that is designed for either 24- or 96-sample format, which requires appropriate cell number and volume adjustments. The Seahorse XF96 Extracellular Flux Analyzer uses a 96-well microplate assay format, and in our experience, the optimal density for human T cells is $2–4 \times 10^5$ cells per well which may need to be adjusted for other cell types. Measuring uncoupled respiration requires the addition of an uncoupler such as FCCP which needs to be titrated depending on the cell type. Detailed guides for using the Seahorse Analyzer can be found at the following Web site (http://www.seahorsebio.com/learning/protocols.php).

1. The day before the metabolic assay, add 200 μl of Seahorse Bioscience XF Calibrant pH 7.4 to each well of the XF-96 utility plate, position the sensor cartridge on top of the utility plate, and incubate at 37 °C overnight.
2. Prepare Cell-Tak solution by adding 73 μl of 1.5 mg/ml stock concentration Cell-Tak to 1.1 ml sterile H_2O. This results in a final concentration of 0.1 mg/ml and is of sufficient volume for one XF96 microplate.
3. Add 10 μl of the Cell-Tak solution to each well of a XF96 cell culture microplate and then add 40 μl of 0.1 *M* $NaHCO_3$ (pH 8.0) to each well to neutralize and promote Cell-Tak adsorption to the microplate.
4. Incubate the microplate at 37 °C for 1 h to overnight.

5. Prewarm the Seahorse XF assay media (supplemented with 25 m*M* glucose, 2 m*M* glutamine, and 1 m*M* sodium pyruvate) to 37 °C and adjust pH to 7.4 with 1 *M* KOH.
6. Count cell number, centrifuge the cells at 300 × *g* for 5 min, and resuspend in the supplemented XF assay media at 2.5×10^6 cells/ml.
7. Wash each well of the Cell-Tak-coated XF96 cell culture microplate with 200 μl of sterile water.
8. Add 120 μl of cell suspension to each well.
9. Centrifuge the microplate containing the cell suspension at 700 × *g* for 2 min to facilitate cell attachment.
10. Incubate at 37 °C for at least 1 h to allow cell attachment.
11. Up to four different compounds can be injected via ports A, B, C, and D at the appropriate concentrations to modulate mitochondrial respiration. It is important to make the appropriate concentrations of injected compounds to adjust for the sequential increase in total volume after each injection. In an example of a four-compound injection experiment, the cells are in a starting volume of 120 μl and the total volume increases to 200 μl with the addition of four sequential 20 μl injections of compounds. To adjust for the increase in volume, the stock concentrations of the injected compounds are appropriately diluted in warm XF medium to provide the indicated final concentration at each of the respective final volume. Hence, the order in which the injections are made cannot be altered after the diluted compounds have been prepared.

Port	Compounds	Stock conc.	Final conc. (vol.)	Injection compound dilution
A	Oligomycin	50 m*M*	2.5 μ*M* (140 μl)	1.5 μl stock + 4.28 ml media
B	FCCP	5 m*M*	1 μ*M* (160 μl)	6 μl stock + 3.75 ml media
C	FCCP	5 m*M*	2 μ*M* (180 μl)	6 μl stock + 3.33 ml media
D	Antimycin A	5 m*M*	1 μ*M* (200 μl)	6 μl stock + 3 ml media

12. Aliquot 20 μl of each compound into the corresponding ports of the sensor cartridge including the control wells for background correction.
13. Reposition the sensor cartridge over the utility plate and place them in the 37 °C incubator for at least 30 min.

14. Load the prewarmed sensor cartridge and utility plate into the instrument and start calibration.
15. After the calibration is complete, replace the calibrant utility plate with the cell-attached XF96 cell culture microplate and run the XF software program. The data analysis is briefly summarized in Section 4.

3. SKELETAL MUSCLE MYOBLAST PROTOCOL

The needle biopsy of the human skeletal muscle must be performed by a trained operator using sterile clinical grade equipments and anesthetics with appropriate approval as required by the investigator's institutional review board. The biopsy technique itself is not described here as it is beyond the scope of an experimental methods chapter. Needle biopsy specimens can be obtained from an accessible skeletal muscle group such as the *vastus lateralis* (lateral quadriceps of the leg). In our experience, a standard skeletal muscle biopsy needle (Ref# 8065, Cadence Science) can yield up to 50–100 mg of tissue from which about 25–50 mg is used for optimal myoblast preparation. The myoblast culture protocol is a modification of a previously described quantitative technique for growing adult skeletal muscle cells (Taanman et al., 1997; Yasin et al., 1977). Briefly, the muscle tissue is enzymatically dissociated using a combination of trypsin and collagenase, and the myoblasts are expanded in medium supplemented with uridine and high concentration of fetal bovine serum under reduced atmospheric oxygen (5% O_2).

3.1. Materials

1. DMEM (supplemented with GlutaMax-I, 4.5 g/l glucose, 1 m*M* pyruvate) (Invitrogen, cat# 10569-010)
2. Fetal bovine serum (FBS) (certified, US origin, Invitrogen, cat# 16000-044)
3. Uridine (20 m*M* stock solution)
4. P/S (10,000 U/ml and 10,000 μg/ml, respectively)
5. Trypsin (2.5%) (Invitrogen-Gibco, cat# 15090-046)
6. Versene (~0.5 m*M* EDTA in PBS) (Invitrogen-Gibco, cat# 15040-066)
7. Collagenase, type II (555 U/mg) (Sigma-Aldrich, cat# C6885)
8. BSA (essentially fatty acid free) (Sigma-Aldrich, cat# A8806)
9. 60 mm tissue culture dish (collagen type I coated) (BD, cat# 354401)
10. 75 cm^2 vented flask (collagen type I coated) (BD, cat# 354485)

11. Cell strainer (40 μm nylon) (BD, cat# 352340)
12. Sterile glass beaker (100 ml)
13. Sterile magnetic stir bar
14. Sterile hypodermic needle
15. Sterile aluminum foil
16. Water bath with shaker
17. Tissue culture incubator with 5% CO_2 and 5% O_2 capability
18. Diluted trypsin solution:

Reagent	Stock conc.	Volume	Final conc.
Trypsin	2.5%	4 ml	0.1%
Versene (EDTA in PBS)	~0.5 m*M*	100 ml	~0.5 m*M*
Total		104 ml	

19. Enzymatic dissociation solution (prepare fresh before each experiment):

Reagent	Stock conc.	Volume/mass	Final conc.
Collagenase, type II	555 U/mg	21.6 mg	240 U/ml
BSA (fatty acid free)		40 mg	
DMEM medium		10 ml	
Filter the above mix through a 0.2-μm filter into a 50-ml sterile tube then add:			
Diluted trypsin solution	0.1%	3.2 ml	0.0064%
DMEM medium		40 ml	
Total		50 ml	

20. Complete myoblast medium:

Reagent	Stock conc.	Volume	Final conc.
DMEM		500 ml	
FBS	100%	100 ml	~20%
P/S		3 ml	50 U/ml and 50 μg/ml
Uridine	20 m*M*	60 μl	2 μ*M*
Total		603 ml	

3.2. Myoblast culture

Note: Care should be taken to split the primary human myoblast cultures before they reach confluency because they can fuse and form myotubes. The myoblasts should be maintained below 70% confluency and can be passaged about 10 times before they lose proliferative capacity. Unless indicated otherwise, it is important to note that human myoblasts are maintained in 5% CO_2 under reduced oxygen tension (5% O_2) to maintain their proliferative potential.

1. Use 25–50 mg of skeletal muscle needle biopsy sample. A minimum of 25 mg is recommended for successfully culturing and expanding human myoblasts.
2. Place the muscle tissue in a sterile 100-ml beaker and add 10 ml of enzymatic dissociation solution. Use two sterile needles mounted on syringes to tease the muscle tissue apart for about 3 min.
3. Cover the beaker with a sterile aluminum foil and incubate it in a water bath at 37 °C with slow (40 rpm) shaking for 12 min.
4. Add 10 ml of complete myoblast medium which inactivates trypsin and triturate the muscle fragments by pipetting up and down 10 times through a 10-ml pipette.
5. Allow the muscle fragments to settle, then use a 10-ml pipette to transfer the supernatant and filter through a cell strainer (40 μm nylon) into a sterile 50-ml conical tube.
6. Add another 10 ml of enzymatic dissociation solution to the beaker with residual muscle fragments and incubate at 37 °C for 12 min.
7. During this time, centrifuge the supernatant from step 5 in the 50-ml tube at 300 × *g* for 10 min to sediment the cells. Aspirate off the supernatant, resuspend the cell pellet in 0.5 ml of complete myoblast medium, and place the loosely capped 50-ml tube in a 37 °C humidified incubator (5% CO_2 and 5% O_2).
8. Perform two to three cycles of steps 2–7 until the muscle tissue fragments are essentially dissociated.
9. Pool the cells suspended in 0.5 ml of complete myoblast medium and transfer to a collagen-coated 60-mm culture dish. Add an additional ~4–6 ml of complete medium to the dish and mix by swirling.
10. Incubate the cells and change the media every 3–4 days until the cells reach ~70% confluency at which time they can be gently trypsinized and replated on larger collagen-coated 75 cm^2 vented flask.

11. Myoblasts can be resuspended in complete myoblast medium containing 7% DMSO and stored in liquid N_2 for future use. (Remember to thaw out and maintain human myoblasts under 5% O_2 condition for improved viability unless 21% O_2 is required for experimental purposes.)

3.3. Measuring respiration in myoblasts using the Seahorse XF24 analyzer

Note:

1. Unlike mononuclear cells that require a smaller surface area for attachment, myoblasts need more space provided by the 24-well format of the Seahorse XF24 Extracellular Flux Analyzer. A uniform distribution of cells over the bottom of the well is also critical for consistent results.
2. Optimal cell numbers should be used for the experiment. Depending on the cell type, the oxygen consumption rate (OCR) should fall within the linear range of 50–500 pmol/min and the extracellular acidification rate should be higher than 20 milli-pH/min.

   1. Gently dispense 100 µl of cell suspension containing around 30,000 myoblasts into the bottom of a XF24 cell culture microplate well. Allow 1–5 h for cell attachment (~80% confluent) and then gently add an additional 150 µl of the complete myoblast medium for overnight culture. Four wells (A1, B4, C3, and D6) containing only medium are used as the temperature correction controls.
   2. Preincubate the sensor cartridge and utility plate containing the calibrant solution overnight in a non-CO_2 room air 37 °C incubator.
   3. Leave the XF24 analyzer with the software running overnight to stabilize the analyzer.
   4. Next day, check that the analyzer temperature is at 37 °C.
   5. Prewarm the Seahorse assay medium and reagents to 37 °C.
   6. Replace the myoblast medium with the Seahorse XF assay medium, which should be made fresh by adding 25 m*M* glucose, 2 m*M* glutamine, and 1 m*M* sodium pyruvate prior to adjusting the pH to 7.4 with 1 *M* KOH. The wells should not be allowed to dry during media changes. Rinse once with 1 ml of the XF assay medium and then add 700 µl of the medium to each well.
   7. Incubate the cell plate in a non-CO_2 incubator at 37 °C for 1 h before starting the metabolic measurements.

8. As described in Section 2.4, step 11, up to four different compounds placed in the sensor cartridge can be injected via ports A, B, C, and D at the appropriate concentrations to modulate mitochondrial respiration. The working stocks of the compounds are prepared as previously described except a larger volume of 75 μl is loaded.
9. Calibrate the analyzer by loading the sensor cartridge and the calibrant solution-containing utility plate into the instrument tray ensuring correct orientation.
10. When complete, replace the calibration utility plate with the cell-containing microplate for the assay.

4. DATA ANALYSIS

In this chapter, we describe an *in vitro* approach for measuring aerobic metabolism in tissue samples obtained from human subjects. We have focused on mitochondrial function using the Seahorse XF Analyzer although it also measures extracellular acidification representing glycolysis. A basic whole cell Seahorse assay provides the rate of oxygen consumption in the following states using specific pharmacologic compounds: state 3 respiration that is coupled to ATP synthesis; state 4 respiration in the absence of ATP synthesis; maximum mitochondrial respiration that can be achieved by uncoupling to ATP synthesis; and nonmitochondrial oxygen consumption (Fig. 11.2). These measurements can be further analyzed with statistics software such as GraphPad Prism or GraphPad InStat (GraphPad Software, Inc.). Detailed guides to data analysis can be found at the following resource Web site (http://www.seahorsebio.com/learning/protocols.php).

Basal ADP-dependent respiration coupled to ATP synthesis (state 3) is affected by many factors such as substrate availability, cell proliferation rate, and cellular energy requirements. In our study of human lymphocytes and myoblasts, we have observed that alterations in respiratory chain components can affect mitochondrial respiration (Wang et al., 2013). For example, the basal OCR values obtained using T-cell samples of different human subject show a significant positive correlation with the levels of mitochondrial respiratory complex V ATP synthase subunit α protein (Fig. 11.3).

Respiration in the absence of ATP synthesis (state 4) is largely determined by mitochondrial membrane proton leakage which can be estimated by the respiratory control ratio (RCR; ratio of state 3 vs. state 4 OCR). A high RCR indicates tighter coupling of respiration to ATP synthesis and may serve as a functional assessment of mitochondrial integrity. With

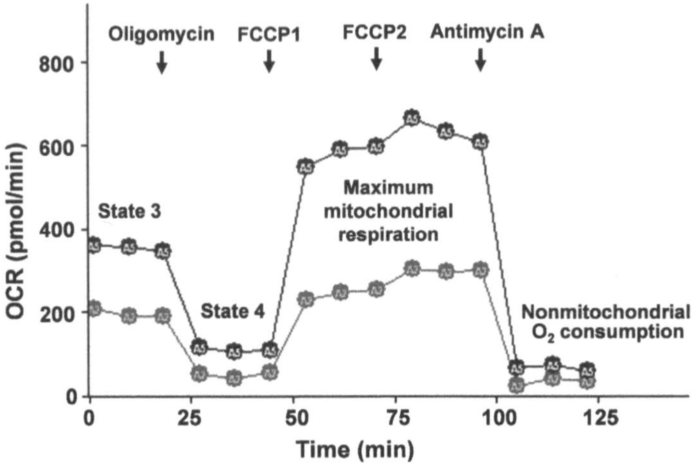

Figure 11.2 *Measuring mitochondrial respiration.* Shown are representative tracings of oxygen consumption rate (OCR) of lymphocytes from two human subjects (red and blue) using the Seahorse XF Analyzer. State 3 respiration in the presence of ADP, state 4 respiration with inhibition of ATP synthesis using oligomycin, uncoupled respiration in the presence of increasing doses of FCCP (1 and 2), and nonmitochondrial oxygen consumption with inhibition of respiration using antimycin A are shown. Arrows indicate the approximate points at which the indicated compounds were added.

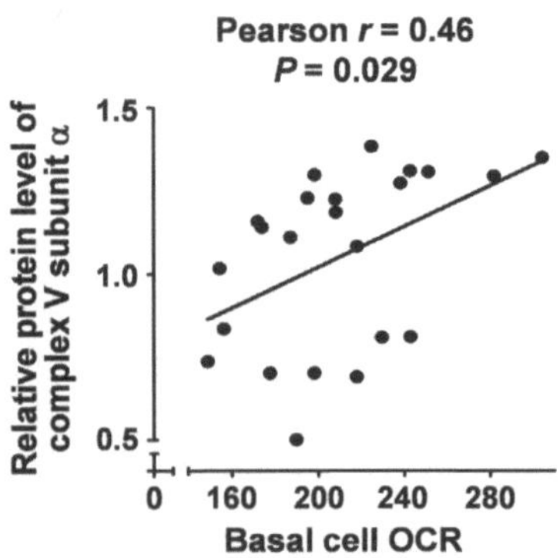

Figure 11.3 *Correlation between mitochondrial function and biogenesis.* A significant positive correlation between basal oxygen consumption rate (OCR) and protein level of mitochondrial respiratory complex V α subunit can be demonstrated in this set of human lymphocyte samples. The protein levels are normalized by a control such as β-actin after quantification using fluorescent Western blot imaging (LI-COR Biosciences).

the addition of the mitochondrial uncoupler FCCP, the maximum respiratory capacity of the cell can be measured (Fig. 11.2). It should be noted that about 10% of the basal OCR represents nonmitochondrial oxygen utilization such as by NADH/NADPH oxidases, various desaturases, and detoxification enzymes (Herst, Tan, Scarlett, & Berridge, 2004). The addition of respiratory complex III inhibitor antimycin A at the end of each Seahorse assay provides an estimate of the oxygen consumption contributed by nonrespiratory chain sources. Such *in vitro* measurements using samples obtained from patients may provide insights into the metabolic changes underlying disease conditions.

ACKNOWLEDGMENTS

This work was supported by the Division of Intramural Research, National Heart, Lung, and Blood Institutes (NHLBI), National Institutes of Health. We thank Martin Pelletier and Kim K. Han for helpful advice with the Seahorse XF Analyzer protocol.

REFERENCES

Caro, P., Kishan, A. U., Norberg, E., Stanley, I. A., Chapuy, B., Ficarro, S. B., et al. (2012). Metabolic signatures uncover distinct targets in molecular subsets of diffuse large B cell lymphoma. *Cancer Cell, 22*, 547–560.

Herst, P. M., Tan, A. S., Scarlett, D. J., & Berridge, M. V. (2004). Cell surface oxygen consumption by mitochondrial gene knockout cells. *Biochimica et Biophysica Acta, 1656*, 79–87.

Jackson, S. H., Devadas, S., Kwon, J., Pinto, L. A., & Williams, M. S. (2004). T cells express a phagocyte-type NADPH oxidase that is activated after T cell receptor stimulation. *Nature Immunology, 5*, 818–827.

Koppenol, W. H., Bounds, P. L., & Dang, C. V. (2011). Otto Warburg's contributions to current concepts of cancer metabolism. *Nature Reviews. Cancer, 11*, 325–337.

Ramanathan, A., Wang, C., & Schreiber, S. L. (2005). Perturbational profiling of a cell-line model of tumorigenesis by using metabolic measurements. *Proceedings of the National Academy of Sciences of the United States of America, 102*, 5992–5997.

Taanman, J. W., Bodnar, A. G., Cooper, J. M., Morris, A. A., Clayton, P. T., Leonard, J. V., et al. (1997). Molecular mechanisms in mitochondrial DNA depletion syndrome. *Human Molecular Genetics, 6*, 935–942.

Wang, P. Y., Ma, W., Park, J. Y., Celi, F. S., Arena, R., Choi, J. W., et al. (2013). Increased oxidative metabolism in the Li-Fraumeni syndrome. *New England Journal of Medicine, 368*, 1027–1032.

Warburg, O. (1956). On the origin of cancer cells. *Science, 123*, 309–314.

Yasin, R., Van Beers, G., Nurse, K. C., Al-Ani, S., Landon, D. N., & Thompson, E. J. (1977). A quantitative technique for growing human adult skeletal muscle in culture starting from mononucleated cells. *Journal of the Neurological Sciences, 32*, 347–360.

Zu, X. L., & Guppy, M. (2004). Cancer metabolism: Facts, fantasy, and fiction. *Biochemical and Biophysical Research Communications, 313*, 459–465.

CHAPTER TWELVE

Use of Chemical Probes to Detect Mitochondrial ROS by Flow Cytometry and Spectrofluorometry

Jing Chen, Clayton E. Mathews[1]
Department of Pathology, Immunology and Laboratory Medicine, The University of Florida College of Medicine, Gainesville, Florida, USA
[1]Corresponding author: e-mail address: clayton.mathews@pathology.ufl.edu

Contents

Abstract

Aerobic respiration is a major source of energy in eukaryotic cells. In this setting, ATP production by the mitochondrial respiratory chain relies on the availability of NADH and $FADH_2$ to donate protons and electrons. The flux of electrons down the electron transport chain, based on a series of oxidation–reduction reactions, releases energy that allow for the transport of H^+ ions across the inner mitochondrial membrane. The resulting proton-motive force is employed to drive ATP synthesis, while the final acceptor of the electrons flowing through the respiratory chain if molecular oxygen. A side effect of this process is the generation of reactive oxygen species (ROS). While mitochondrial ROS (mtROS) production has been linked to many pathological conditions (i.e., aging and tumorigenesis), recent evidence suggests that multiple cells, including

Methods in Enzymology, Volume 542
ISSN 0076-6879
http://dx.doi.org/10.1016/B978-0-12-416618-9.00012-1

malignant cells, employ these by-products of energy production as signals to control various cellular processes. Here, we describe protocols to use chemical probes for measuring mtROS production in intact cells by flow cytometry and spectrofluorometry.

1. INTRODUCTION

The image of the mitochondria (mt) as the powerhouse of the cell has undergone a dramatic transformation. Energy production is a major responsibility of the mt; however, these organelles perform a number of other cellular functions including calcium handling, production of reducing equivalents, cell death through intrinsic apoptosis, and participation in cell signaling. While once viewed in the negative light as damaging by-products of aerobic respiration, reactive oxygen species (ROS) production by mt are now viewed as essential signaling mediators that act in reduction–oxidation signaling pathways. The signaling pathways impacted by ROS generated by the mitochondrial ROS (mtROS) have been linked to both normal as well as aberrant conditions, including tumor growth. Advances in whole-cell detection of ROS will lead to an enhanced understanding of how these molecules influence cellular processes. In this chapter, we describe methods to detect mtROS that can be applied to a wide range of primary and transformed cell types from both human and animal models.

2. MITOCHONDRIAL ROS PRODUCTION

2.1. Physiology

There are four structural entities within the mitochondria: the inner and outer membranes, and the two compartments that they create, the intermembrane space and the matrix. These structural regions are the scaffolds upon which many important metabolic processes take place. Control of mitochondria comes from expression of genes from both the nuclear and mitochondrial genomes. Intergenomic interactions have been implicated in pathophysiology (Chen, Gusdon, Thayer, & Mathews, 2008; Chen, Lu, et al., 2008; Johnson, Zheng, Bykhovskaya, Spirina, & Fischel-Ghodsian, 2001).

Mitochondrial respiration is the process in which water is formed and energy trapped in the bonds of the ATP molecules. The energy that drives respiration is obtained from the oxidation of fatty acids, amino acids, and carbohydrates. The product of macronutrient oxidation and glycolysis is acetyl-CoA which, when joined with oxaloacetate, forms citrate to be processed in the mitochondrial Kreb's cycle. The Kreb's cycle produces reducing equivalents as well as carbon dioxide, in a cycle of enzymatically catalyzed reactions. For every turn of the Kreb's cycle, two carbon dioxide molecules and eight reducing equivalents are produced. The reducing equivalents (NADH and $FADH_2$) are utilized by the respiratory chain to build an electrochemical potential across the mitochondrial inner membrane.

In the electron transport process, the free energy of electron transfer is coupled to ATP synthesis. The free energy necessary to generate ATP is extracted from the oxidation of NADH by NADH dehydrogenase (complex I) and $FADH_2$ by the Kreb's cycle component succinate dehydrogenase (complex II). The electron transport chain (ETC) passes electrons from lower to higher standard reduction potentials and ultimately to oxygen forming water. Oxidative reactions, catalyzed by inner mitochondrial membrane embedded enzyme complexes, generate an electrochemical or pH gradient, by proton translocation, across the mitochondrial inner membrane. ATP can then be made from ADP and Pi by the membrane-bound F_1F_0 ATP synthase using the proton gradient to drive synthesis. The oxidative phosphorylation (OXPHOS) system is composed of five respiratory complexes (complexes I–V) that are assembled from subunits derived from mtDNA and nDNA plus the adenine nucleotide translocator.

Complex I contains one molecule of flavin mononucleotide (FMN), a redox-active prosthetic group, and six to seven iron–sulfur [Fe–S] clusters that participate in the electron transfer (Efremov & Sazanov, 2012; Kim, Khalimonchuk, Smith, & Winge, 2012). Redox-active Fe–S clusters occur as prosthetic groups of nonheme Fe–S proteins. There are two common types, designated [2Fe–2S] and [4Fe–4S] clusters. Each cluster contains an equal number of iron and sulfide ions and is coordinated to four protein-cysteine-sulfhydryl groups, with the iron atoms coordinated between four sulfurs to form a tetrahedron. Electrons pass from the FMN through the multiple Fe–S centers to coenzyme Q [ubiquinone or CoQ]. The reduction of CoQ is coupled to proton translocation (Hatefi, 1985).

Complex II contains three small hydrophobic subunits, all of which are nuclear encoded (Shoffner & Wallace, 1994). The physiological function of complex II is to catalyze the transfer of electrons from succinate to CoQ, with the participation of a covalently bound flavin adenine dinucleotide (FAD), one [4Fe–4S] cluster, two [2Fe–2S] clusters, and one cytochrome *b*560. Electron transfer from succinate to CoQ has a standard redox potential that is insufficient to provide the free energy to drive ATP synthesis, but the complex is important in that it allows these relatively high-potential electrons to enter the electron transport system.

Complex III catalyzes the electron transfer from reduced CoQ to cytochrome *c*. This complex contains two *b*-cytochromes, one cytochrome *c*, and a [2Fe–2S] cluster. Cytochromes are redox-active heme-proteins that reversibly alternate between the Fe(II) and Fe(III) oxidation states. The electrons pass through cytochrome *c* to cytochrome *c* oxidase of complex IV (Wallace, 1992).

The reactions catalyzed by complex IV include the four-electron reductions of O_2 by the single-electron donor, ferrocytochrome *c*. The redox-active centers are two a-type heme groups, which, unlike c-type heme groups contain no sulfur, and two Cu atoms that alternate between their +1 and +2 oxidation states. The identical hemes (a and a3) are each associated with a copper atom (Cua or CuA and Cua3 or CuB, respectively). Heme a and Cua are of low potential and are the recipients of electrons from cytochrome *c*. Heme a3 and CuB are of higher potential and are the sites of oxygen reduction (Hatefi, 1985).

ATP production by the mitochondria occurs through the coupling of the redox free energy to the phosphorylation of ADP (Hatefi, 1985). The free energy of the electrochemical proton gradient derived from the operation of respiratory complexes I, III, and IV, across the inner mitochondrial membrane, is harnessed in the synthesis of ATP from ADP and Pi, by the F_1F_0 ATP synthase (Hatefi, 1985). Also known as complex V of the respiratory chain, this enzyme is also capable of ATP hydrolysis linked to proton translocation from the matrix to the intermembrane space (Hatefi, 1985; Senior, 1988).

Previous reports over the past four decades have provided evidence for ROS production by mitochondria and have further implicated specific sites within the ETC as ROS generators. ETC inhibitors, such as rotenone and antimycin A, have been used extensively to localize mtROS production. Redox centers within Complex I and III have been noted as the major sites of mtROS production (St-Pierre, Buckingham, Roebuck, & Brand, 2002; Sugioka et al., 1988; Turrens & Boveris, 1980). Within complex I, both the FMN and a distal site presumably the ubiquinone-binding site have been shown to be capable of generating ROS with the direction of electron flow dictating the relative contribution from each site (Treberg, Quinlan, & Brand, 2011). ROS from Complex I are released into the mitochondrial matrix. Under ATP-generating conditions, the FMN site within complex I has been shown to be responsible for the majority of ROS production (Quinlan, Treberg, Perevoshchikova, Orr, & Brand, 2012). Complex III supported ROS production can occur at both the Q_i site of the cytochrome bc_1 complex promoted by a partially oxidized ubiquinone pool (Drose & Brandt, 2008) as well as the Qp site that faces the inner membrane space. For Complex III ROS production is directional, with Qi generating ROS in the mitochondrial matrix, while ROS from the Q_p site of Complex III are released into the intermembrane space (Gusdon, Votyakova, & Mathews, 2008; Gusdon, Votyakova, Reynolds, & Mathews, 2007; Muller, Liu, & Van Remmen, 2004; St-Pierre et al., 2002; Tahara, Navarete, & Kowaltowski, 2009).

It is important to note that ROS production is dissimilar when comparing isolated mitochondria from different tissues (St-Pierre et al., 2002). Studies using rat models have demonstrated that during state 4 respiration, mitochondria from brain and heart produce significantly more ROS than kidney, skeletal muscle, or liver mitochondria. During state 3 respiration (after addition of ADP), a significant drop of peroxide production in brain as well as in heart and skeletal muscle mitochondria was observed; however, ROS production from liver was almost unchanged, while kidney mitochondria exhibited only a slight reduction in ROS. Further, mitochondria isolated from brain, cardiac muscle, and kidney produced high levels of ROS when respiring on succinate, while liver mitochondria exhibited elevated ROS production when complex I substrates were added. Skeletal muscle was uniformly low for all substrates tested. These distinctions could result from gene expression differences in mitochondrial proteins that control mitochondrial respiration or other mitochondrial functions (e.g., substrate transporters) among these tissues. These data also suggest that specific mtDNA mutations exhibit tissue-specific phenotypes due to tissue-specific ROS production that when combined can drive pathophysiology.

2.2. Pathophysiology/apoptosis

Mitochondrial dysfunction has been implicated in a growing number of disorders. The manifestations of these syndromes have in many cases been associated with an imbalance in mtROS generation. The ROS produced by the mitochondria principally consist of superoxide and hydrogen peroxide. ROS generated by the mitochondria have also been implicated in cancer (Sohn et al., 2013), aging (Shigenaga, Hagen, & Ames, 1994; Sohal & Weindruch, 1996; Terzioglu & Larsson, 2007; Tong, Schriner, McCleary, Day, & Wallace, 2007), as well as in cardiovascular disorders, including hypertension (Peterson, Sharma, & Davisson, 2006; Puddu, Puddu, Cravero, Rosati, & Muscari, 2008; Zimmerman et al., 2002), atherosclerosis (Bayir et al., 2006; Hsiai & Berliner, 2007; Pandya, 2001; Singh & Jialal, 2006; Steinberg & Witztum, 2002; Zhang et al., 2007), and myocardial infarction (Surekha et al., 2007; Venardos & Kaye, 2007). mtROS has been tightly linked to neurodegenerative conditions, including Alzheimer's disease (Butterfield, Howard, & LaFontaine, 2001; Kennedy, Sandiford, Skerjanc, & Li, 2012; Lin & Beal, 2006), Parkinson's disease (Gautier, Kitada, & Shen, 2008; Kennedy et al., 2012; Lin & Beal, 2006; Tretter, Takacs, Hegedus, & Adam-Vizi, 2007), amyotrophic lateral sclerosis (Lin & Beal, 2006; Panov et al., 2011), and Huntington's disease (Butterfield et al., 2001; Lin & Beal, 2006; Siddiqui, Hanson, & Andersen, 2012). β-Cell dysfunction and death in both type 1 and type 2 diabetes have been linked to mtROS production (Chen, Gusdon, Piganelli, Leiter, & Mathews, 2011; Gusdon et al., 2007; Gusdon et al., 2008; Krauss et al., 2003), and increased superoxide production has been shown to cause DNA damage leading poly(ADP-ribose) polymerase activation, inhibition of glyceraldehyde 3-phosphate dehydrogenase, and induction of the main pathways of hyperglycemia induced pathology (Du et al., 2003).

In some of these pathophysiological cases, the mitochondria play a key role through participation in the intrinsic pathway of apoptosis by producing ROS, undergoing permeability transition, and releasing cytochrome *c* for apoptosome formation. Literature supports the activation of caspase 8 and downstream apoptotic events by H_2O_2 (Aggarwal, Misro, Maheshwari, Sehgal, & Nandan, 2010; Payton et al., 2007; Perez-Cruz, Carcamo, & Golde, 2007; Rigoulet, Yoboue, & Devin, 2011; Shenoy, Wu, & Pervaiz, 2009). Further, upon exposure to peroxide, TNFR-associated factor 2 (TRAF2), receptor-interacting protein (RIP), and FAS-associated death domain (FADD) dissociate from TNFR1 and this new complex is capable of recruiting and activating caspase 8 as well as ROS modulator (Romo1) to promote apoptosis (Kim, Lee, Park, & Yoo, 2010; Shen et al., 2004; Terry Powers et al., 2010). In sum, mtROS can directly damage cells leading to disease onset, the development of sequelae (e.g., Tumorigenesis or Diabetes complications), and impact cellular processes by acting as signaling intermediates to potentiate proliferative or apoptotic signals.

2.3. Redox signaling

ROS act as signaling mediators not only during cell death/apoptosis but also in processes that control cellular proliferation and differentiation. A clear role for cytoplasmic ROS generated by NADPH Oxidase 2 as well as Dual Oxidase 1 has been shown in T cell receptor signaling as well as downstream activation and differentiation of T cells (Jackson, Devadas, Kwon, Pinto, & Williams, 2004; Kwon et al., 2010; Thayer et al., 2011; Tse et al., 2010). Mitochondrial metabolic activity plays a central role in controlling T cell activation, proliferation, and programmed cell death (Chhabra, 2010). Specifically, during the process of T cell activation, mt are recruited to the immunological synapse where through balanced fusion/fission activity, they control transmembrane potential [$\Delta\Psi\mu$], ATP synthesis, and local calcium concentrations (Baixauli et al., 2011; Quintana et al., 2007; Schwindling, Quintana, Krause, & Hoth, 2010). Additionally, calcium-dependent mtROS production is induced. In T cells, ROS production by the Qp site of complex III is required for antigen-induced T cell activation. The loss of Qp generated ROS in T cells results in a failure of transcription factor activation and an inability to produce interleukin-2, a cytokine essential for T cell proliferation (Kaminski et al., 2010; Sena et al., 2013). However, the specific target of mtROS in this process has not yet been identified.

Mitochondria are now recognized as contributors during cell transformation and tumor growth. Production of mtROS in tumors is thought to result from oncometabolism, a term that describes the increased metabolic and energetic demands of transformed cells. Emerging evidence has associated mtROS production by tumors to the dysregulated expression of nuclear-encoded mitochondrial proteins, such as CISD1 (mitoNEET) (Sohn et al., 2013). Overexpression of the CISD1 in tumors results in iron accumulation, elevated mtROS, and heightened tumor growth, while conversely shRNA-mediated suppression of expression led to reduced cell proliferation and inhibited tumor growth. Overall, mtROS are an important characteristic of oncometabolism. By extensively characterizing the production and actions of mtROS in tumors, including the mechanisms of production and cellular targets, therapeutic strategies can be developed that target processes specific to cell transformation.

3. METHODS TO DETECT ROS PRODUCTION

3.1. Reagents

3.1.1. Culture medium

3.1.1.1. Culture medium for mouse beta cell lines (NIT-1 and NIT-4):

- DMEM (Gibco, Catalog number 11885-084) supplemented with:

- 10% FBS (Hyclone, Catalog number SH30071.03, heat inactivated)
- 0.02% BSA (Sigma, Catalog number A8412)
- 0.15 m*M* HEPES (Cellgro, Catalog number 25-060-Cl)
- MEM nonessential amino acid (Gibco, Catalog number 11140)
- Add glucose to a final concentration of 16.5 m*M*
- Penicillin and streptomycin (Gemini Bio-Products, Catalog number 400-109. Penicillin final concentration 100 U/mL and streptomycin final concentration 100 μg/mL)

3.1.1.2. Culture medium for human beta cell line (BetaLox 5):

- DMEM (Cellgro, Catalog number 10-014-CV) supplemented with:
- 10% FBS (Hyclone, Catalog number SH30071.03, heat inactivated)
- 0.02% BSA (Sigma, Catalog number A8412)
- 15 m*M* HEPES (Cellgro, Catalog number 25-060-Cl)
- MEM Nonessential amino acid (Gibco, 11140)
- Penicillin and streptomycin (Gemini Bio-Products, Catalog number 400-109. Penicillin final concentration 100 U/mL and streptomycin final concentration 100 μg/mL)

3.1.1.3. Culture medium for mouse splenocytes and human PBMC, human T cells:

- RPMI1640 (Cellgro, Catalog number 10-040-CV) supplemented with:
- 10% FBS (Heat Inactivated, HyClone, Catalog number SH30071.03)
- 10 m*M* HEPES (Cellgro, Catalog number 25-060-Cl)
- Penicillin and streptomycin (Gemini Bio-Products, Catalog number 400-109. Penicillin final concentration 100 U/mL and streptomycin final concentration 100 μg/mL)
- 1 m*M* sodium pyruvate (Cellgro, Catalog number 25-000-cl)
- 50 μg/mL uridine (Sigma, Catalog number U3003)
- 25 m*M* glucose final concentration

3.1.1.4. Hemolytic Gey's buffer

Stocks of A, B, and C can be made ahead and stored in the refrigerator; however, 1 × Gey's buffer must be made fresh each day by mixing stocks A (4 parts), B (1 part), C (1 part), and water (14 parts).

Solution A	
NH_4Cl	131.08 m*M*
KCl	4.96 m*M*
$Na_2HPO_4 \cdot 12H_2O$	0.85 m*M*
KH_2PO_4	176.00 μ*M*
Glucose	5.55 m*M*
Phenol red	28.22 μ*M*
Dissolve in 1 L of distilled water and autoclave	

Solution B	
$MgCl_2 \cdot 6H_2O$	1.03 m*M*
$MgSO_4 \cdot 6H_2O$	0.29 m*M*
$CaCl_2$	1.16 m*M*
Dissolve in 0.5 L of distilled water and autoclave	

Solution C	
$NaHCO_3$	13.39 m*M*
Dissolve in 0.5 L of distilled water and autoclave	

3.1.2. Fluorescent dyes:

- MitoSox Red (Invitrogen, Catalog number M36008)
- CYTOX Green (Invitrogen Catalog number S-7020)
- Annexin V-APC (Invitrogen, Catalog number A35110)
- $DiOC_6$ (Invitrogen, Catalog number D273)

- CM-H_2DCFDA (Invitrogen, Catalog number C6827)
- H_2DCFDA (Sigma, Catalog number D6883)
- Dihydroethidium (DHE) (Invitrogen, Catalog number D23806)

3.1.3. Buffers:

- PBS (Cellgro, Catalog number 21-040-CV):
- HBSS (Lonza, Catalog number10-527F)
- FACS buffer: 1 × PBS + 2% FBS, phenol red-free

3.1.4. Other reagents:

- Cytokines:
  - **a.** Recombinant human tumor necrosis factor alpha (TNF-α) (R & D Systems, Catalog number 210-TA-010/CF)
  - **b.** Recombinant human interferon gamma (IFN-γ) (BD Biosciences, Catalog number 554617)
  - **c.** Recombinant mouse TNF-α (eBioscience, Catalog number 14-8321-63)
  - **d.** Recombinant mouse IFN-γ (Sigma, Catalog number I4777)
- Ficoll (GE Healthcare, Catalog number 17-1440-03)
- T cell enrichment cocktail (Stemcell, Catalog number 15021/15061)
- 0.25% Trypsin–EDTA (Gemini Bio-product, Catalog number 400-151)

3.2. Cell preparation for flowcytometry analysis

3.2.1. Mouse beta cell lines (NIT-1 and NIT-4):

1. Seed cells in 12-well plates at 5×10^5 cells per well. Allow cells attach overnight.
2. On the second day, treat cells accordingly (see Section 3.3 as an example).
3. On the day of the assay, the dyes should be prepared fresh and the cells stained as described below.

3.2.2. Human beta cell line (BetaLox5):

1. BetaLox5 cells are seeded in 12-well Corning Costar culture plates (Fisher Scientific) at a density of 5×10^4 cells per well in a total of 500 μL and allowed to adhere for 24 h.
2. Treat cells accordingly (see Section 3.3 as an example).

3. On the day cells will be analyzed on a flow cytometer, stain cells with MitoSox Red and $DiOC_6$ dye as described below.

3.2.3. Mouse splenocytes:

1. Harvest mouse spleens.
2. Put one spleen in 5 mL ice-cold HBSS and homogenize using a 7-mL glass Tenbroeck homogenizer (Kontes Glass Company).
3. Collect cells by centrifuging at 350 × *g* for 5 min at 4 °C.
4. Remove red blood cells by incubating cells in hemolytic Gey's solution for 5 min on ice. Stop reaction by adding same volume of isotonic HBSS. Spin down at 350 × *g* for 5 min.
5. Resuspend cells in cold HBSS and pass through cell strainer [Falcon, Catalog number 352340] to remove dead cell clusters.
6. Count cells and proceed to staining.

3.2.4. Human PBMC:

1. Collect human blood samples using a BD Vacutainer with heparin sulfate [Green Top].
2. Leave samples on a rocker overnight.
3. Transfer blood to a 50-mL conical tube. Fill to 17.5 mL per tube.
4. Add DPBS without calcium or magnesium to a total volume of 35 mL. This achieves a 1:1 dilution v/v dilution.
5. Gently add 15 mL Ficoll beneath the sample. We use the following method: Place a 9″ glass Pasteur pipette into the conical so that the small end is at the bottom of the tube. Then use a 5-mL serological pipette to add Ficoll to the large end of Pasteur pipette.
6. Centrifuge at 800 × *g* for 20 min at room temperature with the centrifuge break set to off.
7. Aspirate plasma layer to within 1 cm of interphase layer.
8. Collect interphase layer (buffy coat) using a transfer pipette, slowly sweep top of the desired layer, constantly releasing the pipette bulb and transfer to a new 50-mL conical tube.
9. Wash with PBS up to 50 mL.
10. Centrifuge at 450 × *g* for 10 min at room temperature. Max brake and max acceleration.
11. Resuspend cells in 1 mL PBS. Remove red blood cells using NH_4CL if necessary. Add 9 mL NH_4Cl (0.8% NH_4CL in H_2O, 0.1 m*M* EDTA, buffered with $NaHCO_3$ to

pH 7.2–7.6) to 1 mL cell suspension, incubate on ice for 10 min. Wash twice with PBS as above.

12. Count cells and proceed to staining.

3.2.5. Human T cells:

1. Collect human blood samples using BD Vacutainers with heparin sulfate [Green Top]
2. Leave samples on the rocker for gentle shake overnight.
3. Enrich T cells using RosetteSep Human T cell enrichment cocktail (Stemcell, Catalog number 15021/15061) following the manufacturer's instructions.
4. Treat cells accordingly.
   - We activate T cells *in vitro* by seeding 1×10^6 cells per well of enriched human T cells in 12-well plates precoated with stimulating antibodies or isotype controls. Activation generally follows a time course of 24, 48, and 72 h.
5. Proceed to staining and flow cytometry analysis.

3.3. Induction of apoptosis

Apoptosis is induced in mouse and human beta cell lines by combined inflammatory cytokine treatment: IFN-γ and TNF-α. Cell death including apoptosis is detected using CYTOX Green or Propidium Iodide and Annexin-V, or the mitochondrial membrane potential dye DiOC6. ROS production is detected simultaneously as described below.

3.3.1. For mouse beta cell lines NIT-1 and NIT-4, after allowing cells to attach to the plate for overnight, dilute recombinant mouse TNF-α 1000 U/mL and recombinant mouse IFN-γ 1000 U/mL in culture medium. Remove old medium from wells and feed with fresh medium or medium with cytokines. Cytokines are added at different time points to achieve desired treatment durations. Control wells are fed with fresh medium on the second day. Since cytokines are dissolved in culture medium, this experimental design does not require vehicle control treatment ofbeta cells. If diluent of the treatment reagents is not culture medium, vehicle control wells should be added with correspondent dissolvent at the same concentration as treated wells.

3.3.2. For human beta cell line BetaLox 5: Allow cells to attach to the plate for 24 h. Add recombinant human TNF-α (2000 U/mL) and recombinant human IFN-γ (1000 U/mL). Incubate for 48 h. Proceed to flow cytometer analysis.

3.4. Instruments

- Flow cytometer: We have used both the Accuri 6 as well as BD LSRFortessa cytometers for these protocols
- Plate reader: SpectraMax M5 microplate reader
- Cell culture facility and mouse work station

3.5. Steps using flow cytometer

3.5.1. Prepare MitoSox Red before use, by adding 13.2 μL DMSO to one 50 μg vial to yield a stock solution of 5 m*M*. Dilute the dye to working concentration in culture medium. Avoid light. Working concentrations vary for different cell types, titration prior to experiment is recommended. In our studies, we have used 2.5 μ*M* for mouse and human beta cell lines and 5 μ*M* for human PBMC and T cells.

3.5.2. For beta cell lines seeded in 12-well culture plates:

1. Remove old medium and add 1 mL diluted dye to each well except unstained controls.
2. Incubate at 37 °C for 15 min in the dark.
3. Remove dye, rinse once with HBSS.
4. Harvest cells by incubation with trypsin for 5 min at room temperature. Trypsin is neutralized with an equal volume of culture medium.
5. Wash cells once with phenol red-free FACS buffer (we use PBS + 2% FBS).
6. Add dyes to exclude dead cells. For mouse beta cells, we use CYTOX Green, diluted in FACS buffer to 10 μ*M*, and Annexin-V-APC (Invitrogen, Catalog number A35110) with a 1/5 dilution in FACS buffer. Add 10 μL of each diluted dye to each tube. Stain for 1 min at room temperature and wash with FACS buffer. For human beta cell lines, we use DiOc6 20 n*M*, added together with MitoSox Red staining.

3.5.3. For mouse lymphocytes or human PBMCs:

1. Stain for cell surface markers as usual, if desired. Wash.
2. Add 500 μL culture medium containing dye and incubate at 37 °C for 15 min in dark.
3. Wash with 2-mL of phenol red-free FACS buffer.
4. Resuspend in 300 μL FACS buffer. Add 1:10 diluted Annexin-V 10 μL to stain dead cells.

3.5.4. Proceed to flow cytometer analysis. Depending on the instrument corresponding settings should be chosen. For example, using an Accuri C6 for mouse beta cells, the MitoSox Red signal is detected in both the FL-2 and FL-3 channels, while CYTOX green is collected by the FL-1 channel and Annexin-V-APC in the FL-4 channel. For human beta cells, mouse splenocytes, as well as human PBMCs or T cells, we have used a BD LSR Fortessa. With the Fortessa, Mitosox Red is detected in the PE channel and DiOC6 via the FITC channel.

3.5.5. Results are expressed as ratio to control. Figure 12.1 provides an example of the gating strategy for detection of human T cell mtROS using flow cytometry.

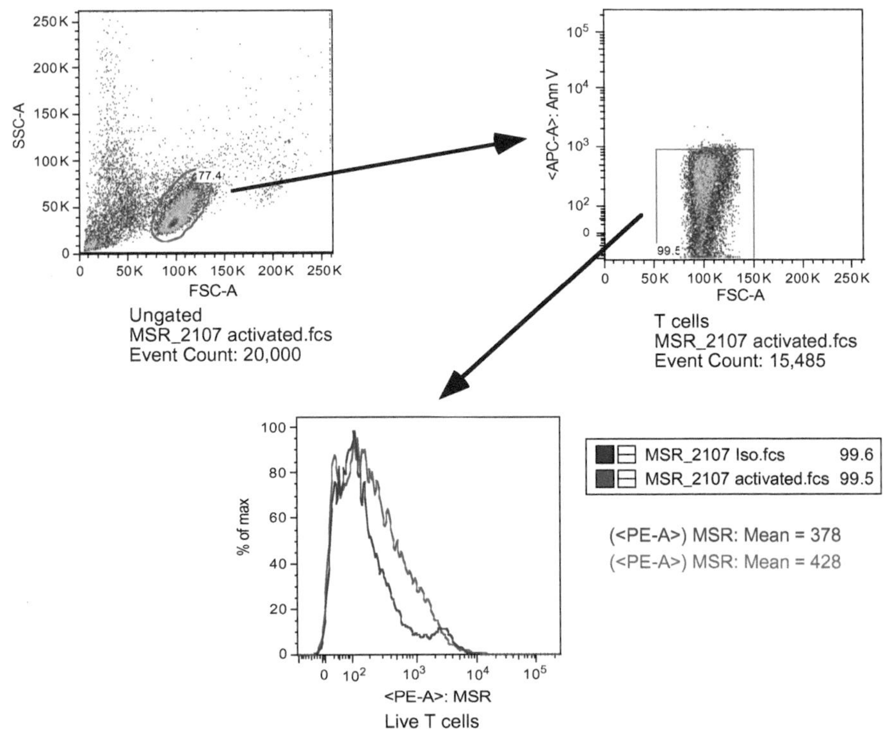

Figure 12.1 *Gating strategy for the cytofluorometic detection of the mitochondrial ROS in human T cells.* Samples are enriched human T cells as mentioned in the text (step 3.2.5). Lower T cell percentage is due to the fact that red blood cells are not lysed in this experiment. The top left panel is a dot plot for the T cell population that was gated on side-scatter and forward-scatter. Live T cells were gated by exclusion of Annexin-V positive cells (top right panel). Lower panel shows comparison of MitoSox Red histogram between activated T cells (red line) and isotype antibody treated controls (blue line). (See the color plate.)

In summary, the mitochondria-specific dye MitoSox Red can be used to detect basal as well as cytokine-induced mtROS production in mouse and human beta cell lines. Additionally, these methods can also be employed for the measurement of mtROS production in primary mouse and human lymphocytes. Selection of fluorescent dyes such as DHE or DCFDA will allow detection of total cell ROS production using this method. For mouse and human cells, the described method allows time course studies of ROS production during both short- (minutes) and long-term (days) experiments.

3.6. Steps using plate reader

3.6.1. Seed cells in 96-well flat bottom plate at 1×10^5 cells per well. Allow cells attach overnight.

3.6.2. On the second day, prewarm the plate reader at 37 °C.

3.6.3. Dilute ROS detection dye to the desired final concentration in culture medium. Remove medium from the wells of the plate and add 200 μL medium with dye. We use three dyes separately: CM-H_2DCFDA at final concentration of 5 μ*M*, DHE at final concentration of 100 *M*, H_2DCFDA at final concentration of 10 μ*M*. Dye concentrations are also cell type dependent and titration is recommended.

3.6.4. Read on a SpectraMax M5 or similar spectrofluorometer. Set at kinetic read mode with a temperature of 37 °C and read for 2 h with an interval of 1 min 20 s with 6 reads per well. Fluorescence reading Ex/Em/Cutoff: Lm1 510/590/570 (for DHE). Lm2 488/517/515 (for both H_2DCFDA and CM-H_2DCFDA). After a 15-min read to establish base line, open spectrofluorometer, add treatments to corresponding wells, reinsert plate into the spectrofluorometer, and continue reading.

In summary, the plate reader method is useful to detect acute ROS production (minutes to hours). While cytokine treatment of beta cells did not elicit ROS production over a short-term experiment, positive control wells where exogenous peroxide was added induced a sharp increase of fluorescence.

4. CONCLUDING REMARKS

ROS play essential roles for cell physiology and participate in many pathological processes. Mitochondria are an important source of cell ROS production. Here, we introduce techniques to detect mitochondrial-specific ROS production in several cell types using flow cytometry and spectrofluorometer plate reader. By selecting different ROS-detecting probes, it is possible to detect either total cellular ROS or mitochondrial-specific ROS. These techniques allow investigators to monitor basal ROS levels as well as dynamic ROS production in hours (using spectrofluorometer plate reader) or in days (using a flow cytometer). In addition to the detection of dynamic ROS production during long-term treatment, these flow cytometry methods also allow for the measurement of ROS in most specific types as well as in cells during different conditions or heterogeneous cell populations by costaining with cell surface markers.

ACKNOWLEDGMENTS

This work was supported by grants from the National Institute of Health (AI056374, AI42288, and DK074656) and the Sebastian Family Endowment for Diabetes Research to C. E. M. as well as grants from the Juvenile Diabetes Research Foundation and the American Diabetes Association to J. C. and C. E. M.

REFERENCES

Aggarwal, A., Misro, M. M., Maheshwari, A., Sehgal, N., & Nandan, D. (2010). N-acetylcysteine counteracts oxidative stress and prevents hCG-induced apoptosis in rat Leydig cells through down regulation of caspase-8 and JNK. *Molecular Reproduction and Development*, *77*, 900–909.

Baixauli, F., Martin-Cofreces, N. B., Morlino, G., Carrasco, Y. R., Calabia-Linares, C., Veiga, E., et al. (2011). The mitochondrial fission factor dynamin-related protein 1 modulates T-cell receptor signalling at the immune synapse. *The EMBO Journal*, *30*, 1238–1250.

Bayir, H., Fadeel, B., Palladino, M. J., Witasp, E., Kurnikov, I. V., Tyurina, Y. Y., et al. (2006). Apoptotic interactions of cytochrome c: Redox flirting with anionic phospholipids within and outside the mitochondria. *Biochimica et Biophysica Acta*, *1757*, 648–659.

Butterfield, D. A., Howard, B. J., & LaFontaine, M. A. (2001). Brain oxidative stress in animal models of accelerated aging and the age-related neurodegenerative disorders, Alzheimer's disease and Huntington's disease. *Current Medicinal Chemistry*, *8*, 815–828.

Chen, J., Gusdon, A. M., Piganelli, J., Leiter, E. H., & Mathews, C. E. (2011). mt-Nd2(a) Modifies resistance against autoimmune type 1 diabetes in NOD mice at the level of the pancreatic beta-cell. *Diabetes*, *60*, 355–359.

Chen, J., Gusdon, A. M., Thayer, T. C., & Mathews, C. E. (2008). Role of increased ROS dissipation in prevention of T1D. *The Annals of the New York Academy of Sciences*, *1150*, 157–166.

Chen, J., Lu, Y., Lee, C. H., Li, R., Leiter, E. H., & Mathews, C. E. (2008). Commonalities of genetic resistance to spontaneous autoimmune and free radical-mediated diabetes. *Free Radical Biology & Medicine*, *45*, 1263–1270.

Chhabra, A. (2010). Mitochondria-centric activation induced cell death of cytolytic T lymphocytes and its implications for cancer immunotherapy. *Vaccine*, *28*, 4566–4572.

Drose, S., & Brandt, U. (2008). The mechanism of mitochondrial superoxide production by the cytochrome bc1 complex. *The Journal of Biological Chemistry*, *283*, 21649–21654.

Du, X., Matsumura, T., Edelstein, D., Rossetti, L., Zsengeller, Z., Szabo, C., et al. (2003). Inhibition of GAPDH activity by poly(ADP-ribose) polymerase activates three major pathways of hyperglycemic damage in endothelial cells. *The Journal of Clinical Investigation*, *112*, 1049–1057.

Efremov, R. G., & Sazanov, L. A. (2012). The coupling mechanism of respiratory complex I—A structural and evolutionary perspective. *Biochimica et Biophysica Acta, 1817*, 1785–1795.

Gautier, C. A., Kitada, T., & Shen, J. (2008). Loss of PINK1 causes mitochondrial functional defects and increased sensitivity to oxidative stress. *Proceedings of the National Academy of Sciences of the United States of America, 105*, 11364–11369.

Gusdon, A. M., Votyakova, T. V., & Mathews, C. E. (2008). mt-Nd2a suppresses reactive oxygen species production by mitochondrial complexes I and III. *The Journal of Biological Chemistry, 283*, 10690–10697.

Gusdon, A. M., Votyakova, T. V., Reynolds, I. J., & Mathews, C. E. (2007). Nuclear and mitochondrial interaction involving mt-Nd2 leads to increased mitochondrial reactive oxygen species production. *The Journal of Biological Chemistry, 282*, 5171–5179.

Hatefi, Y. (1985). The mitochondrial electron transport and oxidative phosphorylation system. *Annual Review of Biochemistry, 54*, 1015–1069.

Hsiai, T., & Berliner, J. A. (2007). Oxidative stress as a regulator of murine atherosclerosis. *Current Drug Targets, 8*, 1222–1229.

Jackson, S. H., Devadas, S., Kwon, J., Pinto, L. A., & Williams, M. S. (2004). T cells express a phagocyte-type NADPH oxidase that is activated after T cell receptor stimulation. *Nature Immunology, 5*, 818–827.

Johnson, K. R., Zheng, Q. Y., Bykhovskaya, Y., Spirina, O., & Fischel-Ghodsian, N. (2001). A nuclear-mitochondrial DNA interaction affecting hearing impairment in mice. *Nature Genetics, 27*, 191–194.

Kaminski, M. M., Sauer, S. W., Klemke, C. D., Suss, D., Okun, J. G., Krammer, P. H., et al. (2010). Mitochondrial reactive oxygen species control T cell activation by regulating IL-2 and IL-4 expression: Mechanism of ciprofloxacin-mediated immunosuppression. *Journal of Immunology, 184*, 4827–4841.

Kennedy, K. A., Sandiford, S. D., Skerjanc, I. S., & Li, S. S. (2012). Reactive oxygen species and the neuronal fate. *Cellular and Molecular Life Sciences, 69*, 215–221.

Kim, H. J., Khalimonchuk, O., Smith, P. M., & Winge, D. R. (2012). Structure, function, and assembly of heme centers in mitochondrial respiratory complexes. *Biochimica et Biophysica Acta, 1823*, 1604–1616.

Kim, J. J., Lee, S. B., Park, J. K., & Yoo, Y. D. (2010). TNF-alpha-induced ROS production triggering apoptosis is directly linked to Romo1 and Bcl-X(L). *Cell Death and Differentiation, 17*, 1420–1434.

Krauss, S., Zhang, C. Y., Scorrano, L., Dalgaard, L. T., St-Pierre, J., Grey, S. T., et al. (2003). Superoxide-mediated activation of uncoupling protein 2 causes pancreatic beta cell dysfunction. *The Journal of Clinical Investigation, 112*, 1831–1842.

Kwon, J., Shatynski, K. E., Chen, H., Morand, S., de Deken, X., Miot, F., et al. (2010). The nonphagocytic NADPH oxidase Duox1 mediates a positive feedback loop during T cell receptor signaling. *Science Signaling, 3*, ra59.

Lin, M. T., & Beal, M. F. (2006). Mitochondrial dysfunction and oxidative stress in neurodegenerative diseases. *Nature, 443*, 787–795.

Muller, F. L., Liu, Y., & Van Remmen, H. (2004). Complex III releases superoxide to both sides of the inner mitochondrial membrane. *The Journal of Biological Chemistry, 279*, 49064–49073.

Pandya, D. P. (2001). Oxidant injury in coronary heart disease (Part-I). *Comprehensive Therapy, 27*, 284–292.

Panov, A., Kubalik, N., Zinchenko, N., Hemendinger, R., Dikalov, S., & Bonkovsky, H. L. (2011). Respiration and ROS production in brain and spinal cord mitochondria of transgenic rats with mutant G93a Cu/Zn-superoxide dismutase gene. *Neurobiology of Disease, 44*, 53–62.

Payton, K. S., Sheldon, R. A., Mack, D. W., Zhu, C., Blomgren, K., Ferriero, D. M., et al. (2007). Antioxidant status alters levels of Fas-associated death domain-like IL-1B-converting enzyme inhibitory protein following neonatal hypoxia-ischemia. *Developmental Neuroscience*, *29*, 403–411.

Perez-Cruz, I., Carcamo, J. M., & Golde, D. W. (2007). Caspase-8 dependent TRAIL-induced apoptosis in cancer cell lines is inhibited by vitamin C and catalase. *Apoptosis*, *12*, 225–234.

Peterson, J. R., Sharma, R. V., & Davisson, R. L. (2006). Reactive oxygen species in the neuropathogenesis of hypertension. *Current Hypertension Reports*, *8*, 232–241.

Puddu, P., Puddu, G. M., Cravero, E., Rosati, M., & Muscari, A. (2008). The molecular sources of reactive oxygen species in hypertension. *Blood Pressure*, *17*, 70–77.

Quinlan, C. L., Treberg, J. R., Perevoshchikova, I. V., Orr, A. L., & Brand, M. D. (2012). Native rates of superoxide production from multiple sites in isolated mitochondria measured using endogenous reporters. *Free Radical Biology & Medicine*, *53*, 1807–1817.

Quintana, A., Schwindling, C., Wenning, A. S., Becherer, U., Rettig, J., Schwarz, E. C., et al. (2007). T cell activation requires mitochondrial translocation to the immunological synapse. *Proceedings of the National Academy of Sciences of the United States of America*, *104*, 14418–14423.

Rigoulet, M., Yoboue, E. D., & Devin, A. (2011). Mitochondrial ROS generation and its regulation: Mechanisms involved in H(2)O(2) signaling. *Antioxidants & Redox Signaling*, *14*, 459–468.

Schwindling, C., Quintana, A., Krause, E., & Hoth, M. (2010). Mitochondria positioning controls local calcium influx in T cells. *Journal of Immunology*, *184*, 184–190.

Sena, L. A., Li, S., Jairaman, A., Prakriya, M., Ezponda, T., Hildeman, D. A., et al. (2013). Mitochondria are required for antigen-specific T cell activation through reactive oxygen species signaling. *Immunity*, *38*, 225–236.

Senior, A. E. (1988). ATP synthesis by oxidative phosphorylation. *Physiological Reviews*, *68*, 177–231.

Shen, H. M., Lin, Y., Choksi, S., Tran, J., Jin, T., Chang, L., et al. (2004). Essential roles of receptor-interacting protein and TRAF2 in oxidative stress-induced cell death. *Molecular and Cellular Biology*, *24*, 5914–5922.

Shenoy, K., Wu, Y., & Pervaiz, S. (2009). LY303511 enhances TRAIL sensitivity of SHEP-1 neuroblastoma cells via hydrogen peroxide-mediated mitogen-activated protein kinase activation and up-regulation of death receptors. *Cancer Research*, *69*, 1941–1950.

Shigenaga, M. K., Hagen, T. M., & Ames, B. N. (1994). Oxidative damage and mitochondrial decay in aging. *Proceedings of the National Academy of Sciences of the United States of America*, *91*, 10771–10778.

Shoffner, J. M., & Wallace, D. C. (1994). Oxidative phosphorylation diseases and mitochondrial DNA mutations: Diagnosis and treatment. *Annual Review of Nutrition*, *14*, 535–568.

Siddiqui, A., Hanson, I., & Andersen, J. K. (2012). Mao-B elevation decreases parkin's ability to efficiently clear damaged mitochondria: Protective effects of rapamycin. *Free Radical Research*, *46*, 1011–1018.

Singh, U., & Jialal, I. (2006). Oxidative stress and atheroschlerosis. *Pathophysiology*, *13*, 129–142.

Sohal, R. S., & Weindruch, R. (1996). Oxidative stress, caloric restriction, and aging. *Science*, *273*, 59–63.

Sohn, Y. S., Tamir, S., Song, L., Michaeli, D., Matouk, I., Conlan, A. R., et al. (2013). NAF-1 and mitoNEET are central to human breast cancer proliferation by maintaining mitochondrial homeostasis and promoting tumor growth. *Proceedings of the National Academy of Sciences of the United States of America*, *110*, 14676–14681.

Steinberg, D., & Witztum, J. L. (2002). Is the oxidative modification hypothesis relevant to human atherosclerosis? Do the antioxidant trials conducted to date refute the hypothesis? *Circulation, 105*, 2107–2111.

St-Pierre, J., Buckingham, J. A., Roebuck, S. J., & Brand, M. D. (2002). Topology of superoxide production from different sites in the mitochondrial electron transport chain. *The Journal of Biological Chemistry, 277*, 44784–44790.

Sugioka, K., Nakano, M., Totsune-Nakano, H., Minakami, H., Tero-Kubota, S., & Ikegami, Y. (1988). Mechanism of O2- generation in reduction and oxidation cycle of ubiquinones in a model of mitochondrial electron transport systems. *Biochimica et Biophysica Acta, 936*, 377–385.

Surekha, R. H., Srikanth, B. B. M. V., Jharna, P., Ramachandra, R. V., Dayasagar, R. V., & Jyothy, A. (2007). Oxidative stress and total antioxidant status in myocardial infarction. *Singapore Medical Journal, 48*, 137–142.

Tahara, E. B., Navarete, F. D., & Kowaltowski, A. J. (2009). Tissue-, substrate-, and site-specific characteristics of mitochondrial reactive oxygen species generation. *Free Radical Biology & Medicine, 46*, 1283–1297.

Terry Powers, J. L., Mace, K. E., Parfrey, H., Lee, S. J., Zhang, G., & Riches, D. W. (2010). TNF receptor-1 (TNF-R1) ubiquitous scaffolding and signaling protein interacts with TNF-R1 and TRAF2 via an N-terminal docking interface. *Biochemistry, 49*, 7821–7829.

Terzioglu, M., & Larsson, N. G. (2007). Mitochondrial dysfunction in mammalian ageing. *Novartis Foundation Symposium, 287*, 197–213.

Thayer, T. C., Delano, M., Liu, C., Chen, J., Padgett, L. E., Tse, H. M., et al. (2011). Superoxide production by macrophages and T cells is critical for the induction of autoreactivity and type 1 diabetes. *Diabetes, 60*, 2144–2151.

Tong, J. J., Schriner, S. E., McCleary, D., Day, B. J., & Wallace, D. C. (2007). Life extension through neurofibromin mitochondrial regulation and antioxidant therapy for neurofibromatosis-1 in Drosophila melanogaster. *Nature Genetics, 39*, 476–485.

Treberg, J. R., Quinlan, C. L., & Brand, M. D. (2011). Evidence for two sites of superoxide production by mitochondrial NADH-ubiquinone oxidoreductase (complex I). *The Journal of Biological Chemistry, 286*, 27103–27110.

Tretter, L., Takacs, K., Hegedus, V., & Adam-Vizi, V. (2007). Characteristics of alpha-glycerophosphate-evoked H2O2 generation in brain mitochondria. *Journal of Neurochemistry, 100*, 650–663.

Tse, H. M., Thayer, T. C., Steele, C., Cuda, C. M., Morel, L., Piganelli, J. D., et al. (2010). NADPH oxidase deficiency regulates Th lineage commitment and modulates autoimmunity. *Journal of Immunology, 185*, 5247–5258.

Turrens, J. F., & Boveris, A. (1980). Generation of superoxide anion by the NADH dehydrogenase of bovine heart mitochondria. *The Biochemical Journal, 191*, 421–427.

Venardos, K. M., & Kaye, D. (2007). Myocardial ischemia-reperfusion injury, antioxidant enzyme systems, and selenium: A review. *Current Medicinal Chemistry, 14*, 1539–1549.

Wallace, D. C. (1992). Diseases of the mitochondrial DNA. *Annual Review of Biochemistry, 61*, 1175–1212.

Zhang, H., Luo, Y., Zhang, W., He, Y., Dai, S., Zhang, R., et al. (2007). Endothelial-specific expression of mitochondrial thioredoxin improves endothelial cell function and reduces atherosclerotic lesions. *The American Journal of Pathology, 170*, 1108–1120.

Zimmerman, M. C., Lazartigues, E., Lang, J. A., Sinnayah, P., Ahmad, I. M., Spitz, D. R., et al. (2002). Superoxide mediates the actions of angiotensin II in the central nervous system. *Circulation Research, 91*, 1038–1045.

CHAPTER THIRTEEN

Methods to Monitor ROS Production by Fluorescence Microscopy and Fluorometry

Aleksandra Wojtala[*,1], **Massimo Bonora**[†,1], **Dominika Malinska**[*,1], **Paolo Pinton**[†], **Jerzy Duszynski**[*], **Mariusz R. Wieckowski**[‡,2]

[*]Nencki Institute of Experimental Biology, Warsaw, Poland
[†]Department of Experimental and Diagnostic Medicine, Section of General Pathology, Interdisciplinary Centre for the Study of Inflammation (ICSI), BioPharmaNet, University of Ferrara, Ferrara, Italy
[‡]Department of Biochemistry, Nencki Institute of Experimental Biology, Warsaw, Poland
[1]These authors contributed equally to this chapter.
[2]Corresponding author: e-mail address: m.wieckowski@nencki.gov.pl

Contents

Abstract

Mitochondria are considered one of the main sources of reactive oxygen species (ROS). The overgeneration of ROS can evoke an intracellular state of oxidative stress, leading to permanent cell damage. Thus, the intracellular accumulation of ROS may not only disrupt the functions of specific tissues and organs but also lead to the premature death of the entire organism. Less severe increases in ROS levels may lead to the nonlethal

Methods in Enzymology, Volume 542
ISSN 0076-6879
http://dx.doi.org/10.1016/B978-0-12-416618-9.00013-3

oxidation of fundamental cellular components, such as proteins, phospholipids, and DNA, hence exerting a mutagenic effect that promotes oncogenesis and tumor progression. Here, we describe the use of chemical probes for the rapid detection of ROS in intact and permeabilized adherent cells by fluorescence microscopy and fluorometry. Moreover, after discussing the limitations described in the literature for the fluorescent probes presented herein, we recommend methods to assess the production of specific ROS in various fields of investigation, including the study of oncometabolism.

1. INTRODUCTION

To date, seven distinct sites of reactive oxygen species (ROS) production have been identified in mammalian mitochondria. The two sites that have been studied most extensively are complexes I and III of the mitochondrial respiratory chain (Brand, 2010; St-Pierre, Buckingham, Roebuck, & Brand, 2002). It has also been suggested that the age-related deterioration of complex II can lead to superoxide production (Ishii, Miyazawa, Hartman, & Ishii, 2011). The involvement of complex II in ROS production was also discussed in the context of diabetes (Nishikawa et al., 2000), and it was suggested that mutations in complex II might also result in superoxide overproduction (Ishii et al., 2005). Additionally, the role of mitochondrial complex II in ROS production during hypoxia has been discussed by Paddenberg et al., (2003) and Yankovskaya et al., (2003). Other documented sources of ROS production in the mitochondria include monoamine oxidase and dihydroorotate dehydrogenase (Cadenas & Davies, 2000; Lambertucci et al., 2008). Furthermore, the flavoproteins acyl-CoA dehydrogenase and glycerol phosphate dehydrogenase can generate ROS in tissues during the oxidation of lipid-derived substrates (Lambertucci et al., 2008; St-Pierre et al., 2002). Both pyruvate and α-ketoglutarate dehydrogenase contain flavoenzyme dihydrolipoyl dehydrogenase subunits and can also participate in ROS production (Starkov et al., 2004; Tahara, Barros, Oliveira, Netto, & Kowaltowski, 2007). However, Brown and Borutaite (2011) present a number of examples supporting the hypothesis that mitochondria cannot be considered as the primary source of ROS. It is important to note that mitochondria are not the only sites of intracellular ROS production. Microsomes (e.g., cytochrome P450 and diamine oxidase), peroxisomes (e.g., enzymes involved in fatty acid oxidation), and some enzymes in the plasma membrane (e.g., NADPH oxidase and lipooxygenase) have been identified as ROS generators as well.

Apart from the aforementioned harmful effects, ROS also can have a beneficial role in cell physiology. Under physiological conditions, ROS can act as mediators and regulators of metabolism. ROS affect the transmission of signals to and throughout the cell, and they can be considered second messengers during growth, differentiation, and cell death. ROS can activate proteins involved in cell division (mitogen-activated proteins) and participate in the body's immune responses, as well influence the expression of many genes or regulate the synthesis of prostanoids (Bartosz, 2009; Dröge, 2002).

Many different methodological approaches have been described for monitoring extra- and intracellular ROS production. For comprehensive reviews and original papers concerning ROS measurement using the fluorescent probes described later, as well as equipment and many other methods, refer the following references: Zielonka and Kalyanaraman (2010), Dikalov and Harrison (2014), Grisham (2013), Halliwell and Whiteman (2004), Kalyanaraman et al. (2012), and Winterbourn (2014).

2. MEASUREMENT OF REACTIVE OXYGEN SPECIES PRODUCTION IN INTACT ADHERENT CELLS USING FLUORESCENT PROBES

2.1. Concept

One of the fastest, easiest, most user-friendly, and accessible methods for monitoring ROS production is based on the detection of ROS-sensitive fluorescent probes using a fluorescence microplate reader. There are several analytical, fluorescence approaches used to detect and characterize different ROS.

Later, we present the application of dihydroethidium (DHE), MitoSOX Red, and 5-(and 6)-chloromethyl-2′,7′-dichlorohydrofluorescein diacetate (CM-H_2DCFDA) to measure cytosolic superoxide, mitochondrial superoxide, and hydrogen peroxide production, respectively, in intact adherent cells. These probes are all oxidized to form intermediate probe-derived radicals that are successively oxidized to generate the corresponding fluorescent products (Winterbourn, 2014). Their reactivity toward different oxygen species is briefly described, as well as the strengths and limitations of their use (for more detailed analysis, refer to the following: Dikalov & Harrison, 2014; Wardman, 2007; Winterbourn, 2014). Despite criticism and the limitations described in the literature, the presented methods can be useful for basic and quick screening experiments to provide an overview of general ROS production.

DHE is the reduced form of the widely used DNA dye ethidium bromide. Due to its desirable quality of passively diffusing into cells, along with its high reactivity, DHE has been commonly used to detect cytosolic superoxide (Gomes, Fernandes, & Lima, 2005; Wardman, 2007; Zhao et al., 2003). Upon its reaction with the superoxide anion, DHE forms a red fluorescent product, 2-hydroxyethidium (Kalyanaraman et al., 2012; Zielonka, Vasquez-Vivar, & Kalyanaraman, 2008), with maximum excitation and emission peaks at 500 and 580 nm, respectively (Table 13.1). DHE itself displays blue fluorescence in the cell cytoplasm; however, the oxidized form 2-hydroxyethidium intercalates into DNA and exhibits red fluorescence (Bucana, Saiki, & Nayar, 1986). DHE is known to be the most specific fluorescent probe for superoxide detection (Wardman, 2007); however, apart from superoxide-mediated oxidation to 2-hydroxyethidium, DHE can also undergo unspecific oxidation by $ONOO^-$ or $^{\bullet}OH$ into ethidium (Gomes et al., 2005; Kalyanaraman et al., 2012; Patsoukis, Papapostolou, & Georgiou, 2005; Wardman, 2007; Zielonka, Hardy, & Kalyanaraman, 2009). Moreover, DHE can be oxidized by cytochrome *c* (Gomes et al., 2005; Zielonka, Srinivasan, et al., 2008). Because the fluorescence spectra of ethidium and 2-dihydroxyethidium differ only slightly, fluorometric measurements are not sufficient to distinguish between these two products, and this should be taken into account when interpreting the results. To verify the superoxide specificity of the signal, other analytical methods should be applied, such as HPLC and mass spectrometry (Zielonka, Vasquez-Vivar, et al., 2008).

MitoSOX Red is a dye used to measure superoxide production in the mitochondrial matrix. Because MitoSOX Red is a cationic derivative of DHE, its reaction with superoxide anions is very similar to that of DHE (Kalyanaraman et al., 2012). However, it reacts with superoxide faster than DHE and is rapidly targeted to the mitochondria (due to the positive charge of the cationic triphenylphosphonium substituent), where it is oxidized by superoxide to form 2-hydroxymitoethidium, which excites and emits at 510 and 580 nm, respectively (Robinson, Janes, & Beckman, 2008; Zielonka, Vasquez-Vivar, et al., 2008). Similar to DHE, MitoSOX Red can also undergo unspecific reactions with other oxidants to form mito-ethidium, which overlaps the fluorescence peak of 2-hydroxymitoethidium (Kalyanaraman et al., 2012; Zielonka & Kalyanaraman, 2010). Interestingly, it has been shown that the superoxide-specific product of MitoSOX Red oxidation has a specific excitation peak at $\sim$400 nm (Robinson et al., 2006). Moreover, the user should be aware of the cyto- and mitotoxicity

Table 13.1 The list of fluorescent probes used to detect ROS with the use of microplate reader

Fluorescent probe	ROS detected	Excitation/emission wavelengths (nm)	Fluorescent product
Dihydroethidium (DHE)	Cytosolic superoxide but also $ONOO^-$ or $^{\bullet}OH$	535/635	2-Hydroxyethidium but also ethidium
MitoSOX Red	Mitochondrial superoxide but also other oxidants	510/595 alternatively: 400/595	2-Hydroxymitoethidium but also mito-ethidium
5-(and 6)-chloromethyl-2′,7′-dichlorohydrofluorescein diacetate (CM-H_2DCFDA)	Hydrogen peroxide but also hydroxyl radical, carbonate radical, and nitrogen dioxide	495/520	Dichlorofluorescein (DCF)

Excitation/emission wavelengths and fluorescent products are presented.

of these probes (Gomes et al., 2005; Kalyanaraman et al., 2012; Zielonka & Kalyanaraman, 2010), the potential modification of mitochondrial morphology, and the redistribution of fluorescence to the nuclei at higher probe concentrations. For more complex and detailed analyses, additional, more supportive methods such as fluorescence spectroscopy are also recommended (Nazarewicz, Bikineyeva, & Dikalov, 2013).

CM-H_2DCFDA is often called a "hydrogen peroxide-detecting probe for the measurement of hydrogen peroxide in intact cells"; however, the range of ROS detected by this probe is much broader. CM-H_2DCFDA freely permeates the plasma membrane and is hydrolyzed in the cytosol to form the DCFH carboxylate anion (Gomes et al., 2005; Kalyanaraman et al., 2012). Oxidation results in the formation of fluorescent DCF, which is maximally excited at 495 nm and emits at 520 nm (Kalyanaraman et al., 2012). The oxidation of H_2DCF to DCF is a two-step process: first, the DCF radical is formed, and then it is further oxidized to DCF in a reaction with molecular oxygen (Wardman, 2007). The first step of H_2DCF oxidation can be mediated by different radical species: hydroxyl radical, carbonate radical, and nitrogen dioxide, as well as by thiyl radicals resulting from thiol oxidation (Halliwell & Whiteman, 2004; Kalyanaraman et al., 2012; Winterbourn, 2014). H_2O_2 does not react with H_2DCF directly but requires the presence of peroxidases or other enzymes containing transition metals. Moreover, an alteration of the signal can be caused by antioxidant enzymes or superoxide that competes with the probe for the ROS (Winterbourn, 2014). Moreover, the superoxide radical is formed in the second step of probe oxidation and can then be dismutated to hydrogen peroxide and cause self-amplification of the signal (Grisham, 2013).

Despite the limitations described earlier, H_2DCFDA, DHE, and MitoSox Red are still the convenient probes for measuring general aspects of intracellular oxidative stress and redox status, provided that the constraints imposed by probe chemistry are taken into account when interpreting the results. Furthermore, to avoid data misinterpretation due to variations in cell numbers, normalization of the data based on protein amount is required, and a sulforhodamine B (SRB)-based method is also presented.

2.2. Experimental setup

The complete measurement buffer consists of the following: phosphate-buffered saline (PBS) with Ca^{2+} and Mg^{2+}, pH 7.4.

Directly before the measurement, 5 m*M* glucose is added.

Stock solutions

- DHE (on the day of the experiment, a working solution is prepared from 10 m*M* DMSO stock (1 mg in 317 μl of DMSO), diluted in DMEM)—protect from light
- MitoSOX Red (on the day of the experiment, a working solution is prepared from 5 m*M* DMSO stock (50 μg in 13 μl of DMSO), diluted in measurement buffer)—protect from light
- CM-H_2DCFDA (on the day of the experiment, a working solution is prepared from 5 m*M* DMSO stock (50 μg in 20 μl of DMSO), diluted in measurement buffer)—protect from light
- Glucose, 1 *M* stock

2.3. Procedure

Preparation of cells

Cells cultured in standard Dulbecco's modified Eagle medium (DMEM) are seeded into multiwell plates (approximately 25,000–30,000 cells per well). The medium should be changed the day before the experiment. Cells should reach 80–90% of confluence on the day of the experiment. The user should adjust the number of seeded cells and the culture duration to suit the specific cell type used.

2.3.1 Measurements

2.3.1.1 Measurement of cytosolic ROS production with the use of DHE

Cytosolic superoxide can be measured using the DHE fluorescent probe. Cells grown in multiwell plates should be incubated for 20 min (needed to allow the probe to enter the cell and start the reaction) at 37 °C in 0.5 ml of DMEM containing 0.5 μ*M* DHE. After the incubation, cells should be washed twice with PBS. The fluorescence can be measured in the measurement buffer using a microplate reader set to 535 nm excitation (Ex bandwidth: 25 nm) and 635 nm emission (Em bandwidth: 35 nm) wavelengths. This is a single-read measurement that does not define the kinetics of the reaction. Figure 13.1A shows the results obtained with this method.

2.3.1.2 Measurement of mitochondrial ROS production with the use of MitoSOX Red

Mitochondrial superoxide production can be measured using the MitoSOX Red fluorescent probe. Cells grown in multiwell plates should be washed twice with PBS to remove the medium and subsequently incubated for 10 min (needed to allow the probe to enter the cell and start the reaction

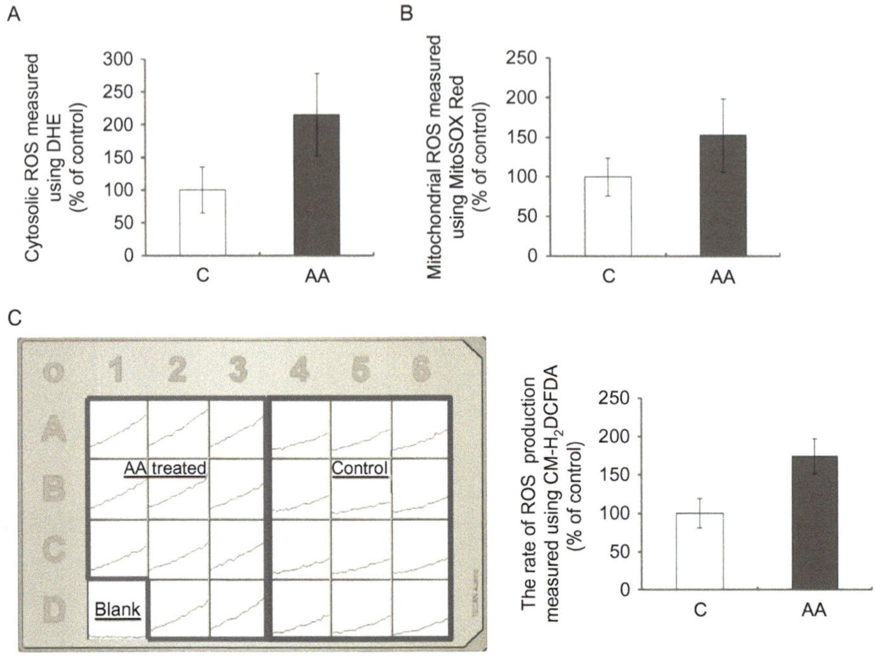

Figure 13.1 *Effect of antimycin A on ROS production in intact human fibroblasts measured with the use of a microplate reader.* C—control fibroblasts, AA—fibroblasts treated with antimycin A; (A) cytosolic ROS production measured using DHE; (B) mitochondrial ROS production measured using MitoSOX Red; (C) ROS generation measured using CM-H_2DCFDA. After the measurements, it is necessary to determine the cellular protein content via the sulforhodamine B (SRB) assay, as described by Vichai and Kirtikara (2006), which enables the standardization of ROS production as a function of cell number (protein level). A defect in the mitochondrial respiratory chain results in higher cytosolic and mitochondrial superoxide production. The rate of H_2O_2 production is also increased.

within the mitochondria) at 37 °C in 0.5 ml of measurement buffer containing 5 μ*M* MitoSOX Red. After the incubation, the cells should be washed twice with PBS. The fluorescence can be monitored in the measurement buffer with a microplate reader set to 510 nm excitation (Ex bandwidth: 10 nm) and 595 nm emission (Em bandwidth: 35 nm) wavelengths. Alternatively, if the signal is high enough, the fluorescence can also be measured using 400 nm excitation to measure only the superoxide-specific product of MitoSOX Red oxidation. This is a single-read measurement that does not define the kinetics of the reaction. Avoid exposing the MitoSOX Red stock solution and dye-loaded cells to light before measuring. Figure 13.1B shows the results obtained with this method.

2.3.1.3 Measurement of ROS production with the use of CM-H_2DCFDA

The rate of H_2O_2 production can be measured using the CM-H_2DCFDA fluorescent probe. Cells grown in multiwell plates should be washed twice with PBS to remove the medium. Next, add 0.5 ml of measurement buffer containing 2 μ*M* CM-H_2DCFDA. Immediately after addition, start measuring the fluorescence in kinetic mode at 485 nm excitation (Ex bandwidth: 20 nm) and 520 nm emission (Em bandwidth: 10 nm) wavelengths. Figure 13.1C shows the results obtained with this method.

3. Quantification of Protein Level with the Use of the SRB Assay

The protein concentration in each well should be determined after the measurements described earlier to standardize ROS level/production as a function of the protein level (i.e., cell number) in each well. This can be performed using the SRB assay, as described by Vichai and Kirtikara (2006).

3.1. Experimental setup

Stock solutions

- Methanol
- Acetic acid
- Tris (10 m*M*), pH 10
- SRB

3.2. Procedure

The following protocol has been optimized for 24-well plates. Directly after the measurement of ROS described in Section 2, fix the cells by gently adding a cold solution of 1% (v/v) acetic acid in methanol and incubating for at least 1 h at −20 °C. Next, dry the plate (e.g., in an incubator). Add 0.5% (w/v) SRB solution in 1% acetic acid to each well and incubate for 1 h at 37 °C (gentle shaking is recommended). Aspirate the SRB solution and gently wash the cells with 1% (v/v) acetic acid. Repeat this step a few times until the remaining SRB solution is removed. Dry the plate (e.g., in an incubator). To the dried wells, add 1 ml of 10 m*M* Tris to dissolve the dye and gently shake for 15 min. Afterward, transfer 200 μl of the solution from each well into a 96-well plate and measure the absorbance at 540 nm using the microplate reader.

4. MEASUREMENT OF MITOCHONDRIAL ROS PRODUCTION IN PERMEABILIZED CELLS WITH THE AMPLEX RED FLUORESCENT PROBE

4.1. Concept

Amplex red (10-acetyl-3,7-dihydroxyphenoxazine) is a nonfluorescent derivative of dihydroresorufin that is used to detect ROS in various experimental systems. This probe is oxidized to resorufin (excitation/emission maxima = 563/587 nm) in a peroxidase-catalyzed reaction with H_2O_2 (Zhou, Diwu, Panchuk-Voloshina, & Haugland, 1997). The probe is characterized by high sensitivity (down to 50 n*M* H_2O_2) and product stability. Amplex red is cell impermeable and is thus mostly applied to measure ROS production in isolated organelles (usually isolated mitochondria because these are considered an important intracellular source of ROS) or to detect extracellular H_2O_2 release (Dikalov, Griendling, & Harrison, 2007). Amplex red can also be used to monitor mitochondrial ROS production *in situ* using permeabilized cells, which allows the investigation of mitochondrial function in a more physiological environment than in the case of isolated organelles because it preserves the interactions of mitochondria with other organelles and avoids the mitochondrial fragmentation that occurs during isolation (Kuznetsov et al., 2008). Additionally, the number of cells used for the assay is much smaller than that needed when isolating mitochondria. Apart from ensuring the access of the probe to the mitochondria, cell permeabilization allows the manipulation of mitochondrial respiratory chain function through the application of appropriate substrates and inhibitors. Additionally, permeabilization leads to the dilution of cytosolic enzymes in the measurement medium, which diminishes the influence of cytosolic ROS-removing enzymes, such as cytosolic superoxide dismutase 1 (SOD1) and catalase, on the amount of ROS detected by the probe. Therefore, the ROS levels measured do not reflect actual intracellular ROS levels. However, this method is very convenient for investigating the relationship between the functional state of the mitochondria and the amount of ROS released by these organelles.

Later, we present the application of an Amplex red–horseradish peroxidase system to compare mitochondrial ROS production in C_2C_{12} myoblasts and myotubes. It should be mentioned that nonmitochondrial ROS sources (such as peroxisomes and plasma membrane NADPH oxidases) can also contribute somewhat to the results. However, the influence of nonmitochondrial ROS is minimized due to the strong dilution of their substrates on cell permeabilization. The addition of particular substrates to the measurement medium (in this case, substrates fuelling mitochondrial respiration) makes it possible to control which source of ROS will contribute the most to the ROS production being measured.

Because Amplex red oxidation is also observed in the presence of superoxide (Malinska, Kudin, Debska-Vielhaber, Vielhaber, & Kunz, 2009), measurements are performed in the presence of exogenous superoxide dismutase to ensure the conversion of mitochondrially released superoxide into H_2O_2 and avoid the bias caused by potential differences in the sensitivity of the probe to these two ROS species. Another source of error, particularly when working with permeabilized cells, could arise from the oxidation of Amplex red in the presence of HRP and NAD(P)H. Such oxidation can also be prevented using superoxide dismutase (Votyakova & Reynolds, 2004). This phenomenon has to be accounted for while working with permeabilized cells because it could interfere with the results.

4.2. Experimental setup

The complete measurement buffer consists of the following: 120 m*M* KCl, 5 m*M* KH_2PO_4, 5 m*M* EGTA, 10 m*M* HEPES, pH 7.4.

Directly before the measurement, add 5 m*M* $MgCl_2$.

Stock solutions

- Amplex red (on the day of the experiment, a working solution is prepared from 20 m*M* DMSO stock diluted 100 times in the measurement medium)
- Horseradish peroxidase (4000 U/ml in measurement medium)
- Superoxide dismutase (7000 U/ml in measurement medium)
- Catalase
- Digitonin (10 mg/ml in water; to dissolve, heating may be necessary)
- H_2O_2 (10 μ*M*, freshly diluted from 30% solution)

Respiratory substrates and inhibitors: 1 *M* potassium succinate and 1 mg/ml antimycin A.

4.3. Procedure

Preparing the cells

C_2C_{12} myoblasts and myotubes are detached by trypsinization and spun down, and the cell pellet is suspended in PBS (pellet from one 10 cm culture plate in 100 μl of PBS).

4.3.1 Measurement

The measurement is performed in the presence of 1 μ*M* Amplex red, 20 U/ml horseradish peroxidase, and 35 U/ml superoxide dismutase. Amplex red oxidation is monitored with a fluorometer at $\lambda_{ex}=560$ nm and $\lambda_{em}=590$ nm. The reaction is initiated by adding the cell suspension (ca. 0.3 mg/ml protein), and the cells are then permeabilized by the addition of 40 μg/ml digitonin. An increase in Amplex red oxidation is observed due to the dilution of cytosolic SOD1, which makes more of the produced H_2O_2 available to react with the probe. Next, the appropriate respiratory substrates and inhibitors are applied to facilitate measurement of the changes in mitochondrial ROS production induced by changes in the redox state of particular respiratory chain reaction centers. Calibration is performed by adding known amounts of H_2O_2 to the cells to account for endogenous catalase activity within the cells. This enables the calculation of the actual amount of ROS released from the mitochondria. The sensitivity of this method to H_2O_2 is verified by catalase addition.

Figure 13.2 shows a comparison of H_2O_2 release from the mitochondria of C_2C_{12} myoblasts and myotubes using succinate as the respiratory substrate

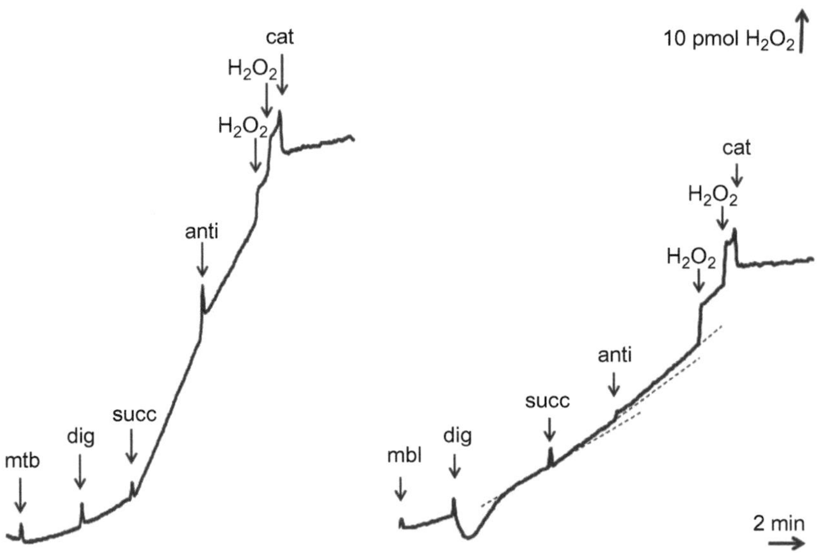

Figure 13.2 *Mitochondrial ROS generation in digitonin-permeabilized C_2C_{12} myotubes (left panel) and myoblasts (right panel) measured with the Amplex red–horseradish peroxidase system.* mbl—myoblasts, 0.3 mg/ml protein, mtb—myotubes, 0.3 mg/ml protein, dig—digitonin, 10 μg/ml, succ—succ, 10 m*M*, anti—antimycin A, 1 μg/ml, H_2O_2—10 pmol of hydrogen peroxide, cat—catalase, 8 kU/ml.

and in the presence of antimycin A, an inhibitor of respiratory complex III. In highly polarized mitochondria, oxidation using succinate as a predominant substrate is known to lead to reverse electron transport from complex II to complex I, resulting in strong reduction of complex I and enhanced superoxide production by this complex (Ksenzenko, Konstantinov, Khomutov, Tikhonov, & Ruuge, 1983). Antimycin A blocks electron transfer at complex III, leading to depolarization of the mitochondria (thereby abolishing reverse electron transport) and to stimulation of superoxide production by respiratory complex III (Votyakova & Reynolds, 2001). The experiment shows generally higher mitochondrial ROS production in C_2C_{12} myotubes in comparison to myoblasts and also shows the higher susceptibility of myotube mitochondria to the occurrence of reverse electron transport.

5. MEASUREMENT OF ROS BY FLUORESCENCE MICROSCOPY

5.1. Concept

Despite the many advantages of the plate reader format in measuring ROS (as described in Section 2), some experimental conditions require a more precise readout, even at the cost of lower throughput. Fluorescence microscopy would then be more appropriate to use in these cases. Microscopy is especially useful in cases with a nonhomogenous sample, in which only some cells needs to be analyzed (i.e., differentiating cells, a sample undergoing gene expression manipulation, a cell cycle phase-specific assay, etc.), or when a nonhomogeneous subcellular compartment needs to be analyzed. Several fluorescence microscopy techniques have been successfully used to investigate ROS, ranging from wide field to spinning disk confocal microscopy. The following procedure can be adapted to measure ROS in adherent cells using most common fluorescence microscopy techniques.

5.2. Experimental setup and procedure

Modified Krebs Ringer Buffer (KRB): 135 m*M* NaCl, 5 m*M* KCl, 1 m*M* $CaCl_2$, 0.4 m*M* KH_2PO_4, 1 m*M* $MgSO_4$, 20 m*M* HEPES, pH 7.4.

Directly before the measurement, 5.5 m*M* glucose should be added.

5.2.1 Cell preparation

Plate cells on cover slips or glass-bottom petri dishes and let the cells grow in complete medium for at least 24 h at a density that allows the cells to reach a

Table 13.2 The list of most common fluorescent reporters compatible with CM-H_2DCFDA, DHE, and MitoSOX RED and dedicated filter sets for laser scanning (LSCM) or wide-field microscopy

		LSCM		Wide-field microscopy	
Dye	**Compatible reporter**	**Excitation**	**Emission**	**Excitation**	**Emission**
CM-H_2DCFDA	RFP, DsRed, mCherry	488—525/20, 543—LLP560		490/20—525/ 20,545/25607/70	
DHE/MitoSOX Red	EGFP, MitoTracker GREEN	514—LP560, 488—525/20		500/20—LP560, 490/20—525/20	
DHE/MitoSOX Red	MitoTracker DEEP RED	514—LP560, 633—LP650		490/20—580/65, 630/20—LP650	

maximum of 80% confluence on the day of the experiment. Proper seeding conditions are extremely dependent on cell type and have to be optimized by each experimenter; nevertheless, a general requirement is that cells should be seeded as uniformly as possible, avoiding the formation of cell aggregates. If the experimental condition requires DNA transfection, a certain degree of toxicity due to the transfection method has to be considered. In some experiments, transfection with fluorescent reporters may be recommended. A list of fluorescent reporters compatible with the discussed probes is presented in Table 13.2.

5.2.2 Counterstaining

To more precisely analyze MitoSOX Red data, mitochondrial counterstaining can be performed. The best counterstaining can be achieved through the use of targeted fluorescent proteins; however, if cells are refractory to transfection, MitoTracker staining is suggested. A large number of fluorescent probes are available for staining mitochondria. Due to the particular properties of the MitoSOX Red spectrum, we suggest MitoTracker Deep Red as the best option for staining mitochondria without interfering with the MitoSOX Red signal. However, if MitoTracker Deep Red cannot be acquired, MitoTracker GREEN should be used instead.

5.2.3 Procedure and signal recording

On the day of the experiment, prepare a loading solution composed of CM-H_2DCFDA, DHE, or MitoSOX Red, depending on the ROS to be measured (see Table 13.1 for working concentrations), 0.2% (v/v) pluronic acid and (in the case of mitochondrial counterstaining) 100 μM

MitoTracker Deep Red/GREEN in modified KRB; the solution can be stored at 4 °C for a few hours. If the stimulation or inhibition of ROS production is part of the experimental design, a 10 × stock of any required compounds (e.g., Antimycin A, Rotenone, or CCCP) should be prepared in modified KRB solution. Wash the cells twice to remove dead cells and cell debris, add 1 ml of the loading solution (prewarmed at room temperature) to the cover slip, and incubate for 10 min at 37 °C. After loading, the cover slip should be mounted in a metal cage or other appropriate support depending on the microscope model and covered with 900 μl of modified KRB. Mount the sample on an inverted microscope equipped with a thermostatic chamber and a magnification ranging from 20 × to 63 ×. CM-H_2DCFDA and DHE will be loaded in several compartments, but the signal will appear mostly cytosolic due to the resolution limit of fluorescence microscopy. As such, a high-resolution setup is not necessary, and a lower magnification is in fact preferred to allow the analysis of a higher number of cells. Excitation sources and emission filters for the ROS probes are suggested in Table 13.3.

ROS-sensitive dyes are irreversibly oxidized by ROS and trapped in the cells. The kinetics of this accumulation directly represents the ROS production rate. Based on this, a time series should be acquired using a delay of at least 15–30 s to avoid photoactivation not related to ROS production. The basal rate could be estimated by acquiring unstimulated cells for 5–10 min. Chemical stimulation can then be performed by adding 100 μl of the 10 × stock solution (e.g., Antimycin A or Rotenone), allowing the substance to diffuse through the chamber and continuing to record for the desired time.

5.2.4 Data analysis

Image analysis can be performed offline using a variety of commercial (e.g., Metamorph, NIS, and Volocity) and open-source (e.g., ImageJ, Bioimage

Table 13.3 The list of commonly used dyes for ROS production measurement with suggested filter sets for confocal (LSCM) and wide-field microscopy as well as their working concentrations

	LSCM		Wide-field microscopy		
Dye	**Excitation**	**Emission**	**Excitation**	**Emission**	**Working concentration (μM)**
CM-H_2DCFDA	488—LP505		490/20—LP510		5
DHE	514—LP530		490/20—LP510		5
MitoSOX Red	514—LP530		490/20—LP510		0.2

XD, and Cell Profiler) bioimaging software. The first step that needs to be performed is background subtraction. Next, a region of interest (ROI) has to be drawn within each cell. If counterstaining is available, the counterstained image can be binarized and used as a mask to filter the original image (see Figure 13.3). Once ROIs have been drawn, the average intensity within each ROI can be measured for each time point and exported to a

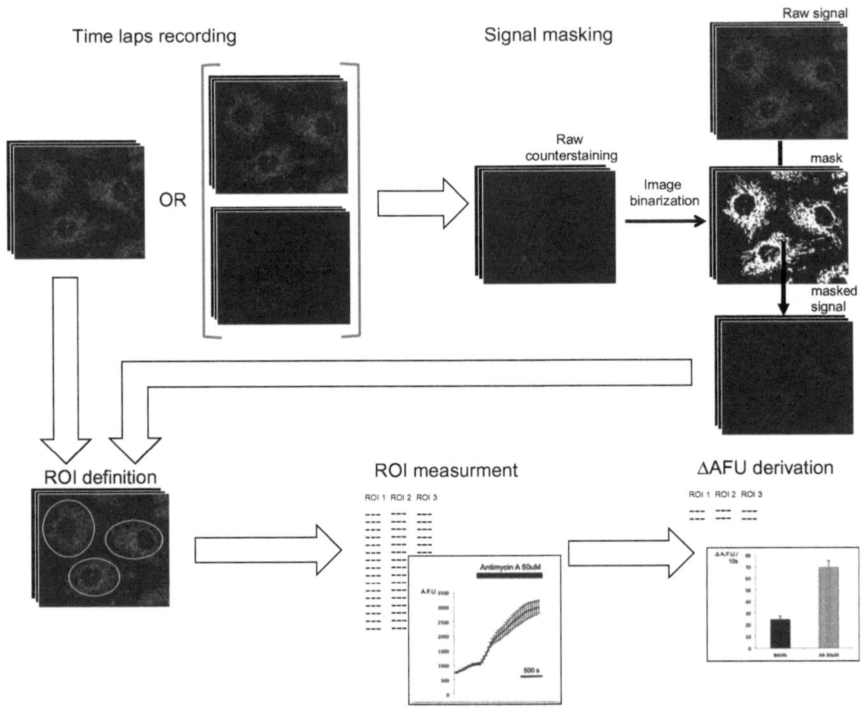

Figure 13.3 *Overview of ROS imaging and analysis workflow.* A time-lapse series is acquired with the preferred dye alone or in combination with counterstaining for the specific subcellular regions (e.g., mitochondria) or with other conditions (e.g., co-transfection with an oligonucleotide or cDNA). For counterstaining, the raw counterstaining images should be converted into a mask by image binarization. Such a mask is used to filter the raw signal images for the ROS probe, thereby allowing only the overlapping regions to be considered in the analysis. Regions of interest (ROIs) should then be defined in the raw or filtered images and the average fluorescence intensity calculated for each ROI at each time point; such data will be collected in a spreadsheet for further analysis. The delta value is then determined by calculating the difference between the intensity at each time point and the intensity of the preceding time point. Finally, the average delta intensity, expressed in arbitrary fluorescence units (ΔAFU), of each ROI has to be calculated for each experimental condition. (See the color plate.)

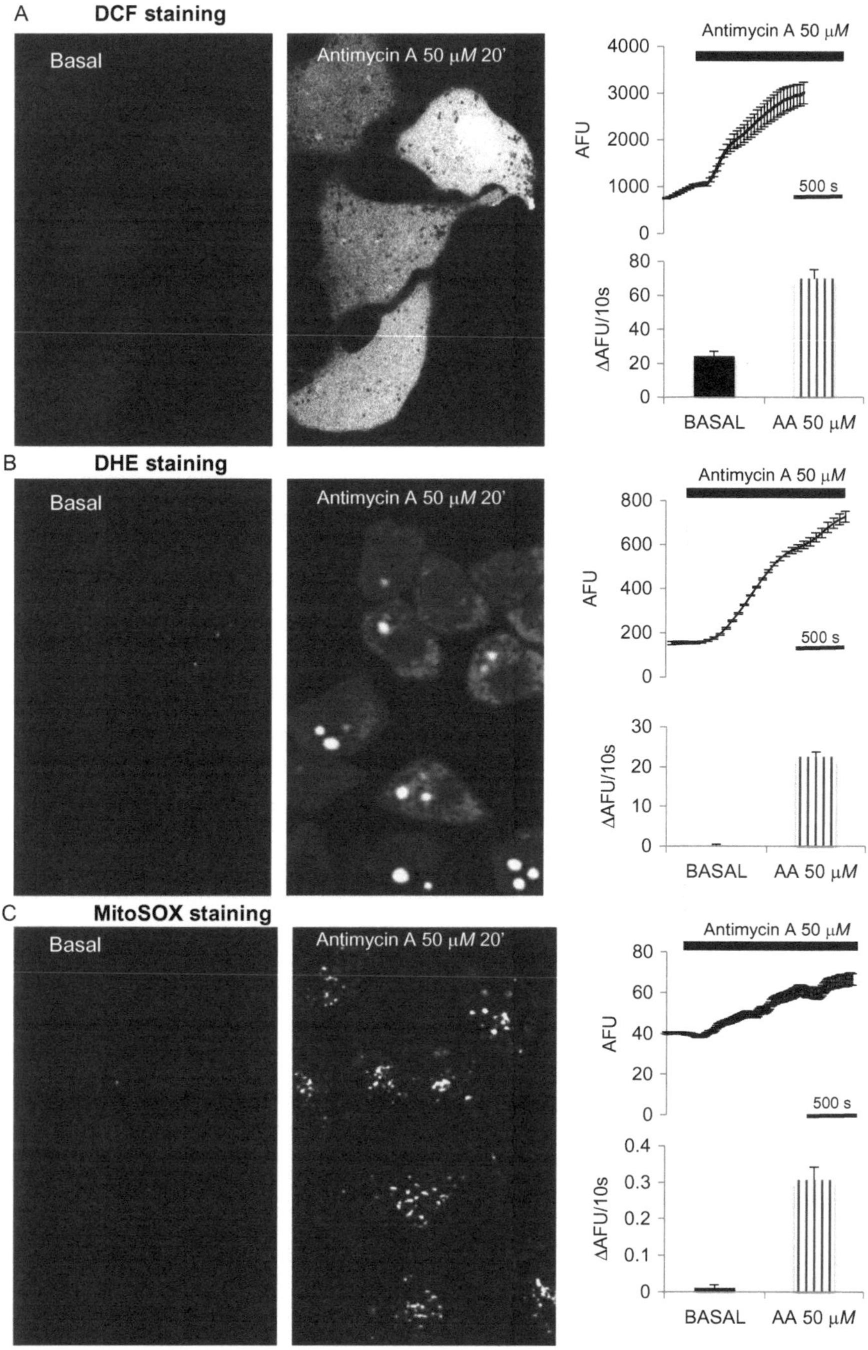

Figure 13.4 *Image-based measurement of ROS production after stimulation.* Representative measurements of cytosolic ROS production by CM-H_2DCFDA (A), cytosolic ROS by DHE (B), and mitochondrial ROS by MitoSOX Red (C) before and after stimulation with 50 μ*M* antimycin A (AA) in HeLa cells by laser scanning confocal microscopy. For each group, representative false-color images are displayed, along with graphs of the kinetics and ROS production rates. (See the color plate.)

spreadsheet. With the use of software such as Microsoft Excel and Libreoffice Calc, the ΔF can be calculated for each cell to represent the rate of ROS production. An average ΔF has to be calculated for each phase of the experiment if different stimulants or treatments were used. Such a value (expressed as ΔAFU) could be collected for different replicates and compared between different conditions (as an example, see Figure 13.4).

ACKNOWLEDGMENTS

This work was supported by the Polish National Science Center (UMO-2011/11/M/NZ3/02128), Statutory Funding from Nencki Institute of Experimental Biology, Polish Ministry of Science and Higher Education grant W100/HFSC/2011 and BIO-IMAGing in Research Innovation and Education (FP7-REGPOT-2010-1) for MRW, JD, AW, and DM. Moreover, PP and MB are financed by AIRC, Telethon (GGP11139B), the Italian Ministry of Education, University and Research and the Italian Ministry of Health.

REFERENCES

Bartosz, G. (2009). Reactive oxygen species: Destroyers or messengers? *Biochemical Pharmacology*, 77(8), 1303–1315.

Brand, M. D. (2010). The sites and topology of mitochondrial superoxide production. *Experimental Gerontology*, *45*, 466–472.

Brown, G. C., & Borutaite, V. (2011). There is no evidence that mitochondria are the main source of reactive oxygen species in mammalian cells. *Mitochondrion*, *12*(1), 1–4.

Bucana, C., Saiki, I., & Nayar, R. (1986). Uptake and accumulation of the vital dye hydroethidine in neoplastic cells. *The Journal of Histochemistry and Cytochemistry*, *34*, 1109–1115.

Cadenas, E., & Davies, K. J. (2000). Mitochondrial free radical generation, oxidative stress, and aging. *Free Radical Biology & Medicine*, *29*, 222–230.

Dikalov, S., Griendling, K. K., & Harrison, D. G. (2007). Measurement of reactive oxygen species in cardiovascular studies. *Hypertension*, *49*(4), 717–727.

Dikalov, S., & Harrison, D. G. (2014). Methods for detection of mitochondrial and cellular reactive oxygen species. *Antioxidants & Redox Signaling*, *20*, 372–382. http://dx.doi.org/10.1089/ars.2012.4886.

Dröge, W. (2002). Free radicals in the physiological control of cell function. *Physiological Reviews*, *82*(1), 47–95.

Gomes, A., Fernandes, E., & Lima, J. (2005). Fluorescence probes used for detection of reactive oxygen species. *Journal of Biochemical and Biophysical Methods*, *65*, 45–80.

Grisham, M. B. (2013). Methods to detect hydrogen peroxide in living cells: Possibilities and pitfalls. *Comparative Biochemistry and Physiology. Part A, Molecular & Integrative Physiology*, *165*(4), 429–438.

Halliwell, B., & Whiteman, M. (2004). Measuring reactive species and oxidative damage in vivo and in cell culture: How should you do it and what do the results mean? *British Journal of Pharmacology*, *142*(2), 231–255.

Ishii, T., Miyazawa, M., Hartman, P. S., & Ishii, N. (2011). Mitochondrial superoxide anion (O(2)(-)) inducible "mev-1" animal models for aging research. *BMB Reports*, *44*(5), 298–305.

Ishii, T., Yasuda, K., Akatsuka, A., Hino, O., Hartman, P. S., & Ishii, N. (2005). A mutation in the SDHC gene of complex II increases oxidative stress, resulting in apoptosis and tumorigenesis. *Cancer Research*, *65*(1), 203–209.

Kalyanaraman, B., Darley-Usmar, V., Davies, K. J., Dennery, P. A., Forman, H. J., Grisham, M. B., et al. (2012). Measuring reactive oxygen and nitrogen species with fluorescent probes: Challenges and limitations. *Free Radical Biology & Medicine*, *52*, 1–6.

Ksenzenko, M., Konstantinov, A. A., Khomutov, G. B., Tikhonov, A. N., & Ruuge, E. K. (1983). Effect of electron transfer inhibitors on superoxide generation in the cytochrome bc1 site of the mitochondrial respiratory chain. *FEBS Letters*, *155*(1), 19–24, PubMed PMID: 6301880.

Kuznetsov, A. V., Veksler, V., Gellerich, F. N., Saks, V., Margreiter, R., & Kunz, W. S. (2008). Analysis of mitochondrial function in situ in permeabilized muscle fibers, tissues and cells. *Nature Protocols*, *3*(6), 965–976.

Lambertucci, R. H., Hirabara, S. M., Silveira Ldos, R., Levada-Pires, A. C., Curi, R., & Pithon-Curi, T. C. (2008). Palmitate increases superoxide production through mitochondrial electron transport chain and NADPH oxidase activity in skeletal muscle cells. *Journal of Cellular Physiology*, *216*, 796–804.

Malinska, D., Kudin, A. P., Debska-Vielhaber, G., Vielhaber, S., & Kunz, W. S. (2009). Chapter 23 Quantification of superoxide production by mouse brain and skeletal muscle mitochondria. *Methods in Enzymology*, *456*, 419–437.

Nazarewicz, R. R., Bikineyeva, A., & Dikalov, S. I. (2013). Rapid and specific measurements of superoxide using fluorescence spectroscopy. *Journal of Biomolecular Screening*, *18*(4), 498–503.

Nishikawa, T., Edelstein, D., Du, X. L., Yamagishi, S., Matsumura, T., Kaneda, Y., et al. (2000). Normalizing mitochondrial superoxide production blocks three pathways of hyperglycaemic damage. *Nature*, *404*, 787–790.

Paddenberg, R., Ishaq, B., Goldenberg, A., Faulhammer, P., Rose, F., Weissmann, N., et al. (2003). Essential role of complex II of the respiratory chain in hypoxia-induced ROS generation in the pulmonary vasculature. *American Journal of Physiology. Lung Cellular and Molecular Physiology*, *284*(5), 710–719.

Patsoukis, N., Papapostolou, I., & Georgiou, C. D. (2005). Interference of non-specific peroxidases in the fluorescence detection of superoxide radical by hydroethidine oxidation: A new assay for H_2O_2. *Analytical and Bioanalytical Chemistry*, *381*(5), 1065–1072.

Robinson, K. M., Janes, M. S., & Beckman, J. S. (2008). The selective detection of mitochondrial superoxide by live cell imaging. *Nature Protocols*, *3*, 941–947.

Robinson, K. M., Janes, M. S., Pehar, M., Monette, J. S., Ross, M. F., Hagen, T. M., et al. (2006). Selective fluorescent imaging of superoxide in vivo using ethidium-based probes. *Proceedings of the National Academy of Sciences of the United States of America*, *103*, 15038–15043.

Starkov, A. A., Fiskum, G., Chinopoulos, C., Lorenzo, B. J., Browne, S. E., Patel, M. S., et al. (2004). Mitochondrial alpha-ketoglutarate dehydrogenase complex generates reactive oxygen species. *The Journal of Neuroscience*, *24*, 7779–7788.

St-Pierre, J., Buckingham, J. A., Roebuck, S. J., & Brand, M. D. (2002). Topology of superoxide production from different sites in the mitochondrial electron transport chain. *The Journal of Biological Chemistry*, *277*, 44784–44790.

Tahara, E. B., Barros, M. H., Oliveira, G. A., Netto, L. E., & Kowaltowski, A. J. (2007). Dihydrolipoyl dehydrogenase as a source of reactive oxygen species inhibited by caloric restriction and involved in Saccharomyces cerevisiae aging. *The FASEB Journal*, *21*, 274–283.

Vichai, V., & Kirtikara, K. (2006). Sulforhodamine B colorimetric assay for cytotoxicity screening. *Nature Protocols*, *3*, 1112–1116.

Votyakova, T. V., & Reynolds, I. J. (2001). DeltaPsi(m)-dependent and -independent production of reactive oxygen species by rat brain mitochondria. *Journal of Neurochemistry*, *79*(2), 266–277.

Votyakova, T. V., & Reynolds, I. J. (2004). Detection of hydrogen peroxide with Amplex Red: Interference by NADH and reduced glutathione auto-oxidation. *Archives of Biochemistry and Biophysics*, *431*, 138–144.

Wardman, P. (2007). Fluorescent and luminescent probes for measurement of oxidative and nitrosative species in cells and tissues: Progress, pitfalls and prospects. *Free Radical Biology & Medicine*, *43*, 995–1022.

Winterbourn, C. C. (2014). The challenges of using fluorescent probes to detect and quantify specific reactive oxygen species in living cells. *Biochimica et Biophysica Acta*, *1840*, 730–738.

Yankovskaya, V., Horsefield, R., Törnroth, S., Luna-Chavez, C., Miyoshi, H., Léger, C., et al. (2003). Architecture of succinate dehydrogenase and reactive oxygen species generation. *Science*, *299*(5607), 700–704.

Zhao, H., Kalivendi, S., Zhang, H., Joseph, J., Nithipatikom, K., Vásquez-Vivar, J., et al. (2003). Superoxide reacts with hydroethidine but forms a fluorescent product that is distinctly different from ethidium: Potential implications in intracellular fluorescence detection of superoxide. *Free Radical Biology & Medicine*, *34*, 1359–1368.

Zhou, M., Diwu, Z., Panchuk-Voloshina, N., & Haugland, R. P. (1997). A stable nonfluorescent derivative of resorufin for the fluorometric determination of trace hydrogen peroxide: Applications in detecting the activity of phagocyte NADPH oxidase and other oxidases. *Analytical Biochemistry*, *253*(2), 162–168.

Zielonka, J., Hardy, M., & Kalyanaraman, B. (2009). HPLC study of oxidation products of hydroethidine in chemical and biological systems: Ramifications in superoxide measurements. *Free Radical Biology & Medicine*, *46*, 329–338.

Zielonka, J., & Kalyanaraman, B. (2010). Hydroethidine- and MitoSOX-derived red fluorescence is not a reliable indicator of intracellular superoxide formation: Another inconvenient truth. *Free Radical Biology & Medicine*, *48*(8), 983–1001.

Zielonka, J., Srinivasan, S., Hardy, M., Ouari, O., Lopez, M., Vasquez-Vivar, J., et al. (2008). Cytochrome c-mediated oxidation of hydroethidine and mito-hydroethidine in mitochondria: Identification of homo- and heterodimers. *Free Radical Biology & Medicine*, *44*, 835–846.

Zielonka, J., Vasquez-Vivar, J., & Kalyanaraman, B. (2008). Detection of 2-hydroxyethidium in cellular systems: A unique marker product of superoxide and hydroethidine. *Nature Protocols*, *3*, 8–21.

CHAPTER FOURTEEN

Genetically Encoded Redox Sensors

Wai Kan Chiu[*,†], **Atif Towheed** [*,†], **Michael J. Palladino**[*,†,1]
[*]Department of Pharmacology & Chemical Biology, University of Pittsburgh School of Medicine, Pittsburgh, Pennsylvania, USA
[†]Pittsburgh Institute for Neurodegenerative Diseases, University of Pittsburgh School of Medicine, Pittsburgh, Pennsylvania, USA
[1]Corresponding author: e-mail address: mjp44@pitt.edu

Contents

Abstract

Endogenous redox sensors detect fluctuations in the intracellular redox equilibrium and are critical for the maintenance of homeostasis. Such systems have been exploited to engineer genetically encoded redox sensors to detect dynamic oxidative changes within a cellular environment. Most genetically encoded redox sensors detect reactive oxygen species (ROS) such as superoxide anion, hydrogen peroxide and hydroxyl radical. Technical hurdles including the limited temporal and spatial resolution as well as tissue heterogeneity have complicated the realization of the full potential of genetically encoded redox sensors in animals until recently. Alterations in the concentration and subcellular localization of ROS are integral to numerous disorders, including neurodegenerative diseases and cancer. Thus, genetically encoded redox sensors are useful for

Methods in Enzymology, Volume 542
ISSN 0076-6879
http://dx.doi.org/10.1016/B978-0-12-416618-9.00014-5

the study of the pathogenesis and progression of multiple diseases. Moreover, the ultimate generation of genetically encoded redox sensors provides substantial advantages over conventional methods such as ROS-sensitive fluorescent probes. Here, we review examples of genetically encoded redox sensors, present their application to various fields of biomedical investigation, including the study of oncometabolism, discuss their drawbacks and explore future developments.

1. INTRODUCTION

In a living cell, redox reactions have an important role in a wide range of biochemical processes such as cell signaling, regulation of transcription (Allen, 1993a, 1993b), oxidative phosphorylation, photosynthesis (Pfannschmidt & Liere, 2005; Pfannschmidt, Schutze, Brost, & Oelmuller, 2001), and immune defense (Rahman, 1999; Rahman & MacNee, 2000a, 2000b). Maintaining cellular redox balance is critical for various normal cellular function. Extreme changes in the oxidative products in a cell can lead to oxidative stress resulting in lesions of proteins, lipids, or DNA (Sedelnikova et al., 2010). These oxidative products can originate in a cell from exposure to either exogenous (X-rays, oxidizing chemicals, UV radiation) or endogenous (metabolic processes, and inflammation) sources (Altieri, Grillo, Maceroni, & Chichiarelli, 2008; Cadet, Douki, & Ravanat, 2010; De Bont & van Larebeke, 2004; Sedelnikova et al., 2010). There are four main types of oxidative products produced as a result of cellular metabolic processes: (1) reactive oxygen species (ROS), (2) reactive nitrogen species (RNS), (3) reactive chloride species, and (4) reactive sulfur species. Among the different types of oxidative products, the most studied species to date are ROS and to a lesser extent RNS. Accumulation or generation of excessive ROS can lead to "oxidative stress" and damage to DNA, RNA, lipids, and proteins in a cell (Darley-Usmar & Halliwell, 1996). This damage contributes significantly to pathogenesis of a variety of conditions such as atherosclerosis (Dhalla, Temsah, & Netticadan, 2000; Harrison, Griendling, Landmesser, Hornig, & Drexler, 2003; Schwartz et al., 1992), diabetes (Piconi, Quagliaro, & Ceriello, 2003), neurodegenerative diseases (Benzi & Moretti, 1995; Sun & Chen, 1998), and cancer (Ames, Gold, & Willett, 1995; Hussain, Hofseth, & Harris, 2003; Malins, Polissar, & Gunselman, 1996; Poulsen, Prieme, & Loft, 1998; Waris & Ahsan, 2006). The role of ROS in a cell is an active area of research and reports suggest that markedly elevated ROS has detrimental effects while less extreme ROS levels are critical to normal cell signaling (Martin & Barrett, 2002). Thus, a balance between pro-oxidants and antioxidants is crucial to maintaining a healthy cellular environment. There are intrinsic mechanisms in a cell to monitor and regulate ROS levels. Redox sensors are encoded naturally in living cells to detect oxidative changes and are crucial to the activation of downstream effectors that facilitate normal biological activities. For example, the σ^R-RsrA system in Gram-positive actinomycetes, which, upon disulfide stress, can undergo structural changes by forming disulfide bonds and release σ^R, a RNA polymerase factor, allowing it to bind to RNA polymerase and activate transcription (Green & Paget, 2004). Additionally, genetically encoded fluorescent redox sensors have been engineered to detect and measure redox activities in cellular compartments *in vivo* and have proven instrumental in elucidating cellular homeostasis and disease pathogenesis.

1.1. Common cellular oxidative products and antioxidants

ROS are radical and nonradical products of incomplete reduction of molecular oxygen. Hydrogen peroxide (H_2O_2), superoxide radicals ($O_2^{\cdot -}$), and hydroxyl radical (.OH) are examples of a few ROS produced in biochemical reactions. RNS include nitric oxide (·NO) and peroxynitrite ($ONOO^-$). Peroxynitrite is produced when ·NO reacts with superoxide ($O_2^{\cdot -}$). Additional RNS products can be formed when peroxynitrite reacts with other molecules to form nitrogen dioxide (·NO_2), dinitrogen trioxide (N_2O_3), and free radicals. Cellular redox status is maintained by a balance between the production of ROS and the antioxidant systems in a cell (Fig. 14.1). Mitochondria are widely known to be a principle source of oxidative products in the cell (Brand et al., 2004; Chance, Sies, & Boveris, 1979; St-Pierre, Buckingham, Roebuck, & Brand, 2002), however, other sources of oxidative stress are known (Brown & Borutaite, 2012; St-Pierre et al., 2002). In addition to mitochondrial derived ROS, enzymes such as xanthine oxidase, nitric oxide synthase, and lipoxygenases produce metabolic oxidative byproducts that are of biomedical importance (Comhair & Erzurum, 2002). Of note, NADPH oxidases (NOX) generate oxidative products as their primary function. Although mitochondria are reported to be the major producer of ROS, other cellular compartments such as endoplasmic reticulum and peroxisomes are actively being debated as significant contributors of ROS as well (Brown & Borutaite, 2012). Most of the oxygen in a cell is reduced to water, but a small percentage is converted to ROS, mainly to

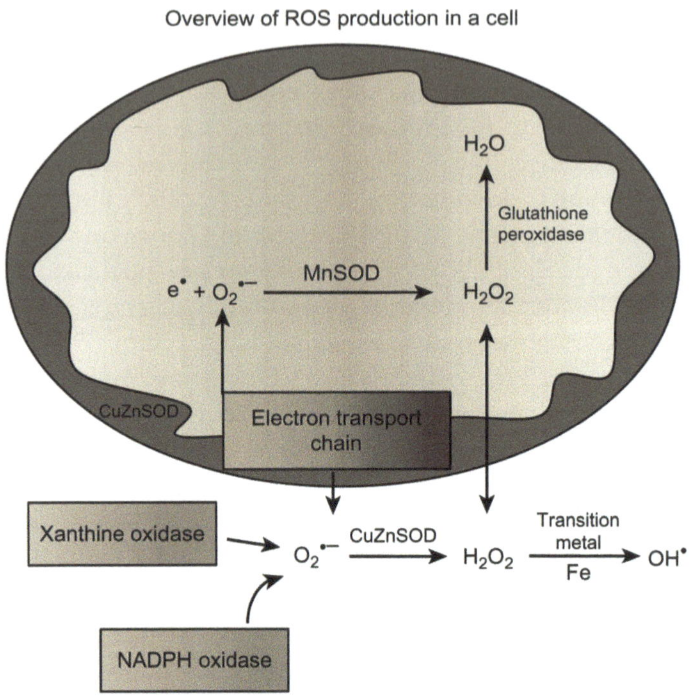

Figure 14.1 *ROS formation and clearance.* Electrons can leak from the respiratory chain and react with an oxygen molecule to form superoxide. Superoxide is reduced to hydrogen peroxide by manganese superoxide dismutase (MnSOD or SOD2) in the matrix or by copper-zinc superoxide dismutase (CuZnSOD or SOD1) in the cytosol and the inner mitochondrial membrane space. Hydrogen peroxide, unlike superoxide, is uncharged and can diffuse in and out of the cell more easily. It can be further reduced to water by glutathione peroxidase, or react with transition metal, such as iron, to form hydroxyl radical. Hydroxyl radical is highly reactive and can damage virtually any macromolecules, including amino acid conversion, nucleic acid mutation, and lipid peroxidation.

superoxide (Boveris & Chance, 1973). Superoxide is formed when oxygen acquires a free electron from the electron transport chain (ETC) (Fridovich, 1978; Halliwell, 1994; Halliwell & Gutteridge, 1990; Powis, 1987). It is the species from which other ROS (hydrogen peroxide, hydroxyl radical) are derived. It is short-lived and, because it is charged, it does not cross lipid bilayers. Superoxide reacts rapidly with dismutase proteins—SOD2 in the mitochondrial matrix or SOD1 in the cytosol and the mitochondrial inner membrane space—to form hydrogen peroxide (Fukai & Ushio-Fukai, 2011). Hydrogen peroxide is also reactive but is more stable and capable of crossing cell membranes. It can be reduced to water by catalase or glutathione peroxidase (GPx) (Antunes, Han, & Cadenas, 2002) or reacts with transition metals to form hydroxyl radical (Comhair & Erzurum, 2002) that is capable of DNA damage. In addition to SODs and catalase, the antioxidant system consists of GPx and thioredoxin reductase enzymes, and thioredoxin, peroxiredoxin, glutathione, and NADPH molecules that maintain equilibrium among the various redox species (Bashan, Kovsan, Kachko, Ovadia, & Rudich, 2009; Turrens, 2003). A recent review has suggested that the immediate precursor of glutathione, γ-glutamylcysteine, can protect the mitochondria from excessive ROS (Quintana-Cabrera & Bolanos, 2013). Other antioxidants such as vitamin C has been shown to play a role in reducing superoxide production (Iino, Iwase, Sonoki, Yoshinari, & Iida, 2005) and vitamin E has been shown to reduce the generation of ROS in the liver of diabetic rats (Rhee, Jeong, & Choi, 2005). However, the utility of pharmacologic antioxidants to prevent disease pathogenesis remains somewhat unclear (Evans, 2008; Vivekananthan, Penn, Sapp, Hsu, & Topol, 2003).

1.2. Genetically encoded redox sensors

The role of ROS in disease pathogenesis and importance of regulating cellular redox homeostasis are of increasing biomedical importance and are fields of intensive study. Numerous techniques have been utilized to detect redox changes including genetically encoded and nongenetic approaches. Some detection methods involve sophisticated tools such as magnetic resonance imaging (Thelwall et al., 2005), positron emission tomography (Ikawa et al., 2009), X-ray synchrotron (Debenham et al., 1996), mass spectrometry (MS) (Fenaille, Guy, & Tabet, 2003), time-of-flight mass spectrometry (Weber, Engelmann, Becher, & Hecker, 2004), and nanostructure initiator mass spectrometry (Greving, Patti, & Siuzdak, 2011). However, there are significant drawbacks associated with these techniques. For example, MS has limited spatial and temporal resolutions and provides a snapshot of the reaction instead of continuous temporal coverage (Okumoto, Jones, & Frommer, 2012). Destruction of tissues and extensive manipulation of the cell required for many of these techniques alter the oxidative status of the cell, potentially leading to erroneous conclusions. In addition, these techniques require expensive equipment that may not be readily available to some laboratories. Reporter-based systems that utilize promoters provide more extensive coverage of reaction changes; however, they usually require transcription and translation and, thus, report oxidative conditions with low temporal resolution.

There are numerous naturally encoded, endogenous redox sensors in biological systems. However, identifying redox-sensitive proteins has been challenging because it typically requires time-consuming biochemical assays. A recently developed proteomic technique termed OxiCAT, however, allows for rapid detection of potential redox proteins *in vivo* (Leichert et al., 2008). Briefly, OxiCAT is a chemical method that labels reduced cysteines with an isotopically light form of a tag and labels oxidized cysteines with an isotopically heavy form of the tag. Following trypsin digestion, quantitative MS determines the extent of thiol modification in the protein. This allows rapid identification of redox-sensitive proteins of interest, the precise cysteine residues that are involved, and the extent of their modification upon oxidative stress. Alternatively, engineered genetically encoded redox sensors can be used that are typically introduced by transfection or transgenesis. These have various sensor designs, including a single-fluorescent protein sensor, Förster resonance energy transfer (FRET)-based sensor, or a chimeric fluorescent protein sensor.

Genetically encoded redox sensors have substantial advantages over conventional tools to measure redox levels *in vivo*. Genetically encoded redox sensors are sensitive, minimally invasive, and have high cellular and subcellular resolution. Moreover, they detect changes taking place in live cells in real time. Continuous development of engineered genetically encoded redox sensors will provide unlimited opportunities in tracking oxidative changes within living cells in real time.

2. ENDOGENOUSLY ENCODED REDOX SENSORS

Several redox sensors have been discovered that detect specific oxidation products demonstrating the importance of regulating cellular redox status. Yeast and bacterial studies have been instrumental in the identification and development of numerous redox sensors. Importantly, these endogenous genetically encoded redox sensors have the potential to be optimized for expression in higher eukaryotic cells and developed into powerful redox reporters.

2.1. Disulfide stress detection

σ^R-RsrA system is specific to Gram-positive actinomycetes (Kang et al., 1999; Manganelli et al., 2002; Paget, Kang, Roe, & Buttner, 1998) and monitors disulfide stress. σ^R is an RNA polymerase factor that binds to the anti-sigma factor RsrA, a small cysteine containing zinc metalloprotein, in an inactive state. RsrA forms at least one intramolecular disulfide bond when under disulfide stress, which causes a structural change and the subsequent release of σ^R (Kang et al., 1999; Li et al., 2003; Paget et al., 2001). Upon release, σ^R associates with RNA polymerase and activates transcription.

2.2. Nitric oxide detection

The versatile *E. coli* SoxRS system not only monitors superoxide stress (Section 3.2) but also detects nitric oxide; however, instead of using the electron transfer system, SoxR is activated by nitrosylation of the Fe–S clusters (Demple, 1999; Ding & Demple, 2000).

Another *E. coli* protein, Fnr, can also react with nitric oxide to form a dinitrosyl-iron–cysteine complex (Cruz-Ramos et al., 2002). It has been shown that Fnr can repress the *hmp* gene, but when nitric oxide is present, *hmp* is derepressed and activated to encode a flavohemoglobin that detoxifies nitric oxide (Cruz-Ramos et al., 2002; Poole & Hughes, 2000).

Soluble guanylate cyclases (sGCs) in mammals are hemoproteins that can bind to both nitric oxide and carbon monoxide but not oxygen, and function as endogenous nitric oxide sensors (Derbyshire & Marletta, 2012; Stuehr, Santolini, Wang, Wei, & Adak, 2004). They exist as heterodimers (α/β) and the heme-binding region is in β_1 (Karow et al., 2005), which has a domain that belongs to the H-NOX (heme-nitric oxide/oxygen binding) family (Derbyshire & Marletta, 2012). Insights into the heme domain structure come mostly from bacterial H-NOX proteins (Derbyshire & Marletta, 2012). At the resting state, the heme tetrapyrrole ring is bound to a ferrous ion with histidine coordination, forming a five-coordinate species. Upon encounter with nitric oxide, the nitric oxide can bind to the heme ring and form a six-coordinate species. Histidine will then dissociate, allowing a low-activity, nitric oxide-bound five-coordinate species to form (Derbyshire & Marletta, 2012; Stone & Marletta, 1996). If nitric oxide is abundant, a second molecule will bind to the heme, displacing the histidine. The second NO will then depart, leaving a high-activity nitric oxide-bound five-coordinate species (Martin, Berka, Sharina, & Tsai, 2012). The two five-coordinate species cannot be distinguished by electronic absorption spectroscopy but do exhibit different signals by electron paramagnetic spectrometry (Derbyshire et al., 2008) Mg^{2+}-GTP and Mg^{2+}/cGMP/PP_i are capable of inducing a high-activity form of sGCs (Cary, Winger, & Marletta, 2005; Russwurm & Koesling, 2004). cGMP is made from GTP by guanylate cyclases and is capable of activating ion channels, phosphodiesterases, and kinases (Girvan & Munro, 2013).

3. ENGINEERED GENETICALLY ENCODED REDOX SENSORS

This class of redox sensor includes those that were engineered and exogenously expressed, typically by transfection into cells or transgenesis in animal systems. These include sensors developed using fluorescent proteins or chimeric fusions of fluorescent proteins and include many that were inspired from an endogenously encoded sensor that was engineered into a useful redox reporting tool.

3.1. Reduction–oxidation-sensitive green fluorescent protein

Reduction–oxidation-sensitive green fluorescent protein (roGFP) detects oxidative changes in the cell (Cannon & Remington, 2006; Hanson et al., 2004; Lohman & Remington, 2008). The fluorescent properties of roGFP in response to oxidation states (reduced dithiol or the oxidized disulfide) is due to the addition two cysteines in the beta barrel structure of the GFP. roGFP has been extensively used in various *in vitro* and *in vivo* studies (Berkelhamer et al., 2013; Celotto, Liu, Vandemark, & Palladino, 2012; Duprez, Roma, Close, & Jonas, 2012; Farrow et al., 2012; Gibbs-Flournoy, Simmons, Bromberg, Dick, & Samet, 2013; Guzman et al., 2010; Heller, Meyer, & Tudzynski, 2012; Liu, Celotto, Romero, Wipf, & Palladino, 2012; Loor et al., 2010; Meyer et al., 2007; Mungai et al., 2011; van Lith, Tiwari, Pediani, Milligan, & Bulleid, 2011; Wedgwood et al., 2011; Yu et al., 2009). roGFPs allow real-time monitoring of redox status in a cell but some have a slow response time to redox changes (Cannon & Remington, 2006; Hanson et al., 2004). Newer versions of roGFP1 with basic residues positioned near key cysteines have been designed resulting in an enhanced response time (Cannon & Remington, 2006). The roGFP2 variant has a pH-dependent fluorescence emission and has a much broader dynamic range (Meyer et al., 2007; Schwarzlander et al., 2008).

Fusion protein Grx1–roGFP2 is different from its predecessors (Gutscher et al., 2008). It has a "built-in glutaredoxin" by fusing glutaredoxin (Grx1) and roGFP2. As the thiol-disulfide exchange between the two is catalyzed by glutaredoxin, having Grx1 fused to roGFP2 eliminates the rate-limiting factor (Lukyanov & Belousov, 2014). It has already shown to be superior to roGFP2 in reaction rate and sensitivity: Grx1–RoGFP2 can respond to reduced glutathione (GSH) or oxidized glutathione (GSSG) in minutes, while roGFP2 is unresponsive under similar conditions. It is also sensitive to determine trace amount of GSSG, making it the best choice currently available for measuring GSH/GSSG ratio in highly reduced environments (Gutscher et al., 2008). The same group fused a yeast peroxidase (Orp1) to roGFP2 to make Orp1-roGFP2 (Gutscher et al., 2009). Like HyPer (Section 3.3), Orp1-roGFP2 reacts with hydrogen peroxide. However, HyPer reacts with hydrogen peroxide directly, while reaction between Orp1-roGFP2 and hydrogen peroxide is mediated by peroxidase, explaining the faster oxidation rate by HyPer (Gutscher et al., 2009).

roGFP variants have been successfully targeted to the mitochondria *in vitro* (Berkelhamer et al., 2013; Duprez et al., 2012; Farrow et al., 2012; Gibbs-Flournoy et al., 2013; Loor et al., 2010; Mungai et al., 2011) and *in vivo* (Celotto et al., 2012; Guzman et al., 2010; Heller et al., 2012; Hrizo et al., 2013; Liu, Celotto, et al., 2012, Liu, Liu, Wang, Yang, & Wang, 2012; Meyer et al., 2007). Although to date the mitochondrion is the most targeted organelle, roGFP variants have also been targeted to endoplasmic reticulum (Birk et al., 2013; van Lith et al., 2011) and have the potential to be targeted to other cellular compartments. Another redox-sensitive fluorescent protein (rxYFP) (Ostergaard, Henriksen, Hansen, & Winther, 2001), the first engineered, genetically encoded redox sensor protein (Lukyanov & Belousov, 2014), has been used to specifically detect glutathione-specific redox state in cellular compartments (Banach-Latapy et al., 2013; Kolossov et al., 2012).

3.2. Superoxide stress detection

The *E. coli* SoxRS system can induce production of 45 proteins that regulate a wide array of responses to superoxide exposure, including DNA repair, cellular reducing power maintenance, and metabolism. Briefly, SoxR normally contains $[2Fe–2S]^{1+}$ clusters and binds to the *soxS* promoter repressing expression (Hidalgo, Bollinger, Bradley, Walsh, & Demple, 1995; Wu, Dunham, & Weiss, 1995). Superoxides change $[2Fe–2S]^{1+}$ into $[2Fe–2S]^{2+}$ clusters altering its structure and *soxS* promoter binding, and activating transcription by allowing RNA polymerase association. Once oxidative stress is removed, $[2Fe–2S]^{2+}$ clusters are reduced back to $[2Fe–2S]^{1+}$ clusters by reductase proteins Rsx and RseC (Ding & Demple, 1997; Koo et al., 2003) and transcription is repressed.

The *E. coli* aconitases AcnA and AcnB are Fe–S-cluster-based superoxide sensors (Gardner & Fridovich, 1992; Tang, Quail, Artymiuk, Guest, & Green, 2002). When oxidative stress is low, AcnB can serve as a repressor of *sodA* (manganese superoxide dismutase) translation. When exposed to superoxide, the Fe–S cluster will disassemble and inactivate AcnB. The more stable AcnA takes over the function of AcnB and allows the inactivated AcnB to continue to block *sodA* translation. However, if the superoxide level continues to rise, SoxRS will be activated to enhance *sodA* transcription.

Recently, a mitochondrial circularly permuted yellow fluorescent protein (mt-cpYFP) has been reported to be a novel superoxide indicator (Wang et al., 2008). The paper shows that superoxide production is controlled by mitochondrial permeability transition pore openings and ETC. Most interestingly, an individual mitochondrion can produce highly localized spikes of superoxide that are detectable as flashes of mt-cpYFP fluorescence, even in resting conditions (Wang et al., 2008).

3.3. Peroxide stress detection

The *E. coli* transcriptional activator OxyR is a well-studied thiol-based redox sensor that is prevalent in both Gram-positive and Gram-negative bacteria. OxyR has six cysteine thiols, which allows OxyR to form several reversible bonds, including disulfide (R-S-S-R′), sulfenic acid (R-SOH), or S-nitrosothiol (R-SNO) bonds, the latter of which occurs when RNS are encountered. Additionally, it can also form irreversible bonds to produce sulfinic acid (R-SO_2H) and sulfonic acid (R-SO_3H) (Green & Paget, 2004). With six cysteine thiols, the OxyR is well equipped to sense oxidative stress. However, the general consensus is that only one cysteine (Cys 199) is involved in hydrogen peroxide-mediated activation of OxyR, while the role of another cysteine (Cys 208) is controversial (Kim et al., 2002). In the presence of peroxide, the expression of detoxifying enzymes such as hydroperoxidase I and alkylhydroperoxide reductase is activated by OxyR (Zheng et al., 2001).

The *E. coli* protein IscR also responds to hydrogen peroxide stress; however, its sensor is Fe–S-based instead of thiol-based. It has a [2Fe–2S] cluster (Schwartz, Djaman, Imlay, & Kiley, 2000) and is a repressor of *isc* gene expression. When exposed to hydrogen peroxide, IscR is inactivated with allows for *isc* gene transcription (Zheng et al., 2001).

The yeast *Schizosaccharomyces pombe* activates two different pathways in response to different levels of hydrogen peroxide stress (Vivancos, Castillo, Jones, Ayte, & Hidalgo, 2004). Yeast can activate and cause nuclear translocation of Pap1, a homolog of the mammalian transcription factor c-Jun (Toda, Shimanuki, & Yanagida, 1991), at low hydrogen peroxide concentrations (~0.2 m*M*). However, at higher hydrogen peroxide concentrations (~1 m*M*), either Styl1, a mitogen-activated protein kinase (MAPK), has to first be activated (Nguyen, Lee, Place, & Shiozaki, 2000; Quinn et al., 2002) or H_2O_2-decomposing enzymes (Degols, Shiozaki, & Russell, 1996; Wilkinson et al., 1996) have to be overexpressed before Pap1 can be translocated to the nucleus.

HyPer, a chromosomally encoded ratiometric hydrogen peroxide sensor protein (Belousov et al., 2006; Markvicheva, Bogdanova, Staroverov, Lukyanov, & Belousov, 2008), has been utilized *in vivo* to examine H_2O_2 levels throughout the developmental stages and lifespan of *C. elegans* (Knoefler et al., 2012). HyPer-2 (Markvicheva et al., 2011) and HyPer-3 (Bilan et al., 2013), improved versions of HyPer with more extended dynamic ranges, are used to examine intracellular changes in H_2O_2 levels. It has been shown that *C. elegans* have significant levels of oxidants during larval development, but these levels drop as the worms enter reproductive age and then rise again with animal aging (Knoefler et al., 2012). It has also been demonstrated that long-lived mutants transition to reducing conditions faster during larval stages while short-lived mutants have consistently higher levels of oxidation throughout their lives, suggesting that decreasing oxidant levels early and keeping oxidant levels low provide significant survival advantages (Knoefler et al., 2012).

3.4. Transition metal detection

It has recently been demonstrated that molybdate can cause G2/M cell arrest in the human prostrate cancer cell line PC-3 (Liu, Liu, et al., 2012). Molybdenum is an essential transitional metal for living organisms and tracking it in real time will be of interest in understanding cancer pathogenesis (Nakanishi et al., 2013). Various enzymes, such as xanthine oxidase, sulfite oxidase, aldehyde oxidase, and nitrate reductase, utilize molybdenum as a cofactor to oxidize and reduce carbon, nitrogen, and sulfur metabolites. It has a range of oxidation states from +2 to +6. Most of our knowledge about its transport, storage, and transcription regulation is derived from bacteria. Recently, a FRET-based genetically encoded redox sensor for eukaryotes was developed and a novel oxalate-sensitive molybdate transporter was discovered in animal cells (Nakanishi et al., 2013). Further investigation will be needed to fully understand the role of molybdate in redox biology and cellular metabolism. The development of this genetically encoded redox sensor provides a crucial first step toward that goal.

3.5. NAD^+/NADH detection

NAD^+/NADH redox pair is critical to oxidation reactions as it supplies electrons to the mitochondria ETC (Pollak, Dolle, & Ziegler, 2007). Electron leakage from NADH can also contribute to increased ROS levels in cells (Holmgren et al., 2005; Starkov, 2008). Recently, Peredox and Frex proteins (Hung, Albeck, Tantama, & Yellen, 2011; Zhao et al., 2011) have been developed to measure NAD^+/NADH ratios and NADH levels, respectively. Peredox was based on protein Rex from *Thermus aquaticus* (Hung et al., 2011) and Frex was based on protein Rex from *Bacillus subtilis* (Zhao et al., 2011). Rex is a repressor that can lose its affinity to its target DNA when NADH binds to it. When oxygen is limited or when respiration is inhibited, the NAD^+/NADH ratio decreases, which increases the probability of NADH Rex binding. This unique property of Rex was capitalized upon and used to engineer a specific nutrient sensor that reports the metabolic state of transgenic CHO cells (Weber, Link, & Fussenegger, 2006). However, currently their use is somewhat limited due to high sensitivity of Peredox to NADH and pH sensitivity of Frex (Zhao et al., 2011).

4. OXIDATIVE STRESS IN CANCER AND POTENTIAL APPLICATIONS OF REDOX SENSORS IN ONCOLOGY RESEARCH

It is well known that ROS can cause DNA damage (Veskoukis, Tsatsakis, & Kouretas, 2012). Elevated ROS can inactivate antioxidant proteins as well as DNA repair enzymes (Eiberger et al., 2008; Gillard, Begusova, Castaing, & Spotheim-Maurizot, 2004). Elevated ROS has been detected not only within the tumor microenvironment but also in distant organs through a bystander effect (Sedelnikova et al., 2010). The role of ROS in tumorigenesis is extraordinarily complex and incompletely understood. It has been reported that ROS can simultaneously stimulate the oncogenic Ras–Raf–MEK1/2–ERK1/2 pathway and the cancer suppression p38 MAPK pathway (Pan, Hong, & Ren, 2009). More research is needed to determine under what conditions one pathway becomes dominant. Factors such as p53 are also closely involved in ROS regulation. Under low-intensity stress, p53 activates antioxidant genes (Hussain et al., 2004; Sablina et al., 2005; Tan et al., 1999), such as *sestrins* (Lee, Bodmer,

Bier, & Karin, 2010; Lee, Budanov, et al., 2010). However, under high-intensity stress, p53 will activate pro-oxidant redox proteins in an attempt to induce apoptosis (Liu, Chen, & St Clair, 2008; Polyak, Xia, Zweier, Kinzler, & Vogelstein, 1997; Rivera & Maxwell, 2005). Thus, p53 dynamically responds to ROS with varying levels of activation resulting in a wide range of transcriptional target activation. The ataxia-telangiectasia-mutated (ATM) protein kinase is well known for its DNA damage response, however, recently it has been reported that ATM can also be activated by direct oxidation (Guo, Deshpande, & Paull, 2010; Guo, Kozlov, Lavin, Person, & Paull, 2010). Once ATM is activated, it can phosphorylate the tumor suppressor p53 (Canman et al., 1998).

The role of ROS in tumor development is just beginning to be understood. It has been shown that in mouse models, ROS serve as mediators of macrophages' antitumor activities (Nathan & Cohn, 1980; Nathan & Klebanoff, 1982; Nathan, Silverstein, Brukner, & Cohn, 1979). Additionally, ROS produced by neutrophils have been shown to contribute to tumorigenesis (Weitzman, Weitberg, Clark, & Stossel, 1985). ROS are also produced by tumor cells themselves affecting cancer pathogenesis (Liou & Storz, 2010). For example, elevated ROS can induce mitochondrial DNA mutations that enhance the metastatic potential of tumor cells (Ishikawa et al., 2008) or lead to genomic instability (Lonkar & Dedon, 2011) and produce angiogenic factors (Liou & Storz, 2010).

Thioredoxins (Trxs) and glutaredoxins (Grxs) are thiol-disulfide oxidoreductases, major human redox sensors, and play an important role in cellular redox homeostasis (Holmgren, 1976, 1979a, 1979b; Laurent, Moore, & Reichard, 1964). Trxs have a conserved Cys-Gly-Pro-Cys motif (Holmgren, 1968) that is conserved in most organisms. There are two Trx systems in mammals: a cytosolic Trx1 and a mitochondrial Trx2. Trx1 is upregulated in liver (Nakamura et al., 1992), lung (Gasdaska, Oblong, Cotgreave, & Powis, 1994), and colon (Berggren et al., 1996) cancers. It has been shown that in normal tissue Grx2b and Grx2c are only expressed in testes, but they have been detected in various cancer cell lines as well (Lonn et al., 2008). Additionally, a strong correlation between the degree of differentiation of adenocarcinomas and Grx1 and Grx2a expression has been demonstrated (Fernandes et al., 2009). The Grx field has been expanded rapidly, with about two-thirds of the articles for "glutaredoxins" being published in the last decade. However, more research is needed to determine its role in tumorigenesis. Developing sensors using Trxs and Grxs to visualize the process may be particularly useful.

While it is well established that oxidative stress is important in oncopathogenesis, our current understanding is incomplete and genetically encoded redox sensors are poised to make important contributions to this field.

5. LIMITATIONS AND FUTURE OF GENETICALLY ENCODED REDOX SENSORS

Although genetically encoded redox sensors possess substantial advantages over other conventional methods to measure redox reactions, they are not without limitations. First, they are sensitive to pH (Lukyanov & Belousov, 2014) and may have limited usage in highly acidified environments, such as lysosomes. It is stated that measurement with Frex, a genetically encoded redox sensor for NADH, would require corrections when used in mitochondria because of a wider range of pH within the organelle (Zhao et al., 2011). Frex is also incapable of measuring NAD^+/NADH ratio; however, new genetically encoded redox sensors are being developed to address this issue (Zhao et al., 2011). Another challenge is to make sure that genetically encoded redox sensors do not interfere with normal cellular processes. For example, questions have been raised regarding the ability of locally expressed HyPer to alter hydrogen peroxide concentration in subcellular compartments (Lukyanov & Belousov, 2014). With increasing interests in discovering and developing genetically encoded redox sensors, it is expected that future generations of genetically encoded redox sensors will have further refined temporal and spatial resolution and improved targeting to specific tissues and organelles.

6. SUMMARY

Redox reactions are essential to normal cellular function and the ability to measure oxidative changes is crucial in understanding how such changes affect metabolic pathways, cellular repair, and other important cellular mechanisms. An ideal genetically encoded redox sensor should be selective for the type of oxidant under investigation, have high signal to noise ratio, show site specificity for compartment localization, and have quick response time. Additionally, it is extraordinarily helpful if a sensor is ratiometric, resistant to pH variations, and nontoxic to the cell (Wang et al., 2013).

Some of the commonly used probes produce results that are complicated to interpret due to the limitations of selectivity and sensitivity. A formidable challenge in detecting the oxidative product is to trap it at the site of its origin as they have a propensity to diffuse away (Winterbourn, 2014). An ideal probe should be able to detect the oxidant before allowing it to diffuse. Genetically encoded redox sensors have excellent spatial and temporal resolution and allow real-time measurements of oxidative changes in living cells. Ratiometric detection, site-specific targeting, and broad ranges of sensitivity make these sensors advantageous over conventional probes. Although most of the genetically encoded redox sensors were identified in microbes, their discoveries and applications in higher organisms have increased recently. With the recent advances in imaging technology, such as time-lapse imaging in intact cells or *in vivo* imaging in animal models, genetically encoded ROS sensors will play an important role in the understanding of the complexity of *in vivo* redox dynamics. As it has already been established that oxidative stress plays an important role in tumor formation, further developments and applications of these sensors will substantially improve our understanding of oncopathogenesis *in vivo*.

REFERENCES

Allen, J. F. (1993a). Control of gene expression by redox potential and the requirement for chloroplast and mitochondrial genomes. *Journal of Theoretical Biology*, *165*(4), 609–631. http://dx.doi.org/10.1006/jtbi.1993.1210.

Allen, J. F. (1993b). Redox control of transcription: Sensors, response regulators, activators and repressors. *FEBS Letters*, *332*(3), 203–207.

Altieri, F., Grillo, C., Maceroni, M., & Chichiarelli, S. (2008). DNA damage and repair: From molecular mechanisms to health implications. *Antioxidants & Redox Signaling*, *10*(5), 891–937. http://dx.doi.org/10.1089/ars.2007.1830.

Ames, B. N., Gold, L. S., & Willett, W. C. (1995). The causes and prevention of cancer. *Proceedings of the National Academy of Sciences of the United States of America*, *92*(12), 5258–5265.

Antunes, F., Han, D., & Cadenas, E. (2002). Relative contributions of heart mitochondria glutathione peroxidase and catalase to H(2)O(2) detoxification in in vivo conditions. *Free Radical Biology & Medicine*, *33*(9), 1260–1267.

Banach-Latapy, A., He, T., Dardalhon, M., Vernis, L., Chanet, R., & Huang, M. E. (2013). Redox-sensitive YFP sensors for monitoring dynamic compartment-specific glutathione redox state. *Free Radical Biology & Medicine*, *65C*, 436–445. http://dx.doi.org/10.1016/j.freeradbiomed.2013.07.033.

Bashan, N., Kovsan, J., Kachko, I., Ovadia, H., & Rudich, A. (2009). Positive and negative regulation of insulin signaling by reactive oxygen and nitrogen species. *Physiological Reviews*, *89*(1), 27–71. http://dx.doi.org/10.1152/physrev.00014.2008.

Belousov, V. V., Fradkov, A. F., Lukyanov, K. A., Staroverov, D. B., Shakhbazov, K. S., Terskikh, A. V., et al. (2006). Genetically encoded fluorescent indicator for intracellular hydrogen peroxide. *Nature Methods*, *3*(4), 281–286. http://dx.doi.org/10.1038/nmeth866.

Benzi, G., & Moretti, A. (1995). Are reactive oxygen species involved in Alzheimer's disease? *Neurobiology of Aging*, *16*(4), 661–674.

Berggren, M., Gallegos, A., Gasdaska, J. R., Gasdaska, P. Y., Warneke, J., & Powis, G. (1996). Thioredoxin and thioredoxin reductase gene expression in human tumors and cell lines, and the effects of serum stimulation and hypoxia. *Anticancer Research*, *16*(6B), 3459–3466.

Berkelhamer, S. K., Kim, G. A., Radder, J. E., Wedgwood, S., Czech, L., Steinhorn, R. H., et al. (2013). Developmental differences in hyperoxia-induced oxidative stress and cellular responses in the murine lung. *Free Radical Biology & Medicine*, *61C*, 51–60. http://dx.doi.org/10.1016/j.freeradbiomed.2013.03.003.

Bilan, D. S., Pase, L., Joosen, L., Gorokhovatsky, A. Y., Ermakova, Y. G., Gadella, T. W., et al. (2013). HyPer-3: A genetically encoded H(2)O(2) probe with improved performance for ratiometric and fluorescence lifetime imaging. *ACS Chemical Biology*, *8*(3), 535–542. http://dx.doi.org/10.1021/cb300625g.

Birk, J., Meyer, M., Aller, I., Hansen, H. G., Odermatt, A., Dick, T. P., et al. (2013). Endoplasmic reticulum: Reduced and oxidized glutathione revisited. *Journal of Cell Science*, *126*(Pt. 7), 1604–1617. http://dx.doi.org/10.1242/jcs.117218.

Boveris, A., & Chance, B. (1973). The mitochondrial generation of hydrogen peroxide. General properties and effect of hyperbaric oxygen. *The Biochemical Journal*, *134*(3), 707–716.

Brand, M. D., Affourtit, C., Esteves, T. C., Green, K., Lambert, A. J., Miwa, S., et al. (2004). Mitochondrial superoxide: Production, biological effects, and activation of uncoupling proteins. *Free Radical Biology & Medicine*, *37*(6), 755–767. http://dx.doi.org/10.1016/j.freeradbiomed.2004.05.034.

Brown, G. C., & Borutaite, V. (2012). There is no evidence that mitochondria are the main source of reactive oxygen species in mammalian cells. *Mitochondrion*, *12*(1), 1–4. http://dx.doi.org/10.1016/j.mito.2011.02.001.

Cadet, J., Douki, T., & Ravanat, J. L. (2010). Oxidatively generated base damage to cellular DNA. *Free Radical Biology & Medicine*, *49*(1), 9–21. http://dx.doi.org/10.1016/j.freeradbiomed.2010.03.025.

Canman, C. E., Lim, D. S., Cimprich, K. A., Taya, Y., Tamai, K., Sakaguchi, K., et al. (1998). Activation of the ATM kinase by ionizing radiation and phosphorylation of p53. *Science*, *281*(5383), 1677–1679.

Cannon, M. B., & Remington, S. J. (2006). Re-engineering redox-sensitive green fluorescent protein for improved response rate. *Protein Science*, *15*(1), 45–57. http://dx.doi.org/10.1110/ps.051734306.

Cary, S. P., Winger, J. A., & Marletta, M. A. (2005). Tonic and acute nitric oxide signaling through soluble guanylate cyclase is mediated by nonheme nitric oxide, ATP, and GTP. *Proceedings of the National Academy of Sciences of the United States of America*, *102*(37), 13064–13069. http://dx.doi.org/10.1073/pnas.0506289102.

Celotto, A. M., Liu, Z., Vandemark, A. P., & Palladino, M. J. (2012). A novel Drosophila SOD2 mutant demonstrates a role for mitochondrial ROS in neurodevelopment and disease. *Brain & Behavior*, *2*(4), 424–434. http://dx.doi.org/10.1002/brb3.73.

Chance, B., Sies, H., & Boveris, A. (1979). Hydroperoxide metabolism in mammalian organs. *Physiological Reviews*, *59*(3), 527–605.

Comhair, S. A., & Erzurum, S. C. (2002). Antioxidant responses to oxidant-mediated lung diseases. *American Journal of Physiology: Lung Cellular and Molecular Physiology*, *283*(2), L246–L255. http://dx.doi.org/10.1152/ajplung.00491.2001.

Cruz-Ramos, H., Crack, J., Wu, G., Hughes, M. N., Scott, C., Thomson, A. J., et al. (2002). NO sensing by FNR: Regulation of the Escherichia coli NO-detoxifying flavohaemoglobin, Hmp. *The EMBO Journal*, *21*(13), 3235–3244. http://dx.doi.org/10.1093/emboj/cdf339.

Darley-Usmar, V., & Halliwell, B. (1996). Blood radicals: Reactive nitrogen species, reactive oxygen species, transition metal ions, and the vascular system. *Pharmaceutical Research*, *13*(5), 649–662.

De Bont, R., & van Larebeke, N. (2004). Endogenous DNA damage in humans: A review of quantitative data. *Mutagenesis*, *19*(3), 169–185.

Debenham, M. J., Hao, Q., Hasnain, S. S., Dodd, F. E., Abraham, Z. H., & Eady, R. R. (1996). Structure solution of azurin II from Alcaligenes xylosoxidans using the Laue method: Possibility of studying in situ redox changes using X-rays. *Journal of Synchrotron Radiation*, *3*(Pt. 1), 14–19. http://dx.doi.org/10.1107/S0909049595013240.

Degols, G., Shiozaki, K., & Russell, P. (1996). Activation and regulation of the Spc1 stress-activated protein kinase in Schizosaccharomyces pombe. *Molecular and Cellular Biology*, *16*(6), 2870–2877.

Demple, B. (1999). Genetic responses against nitric oxide toxicity. *Brazilian Journal of Medical and Biological Research*, *32*(11), 1417–1427.

Derbyshire, E. R., Gunn, A., Ibrahim, M., Spiro, T. G., Britt, R. D., & Marletta, M. A. (2008). Characterization of two different five-coordinate soluble guanylate cyclase ferrous-nitrosyl complexes. *Biochemistry*, *47*(12), 3892–3899. http://dx.doi.org/10.1021/bi7022943.

Derbyshire, E. R., & Marletta, M. A. (2012). Structure and regulation of soluble guanylate cyclase. *Annual Review of Biochemistry*, *81*, 533–559. http://dx.doi.org/10.1146/annurev-biochem-050410-100030.

Dhalla, N. S., Temsah, R. M., & Netticadan, T. (2000). Role of oxidative stress in cardiovascular diseases. *Journal of Hypertension*, *18*(6), 655–673.

Ding, H., & Demple, B. (1997). In vivo kinetics of a redox-regulated transcriptional switch. *Proceedings of the National Academy of Sciences of the United States of America*, *94*(16), 8445–8449.

Ding, H., & Demple, B. (2000). Direct nitric oxide signal transduction via nitrosylation of iron-sulfur centers in the SoxR transcription activator. *Proceedings of the National Academy of Sciences of the United States of America*, *97*(10), 5146–5150.

Duprez, J., Roma, L. P., Close, A. F., & Jonas, J. C. (2012). Protective antioxidant and antiapoptotic effects of ZnCl2 in rat pancreatic islets cultured in low and high glucose concentrations. *PLoS One*, 7(10), e46831. http://dx.doi.org/10.1371/journal.pone.0046831.

Eiberger, W., Volkmer, B., Amouroux, R., Dherin, C., Radicella, J. P., & Epe, B. (2008). Oxidative stress impairs the repair of oxidative DNA base modifications in human skin fibroblasts and melanoma cells. *DNA Repair (Amst)*, 7(6), 912–921. http://dx.doi.org/10.1016/j.dnarep.2008.03.002.

Evans, J. (2008). Antioxidant supplements to prevent or slow down the progression of AMD: A systematic review and meta-analysis. *Eye (London, England)*, *22*(6), 751–760. http://dx.doi.org/10.1038/eye.2008.100.

Farrow, K. N., Lee, K. J., Perez, M., Schriewer, J. M., Wedgwood, S., Lakshminrusimha, S., et al. (2012). Brief hyperoxia increases mitochondrial oxidation and increases phosphodiesterase 5 activity in fetal pulmonary artery smooth muscle cells. *Antioxidants & Redox Signaling*, *17*(3), 460–470. http://dx.doi.org/10.1089/ars.2011.4184.

Fenaille, F., Guy, P. A., & Tabet, J. C. (2003). Study of protein modification by 4-hydroxy-2-nonenal and other short chain aldehydes analyzed by electrospray ionization tandem mass spectrometry. *Journal of the American Society for Mass Spectrometry*, *14*(3), 215–226. http://dx.doi.org/10.1016/S1044-0305(02)00911-X.

Fernandes, A. P., Capitanio, A., Selenius, M., Brodin, O., Rundlof, A. K., & Bjornstedt, M. (2009). Expression profiles of thioredoxin family proteins in human lung cancer tissue: Correlation with proliferation and differentiation. *Histopathology*, *55*(3), 313–320. http://dx.doi.org/10.1111/j.1365-2559.2009.03381.x.

Fridovich, I. (1978). The biology of oxygen radicals. *Science*, *201*(4359), 875–880.
Fukai, T., & Ushio-Fukai, M. (2011). Superoxide dismutases: Role in redox signaling, vascular function, and diseases. *Antioxidants & Redox Signaling*, *15*(6), 1583–1606. http://dx.doi.org/10.1089/ars.2011.3999.
Gardner, P. R., & Fridovich, I. (1992). Inactivation-reactivation of aconitase in Escherichia coli. A sensitive measure of superoxide radical. *The Journal of Biological Chemistry*, *267*(13), 8757–8763.
Gasdaska, P. Y., Oblong, J. E., Cotgreave, I. A., & Powis, G. (1994). The predicted amino acid sequence of human thioredoxin is identical to that of the autocrine growth factor human adult T-cell derived factor (ADF): Thioredoxin mRNA is elevated in some human tumors. *Biochimica et Biophysica Acta*, *1218*(3), 292–296.
Gibbs-Flournoy, E. A., Simmons, S. O., Bromberg, P. A., Dick, T. P., & Samet, J. M. (2013). Monitoring intracellular redox changes in ozone-exposed airway epithelial cells. *Environmental Health Perspectives*, *121*(3), 312–317. http://dx.doi.org/10.1289/ehp.1206039.
Gillard, N., Begusova, M., Castaing, B., & Spotheim-Maurizot, M. (2004). Radiation affects binding of Fpg repair protein to an abasic site containing DNA. *Radiation Research*, *162*(5), 566–571.
Girvan, H. M., & Munro, A. W. (2013). Heme sensor proteins. *The Journal of Biological Chemistry*, *288*(19), 13194–13203. http://dx.doi.org/10.1074/jbc.R112.422642.
Green, J., & Paget, M. S. (2004). Bacterial redox sensors. *Nature Reviews Microbiology*, *2*(12), 954–966. http://dx.doi.org/10.1038/nrmicro1022.
Greving, M. P., Patti, G. J., & Siuzdak, G. (2011). Nanostructure-initiator mass spectrometry metabolite analysis and imaging. *Analytical Chemistry*, *83*(1), 2–7. http://dx.doi.org/10.1021/ac101565f.
Guo, Z., Deshpande, R., & Paull, T. T. (2010). ATM activation in the presence of oxidative stress. *Cell Cycle*, *9*(24), 4805–4811.
Guo, Z., Kozlov, S., Lavin, M. F., Person, M. D., & Paull, T. T. (2010). ATM activation by oxidative stress. *Science*, *330*(6003), 517–521. http://dx.doi.org/10.1126/science.1192912.
Gutscher, M., Pauleau, A. L., Marty, L., Brach, T., Wabnitz, G. H., Samstag, Y., et al. (2008). Real-time imaging of the intracellular glutathione redox potential. *Nature Methods*, *5*(6), 553–559. http://dx.doi.org/10.1038/nmeth.1212.
Gutscher, M., Sobotta, M. C., Wabnitz, G. H., Ballikaya, S., Meyer, A. J., Samstag, Y., et al. (2009). Proximity-based protein thiol oxidation by H2O2-scavenging peroxidases. *The Journal of Biological Chemistry*, *284*(46), 31532–31540. http://dx.doi.org/10.1074/jbc.M109.059246.
Guzman, J. N., Sanchez-Padilla, J., Wokosin, D., Kondapalli, J., Ilijic, E., Schumacker, P. T., et al. (2010). Oxidant stress evoked by pacemaking in dopaminergic neurons is attenuated by DJ-1. *Nature*, *468*(7324), 696–700. http://dx.doi.org/10.1038/nature09536.
Halliwell, B. (1994). Free radicals, antioxidants, and human disease: Curiosity, cause, or consequence? *Lancet*, *344*(8924), 721–724.
Halliwell, B., & Gutteridge, J. M. (1990). Role of free radicals and catalytic metal ions in human disease: An overview. *Methods in Enzymology*, *186*, 1–85.
Hanson, G. T., Aggeler, R., Oglesbee, D., Cannon, M., Capaldi, R. A., Tsien, R. Y., et al. (2004). Investigating mitochondrial redox potential with redox-sensitive green fluorescent protein indicators. *The Journal of Biological Chemistry*, *279*(13), 13044–13053. http://dx.doi.org/10.1074/jbc.M312846200.
Harrison, D., Griendling, K. K., Landmesser, U., Hornig, B., & Drexler, H. (2003). Role of oxidative stress in atherosclerosis. *The American Journal of Cardiology*, *91*(3A), 7A–11A.
Heller, J., Meyer, A. J., & Tudzynski, P. (2012). Redox-sensitive GFP2: Use of the genetically encoded biosensor of the redox status in the filamentous fungus Botrytis cinerea.

Molecular Plant Pathology, *13*(8), 935–947. http://dx.doi.org/10.1111/j.1364-3703.2012.00802.x.
Hidalgo, E., Bollinger, J. M., Jr., Bradley, T. M., Walsh, C. T., & Demple, B. (1995). Binuclear [2Fe-2S] clusters in the Escherichia coli SoxR protein and role of the metal centers in transcription. *The Journal of Biological Chemistry*, *270*(36), 20908–20914.
Holmgren, A. (1968). Thioredoxin. 6. The amino acid sequence of the protein from escherichia coli B. *European Journal of Biochemistry*, *6*(4), 475–484.
Holmgren, A. (1976). Hydrogen donor system for Escherichia coli ribonucleoside-diphosphate reductase dependent upon glutathione. *Proceedings of the National Academy of Sciences of the United States of America*, *73*(7), 2275–2279.
Holmgren, A. (1979a). Glutathione-dependent synthesis of deoxyribonucleotides. Characterization of the enzymatic mechanism of Escherichia coli glutaredoxin. *The Journal of Biological Chemistry*, *254*(9), 3672–3678.
Holmgren, A. (1979b). Glutathione-dependent synthesis of deoxyribonucleotides. Purification and characterization of glutaredoxin from Escherichia coli. *The Journal of Biological Chemistry*, *254*(9), 3664–3671.
Holmgren, A., Johansson, C., Berndt, C., Lonn, M. E., Hudemann, C., & Lillig, C. H. (2005). Thiol redox control via thioredoxin and glutaredoxin systems. *Biochemical Society Transactions*, *33*(Pt. 6), 1375–1377. http://dx.doi.org/10.1042/BST20051375.
Hrizo, S. L., Fisher, I. J., Long, D. R., Hutton, J. A., Liu, Z., & Palladino, M. J. (2013). Early mitochondrial dysfunction leads to altered redox chemistry underlying pathogenesis of TPI deficiency. *Neurobiology of Disease*, *54*, 289–296. http://dx.doi.org/10.1016/j.nbd.2012.12.020.
Hung, Y. P., Albeck, J. G., Tantama, M., & Yellen, G. (2011). Imaging cytosolic NADH-NAD(+) redox state with a genetically encoded fluorescent biosensor. *Cell Metabolism*, *14*(4), 545–554. http://dx.doi.org/10.1016/j.cmet.2011.08.012.
Hussain, S. P., Amstad, P., He, P., Robles, A., Lupold, S., Kaneko, I., et al. (2004). p53-induced up-regulation of MnSOD and GPx but not catalase increases oxidative stress and apoptosis. *Cancer Research*, *64*(7), 2350–2356.
Hussain, S. P., Hofseth, L. J., & Harris, C. C. (2003). Radical causes of cancer. *Nature Reviews Cancer*, *3*(4), 276–285. http://dx.doi.org/10.1038/nrc1046.
Iino, K., Iwase, M., Sonoki, K., Yoshinari, M., & Iida, M. (2005). Combination treatment of vitamin C and desferrioxamine suppresses glomerular superoxide and prostaglandin E production in diabetic rats. *Diabetes, Obesity & Metabolism*, 7(1), 106–109. http://dx.doi.org/10.1111/j.1463-1326.2005.00371.x.
Ikawa, M., Okazawa, H., Arakawa, K., Kudo, T., Kimura, H., Fujibayashi, Y., et al. (2009). PET imaging of redox and energy states in stroke-like episodes of MELAS. *Mitochondrion*, *9*(2), 144–148. http://dx.doi.org/10.1016/j.mito.2009.01.011.
Ishikawa, K., Takenaga, K., Akimoto, M., Koshikawa, N., Yamaguchi, A., Imanishi, H., et al. (2008). ROS-generating mitochondrial DNA mutations can regulate tumor cell metastasis. *Science*, *320*(5876), 661–664. http://dx.doi.org/10.1126/science.1156906.
Kang, J. G., Paget, M. S., Seok, Y. J., Hahn, M. Y., Bae, J. B., Hahn, J. S., et al. (1999). RsrA, an anti-sigma factor regulated by redox change. *The EMBO Journal*, *18*(15), 4292–4298. http://dx.doi.org/10.1093/emboj/18.15.4292.
Karow, D. S., Pan, D., Davis, J. H., Behrends, S., Mathies, R. A., & Marletta, M. A. (2005). Characterization of functional heme domains from soluble guanylate cyclase. *Biochemistry*, *44*(49), 16266–16274. http://dx.doi.org/10.1021/bi051601b.
Kim, S. O., Merchant, K., Nudelman, R., Beyer, W. F., Jr., Keng, T., DeAngelo, J., et al. (2002). OxyR: A molecular code for redox-related signaling. *Cell*, *109*(3), 383–396.
Knoefler, D., Thamsen, M., Koniczek, M., Niemuth, N. J., Diederich, A. K., & Jakob, U. (2012). Quantitative in vivo redox sensors uncover oxidative stress as an early event in life. *Molecular Cell*, *47*(5), 767–776. http://dx.doi.org/10.1016/j.molcel.2012.06.016.

Kolossov, V. L., Leslie, M. T., Chatterjee, A., Sheehan, B. M., Kenis, P. J., & Gaskins, H. R. (2012). Forster resonance energy transfer-based sensor targeting endoplasmic reticulum reveals highly oxidative environment. *Experimental Biology and Medicine (Maywood, N.J.)*, *237*(6), 652–662. http://dx.doi.org/10.1258/ebm.2012.011436.

Koo, M. S., Lee, J. H., Rah, S. Y., Yeo, W. S., Lee, J. W., Lee, K. L., et al. (2003). A reducing system of the superoxide sensor SoxR in Escherichia coli. *The EMBO Journal*, *22*(11), 2614–2622. http://dx.doi.org/10.1093/emboj/cdg252.

Laurent, T. C., Moore, E. C., & Reichard, P. (1964). Enzymatic synthesis of deoxyribonucleotides. IV. Isolation and characterization of thioredoxin the hydrogen donor from Escherichia coli B. *The Journal of Biological Chemistry*, *239*, 3436–3444.

Lee, J. H., Bodmer, R., Bier, E., & Karin, M. (2010). Sestrins at the crossroad between stress and aging. *Aging (Albany NY)*, *2*(6), 369–374.

Lee, J. H., Budanov, A. V., Park, E. J., Birse, R., Kim, T. E., Perkins, G. A., et al. (2010). Sestrin as a feedback inhibitor of TOR that prevents age-related pathologies. *Science*, *327*(5970), 1223–1228. http://dx.doi.org/10.1126/science.1182228.

Leichert, L. I., Gehrke, F., Gudiseva, H. V., Blackwell, T., Ilbert, M., Walker, A. K., et al. (2008). Quantifying changes in the thiol redox proteome upon oxidative stress in vivo. *Proceedings of the National Academy of Sciences of the United States of America*, *105*(24), 8197–8202. http://dx.doi.org/10.1073/pnas.0707723105.

Li, W., Bottrill, A. R., Bibb, M. J., Buttner, M. J., Paget, M. S., & Kleanthous, C. (2003). The Role of zinc in the disulphide stress-regulated anti-sigma factor RsrA from Streptomyces coelicolor. *Journal of Molecular Biology*, *333*(2), 461–472.

Liou, G. Y., & Storz, P. (2010). Reactive oxygen species in cancer. *Free Radical Research*, *44*(5), 479–496. http://dx.doi.org/10.3109/10715761003667554.

Liu, Z., Celotto, A. M., Romero, G., Wipf, P., & Palladino, M. J. (2012). Genetically encoded redox sensor identifies the role of ROS in degenerative and mitochondrial disease pathogenesis. *Neurobiology of Disease*, *45*(1), 362–368. http://dx.doi.org/10.1016/j.nbd.2011.08.022.

Liu, B., Chen, Y., & St Clair, D. K. (2008). ROS and p53: A versatile partnership. *Free Radical Biology & Medicine*, *44*(8), 1529–1535. http://dx.doi.org/10.1016/j.freeradbiomed.2008.01.011.

Liu, T. T., Liu, Y. J., Wang, Q., Yang, X. G., & Wang, K. (2012). Reactive-oxygen-species-mediated Cdc25C degradation results in differential antiproliferative activities of vanadate, tungstate, and molybdate in the PC-3 human prostate cancer cell line. *Journal of Biological Inorganic Chemistry*, *17*(2), 311–320. http://dx.doi.org/10.1007/s00775-011-0852-1.

Lohman, J. R., & Remington, S. J. (2008). Development of a family of redox-sensitive green fluorescent protein indicators for use in relatively oxidizing subcellular environments. *Biochemistry*, *47*(33), 8678–8688. http://dx.doi.org/10.1021/bi800498g.

Lonkar, P., & Dedon, P. C. (2011). Reactive species and DNA damage in chronic inflammation: Reconciling chemical mechanisms and biological fates. *International Journal of Cancer*, *128*(9), 1999–2009. http://dx.doi.org/10.1002/ijc.25815.

Lonn, M. E., Hudemann, C., Berndt, C., Cherkasov, V., Capani, F., Holmgren, A., et al. (2008). Expression pattern of human glutaredoxin 2 isoforms: Identification and characterization of two testis/cancer cell-specific isoforms. *Antioxidants & Redox Signaling*, *10*(3), 547–557. http://dx.doi.org/10.1089/ars.2007.1821.

Loor, G., Kondapalli, J., Schriewer, J. M., Chandel, N. S., Vanden Hoek, T. L., & Schumacker, P. T. (2010). Menadione triggers cell death through ROS-dependent mechanisms involving PARP activation without requiring apoptosis. *Free Radical Biology & Medicine*, *49*(12), 1925–1936. http://dx.doi.org/10.1016/j.freeradbiomed.2010.09.021.

Lukyanov, K. A., & Belousov, V. V. (2014). Genetically encoded fluorescent redox sensors. *Biochimica et Biophysica Acta*, *1840*(2), 745–756. http://dx.doi.org/10.1016/j.bbagen.2013.05.030.

Malins, D. C., Polissar, N. L., & Gunselman, S. J. (1996). Progression of human breast cancers to the metastatic state is linked to hydroxyl radical-induced DNA damage. *Proceedings of the National Academy of Sciences of the United States of America*, *93*(6), 2557–2563.

Manganelli, R., Voskuil, M. I., Schoolnik, G. K., Dubnau, E., Gomez, M., & Smith, I. (2002). Role of the extracytoplasmic-function sigma factor sigma(H) in Mycobacterium tuberculosis global gene expression. *Molecular Microbiology*, *45*(2), 365–374.

Markvicheva, K. N., Bilan, D. S., Mishina, N. M., Gorokhovatsky, A. Y., Vinokurov, L. M., Lukyanov, S., et al. (2011). A genetically encoded sensor for H2O2 with expanded dynamic range. *Bioorganic & Medicinal Chemistry*, *19*(3), 1079–1084. http://dx.doi.org/10.1016/j.bmc.2010.07.014.

Markvicheva, K. N., Bogdanova, E. A., Staroverov, D. B., Lukyanov, S., & Belousov, V. V. (2008). Imaging of intracellular hydrogen peroxide production with HyPer upon stimulation of HeLa cells with epidermal growth factor. *Methods in Molecular Biology*, *476*, 79–86.

Martin, K. R., & Barrett, J. C. (2002). Reactive oxygen species as double-edged swords in cellular processes: Low-dose cell signaling versus high-dose toxicity. *Human & Experimental Toxicology*, *21*(2), 71–75.

Martin, E., Berka, V., Sharina, I., & Tsai, A. L. (2012). Mechanism of binding of NO to soluble guanylyl cyclase: Implication for the second NO binding to the heme proximal site. *Biochemistry*, *51*(13), 2737–2746. http://dx.doi.org/10.1021/bi300105s.

Meyer, A. J., Brach, T., Marty, L., Kreye, S., Rouhier, N., Jacquot, J. P., et al. (2007). Redox-sensitive GFP in Arabidopsis thaliana is a quantitative biosensor for the redox potential of the cellular glutathione redox buffer. *The Plant Journal*, *52*(5), 973–986. http://dx.doi.org/10.1111/j.1365-313X.2007.03280.x.

Mungai, P. T., Waypa, G. B., Jairaman, A., Prakriya, M., Dokic, D., Ball, M. K., et al. (2011). Hypoxia triggers AMPK activation through reactive oxygen species-mediated activation of calcium release-activated calcium channels. *Molecular and Cellular Biology*, *31*(17), 3531–3545. http://dx.doi.org/10.1128/MCB.05124-11.

Nakamura, H., Masutani, H., Tagaya, Y., Yamauchi, A., Inamoto, T., Nanbu, Y., et al. (1992). Expression and growth-promoting effect of adult T-cell leukemia-derived factor. A human thioredoxin homologue in hepatocellular carcinoma. *Cancer*, *69*(8), 2091–2097.

Nakanishi, Y., Iida, S., Ueoka-Nakanishi, H., Niimi, T., Tomioka, R., & Maeshima, M. (2013). Exploring dynamics of molybdate in living animal cells by a genetically encoded FRET nanosensor. *PLoS One*, *8*(3), e58175. http://dx.doi.org/10.1371/journal.pone.0058175.

Nathan, C., & Cohn, Z. (1980). Role of oxygen-dependent mechanisms in antibody-induced lysis of tumor cells by activated macrophages. *The Journal of Experimental Medicine*, *152*(1), 198–208.

Nathan, C. F., & Klebanoff, S. J. (1982). Augmentation of spontaneous macrophage-mediated cytolysis by eosinophil peroxidase. *The Journal of Experimental Medicine*, *155*(5), 1291–1308.

Nathan, C. F., Silverstein, S. C., Brukner, L. H., & Cohn, Z. A. (1979). Extracellular cytolysis by activated macrophages and granulocytes. II. Hydrogen peroxide as a mediator of cytotoxicity. *The Journal of Experimental Medicine*, *149*(1), 100–113.

Nguyen, A. N., Lee, A., Place, W., & Shiozaki, K. (2000). Multistep phosphorelay proteins transmit oxidative stress signals to the fission yeast stress-activated protein kinase. *Molecular Biology of the Cell*, *11*(4), 1169–1181.

Okumoto, S., Jones, A., & Frommer, W. B. (2012). Quantitative imaging with fluorescent biosensors. *Annual Review of Plant Biology*, *63*, 663–706. http://dx.doi.org/10.1146/annurev-arplant-042110-103745.

Ostergaard, H., Henriksen, A., Hansen, F. G., & Winther, J. R. (2001). Shedding light on disulfide bond formation: Engineering a redox switch in green fluorescent protein. *The EMBO Journal*, *20*(21), 5853–5862. http://dx.doi.org/10.1093/emboj/20.21.5853.

Paget, M. S., Bae, J. B., Hahn, M. Y., Li, W., Kleanthous, C., Roe, J. H., et al. (2001). Mutational analysis of RsrA, a zinc-binding anti-sigma factor with a thiol-disulphide redox switch. *Molecular Microbiology*, *39*(4), 1036–1047.

Paget, M. S., Kang, J. G., Roe, J. H., & Buttner, M. J. (1998). sigmaR, an RNA polymerase sigma factor that modulates expression of the thioredoxin system in response to oxidative stress in Streptomyces coelicolor A3(2). *The EMBO Journal*, *17*(19), 5776–5782. http://dx.doi.org/10.1093/emboj/17.19.5776.

Pan, J. S., Hong, M. Z., & Ren, J. L. (2009). Reactive oxygen species: A double-edged sword in oncogenesis. *World Journal of Gastroenterology*, *15*(14), 1702–1707.

Pfannschmidt, T., & Liere, K. (2005). Redox regulation and modification of proteins controlling chloroplast gene expression. *Antioxidants & Redox Signaling*, 7(5–6), 607–618. http://dx.doi.org/10.1089/ars.2005.7.607.

Pfannschmidt, T., Schutze, K., Brost, M., & Oelmuller, R. (2001). A novel mechanism of nuclear photosynthesis gene regulation by redox signals from the chloroplast during photosystem stoichiometry adjustment. *The Journal of Biological Chemistry*, *276*(39), 36125–36130. http://dx.doi.org/10.1074/jbc.M105701200.

Piconi, L., Quagliaro, L., & Ceriello, A. (2003). Oxidative stress in diabetes. *Clinical Chemistry and Laboratory Medicine*, *41*(9), 1144–1149. http://dx.doi.org/10.1515/CCLM.2003.177.

Pollak, N., Dolle, C., & Ziegler, M. (2007). The power to reduce: Pyridine nucleotides–small molecules with a multitude of functions. *The Biochemical Journal*, *402*(2), 205–218. http://dx.doi.org/10.1042/BJ20061638.

Polyak, K., Xia, Y., Zweier, J. L., Kinzler, K. W., & Vogelstein, B. (1997). A model for p53-induced apoptosis. *Nature*, *389*(6648), 300–305. http://dx.doi.org/10.1038/38525.

Poole, R. K., & Hughes, M. N. (2000). New functions for the ancient globin family: Bacterial responses to nitric oxide and nitrosative stress. *Molecular Microbiology*, *36*(4), 775–783.

Poulsen, H. E., Prieme, H., & Loft, S. (1998). Role of oxidative DNA damage in cancer initiation and promotion. *European Journal of Cancer Prevention*, 7(1), 9–16.

Powis, G. (1987). Metabolism and reactions of quinoid anticancer agents. *Pharmacology & Therapeutics*, *35*(1–2), 57–162.

Quinn, J., Findlay, V. J., Dawson, K., Millar, J. B., Jones, N., Morgan, B. A., et al. (2002). Distinct regulatory proteins control the graded transcriptional response to increasing H(2)O(2) levels in fission yeast Schizosaccharomyces pombe. *Molecular Biology of the Cell*, *13*(3), 805–816. http://dx.doi.org/10.1091/mbc.01-06-0288.

Quintana-Cabrera, R., & Bolanos, J. P. (2013). Glutathione and gamma-glutamylcysteine in the antioxidant and survival functions of mitochondria. *Biochemical Society Transactions*, *41*(1), 106–110. http://dx.doi.org/10.1042/BST20120252.

Rahman, I. (1999). Inflammation and the regulation of glutathione level in lung epithelial cells. *Antioxidants & Redox Signaling*, *1*(4), 425–447.

Rahman, I., & MacNee, W. (2000a). Oxidative stress and regulation of glutathione in lung inflammation. *The European Respiratory Journal*, *16*(3), 534–554.

Rahman, I., & MacNee, W. (2000b). Regulation of redox glutathione levels and gene transcription in lung inflammation: Therapeutic approaches. *Free Radical Biology & Medicine*, *28*(9), 1405–1420.

Rhee, S. J., Jeong, Y. C., & Choi, J. H. (2005). Effects of vitamin E on phospholipase A2 activity and oxidative damage to the liver in streptozotocin-induced diabetic rats. *Annals of Nutrition & Metabolism*, *49*(6), 392–396. http://dx.doi.org/10.1159/000088930.

Rivera, A., & Maxwell, S. A. (2005). The p53-induced gene-6 (proline oxidase) mediates apoptosis through a calcineurin-dependent pathway. *The Journal of Biological Chemistry*, *280*(32), 29346–29354. http://dx.doi.org/10.1074/jbc.M504852200.

Russwurm, M., & Koesling, D. (2004). NO activation of guanylyl cyclase. *The EMBO Journal*, *23*(22), 4443–4450. http://dx.doi.org/10.1038/sj.emboj.7600422.

Sablina, A. A., Budanov, A. V., Ilyinskaya, G. V., Agapova, L. S., Kravchenko, J. E., & Chumakov, P. M. (2005). The antioxidant function of the p53 tumor suppressor. *Nature Medicine*, *11*(12), 1306–1313. http://dx.doi.org/10.1038/nm1320.

Schwartz, C. J., Djaman, O., Imlay, J. A., & Kiley, P. J. (2000). The cysteine desulfurase, IscS, has a major role in in vivo Fe-S cluster formation in Escherichia coli. *Proceedings of the National Academy of Sciences of the United States of America*, *97*(16), 9009–9014. http://dx.doi.org/10.1073/pnas.160261497.

Schwartz, C. J., Valente, A. J., Sprague, E. A., Kelley, J. L., Cayatte, A. J., & Rozek, M. M. (1992). Pathogenesis of the atherosclerotic lesion. Implications for diabetes mellitus. *Diabetes Care*, *15*(9), 1156–1167.

Schwarzlander, M., Fricker, M. D., Muller, C., Marty, L., Brach, T., Novak, J., et al. (2008). Confocal imaging of glutathione redox potential in living plant cells. *Journal of Microscopy*, *231*(2), 299–316. http://dx.doi.org/10.1111/j.1365-2818.2008.02030.x.

Sedelnikova, O. A., Redon, C. E., Dickey, J. S., Nakamura, A. J., Georgakilas, A. G., & Bonner, W. M. (2010). Role of oxidatively induced DNA lesions in human pathogenesis. *Mutation Research*, *704*(1–3), 152–159. http://dx.doi.org/10.1016/j.mrrev.2009.12.005.

Starkov, A. A. (2008). The role of mitochondria in reactive oxygen species metabolism and signaling. *Annals of the New York Academy of Sciences*, *1147*, 37–52. http://dx.doi.org/10.1196/annals.1427.015.

Stone, J. R., & Marletta, M. A. (1996). Spectral and kinetic studies on the activation of soluble guanylate cyclase by nitric oxide. *Biochemistry*, *35*(4), 1093–1099. http://dx.doi.org/10.1021/bi9519718.

St-Pierre, J., Buckingham, J. A., Roebuck, S. J., & Brand, M. D. (2002). Topology of superoxide production from different sites in the mitochondrial electron transport chain. *The Journal of Biological Chemistry*, *277*(47), 44784–44790. http://dx.doi.org/10.1074/jbc.M207217200.

Stuehr, D. J., Santolini, J., Wang, Z. Q., Wei, C. C., & Adak, S. (2004). Update on mechanism and catalytic regulation in the NO synthases. *The Journal of Biological Chemistry*, *279*(35), 36167–36170. http://dx.doi.org/10.1074/jbc.R400017200.

Sun, A. Y., & Chen, Y. M. (1998). Oxidative stress and neurodegenerative disorders. *Journal of Biomedical Science*, *5*(6), 401–414.

Tan, M., Li, S., Swaroop, M., Guan, K., Oberley, L. W., & Sun, Y. (1999). Transcriptional activation of the human glutathione peroxidase promoter by p53. *The Journal of Biological Chemistry*, *274*(17), 12061–12066.

Tang, Y., Quail, M. A., Artymiuk, P. J., Guest, J. R., & Green, J. (2002). Escherichia coli aconitases and oxidative stress: Post-transcriptional regulation of sodA expression. *Microbiology*, *148*(Pt. 4), 1027–1037.

Thelwall, P. E., Yemin, A. Y., Gillian, T. L., Simpson, N. E., Kasibhatla, M. S., Rabbani, Z. N., et al. (2005). Noninvasive in vivo detection of glutathione metabolism in tumors. *Cancer Research*, *65*(22), 10149–10153. http://dx.doi.org/10.1158/0008-5472.CAN-05-1781.

Toda, T., Shimanuki, M., & Yanagida, M. (1991). Fission yeast genes that confer resistance to staurosporine encode an AP-1-like transcription factor and a protein kinase related to the

mammalian ERK1/MAP2 and budding yeast FUS3 and KSS1 kinases. *Genes & Development, 5*(1), 60–73.
Turrens, J. F. (2003). Mitochondrial formation of reactive oxygen species. *The Journal of Physiology, 552*(Pt. 2), 335–344. http://dx.doi.org/10.1113/jphysiol.2003.049478.
van Lith, M., Tiwari, S., Pediani, J., Milligan, G., & Bulleid, N. J. (2011). Real-time monitoring of redox changes in the mammalian endoplasmic reticulum. *Journal of Cell Science, 124*(Pt. 14), 2349–2356. http://dx.doi.org/10.1242/jcs.085530.
Veskoukis, A. S., Tsatsakis, A. M., & Kouretas, D. (2012). Dietary oxidative stress and antioxidant defense with an emphasis on plant extract administration. *Cell Stress & Chaperones, 17*(1), 11–21. http://dx.doi.org/10.1007/s12192-011-0293-3.
Vivancos, A. P., Castillo, E. A., Jones, N., Ayte, J., & Hidalgo, E. (2004). Activation of the redox sensor Pap1 by hydrogen peroxide requires modulation of the intracellular oxidant concentration. *Molecular Microbiology, 52*(5), 1427–1435. http://dx.doi.org/10.1111/j.1365-2958.2004.04065.x.
Vivekananthan, D. P., Penn, M. S., Sapp, S. K., Hsu, A., & Topol, E. J. (2003). Use of antioxidant vitamins for the prevention of cardiovascular disease: Meta-analysis of randomised trials. *Lancet, 361*(9374), 2017–2023. http://dx.doi.org/10.1016/S0140-6736(03)13637-9.
Wang, W., Fang, H., Groom, L., Cheng, A., Zhang, W., Liu, J., et al. (2008). Superoxide flashes in single mitochondria. *Cell, 134*(2), 279–290. http://dx.doi.org/10.1016/j.cell.2008.06.017.
Wang, X., Fang, H., Huang, Z., Shang, W., Hou, T., Cheng, A., et al. (2013). Imaging ROS signaling in cells and animals. *Journal of Molecular Medicine (Berlin, Germany), 91*(8), 917–927. http://dx.doi.org/10.1007/s00109-013-1067-4.
Waris, G., & Ahsan, H. (2006). Reactive oxygen species: Role in the development of cancer and various chronic conditions. *Journal of Carcinogenesis, 5*, 14. http://dx.doi.org/10.1186/1477-3163-5-14.
Weber, H., Engelmann, S., Becher, D., & Hecker, M. (2004). Oxidative stress triggers thiol oxidation in the glyceraldehyde-3-phosphate dehydrogenase of Staphylococcus aureus. *Molecular Microbiology, 52*(1), 133–140. http://dx.doi.org/10.1111/j.1365-2958.2004.03971.x.
Weber, W., Link, N., & Fussenegger, M. (2006). A genetic redox sensor for mammalian cells. *Metabolic Engineering, 8*(3), 273–280. http://dx.doi.org/10.1016/j.ymben.2005.12.004.
Wedgwood, S., Lakshminrusimha, S., Fukai, T., Russell, J. A., Schumacker, P. T., & Steinhorn, R. H. (2011). Hydrogen peroxide regulates extracellular superoxide dismutase activity and expression in neonatal pulmonary hypertension. *Antioxidants & Redox Signaling, 15*(6), 1497–1506. http://dx.doi.org/10.1089/ars.2010.3630.
Weitzman, S. A., Weitberg, A. B., Clark, E. P., & Stossel, T. P. (1985). Phagocytes as carcinogens: Malignant transformation produced by human neutrophils. *Science, 227*(4691), 1231–1233.
Wilkinson, M. G., Samuels, M., Takeda, T., Toone, W. M., Shieh, J. C., Toda, T., et al. (1996). The Atf1 transcription factor is a target for the Sty1 stress-activated MAP kinase pathway in fission yeast. *Genes & Development, 10*(18), 2289–2301.
Winterbourn, C. C. (2014). The challenges of using fluorescent probes to detect and quantify specific reactive oxygen species in living cells. *Biochimica et Biophysica Acta, 1840*(2), 730–738. http://dx.doi.org/10.1016/j.bbagen.2013.05.004.
Wu, J., Dunham, W. R., & Weiss, B. (1995). Overproduction and physical characterization of SoxR, a [2Fe-2S] protein that governs an oxidative response regulon in Escherichia coli. *The Journal of Biological Chemistry, 270*(17), 10323–10327.
Yu, S., Qin, W., Zhuang, G., Zhang, X., Chen, G., & Liu, W. (2009). Monitoring oxidative stress and DNA damage induced by heavy metals in yeast expressing a redox-sensitive

green fluorescent protein. *Current Microbiology*, *58*(5), 504–510. http://dx.doi.org/10.1007/s00284-008-9354-y.

Zhao, Y., Jin, J., Hu, Q., Zhou, H. M., Yi, J., Yu, Z., et al. (2011). Genetically encoded fluorescent sensors for intracellular NADH detection. *Cell Metabolism*, *14*(4), 555–566. http://dx.doi.org/10.1016/j.cmet.2011.09.004.

Zheng, M., Wang, X., Templeton, L. J., Smulski, D. R., LaRossa, R. A., & Storz, G. (2001). DNA microarray-mediated transcriptional profiling of the Escherichia coli response to hydrogen peroxide. *Journal of Bacteriology*, *183*(15), 4562–4570. http://dx.doi.org/10.1128/JB.183.15.4562-4570.2001.

CHAPTER FIFTEEN

Use of Genetically Encoded Sensors to Monitor Cytosolic ATP/ADP Ratio in Living Cells

Andrei I. Tarasov*, Guy A. Rutter†,1
*Oxford Centre for Diabetes, Endocrinology and Metabolism, University of Oxford, Oxford, United Kingdom
†Section of Cell Biology, Department of Medicine, Imperial College London, London, United Kingdom
[1]Corresponding author: e-mail address: g.rutter@imperial.ac.uk

Contents

Abstract

ATP is not only recognized as the universal energy "currency" in most cells but also plays a less well-known role as an intracellular and extracellular messenger. Here, we review

Methods in Enzymology, Volume 542
ISSN 0076-6879
http://dx.doi.org/10.1016/B978-0-12-416618-9.00015-7

novel approaches for measuring free ATP (or ATP/ADP ratios) in living mammalian cells by using genetically encoded sensors. We also discuss the key technical aspects of routine real-time ATP/ADP monitoring using as a model one of the last-generation fluorescent probes, a fusion protein commonly known as "Perceval." Finally, we present detailed guidelines for the simultaneous measurement of cytosolic ATP/ADP ratios and Ca^{2+} concentrations alongside electrical parameters in individual pancreatic β cells, in which energy metabolism is tightly linked to plasma membrane excitability to control the secretion of insulin. With appropriate variations, this approach can be adapted to the study of cytosolic ATP/ADP ratios and Ca^{2+} concentrations in malignant cells, two important aspects of oncometabolism.

ABBREVIATIONS

$[ATP]_{cyt}$, $[ATP/ADP]_{cyt}$, $[Ca^{2+}]_{cyt}$ cytosolic-free concentrations of ATP (or ATP/ADP ratio) and Ca^{2+}, respectively

FRET Förster resonance energy transfer

K_{ATP} ATP-sensitive K^+ channel

SNR signal-to-noise ratio

V_m membrane potential

1. INTRODUCTION

1.1. Cytosolic ATP: Energy equivalent and signaling molecule

The production and utilization of adenosine-5′-triphosphate (ATP) is a near-universal feature of living cells. In line with this, the commercial-scale detection of ATP (known as the "ATP test") is employed to report the presence of living microorganisms in industrially used liquids (Griffiths, 1993). More relevant to biomedical science applications of ATP detection are assays for cell viability (Riss et al., 2004) and transplant quality (Goto, Holgersson, Kumagai-Braesch, & Korsgren, 2006). The topic of this chapter, monitoring ATP levels in living cells, represents an expansion of the commercial methods of ATP detection to probe cellular physiology, by means of molecular biology and optical imaging.

Evolution has established ATP as a bioenergetic "currency," that is, a means of efficient transfer of energy equivalents between different systems and processes within the living cell. Thus, the high-energy phosphate bonds in ATP serve as a convenient intermediate between on the one hand exogenous fuels (or energy stores such as fat and glycogen), and energy-requiring processes such as DNA and protein synthesis or the generation of electrochemical gradients across cellular membranes.

At the same time, and in certain specialized (usually eukaryotic) cell types, ATP (or its breakdown products, ADP and AMP) can play important signaling roles, indicating the energetic status of the cell, for example, in response to altered fuel supply or increased workload. In this capacity, ATP is able to bind specifically to and regulate the activity of a number of intracellular and extracellular proteins, without undergoing hydrolysis.

The intracellular targets of ATP represent a very heterogeneous group ranging from protein kinases such as AMP-activated protein kinase (AMPK; also regulated by ADP and AMP; Hardie, Ross, & Hawley, 2012; Rutter & Leclerc, 2009) to ion channels (e.g., the cystic fibrosis transmembrane conductance regulator (Billet & Hanrahan, 2013) and ATP-sensitive K^+ channels (K_{ATP}; Tarasov, Dusonchet, & Ashcroft, 2004). Of note, the above ATP targets and as well as many others have a proven role of metabolic sensors, that is, they serve to link a change in energy metabolism to the activation of various cellular functions such as secretion, etc.

Here, we discuss the factors which control ATP concentration in cells and our own and others' approaches to imaging the dynamics of these changes in real time.

1.2. Factors affecting $[ATP]_{cyt}$ and $[ATP/ADP]_{cyt}$

The pool of cytosolic ATP is replenished by glycolysis or β-oxidation of fatty acids, followed by oxidative phosphorylation, that is, the synthesis of ATP from ADP plus inorganic phosphate. ATP consumption (hydrolysis to ADP) fuels a plethora of processes critical for cell physiology and homeostasis, as discussed above. While ATP synthesis occurs within minutes, the timescales of ATP expenditure may range from hours (biosynthesis) to seconds (ion handling by ATPases).

It is important to remember, though, that the duality of function defined for ATP above is deceptive. In reality, just like the voltage in domestic power supply sockets, free cytosol ATP concentration ($[ATP]_{cyt}$) and the ratio of ATP/ADP ($[ATP/ADP]_{cyt}$) are tightly controlled over quite a narrow range in the vast majority of cell types, reflecting the vital role of the nucleotide in so many biochemical processes. $[ATP]_{cyt}$ is therefore a result of the dynamic equilibrium between the synthesis and the expenditure. Changes in $[ATP]_{cyt}$ can occur as a result of pathological processes, for example, the Warburg effect, toxicity, etc. By contrast, the signaling function of ATP is limited to only few cell types and tissues that can readily vary intracellular ATP in a physiological way, while not compromising normal function: note that, even in these, "basal" levels of ATP cannot fall below a certain threshold level required for cell survival, for example, for the maintenance of ion gradients. A classic example of those in which ATP levels can change in response to an external stimulus (a change in the supply of fuel) are pancreatic islet cells in higher animals. In particular, islet β-cells employ a physiological elevation of $[ATP]_{cyt}$ as the first step in a cascade linking glucose stimulation to insulin secretion (Rutter, 2001).

In practice, the fact that ATP is constantly, and quite rapidly, being turned over in cells means that extracellular factors may exert only a transient effect on $[ATP/ADP]_{cyt}$. By limiting the supply of substrate or exposing the cell to subphysiological temperatures, one may achieve a decrease in ATP synthesis. This, however, may be followed by a decrease in ATP expenditure induced by a feedback loop (e.g., via AMPK; Hardie et al., 2012). Likewise, an increase in ATP expenditure, for example, due to Ca^{2+} influx into the cytosol and the activation of ionic pumps leads, in turn, to a stimulation of oxidative metabolism and ATP production (Tarasov et al., 2013). Importantly, therefore, only the real-time observation of ATP dynamics can provide reliable information about the state of cellular energetics within a given single cell.

1.3. Methodology of $[ATP]_{cyt}$ and $[ATP/ADP]_{cyt}$ measurement

A short history of ATP measurement features several "revolutionary" points that reflect progress in developing key technologies. However, independently of the methodology of the ATP measurement, the key criterion of the biological relevance is the ability to image ATP real time in living systems.

Differently to other markers of a cell's energetic status (such as the reduced state of flavin adenine dinucleotide, FAD, or reduced nicotine adenine dinucleotide, NADH), ATP does not absorb or emit in the visual part of the spectrum (Ito & Ito, 1986). Moreover, given the tight involvement of ATP in almost every aspect of cellular homeostasis, it is easy to imagine how challenging ATP measurement in living cells is; perturbations which deplete ATP levels are likely to have severe consequences.

A straightforward and noninvasive way to monitor ATP levels is to follow the activity of a surrogate molecule, a binding partner of ATP, which is naturally available within the cell. The elegant implementations of this strategy included measurement of cytosolic [Mg^{2+}], a high-affinity ATP ligand (Leyssens, Nowicky, Patterson, Crompton, & Duchen, 1996), or conductance of K_{ATP} channels (Gribble et al., 2000; Tarasov, Girard, & Ashcroft, 2006; Tsuboi, Lippiat, Ashcroft, & Rutter, 2004). Despite the outstanding resolving power that such single-molecule techniques as patch-clamp possess, these indirect approaches suffer from a number of common restrictions that limit the dynamic range and/or the spatial component of ATP measurement. Recently, more *direct* approaches have been based on the engineering of molecular ATP sensors with the light output in the visual range.

The first such direct ATP measurements in living cells were based around the introduction of cDNA encoding firefly (*Photinus pyralis*) luciferase, with subsequent quantification of the resulting luminescence (Kennedy et al., 1999; Ow et al., 1986). Remarkably, the kinetic properties of firefly luciferase with respect to ATP concentration (the enzyme is relatively insensitive to ADP) appear to differ in living cells from those in the test tube, with the half-maximal activating concentration of ATP ($K_{0.5}$) falling within the normal physiological range of ATP (1–10 m*M*; Kennedy et al., 1999). While measurements in populations of cells are reasonably straightforward and can be done using very simple photomultiplier devices (Jouaville, Pinton, Bastianutto, Rutter, & Rizzuto, 1999), measurements in single cells are technically more challenging. Thus, the intensity of luminescence from a single cell is generally very low, and practically any stray photon can affect the recordings which require highly sensitive (usually multiply-intensified) photon-counting cameras (Rutter, Kennedy, Wood, White, & Tavare, 1998). Specialized dark rooms are therefore indispensable for imaging of this sort. Nonetheless, it has been possible with this approach for us to image, at cellular resolution, glucose-dependent changes in free ATP in pancreatic islet beta- (Ainscow & Rutter, 2001) and alpha- (Ravier & Rutter, 2005) cells as well as neurons (Ainscow, Mirshamsi, Tang, Ashford, & Rutter, 2002), intestinal L-cells (Reimann, Williams, da Silva, Rutter, & Gribble, 2004), and cardiac myocytes (Bell, Bright, Rutter, & Griffiths, 2006). Importantly, the use of this and other recombinant probes allows the use of signal peptides facilitating recordings to be made from a variety of subcellular locales, without the need for sophisticated optical (i.e., super-resolution) approaches: examples include the mitochondrial matrix and the subplasma membrane domains (Kennedy et al., 1999).

Although the luminescence approach above has some important advantages (e.g., the lack of any endogenous biologically derived background signal), in practice the weakness of the signal prevents it from being easily adapted for live-cell imaging with adequate spatial and temporal resolution, that is, data capture at the rate of 1 Hz or more and at the level of a few microns square. Assays based on fluorescence imaging, involving drastically greater photon fluxes, are therefore usually required for live-cell imaging. On the down side, the greater brightness of fluorescent probes (excited by a light source) mean that the optical setup needs to be more complex (excitation and emission filters and beam splitters are now required). Nonetheless, with these in place, one can obtain signals with both high temporal and spatial resolutions. Importantly, it now becomes possible to combine recordings of ATP changes (or more precisely ATP/ADP ratio given the sensitivity of these probes to ADP) with that of other signals, both using spectrally shifted dyes for other small molecules and by using techniques such as patch-clamp electrophysiology (Tarasov et al., 2012, 2013).

At present, there exist two generic technologies of engineering the fluorescent sensors specific for a particular analyte. The first one is based on Förster resonance energy transfer (FRET) between two fluorophores (e.g., cyan and yellow fluorescent proteins, CFP and YFP). The sensitivity of the system to the analyte is achieved by placing a specific analyte-binding linker between the two fluorophores that effectively regulates the energy transfer between them. The second approach is based on circular permutation (CP) of a single fluorophore molecule and subsequent insertion of the analyte-sensitive domain inside. The conformational change upon binding the analyte then increases the intensity of fluorescence of the CP fluorophore. Although the two technologies have been used for over 10 years, recombinant sensors reporting ATP reliably in living systems are a fairly recent development and the range of available probes is still not particularly broad. The two most important probes have been generated by the groups of Noji (Imamura et al., 2009; Nakano, Imamura, Nagai, & Noji, 2011) and Yellen (Berg, Hung, & Yellen, 2009). The former is a FRET sensor, while the latter is based on CP of Venus, a GFP variant (Nagai et al., 2002).

While protein engineering approaches have sought to transform the dynamics of $[ATP]_{cyt}$ or $[ATP/ADP]_{cyt}$ into a visual signal, the nonoptical spectroscopic approaches have also been used for ATP imaging. The nuclear magnetic resonance spectroscopy based on ^{31}P reliably resolves the concentrations of ATP, ADP, and creatine phosphate (Kemp & Brindle, 2012). The NMR approach is, however, too insensitive for single-cell imaging, and the ability to measure the $[ATP]_{cyt}$ or $[ATP/ADP]_{cyt}$

dynamics is particularly limited. A distinct avenue, which seemed unthinkably distant only some 20 years ago, involves infrared (IR) spectroscopy. A much needed boost of the sensitivity of the IR spectroscopy has been provided thanks to the development of the multibounce FTIR–IR (Kazarian & Chan, 2013). Recent reports of ATP and ADP measurement in living cells (Yamaguchi et al., 2011) may mark a dawn of a new era in $[ATP]_{cyt}$ and $[ATP/ADP]_{cyt}$ measurements and possibly a new era in metabolomics too. This, however, will require further efforts in technology development, so it is probably safe to say, at the present time, that fluorescence imaging of the $[ATP]_{cyt}$ levels represents the most precise and informative technology available.

The remainder of this chapter describes the details of applying this method to study the glucose-induced changes in $[ATP/ADP]_{cyt}$ in pancreatic β-cells.

2. EXPERIMENTAL COMPONENTS AND CONSIDERATIONS

2.1. Choice of recombinant sensor

The choice of recombinant sensor remains a matter of throwing a dice as the two groups of probes above have been reported to cover a broad range of ATP concentrations (Goehring et al., 2012; Li, Shuai, Gylfe, & Tengholm, 2013; Nakano et al., 2011; Tarasov et al., 2012). However, the available sensors are based on different technologies that have their own pros and cons. CP sensors, as a rule, have a broader dynamic range and better signal-to-noise ratio (SNR). The biggest advantage of the FRET sensors is that they offer a decoupling from the fluorescence intensity, for example, by using fluorescent lifetime imaging or spectral measurements. The important application is measurements in tissues, where the intensity measurements are frequently linked to various motion artifacts. Also, FRET sensors provide meaningful information even at the steady state, allowing cell-to-cell comparisons (though calibrations are generally advisable where experimentally accessible). Most of the studies below were performed using the CP-based reporter based on GFP fusion with the bacterial regulatory protein, GlnK1, from *Methanococcus jannaschii* ("Perceval"; Berg et al., 2009). Note that this probe, unlike luciferase, senses ATP/ADP ratio, rather than free ATP concentration *per se*; importantly, this represents the most likely parameter to affect intracellular targets (e.g., K_{ATP} channels; Tsuboi et al., 2004).

2.2. Controls and multiparameter measurements

Imaging experiments are frequently a source of artifacts and are naturally, therefore, subject to the need for a rigorous set of controls. These controls are required first to determine whether the changes in fluorescence are due to fluctuations in $[ATP/ADP]_{cyt}$ or instead reflect an action on the probe of other factors such as bleaching, motion artifacts (e.g., due to perifusion), and pH. While the controls are easy to set for the case of ions and signaling molecules, calibrating the signal in terms of ATP/ADP ratio is more complex. In our experience, the best way to achieve this is to use multiparameter imaging.

Typically, an experimenter performs negative and positive controls in the end of the experimental run, effectively providing an estimate of the signal in the absence and presence of ATP and indicating the dynamic range of the recorded signal. Lowering $[ATP/ADP]_{cyt}$ is best achieved by inhibiting the major mechanism of ATP generation in the tissue studied. In the case of pancreatic β-cells, most efficient was the application of the mitochondrial uncoupler carbonyl cyanide 4-(trifluoromethoxy)phenylhydrazone (FCCP). The marked advantage of this compound is that, in addition to disruption of ATP production, this metabolic poison also stimulates ATP expenditure by the F_0F_1 synthase (Nicholls, 1977).

Generating a "positive control" for ATP (i.e., achieving saturation of the probe at a maximal ATP/ADP ratio) is generally more of a problem as there is no specific way to transfer ATP or ATP analogue across the cell membrane. Direct permeabilization of the plasma membrane, for example, by using digitonin or *Staphylococcus aureus* α-toxin will inevitably decrease the fluorescence intensity, may lead to probe leakage, and can cause a change in cell shape. In some tissues, there is an option of using a metabolic stimulus, for example, the application of a high (supraphysiological) concentration of glucose in pancreatic β-cells. However, a convenient alternative approach, applicable to many if not all cell types, is to monitor a complementary signal-like cytosolic $[Ca^{2+}]$ or the conductance of K_{ATP} channels. The above signals change simultaneously with $[ATP/ADP]_{cyt}$, thereby confirming the significance of the ATP measurements. In the example below, we utilize parallel recording of both $[Ca^{2+}]_{cyt}$ and plasma membrane potential (V_m) simultaneously with $[ATP/ADP]_{cyt}$.

For the application below, the sensors reporting both $[ATP/ADP]_{cyt}$ (Perceval) and $[Ca^{2+}]_{cyt}$ (Fura-Red) are used in single-wavelength excitation mode, rather than in ratiometric mode. As described earlier (Berg et al., 2009), Perceval green emission could in theory be imaged with excitation at 490 and 405 nm (or 430 nm, the isosbestic point) to obtain a ratiometric signal, while Fura-Red emission can be imaged with excitation at 480 and 420 nm (Kurebayashi, Harkins, & Baylor, 1993). Indeed, ratiometric measurements are more powerful in rejecting artifacts related to shape changes, sensor concentration, etc., and are likely to provide higher SNR. However, for multiparameter imaging, use of the probes in ratiometric mode would come at a high cost. First, the doubled exposure time is likely to affect cell viability during a long recording. Second, the absorption spectrum of Perceval is substantially skewed toward 490 nm (Berg et al., 2009), which is reflected in a significant difference of SNR between the signals excited at 490 and 405 nm (or 430 nm). Using the ratio signal instead of a single-wavelength one (ex 490 nm) would therefore result in a substantial decrease in SNR. Finally, despite the seemingly large gap between the emission maxima of Perceval and Fura-Red (535 and 630 nm), the spectral compatibility of the two sensors is not ideal. While the interference between the signals excited at 490 nm is negligible, the low absorption at 430 and 405 nm reduces Perceval fluorescence to the level of the "bleed-through" from Fura-Red channel.

2.3. Effect of pH

pH dependence is a common feature of recombinant fluorescent sensors that use GFP variants as a backbone. Even those branded as "pH insensitive" (e.g., Venus; Nagai et al., 2002) exhibit strong changes in fluorescence as a function of pH and are, in fact, better pH sensors than the commercially available trappable dyes such as 2′,7′-bis-(2-carboxyethyl)-5-(and-6)-carboxyfluorescein (BCECF). By and large, the fluorescence of the recombinant proteins tends to increase with deprotonation as the pH rises. Note that the use of ratiometric probes does not circumvent this limitation as different moieties or individual fluorescent proteins within a chimeric molecule often have different sensitivities to pH. Thus, the effects of pH are something that cannot be avoided but may be accounted for during $[ATP/ADP]_{cyt}$ measurements, as described below.

Should a particular manipulation be likely to cause changes in intracellular pH (as is the case for the effects of glucose (Shepherd & Henquin, 1995) and KCl (Varadi & Rutter, 2004) in pancreatic β-cells), direct measurements of intracellular probes using trappable intracellular pH sensors (BCECF, SNARF, etc.) are advisable. Should these be confirmed, we recommend either concurrent measurements of pH as proposed by Yellen and colleagues (Berg et al., 2009) or using a proof-by-contradiction approach to show that the observed effect of $[ATP/ADP]_{cyt}$ is not induced by pH. That is, we first assume that the observed changes in $[ATP/ADP]_{cyt}$ are solely due to an alteration in the pH. Recordings of the ATP signal are thereby subject to a pH calibration that allows recalculation of the observed changes in $[ATP/ADP]_{cyt}$ in pH units. This calibration thus demonstrates how large the change in the cytosolic pH would have to be to explain the observed changes in $[ATP/ADP]_{cyt}$ induced by various factors. After this, the true changes in cytosolic pH triggered by these factors are measured in the same experimental system, using a trappable pH sensor (BCECF, SNARF, etc.). The real pH changes are then compared with the hypothetical pH changes to reveal the degree to which pH interferes with the ATP recordings. As a rule of thumb, pH-dependent changes at the level of 10–20% of the ATP-dependent signal are acceptable, while anything beyond these figures would require additional careful evaluation. Generally speaking, a reliable measurement of $[ATP/ADP]_{cyt}$ (or other signals) using recombinant fluorescent probes in a cell or tissue with high alteration of pH is challenging. Further details are provided in Tarasov et al. (2012).

2.4. Choice of perifusion system

Perifusion of the experimental solution is frequently an absolute necessity for imaging ATP changes where normal physiological responses are sought. At the same time, the perifusion system is often at the heart of motion artifacts that tend to confound fluorescence intensity measurements. Any changes in the level (i.e., the depth) of the bath solution are immediately reflected by the fluorescence intensity, thereby "modulating" the apparent kinetics of the measured signal. In the first instance, a careful positioning of the outflow of the perifusion system is important.

Perifusion-induced motion artifacts are easily accountable for when imaging cells grown on cover slips as they represent a mechanically rigid system. A very different story unfolds in the case of measurements in tissues that are not grown on a cover slip and therefore are immobilized under the objective of the microscope mechanically prior to the experiment. In this case, a conventionally used perifusion system based around a peristaltic pump easily introduces periodical oscillations in the intensity of the fluorescence that may appear to look absolutely "genuine" (i.e., biologically derived). The use of a ratiometric probe or set of probes can reveal this artifact and thus is highly recommended in this case (Speier et al., 2008). As a rule, the first thing to do upon discovering the oscillatory pattern in the measured signal is increasing/decreasing the rate of the perifusion. This maneuver should only change the intensity of the basal fluorescence but never the frequency of the genuine oscillations, which is independent of the frequency of perifusion. Generally, the approaches to avoid the perifusion-related artifacts are either linked to attenuating the flow turbulence (using a gravity-fed system, bubble trappers, etc.) or moving off the intensity measurements toward spectral or lifetime imaging.

2.5. Temperature control

The other important parameter to consider during the live-cell measurements is the temperature of the bath solution. The temperature needs to be stable in the physiological range (usually over 30 °C for mammalian cells although oxidative metabolism is active even at 21 °C). The fluctuations in the bath temperature are likely to introduce changes in ATP production and/or expenditure but will also inevitably lead to the mechanical drifts and loss of focus. Furthermore, temperature may also affect the properties of fluorescent biosensors. For example, it has been reported that the dissociation constant of FRET-based ATP biosensor ATeam is altered by temperature (Imamura et al., 2009).The simple conventional approaches to stabilize the temperature are heated stage or an imaging chamber or a flow heater positioned close to the perifusion inflow. An ideal solution, which is unfortunately not always easy to implement when imaging is combined with other technologies like patch-clamp, is a climate-controlled chamber for the microscope. As a simple alternative, keeping the laboratory warm is a good idea. Please note that prewarming the bath solution using a water bath is unlikely to help, as long as there is usually a tubing connection between the bath and the imaging chamber which dissipates heat: polyethylene and silicone tubing have an ideal form factor for heat exchange and therefore the solution that travels along the tubing pipeline gets cooled to the ambient temperature of the room practically instantly.

2.6. Imaging system and filter sets

The essential components of the simplest wide-field imaging system are a microscope with a reflector block, a camera, a light source, and a shutter. As the ATP probes display a reasonably high dynamic range and rely on the detection of bright proteins (such as Venus), there is substantial leeway for choosing the optical hardware. Although routinely we have used a 40×/1.3

objective for imaging, we are also able to record the ATP signal reliably even using a 10 ×/0.3 air objective. As ATP signal changes within tens of seconds, as the fastest, an average charge-coupled device (CCD) camera would suffice. New-generation cMOS cameras are a good choice when a higher temporal or spatial resolution is needed, and can offer significant improvements in SNR over many older generation CCD cameras.

Measuring $[ATP/ADP]_{cyt}$ and $[Ca^{2+}]_{cyt}$ simultaneously is relatively easy when Perceval and Fura-Red are used in combination (Tarasov et al., 2012). Although the emission spectra of the two fluorophores slightly overlap, they are well separated with minimal bleed-through by using a beamsplitter placed in the emission path (565, 535, and 630 nm). Using Fura-2 or Indo-1 with Perceval is more complex and may require spectral unmixing. The problem of combining $[Ca^{2+}]_{cyt}$ with $[ATP/ADP]_{cyt}$ measurements for the ATeam series of probes was resolved by using a different, red-shifted FRET pair, GFP–OFP (Nakano et al., 2011). This excellent step not only allowed an easy combination with Fura-2 but, very importantly, also eliminated an overlap with autofluorescence from FAD that typically plagues every recording utilizing CFP–YFP FRET.

For the light source, an arc lamp or a set of LEDs can be used. Again, due to the fairly slow kinetics of ATP, a simple filter wheel like MT-20 (Olympus) would suffice. The filter changer and the shutter need to be situated off the microscope to reduce the vibration that may have deleterious effects for the patch-clamp experiments. The light source is therefore connected to the microscope via a lightguide. The microscope is positioned on a floating table inside the Faraday cage. The cage is covered with blinds to prevent the light entry. The light blocking is also a fairly liberal moment unless an objective with high numerical aperture is used for imaging.

Imaging software has in the past frequently been supplied together with the hardware system, in one package. These days, with development of the open-source Micromanager platform (http://www.micro-manager.org), things are changing for the better. Two critical points that one needs to remember though are that (i) the imaging system needs to be able to send or receive TTL pulses from/to the patch-clamp system and (ii) online charting capability is essential.

2.7. Patch-clamp system for combined experiments

The patch-clamp system requires very careful grounding as complex imaging systems guarantee an abundance of multiple sources of electrical noise.

As we use the perforated-patch configuration on small pancreatic β-cells, a critical aspect is the stability of the micromanipulator. We therefore advise to avoid using the hydraulic models as these are known to have a substantial drift. Literally, any low-noise amplifier can be used for these recordings, the only critical point being the synchronization between the imaging and electrophysiological systems. However, this puts very little pressure on the choice of the electrophysiology hardware that traditionally has all kinds of triggering capabilities. Due to the slow time constant of the ATP-dependent processes, the digitization at as low as 1 kHz is sufficient, which puts a very little demand on the PC for recording the patch-clamp data.

We have used the flagship software packages from both main vendors, Patchmaster (HEKA, http://www.heka.com) and pClamp (Molecular Devices, http://www.moleculardevices.com/products/software/pclamp.html). Although both packages perform well, in our hands, the long continuous recordings seemed to be better supported in Clampex. A good alternative for the combined measurements seems to be a free WinFluor package from J. Dempster (Strathclyde University, Glasgow, UK, http://spider.science.strath.ac.uk/sipbs/software_imaging.htm), which requires a cheap converter card from National Instruments. The compatibility of the available hardware with this software, however, needs to be checked. We recommend that, for analysis, the patch-clamp data should be exported into Igor Pro (Wavemetrics, http://www.wavemetrics.com) or equivalent software, in order to be combined with the imaging data.

3. TYPICAL EXPERIMENT: $[ATP/ADP]_{CYT}$ IN PANCREATIC β-CELLS

3.1. Cell isolation, culturing, and expression of the recombinant probe

Pancreatic islets are isolated by injection of the collagenase solution into the bile duct, with subsequent digestion of the pancreatic tissue (Ravier & Rutter, 2010; Tarasov et al., 2012). For single-cell studies, islets are mechanically disrupted by trituration using a P1000 pipette. The cells are grown on glass cover slips but plastic can be used too, depending on the demands of the imaging system.

cDNA encoding the probe is added to the cells usually in the form of an adenovirus (plasmid DNA, delivered with a suitable transfection reagent such as lipofectamine, can be used for easily transfected, e.g., clonal cells).

Needless to say, it is critical that the exposure of the cells to adenovirus is minimal. Typically, achieving multiplicity of infection of two (i.e., giving >85% of cells visibly expressing the probe) is more than sufficient. For single pancreatic cells, we were able to achieve this by exposing cells to 1.2×10^5 plaque-forming units (pfu)/cell/μl for 3–4 h. These values will be different for different cell types, being higher for the case when tissue is used. In some cases, the use of different vector is recommended and therefore the best approach to get high rates of expression in a particular tissue would be to clarify the best compatible vector. The cells can be assayed as soon as the expression of the probe is observable (typically 24–48 h for adenovirus).

3.2. Loading cells with [Ca^{2+}] dye

The stock solution of Fura-Red (2 m*M*; Invitrogen) is prepared by dissolving the powder in dimethylsulfoxide (DMSO) supplemented with 20% Pluronic F-127. The working solution (2 μ*M*) is prepared by thoroughly dissolving the stock solution in the extracellular bath solution (see below). Cells are loaded with Fura-Red dye by incubating in 2 μ*M* solution of the dye for 30 min, at room temperature. When viewed under the microscope, the cells should have distinct red fluorescence (ex 440–490 nm, em 630–660 nm). The extracellular solution should not contain micelles of the dye that look like multiple fluorescent dots. Use of freshly prepared dye or sonication is a good way to avoid this.

3.3. Imaging setup

Our imaging setups have been based around an Olympus IX-71 inverted microscope equipped with a 40×/1.3 apochromat oil immersion objective. The excitation, light control, and filtering are provided by an Olympus MT-20 light source connected to the microscope via a light guide. The 490/20 and 440/20 filters (Chroma) were preloaded into the filter wheel of the MT-20. Images are collected using an Olympus F-View-2 CCD camera (Olympus), or a cMOS camera (Andor), either being located in the side port of the microscope. To separate the emission of the two fluorophores, Perceval and Fura-Red, a beam splitter is introduced into the emitted light path (Dual-View; Photometrics). The beam splitter uses a 560-nm dichroic mirror and 535- and 630-nm emission filters. The imaging equipment is controlled by Cell^R software (Olympus) or Micromanager. The acquisition of the first image triggers the patch-clamp acquisition via a TTL pulse.

The imaging chamber is comprised of a simple metal frame that is able to hold the cover slip in a watertight way. The heating of the chamber (at +33 to 35 °C) is accomplished using THO-60 heating stage (Linkam Scientific Instruments). The perifused solution is gravity-fed into the imaging chamber and removed using a peristaltic pump.

3.4. Setting up patch-clamp measurements

Electrophysiological recordings and stimulation are done in whole-cell perforated-patch configuration, using an EPC9 patch-clamp amplifier controlled by Patchmaster acquisition software (HEKA Elektronik). The headstage of the amplifier was mounted onto a Burleigh PCS-5000 piezoelectric micromanipulator. The pipettes were pulled of borosilicate glass. Conventionally for the perforated-patch configuration, the pipette tip was dipped into pipette solution and then backfilled with the same solution containing 0.2 μg/ml amphotericin B. Series resistance and cell capacitance were compensated automatically by the acquisition software. Data were filtered at 1 kHz, digitized at 2 kHz. The conductance of the K_{ATP} channels was normalized to cell capacitance to account for the difference in cell size.

3.5. Experimental solutions

The extracellular bath solution typically contains (m*M*): 120 NaCl, 4.8 KCl, 24 $NaHCO_3$ (saturated with CO_2), 5 HEPES (pH 7.4 with NaOH), 2.5 $CaCl_2$, 1.2 $MgCl_2$. The patch pipette solution contains (m*M*): 76 K_2SO_4, 10 NaCl, 10 KCl, 1 $MgCl_2$, 5 HEPES (pH 7.35 with KOH). The stock solutions of amphotericin B (60 μg/ml) and FCCP (20 m*M*) are prepared in DMSO. The high-K^+ solution is identical to the extracellular bath solution in all but had no NaCl and 125 m*M* KCl. The pH of the high-K^+ solution is adjusted with KOH.

3.6. Recording the $[ATP/ADP]_{cyt}$ signal

A cell of interest expressing the recombinant probe (Perceval) and exhibiting Fura-Red fluorescence is identified. Cells with extremely high levels of fluorescence of any of the two dyes, or with markedly defective morphology, are not considered further. A giga-seal is established via a patch-clamp pipette in the bath solution containing 3 m*M* glucose, and an electrophysiological recording is initiated in the current-clamp mode. Cell plasma membrane potential is monitored continuously at low (3 m*M*) glucose and the signs of hyperpolarization (plasma membrane potential, V_m, around −70 mV with stable dynamics) marked. The patch clamp can also be used as an additional criterion of cell "fitness" and cells displaying non-β-cell or pathological behavior can be discarded.

If all the patterns of a "healthy" β-cell are matched by the patched cell, imaging recordings are initiated. The recordings feature sequential images of Perceval and Fura-Red fluorescence upon short exposures (10–20 ms) to the excitation light (490 and 440 nm). Images are acquired at a reasonably low frequency (0.2 Hz). The patch-clamp recording is triggered by the synchronizing TTL pulse from the imaging system. Patch clamp is started in current-clamp mode and all three parameters (V_m, $[ATP/ADP]_{cyt}$, and $[Ca^{2+}]_{cyt}$) can be monitored simultaneously (Fig. 15.1). The bath solution may then be changed to one containing high (17 m*M*) glucose, which within minutes is expected to cause the elevation of $[ATP/ADP]_{cyt}$ followed by the depolarization of the plasma membrane (Fig. 15.2A). Electrical activity usually follows shortly afterward, and $[Ca^{2+}]_{cyt}$ dramatically increases. This, in turn, typically causes a transient decrease in $[ATP/ADP]_{cyt}$, followed by a potent recovery due to the stimulation of oxidative metabolism by Ca^{2+}

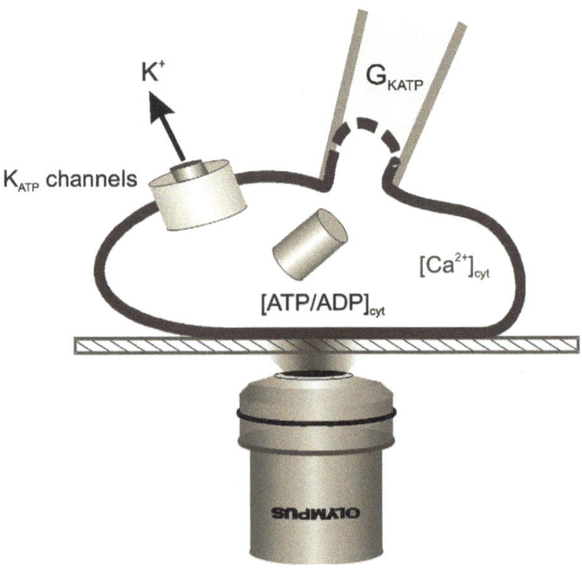

Figure 15.1 *Experimental setup.* A typical experimental setup for combined imaging and patch-clamp electrophysiology that we use to measure $[ATP/ADP]_{cyt}$ alongside $[Ca^{2+}]_{cyt}$ and V_m. The setup is based around an inverted microscope equipped with 40×/1.3 objective (Olympus). The ATP sensor Perceval is delivered adenovirally, $[Ca^{2+}]_{cyt}$ is reported by a trappable dye Fura-Red, and V_m (or $G_{K_{ATP}}$) is measured using a noninvasive perforated-patch whole-cell electrophysiology.

(Tarasov et al., 2012). Further application of 2 μ*M* of FCCP usually lowers $[ATP/ADP]_{cyt}$, repolarizes the plasma membrane, and leaves $[Ca^{2+}]_{cyt}$ somewhat elevated. We note that FCCP can affect intracellular pH in some cell types and changes in this parameter, which may affect the $[ATP/ADP]_{cyt}$ calibration, should be checked.

Note that the patch-clamp configuration can also be switched to voltage clamp before exposing the cell to the high glucose (Fig. 15.2B). By holding the membrane at −70 mV and applying small voltage pulses (10–20 mV pulses at 0.5 Hz), we are then able to monitor K_{ATP} channel conductance, $G_{K_{ATP}}$. The application of high glucose usually leads to an increase in $[ATP/ADP]_{cyt}$ and a decrease in $G_{K_{ATP}}$ (Fig. 15.2B). However, a dramatic increase in $[Ca^{2+}]_{cyt}$ is not registered in the cell shown in the figure as the voltage-gated Ca^{2+} channels were kept closed.

In summary, we are thus able to monitor the dynamics of $[ATP/ADP]_{cyt}$ in pancreatic β-cells, induced by the addition of high glucose, while simultaneously recording both changes in cytosolic-free Ca^{2+} concentration and electrical activity, that is, a "multiparametric" recording from a single cell. An important point in these recordings is that the changes $[ATP/ADP]_{cyt}$ signal agree well with changes in other cellular characteristics. Most importantly, they correlate tightly with the drop in $G_{K_{ATP}}$, which provides an excellent, independent confirmation of the authenticity of the $[ATP/ADP]_{cyt}$ recording.

3.7. Accounting for pH changes

As outlined above, the fluorescence of all GFP-based sensors is dramatically affected by the intracellular pH, resulting in very substantial artifacts. To confirm the authenticity of the $[ATP/ADP]_{cyt}$ dynamics measured in response to glucose, we perform the calibration of the response with high-K^+ nigericin solution. This technique is based on the ability of nigericin to exchange K^+ for H^+ in the absence of Na^+ (Bevensee & Boron, 1998). By exposing the cell to the high-K^+ solution with nigericin (10 μ*M*), we adjust the intracellular pH to the value of the extracellular pH times the ratio of intracellular to extracellular $[K^+]$.

We perform three pH steps in the end of our recordings to assess the response of the Perceval signal to pH (Fig. 15.3). The exposure to the high-K^+ nigericin solution at pH 7.2 slightly shifts the fluorescence down. This does not reflect the fact that the intracellular pH exceeded 7.2 in our experiment but is likely to result from the nonideal adjustment of $[K^+]$

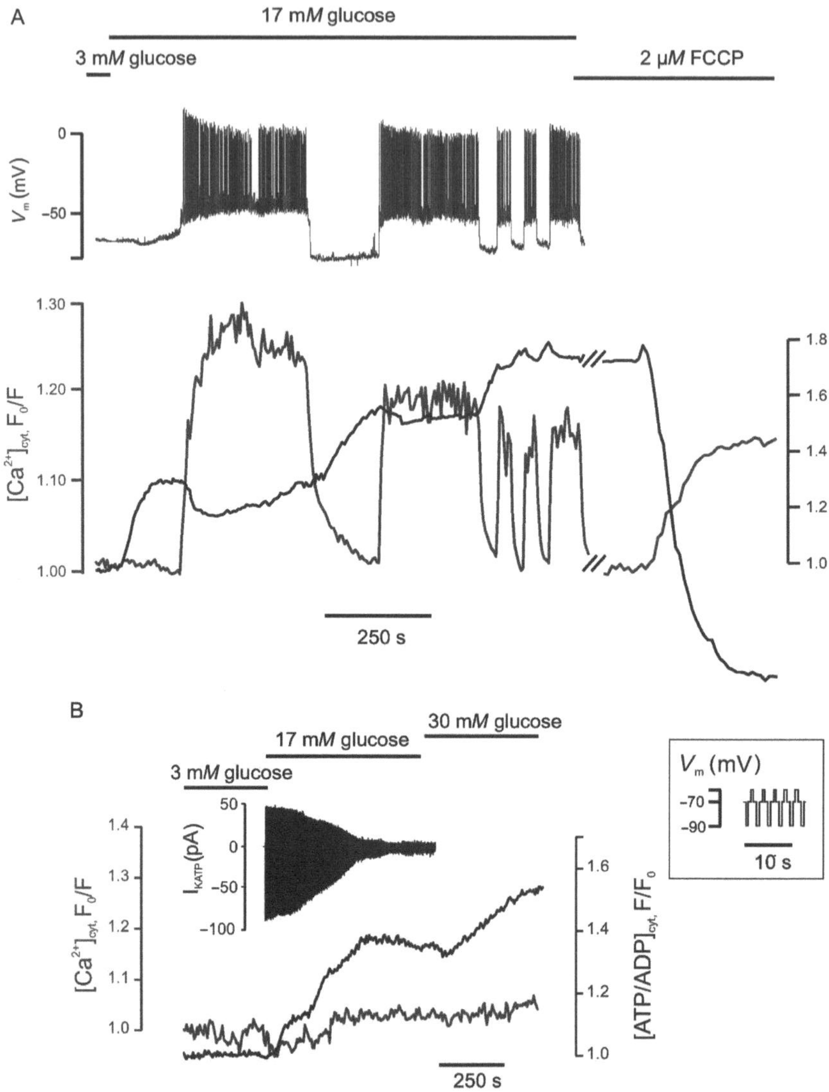

Figure 15.2 *Changes in $[ATP/ADP]_{cyt}$, $[Ca^{2+}]_{cyt}$, and V_m induced by glucose in pancreatic β-cells.* (A) Typical response in $[ATP/ADP]_{cyt}$, $[Ca^{2+}]_{cyt}$, and V_m to the glucose step from 3 to 17 m*M* recorded in a single pancreatic β-cell. The cell expressing Perceval was loaded with Fura-Red and studied using the setup in current-clamp mode (Fig. 15.1). Note the transient decreases in $[ATP/ADP]_{cyt}$ following the $[Ca^{2+}]_{cyt}$ increases induced by the bursts of electrical activity. FCCP was applied at the end of the recording as a negative control. (B) Voltage clamping of the cell at −70 mV keeps the voltage-gated Ca^{2+} channels shut, thereby preventing Ca^{2+} entry into the cytosol. Small depolarizations (inset) applied to the voltage-clamp cell induced K^+ current, mainly through K_{ATP} channels. Note a close correlation of the K_{ATP} channel current ($I_{K_{ATP}}$) with $[ATP/ADP]_{cyt}$. Only a

between the extracellular and intracellular solutions and can be ignored as we are measuring the differential effects of the pH steps. The application of pH 8.2 extracellularly dramatically and reversibly elevates the Perceval fluorescence (Fig. 15.3). The pH 6.2 results, respectively, in a drop of the probe fluorescence (Fig. 15.3).

Implementing the proof-by-contradiction logic, we assume that the glucose-induced effect on Perceval fluorescence is mediated by pH and use the three-step calibration (as above) to assess the pH shift induced by glucose. This assessment gives us a number exceeding 1.1 pH units which is 11 times higher than the assessment of pH changes in the β-cells done with trappable dyes (Shepherd & Henquin, 1995). This means that over 90% of the effect measured in our experiment is definitely not produced by the changes in pH. (This does not confirm that the remaining <10% of the signal changes are pH-dependent, as the changes in cytosolic pH may be caused by the elevation of $[ATP/ADP]_{cyt}$.)

3.8. Analyzing the data

Imaging data are typically analyzed using Cell^R (Olympus), ImageJ (Wayne Rasband, NIMH), and Microsoft Excel. Simultaneous recordings are combined together and analyzed using Igor Pro (Wavemetrics). The results are usually presented as mean ± SEM. Mann–Whitney *U*-test was used to assess the statistical significance of the differences between the independent samples, and Wilcoxon's paired test was used in case of dependent samples. The significance values are specified in each figure; as a rule, $p < 0.05$ differences were considered significant.

4. SUMMARY

Technologies for imaging ATP and ATP/ADP ratio at the level of the single living cell have advanced considerably over the past 15 years, moving from the use of bioluminescent firefly luciferase-based probes to the highly fluorescent Perceval and similar probes which allow high resolution in both time and space. Excitingly, this now allows the operator to perform multiple

fragment of the $I_{K_{ATP}}$ trace, including the whole range of $I_{K_{ATP}}$ variation, is shown for convenience. The step from 3 to 17 m*M* glucose induces an elevation on $[ATP/ADP]_{cyt}$ and corresponding decrease in $I_{K_{ATP}}$. No Ca^{2+}-induced potentiation of $[ATP/ADP]_{cyt}$ is observed. Interestingly, further elevation of the substrate concentration to 30 m*M* prompts a respective increase in $[ATP/ADP]_{cyt}$. (See the color plate.)

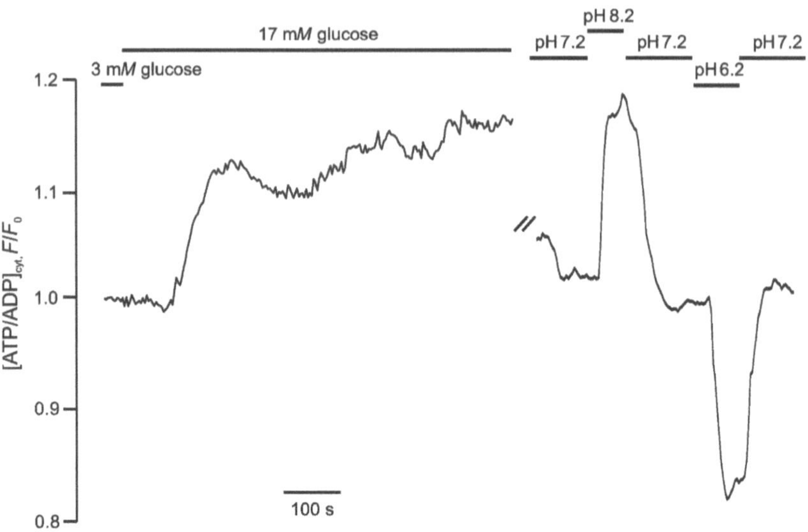

Figure 15.3 *Correcting for pH changes.* After measuring the response of $[ATP/ADP]_{cyt}$ to the glucose stimulus, the pH calibration is performed in the same cell. The cell is exposed to a high-K^+ extracellular solution of predefined pH that includes 10 μ*M* of nigericin. This maneuver makes it possible to equilibrate the pH across the plasma membrane. The exposure to pH 7.2 solution inevitably causes a shift in fluorescence, due to nonideal equilibration of $[K^+]$ across the membrane. The pH steps to 8.2 and 6.2 produce rapid changes in the Perceval fluorescence, both comparable to the effect that the glucose step had on Perceval. However, remembering that the glucose-induced changes in cytosolic pH of β-cells were reported to be around 0.1 of pH unit, we conclude that the effect of glucose on Perceval is authentically mediated by $[ATP/ADP]_{cyt}$.

measurements simultaneously within the same individual cell, removing assumptions which have long-dogged biochemical measurements of ATP and related parameters (e.g., $[Ca^{2+}]_{cyt}$ and membrane potential in excitable cells). As a demonstration of the power of this approach, we have been able, for the first time, to record the previously assumed, but never demonstrated, temporal sequence of events by which glucose, initially by increasing ATP/ADP ratios, leads to the stimulation of electrical activity of pancreatic β-cells and thence to the release of insulin. We expect that continuous improvements in recombinant probes, their targeting to alternative intracellular sites, and the development of near-IR versions, as well as refinements in imaging technology (including the use of super-resolution of or near-field (evanescent wave) microscopes), will allow us to dig deeper into the mechanisms through which ATP controls cellular activity in a range of cell types and settings.

ACKNOWLEDGMENTS

A. I. T. holds an OxBRC fellowship. G. A. R. is supported by a Wellcome Trust Senior Investigator Award (WT098424AIA), MRC Programme (MR/J0003042/1), BBSRC (BB/J015873/1), and Diabetes UK (11/0004210) Project Grants and a Royal Society Wolfson Research Merit Award. The work leading to this publication has received support from the Innovative Medicines Initiative Joint Undertaking under grant agreement no. 155005 (IMIDIA), resources of which are composed of financial contribution from the European Union's Seventh Framework Programme (FP7/2007–2013) and EFPIA companies' in-kind contribution (G. A. R.). We thank Dr. Gary Yellen (Harvard University) and members of the Rutter laboratory for useful discussion.

REFERENCES

Ainscow, E. K., Mirshamsi, S., Tang, T., Ashford, M. L., & Rutter, G. A. (2002). Dynamic imaging of free cytosolic ATP concentration during fuel sensing by rat hypothalamic neurones: Evidence for ATP-independent control of ATP-sensitive K(+) channels. *The Journal of Physiology*, *544*, 429–445.

Ainscow, E. K., & Rutter, G. A. (2001). Mitochondrial priming modifies Ca^{2+} oscillations and insulin secretion in pancreatic islets. *The Biochemical Journal*, *353*, 175–180.

Bell, C. J., Bright, N. A., Rutter, G. A., & Griffiths, E. J. (2006). ATP regulation in adult rat cardiomyocytes: Time-resolved decoding of rapid mitochondrial calcium spiking imaged with targeted photoproteins. *The Journal of Biological Chemistry*, *281*, 28058–28067.

Berg, J., Hung, Y. P., & Yellen, G. (2009). A genetically encoded fluorescent reporter of ATP:ADP ratio. *Nature Methods*, *6*, 161–166.

Bevensee, M. O., & Boron, W. F. (1998). Fluorescence indicators. In K. Kaila & B. R. Ransom (Eds.), *pH and brain function* (pp. 129–151). New-York: Wiley-IEEE ISBN: 0471118389.

Billet, A., & Hanrahan, J. W. (2013). The secret life of CFTR as a calcium-activated chloride channel. *The Journal of Physiology*, *591*(Pt 21), 5273–5278.

Goehring, I., Gerencser, A. A., Schmidt, S., Brand, M. D., Mulder, H., & Nicholls, D. G. (2012). Plasma membrane potential oscillations in insulin secreting Ins-1 832/13 cells do not require glycolysis and are not initiated by fluctuations in mitochondrial bioenergetics. *The Journal of Biological Chemistry*, *287*, 15706–15717.

Goto, M., Holgersson, J., Kumagai-Braesch, M., & Korsgren, O. (2006). The ADP/ATP ratio: A novel predictive assay for quality assessment of isolated pancreatic islets. *American Journal of Transplantation*, *6*, 2483–2487.

Gribble, F. M., Loussouarn, G., Tucker, S. J., Zhao, C., Nichols, C. G., & Ashcroft, F. M. (2000). A novel method for measurement of submembrane ATP concentration. *The Journal of Biological Chemistry*, *275*, 30046–30049.

Griffiths, M. W. (1993). Applications of bioluminescence in the dairy industry. *Journal of Dairy Science*, *76*, 3118–3125.

Hardie, D. G., Ross, F. A., & Hawley, S. A. (2012). AMP-activated protein kinase: A target for drugs both ancient and modern. *Chemistry & Biology*, *19*, 1222–1236.

Imamura, H., Nhat, K. P., Togawa, H., Saito, K., Iino, R., Kato-Yamada, Y., et al. (2009). Visualization of ATP levels inside single living cells with fluorescence resonance energy transfer-based genetically encoded indicators. *Proceedings of the National Academy of Sciences of the United States of America*, *106*, 15651–15656.

Ito, A., & Ito, T. (1986). Absorption spectra of deoxyribose, ribosephosphate, ATP and DNA by direct transmission measurements in the vacuum-UV (150-190 nm) and far-UV

(190-260 nm) regions using synchrotron radiation as a light source. *Photochemistry and Photobiology*, *44*, 355–358.

Jouaville, L. S., Pinton, P., Bastianutto, C., Rutter, G. A., & Rizzuto, R. (1999). Regulation of mitochondrial ATP synthesis by calcium: Evidence for a long-term metabolic priming. *Proceedings of the National Academy of Sciences of the United States of America*, *96*, 13807–13812.

Kazarian, S. G., & Chan, K. L. (2013). ATR-FTIR spectroscopic imaging: Recent advances and applications to biological systems. *Analyst*, *138*, 1940–1951.

Kemp, G. J., & Brindle, K. M. (2012). What do magnetic resonance-based measurements of Pi→ATP flux tell us about skeletal muscle metabolism? *Diabetes*, *61*, 1927–1934.

Kennedy, H. J., Pouli, A. E., Ainscow, E. K., Jouaville, L. S., Rizzuto, R., & Rutter, G. A. (1999). Glucose generates sub-plasma membrane ATP microdomains in single islet beta-cells. Potential role for strategically located mitochondria. *The Journal of Biological Chemistry*, *274*, 13281–13291.

Kurebayashi, N., Harkins, A. B., & Baylor, S. M. (1993). Use of fura red as an intracellular calcium indicator in frog skeletal muscle fibers. *Biophysical Journal*, *64*, 1934–1960.

Leyssens, A., Nowicky, A. V., Patterson, L., Crompton, M., & Duchen, M. R. (1996). The relationship between mitochondrial state, ATP hydrolysis, $[Mg^{2+}]_i$ and $[Ca^{2+}]_i$ studied in isolated rat cardiomyocytes. *The Journal of Physiology*, *496*(Pt. 1), 111–128.

Li, J., Shuai, H. Y., Gylfe, E., & Tengholm, A. (2013). Oscillations of sub-membrane ATP in glucose-stimulated beta cells depend on negative feedback from Ca(2+). *Diabetologia*, *56*, 1577–1586.

Nagai, T., Ibata, K., Park, E. S., Kubota, M., Mikoshiba, K., & Miyawaki, A. (2002). A variant of yellow fluorescent protein with fast and efficient maturation for cell-biological applications. *Nature Biotechnology*, *20*, 87–90.

Nakano, M., Imamura, H., Nagai, T., & Noji, H. (2011). Ca^{2+} regulation of mitochondrial ATP synthesis visualized at the single cell level. *ACS Chemical Biology*, *6*, 709–715.

Nicholls, D. G. (1977). Stoichiometries of proton translocation by mitochondria. *Biochemical Society Transactions*, *5*, 200–203.

Ow, D. W., De Wet, J. R., Helinski, D. R., Howell, S. H., Howell, S. H., Wood, K. V., et al. (1986). Transient and stable expression of the firefly luciferase gene in plant cells and transgenic plants. *Science*, *234*, 856–859.

Ravier, M. A., & Rutter, G. A. (2005). Glucose or insulin, but not zinc ions, inhibit glucagon secretion from mouse pancreatic alpha-cells. *Diabetes*, *54*, 1789–1797.

Ravier, M. A., & Rutter, G. A. (2010). Isolation and culture of mouse pancreatic islets for ex vivo imaging studies with trappable or recombinant fluorescent probes. *Methods in Molecular Biology*, *633*, 171–184.

Reimann, F., Williams, L., da Silva, X. G., Rutter, G. A., & Gribble, F. M. (2004). Glutamine potently stimulates glucagon-like peptide-1 secretion from GLUTag cells. *Diabetologia*, *47*, 1592–1601.

Riss, T. L., Moravec, R. A., Niles, A. L., Benink, H. A., Worzella, T. J., & Minor, L. (2004). *A cell viability assays*. http://www.ncbi.nlm.nih.gov/books/NBK144065/.

Rutter, G. A. (2001). Nutrient-secretion coupling in the pancreatic islet beta-cell: Recent advances. *Molecular Aspects of Medicine*, *22*, 247–284.

Rutter, G. A., Kennedy, H. J., Wood, C. D., White, M. R., & Tavare, J. M. (1998). Real-time imaging of gene expression in single living cells. *Chemistry & Biology*, *5*, R285–R290.

Rutter, G. A., & Leclerc, I. (2009). The AMP-regulated kinase family: Enigmatic targets for diabetes therapy. *Molecular and Cellular Endocrinology*, *297*, 41–49.

Shepherd, R. M., & Henquin, J. C. (1995). The role of metabolism, cytoplasmic Ca^{2+}, and pH-regulating exchangers in glucose-induced rise of cytoplasmic pH in normal mouse pancreatic islets. *The Journal of Biological Chemistry*, *270*, 7915–7921.

Speier, S., Nyqvist, D., Kohler, M., Caicedo, A., Leibiger, I. B., & Berggren, P. O. (2008). Noninvasive high-resolution in vivo imaging of cell biology in the anterior chamber of the mouse eye. *Nature Protocols*, *3*, 1278–1286.

Tarasov, A., Dusonchet, J., & Ashcroft, F. (2004). Metabolic regulation of the pancreatic beta-cell ATP-sensitive K^+ channel: A pas de deux. *Diabetes*, *53*(Suppl. 3), S113–S122.

Tarasov, A. I., Girard, C. A., & Ashcroft, F. M. (2006). ATP sensitivity of the ATP-sensitive K^+ channel in intact and permeabilized pancreatic beta-cells. *Diabetes*, *55*, 2446–2454.

Tarasov, A. I., Semplici, F., Li, D., Rizzuto, R., Ravier, M. A., Gilon, P., et al. (2013). Frequency-dependent mitochondrial Ca^{2+} accumulation regulates ATP synthesis in pancreatic beta cells. *Pflügers Archiv*, *465*, 543–554.

Tarasov, A. I., Semplici, F., Ravier, M. A., Bellomo, E. A., Pullen, T. J., Gilon, P., et al. (2012). The mitochondrial Ca^{2+} uniporter MCU is essential for glucose-induced ATP increases in pancreatic beta-cells. *PLoS One*, 7, e39722.

Tsuboi, T., Lippiat, J. D., Ashcroft, F. M., & Rutter, G. A. (2004). ATP-dependent interaction of the cytosolic domains of the inwardly rectifying K+ channel Kir6.2 revealed by fluorescence resonance energy transfer. *Proceedings of the National Academy of Sciences of the United States of America*, *101*, 76–81.

Varadi, A., & Rutter, G. A. (2004). Ca^{2+}-induced Ca^{2+} release in pancreatic islet beta-cells: Critical evaluation of the use of endoplasmic reticulum-targeted "cameleons" *Endocrinology*, *145*, 4540–4549.

Yamaguchi, R. T., Ayumi, H. I., Aonuma, Y., Yoshimura, Y., Shinohara, Y., Kimura, Y., et al. (2011). Real-time monitoring of mitochondrial adenosine 5-triphosphate synthesis and hydrolysis by surface infrared spectroscopy. *Applied Physics Letters*, *98*, 133703.

CHAPTER SIXTEEN

Methods to Monitor and Compare Mitochondrial and Glycolytic ATP Production

Simone Patergnani*, Federica Baldassari*, Elena De Marchi*, Agnieszka Karkucinska-Wieckowska†, Mariusz R. Wieckowski‡, Paolo Pinton*,[1]

*Section of Pathology, Oncology and Experimental Biology, Laboratory for Technologies of Advanced Therapies (LTTA), Department of Morphology, Surgery and Experimental Medicine, Ferrara, Italy
†Department of Pathology, The Children's Memorial Health Institute, Warsaw, Poland
‡Department of Biochemistry, Nencki Institute of Experimental Biology, Warsaw, Poland
[1]Corresponding author: e-mail address: pnp@unife.it

Contents

Methods in Enzymology, Volume 542
ISSN 0076-6879
http://dx.doi.org/10.1016/B978-0-12-416618-9.00016-9

Abstract

ATP is commonly considered as the main energy unit of the cell and participates in a variety of cellular processes. Thus, intracellular ATP concentrations rapidly vary in response to a wide variety of stimuli, including nutrients, hormones, cytotoxic agents, and hypoxia. Such alterations not necessarily affect cytosolic and mitochondrial ATP to similar extents. From an oncological perspective, this is particularly relevant in the course of tumor progression as well as in the response of cancer cells to therapy. In normal cells, mitochondrial oxidative phosphorylation (OXPHOS) is the predominant source of ATP. Conversely, many cancer cells exhibit an increased flux through glycolysis irrespective of oxygen tension. Assessing the relative contribution of glycolysis and OXPHOS to intracellular ATP production is fundamental not only for obtaining further insights into the peculiarities and complexities of oncometabolism but also for developing therapeutic and diagnostic tools. Several techniques have been developed to measure intracellular ATP levels including enzymatic methods based on hexokinase, glucose-6-phosphate dehydrogenase, and firefly luciferase. Here, we summarize conventional methods for measuring intracellular ATP levels and we provide a detailed protocol based on cytosol- and mitochondrion-targeted variants of firefly luciferase to determine the relative contribution of glycolysis and OXPHOS to ATP synthesis.

1. INTRODUCTION

The mitochondria are part of a complex cellular signaling network that play a central role in several physiological processes (Bonora, Bononi, et al., 2013; Marchi, Patergnani, & Pinton, 2014; Patergnani et al., 2013), in cellular processes (Houtkooper, Pirinen, & Auwerx, 2012), during the stress response (Manoli et al., 2007), and in regulating the homeostasis of second messengers (Giorgi et al., 2012; Marchi & Pinton, 2014). These highly specialized organelles are also the primary source of ATP: during the mitochondrial tricarboxylic acid cycle, the NADH (nicotinamide adenine dinucleotide reduced) produced by the oxidation of glycolytic pyruvate is transferred along the electron transport chain across the inner mitochondrial membrane to boost ATP production. ATP acts as an instant source of energy within the cells of all living organisms, and it participates in a variety of cellular processes. Most importantly, ATP is highly involved in tumorigenesis (Rodriguez-Enriquez et al., 2010; Shackelford & Shaw, 2009).

For example, tumor cells are able to sustain high rates of glycolysis for ATP/energy generation, regardless of oxygen availability (Scatena, 2012). Studies have clearly demonstrated that the metabolic shift from oxidative phosphorylation (OXPHOS) to aerobic glycolysis is a common hallmark of cancer and that the signaling pathways controlling energy production in cancer cells are greatly altered during tumor progression (Gatenby & Gillies, 2004; Pollak, 2013).

A key molecular player that is highly interconnected to ATP synthesis and storage, the 5′ AMP-activated protein kinase (AMPK), has been linked to cancer metabolism and progression. AMPK is a Ser/Thr protein kinase that is sensitive to the cellular AMP/ATP ratio. AMPK acts to phosphorylate several downstream substrates in response to metabolic and energy crises caused by significant alterations in ATP synthesis. The overall effect of AMPK activation promotes ATP synthesis and consequential inhibition of anabolic (ATP-utilizing) processes such as protein and lipid synthesis.

Another critical regulator of ATP homeostasis involved in tumorigenesis is the mitochondrial ADP/ATP carrier ANT, which is responsible for the exchange of ADP for ATP in a 1:1 stoichiometry. Because of its critical role in energy metabolism, defects in the ADP/ATP transport system result in severe disorders. Studies have shown that neoplastic transformations are often associated with mutated expression of members of the ANT family and that ADP/ATP exchange is compromised of different pathological conditions.

Thus, understanding the importance of intracellular ATP levels, the rate of ADP/ATP exchange, and the intracellular ADP levels will undoubtedly be of great importance in elucidating the metabolic behavior of tumor cells.

The most commonly used methods to determine ADP levels are based on the activity of pyruvate kinase (PK). The production of ADP is evaluated by measuring the decrease in absorbance or fluorescence concomitant with the oxidation of NADH to NAD^+ by PK in the presence of phosphoenolpyruvate (Shapiro, Hajec, & Gao, 2011). One limitation of this approach is that the use of UV absorbance or excitation renders the assay prone to interference by the absorbance or fluorescence of the samples being tested. On the other hand, these methods for determining ADP/ATP exchange have been well developed (Duee & Vignais, 1969). Recently, a novel technique to measure this mitochondrial exchange based on the Mg^{2+}-sensitive fluorescent probe Magnesium Green has been described. Using this method, it is now possible to determine the ATP–ADP exchange rate mediated by the ANT and to estimate the molecular turnover of this mitochondrial carrier (Chinopoulos et al., 2009; Kawamata, Starkov, Manfredi, & Chinopoulos, 2010).

Similarly, the assessment of cellular ATP content is a critical measure in many types of experiments, especially in those involving hypoxic, metabolic, or cytotoxic cell injury. Several techniques have been developed. Some approaches are based on reactions catalyzed by different kinases, such as adenylate and PK, and permit an excellent measurement of the total ATP levels. Nevertheless, these methods require a permeabilization step that allows the hydrophilic substrates to penetrate biological membranes. Using an FRET-based sensor, it is possible to measure the level of ATP in intact cells and to determine its cellular distribution and dynamics. Unfortunately, this approach possess critical pitfalls (such as ATP buffering or a substantial modification of ATP homeostasis; Liemburg-Apers et al., 2011). The best method for measuring intracellular ATP employs firefly luciferase, an enzyme that causes the oxidation of luciferin (an oxidizable substrate). This reaction can be quantified because the energy produced releases a photon of light (bioluminescence).

In the following sections, we review the luciferase–luciferin systems currently used to measure intracellular ATP concentrations. We describe accurate methods for the detection of ATP in different cellular compartments using recombinant-targeted chimeras of luciferase. Finally, we describe an enzymatic-based method for accurately measuring ATP levels in cultured cells.

2. MONITORING INTRACELLULAR ATP USING LUCIFERASE-BASED TECHNIQUES

The luciferase isolated from the firefly *Photinus pyralis* is a monomeric enzyme that catalyzes light production in bioluminescent organisms, requiring luciferin, ATP, and O_2 as substrates (DeLuca & McElroy, 1974; de Wet, Wood, DeLuca, Helinski, & Subramani, 1987). The reactions catalyzed by firefly luciferase are:

$$\text{Luciferase} + \text{luciferin} + \text{ATP} \xleftrightarrow{\text{Mg}^{2+}} \text{Luciferase} - \text{luciferyl} - \text{AMP} + \text{PP}_i$$

$$\text{Luciferase} - \text{luciferyl} - \text{AMP} + \text{O}_2 \rightarrow \text{Luciferase} + \text{oxyluciferin} + \text{AMP} + \text{CO}_2 + h\nu$$

Because of its great sensitivity, the firefly luciferase assay has been widely employed to measure intracellular ATP content, and different luciferin–luciferase kit assays have been engineered.

Again, a simple microinjection of purified luciferase lacking a targeting sequence into single isolated cells permits effective assessment of cytosolic ATP levels (Bowers, Allshire, & Cobbold, 1992).

Luciferase can be expressed in the cytosol by introducing an appropriate cDNA into the cell. An example of a luciferase vector is pcDNA3LuciferaseLL/V, a mutagenized luciferase with the C-terminal amino acid substitution L550V, which eliminates the peroxisomal targeting sequence (Bell, Manfredi, Griffiths, & Rutter, 2007).

Similarly, the cytosolic ATP luminescence of luciferase-expressing HeLa cells (HeLa-LUC cells) was continuously monitored by integrating the photon counts over 10-s intervals at 37 °C. The sample was rotated at a frequency of approximately 100 rpm, and the direction was changed every 3–5 s to ensure proper mixing of the solution (Zamaraeva, Sabirov, Manabe, & Okada, 2007). Using the adenoviral-driven expression of recombinant firefly luciferase (AdCMVcLUC), Ainscow and collaborators monitored cytosolic ATP concentrations in real time in isolated human and mouse pancreatic β-cells (Ainscow & Rutter, 2002).

A recent study used a protein transduction domain (PTD)-conjugated luciferase (LUC) to assay intracellular ATP levels in live cells. PTD-LUC can penetrate the cell membrane without affecting its integrity or interfering with cellular metabolism, and its luciferase activity is proportional to the ATP concentration (Lee et al., 2012).

Specifically targeted luciferase chimeras have been engineered based on recombinant DNA technology. A cDNA encoding luciferase can be introduced and expressed in cells and, with the addition of signal peptides, luciferase constructs have been targeted to subcellular compartments (Jouaville, Pinton, Bastianutto, Rutter, & Rizzuto, 1999; Kennedy et al., 1999). Furthermore, recombinant luciferase has been used to study variations in ATP concentration after cell stimulation by agonists that mobilize intracellular calcium (Bonora, Bononi, et al., 2013; Bonora et al., 2012; Brini et al., 1999; Giorgi et al., 2010; Giorgi, Romagnoli, Pinton, & Rizzuto, 2008; Rimessi, Giorgi, Pinton, & Rizzuto, 2008; Rimessi et al., 2009; Voronina et al., 2010). Jouaville and collaborators demonstrated that mitochondrial Ca^{2+} accumulation triggers increased ATP synthesis in the mitochondria and enhanced ATP levels in the cytosol. In contrast, a decrease in mitochondrial calcium accumulation leads to diminished mitochondrial ATP concentrations (Jouaville et al., 1999).

We present a detailed protocol below that uses recombinant luciferase chimeras targeted to discrete intracellular domains and a luminescent reader (luminometer) to monitor basal and induced levels of intracellular ATP.

2.1. Measuring the intracellular ATP concentration with a luminescence reader

In this method, the luminescence is dependent on the presence of luciferin, and it is proportional to the perfused luciferin concentration, between

20 and 200 μ*M*. The light emission is proportional to the ATP concentration in the sample.

The protocol allows the measurement of ATP levels with a recombinant-targeted luciferase in a wide variety of cell types (e.g., HeLa, HEK 293T, CHO, COS-7, SH-SY5Y, A7r5, PC12, mouse embryonic fibroblasts (MEFs), as well as primary cultures of neurons and skeletal muscle myotubes).

2.2. Equipment setup

Luminescence reader (luminometer) for cell population analyses. The photons emitted during the experimental procedure can be recorded and analyzed using relatively inexpensive equipment. The instrument is constructed as follows:

- Photon-counting head H7360-01 (Hamamatsu Photonics)
- Perfusion chamber (Elettrofor)
- Elettrofor Termostitc group GT-80/P (Elettrofor)
- Photon counter C8855-01 (Hamamatsu Photonics)
- Peristaltic pump

The luminometer is composed of an upper part containing a low-noise photomultiplier inserted into a dark box housing a manual shutter (to protect it from light exposure) and a removable temperature-controlled chamber to hold the sample. During ATP measurements, the shutter is open and the perfusion chamber (13 mm in diameter, 2 mm in height, with a volume capacity of 265 μl) is placed in close proximity (<3 mm) to the photomultiplier. The cells are continuously perfused with temperature-controlled medium (water bath, GT-80/P) at 37 °C. A peristaltic pump allows liquid perfusion. The signals generated by the phototube are collected and converted by a digital counting unit connected to a computer (Bonora, Giorgi, et al., 2013; Chiesa et al., 2001; Cobbold, Cuthbertson, Goyns, & Rice, 1983; Webb, Rogers, Karplus, & Miller, 2010).

2.3. Reagent preparation

1. Modified Krebs–Ringer solution (KRB): 125 m*M* NaCl, 5 m*M* KCl, 1 m*M* KH_2PO_4, 1 m*M* $MgSO_4$, 1 m*M* $CaCl_2$, 5.5 m*M* glucose, and 20 m*M* HEPES.
2. Luciferin solution (Luciferin): 20 μ*M* luciferin (Promega, E1605) in KRB solution.

3. Ca^{2+}-mobilizing solution (CMS): Dissolve the Ca^{2+}-perturbing agent in the luciferin solution; different cell types require different agents (e.g., histamine for HeLa cells, bradykinin for human fibroblasts, ATP for MEFs and HEK293T cells, and carbachol for SH-SY5Y cells).

2.4. Sample preparation and transfection

The cells are seeded on glass coverslips (13 mm in diameter) for single-sample luminescence measurements, and they are allowed to grow until they reach 50% confluence. After seeding the cells, wait at least 24 h. The cells are then transfected with a cytosolic (untargeted) firefly luciferase construct (cytLUC) and a mitochondrial-targeted HA1-tagged luciferase construct (mtLUC) using an appropriate transfection method. Cancer and immortalized cell lines are easily transfected using Ca^{2+} phosphate or lipoamine transfection methods. Primary cultures that are notably "hard-to-transfect" require electroporation or infection with adenoviral vectors. After transfection, wait for 36–48 h.

An 876-nucleotide *Bgl*II*Hin*dIII fragment of the cytosolic (untargeted) firefly luciferase was transferred from the pGL3 basic plasmid (Promega) to the pcDNA3 plasmid (Invitrogen), and it was then transferred to VR1012 to generate the cytLUC plasmid.

To obtain the mitochondrial luciferase construct, the photoprotein was engineered with DNA sequences encoding the mitochondrial protein subunit VIII of cytochrome *c* oxidase (COX8) and the hemagglutinin HA1 tag. The entire final construct (mtLUC) was cloned into the expression vectors pcDNAI and VR1012 (Jouaville et al., 1999; Kennedy et al., 1999; Porcelli et al., 2001).

2.5. Measurements

1. Remove the cells from the incubator and place the 13-mm coverslip inside the temperature-controlled (37 °C) perfusion chamber.
2. Begin perfusing the cell monolayer with KRB using a peristaltic pump (standard flow rate: 2.5 ml/min) and place the chamber in close proximity (2–3 mm) to the surface of the low-noise photomultiplier tube.
3. Measure the background luminescence. Begin by recording the basal cps (counts per second) luminescence value for at least 30 s to obtain a stable signal, and then record the lower value as the background (see Fig. 16.1A).

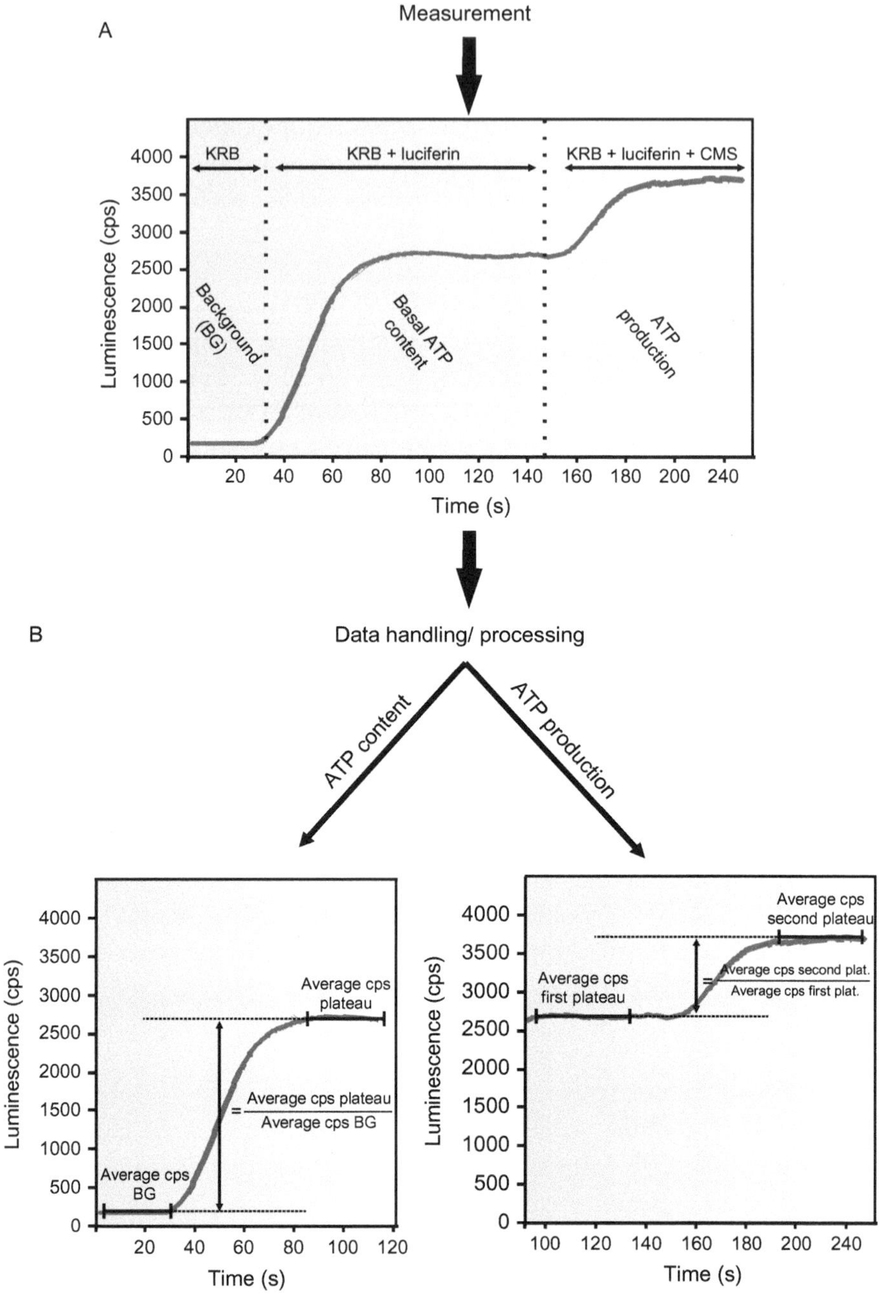

Figure 16.1 *Calibration and measurement of intracellular ATP levels using a luminometer.* (A) A representative luminescence trace employed to measure free [ATP] in intracellular compartments using a specifically targeted chimera of the ATP-sensitive photoprotein luciferase. The resting luminescence (background, BG) was recorded before the addition of luciferin to assess the basal ATP levels. Next, the cells were perfused with a Ca^{2+}-mobilizing solution (CMS) to determine the amount of ATP production. (B) The data were processed to estimate the basal ATP content and the levels of ATP produced. KRB, modified Krebs–Ringer solution; cps, counts per second. (See the color plate.)

4. To record the basal ATP content, pause the perfusion, change from the modified KRB to the luciferin solution, and resume the perfusion. Following the addition of luciferin, light emission (and thus cps values) rapidly increases and reaches a plateau within approximately 2 min (see Fig. 16.1A).
5. Record the ATP production. Wait until the cps values remain unchanged for approximately 15–30 s. Pause the perfusion, change the medium to the CMS solution, and resume the perfusion. Under these conditions, the light emission (and therefore, the cps values) rapidly increases and reaches a second plateau within approximately 1–1.5 minutes. Wait until the cps values remain unchanged for approximately 20–30 s (see Fig. 16.1A).
6. Stop the analysis system, switch off the power supply and the perfusion, and remove the 13-mm coverslip from the chamber notch. Reseal the chamber and wash the entire system extensively by perfusion with distilled water, thus avoiding the occurrence of spurious signals in later experiments

2.6. Data handling/processing

To calculate the basal ATP content, estimate the luminescence values of the plateau generated after the addition of the luciferin solution (see Fig. 16.1B). As the basal ATP content is dependent on the abundance of luciferase transfected, an immunoblot assay with an antibody against luciferase is recommended. To calculate the amount of ATP produced, divide the luminescence values of the plateau generated after the addition of the luciferin solution by the luminescence values of the second plateau generated after the addition of the CMS solution (see Fig. 16.1B).

2.7. Discrimination of glycolytic versus OXPHOS ATP generation

By using both luciferase constructs, it is possible to dynamically monitor the effect of OXPHOS and glycolysis inhibitors to capture an accurate measurement of the contribution of glycolysis and oxidative phosphorylation to the intracellular ATP production.

To mimic OXPHOS suppression, use uncouplers of mitochondrial OXPHOS, such as pentachlorophenol or carbonyl cyanide-4-(trifluoromethoxy)phenylhydrazone. Alternatively, the same results may be obtained using inhibitors of the respiratory chain, such as cyanide

(NaCN, an inhibitor of complex IV). However, the most commonly used inhibitor is oligomycin, a potent inhibitor of the mitochondrial ATP synthase.

Different methods may be used to inhibit ATP production by the glycolytic pathway. If it is available, we recommended utilizing the non-metabolizable analog of glucose, 2-deoxyglucose (2DG); if 2DG is not available, it is possible to block glycolytic ATP synthesis with iodoacetamide (a glyceraldehyde 3-phosphate blocker) or fluoride (NaF, an enolase inhibitor).

Considering the fact that different cell lines display different respiration levels and glycolysis rates, the specific pharmacological effects of the compound used should be evaluated initially to determine the ideal experimental conditions. In the next sections (Sections 2.8–2.11), we describe how to discriminate between glycolytic and OXPHOS ATP generation and discuss the optimal inhibitor dosages for an immortalized and cancer cell line (HeLa cells) and for primary cells (MEFs).

2.8. Reagent setup

1. Modified Krebs–Ringer solution (KRB): 125 m*M* NaCl, 5 m*M* KCl, 1 m*M* KH_2PO_4, 1 m*M* $MgSO_4$, 1 m*M* $CaCl_2$, 5.5 m*M* glucose and 20 m*M* HEPES.
2. Luciferin solution (luciferin): 20 μ*M* luciferin in KRB solution.
3. 2DG solution: 5 m*M* (for immortalized or cancer cell lines) or 20 m*M* (primary cell cultures) deoxyglucose (Sigma-Aldrich, D8375) in luciferin solution.
4. Oligomycin solution (oligomycin): 1 μ*M* (for immortalized or cancer cell lines) or 5 μ*M* (primary cell cultures) oligomycin (Sigma-Aldrich, 75351) in 2DG solution.

2.9. Measurement

1. After seeding the cells as described in Section 2.4, transfect the cells with cytLUC or mtLUC. HeLa cells are easily transfected using Ca^{2+}-phosphate or lipoamines. Primary cells are less sensitive to standard methods and require electroporation or adenoviral vector infection.
2. After transfection, wait for 36–48 h.
3. Begin the measurement by following the instructions provided in Section 2.5 until step 4 (see Fig. 16.2A).

4. Wait until the cps values remain unchanged for approximately 15–30 s. Pause the perfusion, change the medium to the 2DG solution, and resume the perfusion. Under these conditions, the nonmetabolizable analog of glucose triggers a large decrease in the luciferase signal. The light emission (and therefore cps values) decreases until reaching a plateau after approximately 1 min. At this point, wait until the light emission does not change further (approximately 30 s; see Fig. 16.2A).
5. Stop the perfusion system and change the medium to the oligomycin solution. Restart the perfusion. As a consequence of adding the mitochondrial ATP synthase inhibitor, OXPHOS is suppressed and the cps values decrease further until they are comparable to resting luminescence. Wait until the values reach a plateau (approximately 30–60 s; see Fig. 16.2A).
6. Repeat step 6 from Section 2.5.

2.10. Data handling/processing

To calculate the total ATP levels, divide the luminescence values of the plateau generated after the addition of the luciferin solution by the luminescence values of the plateau generated after the addition of the oligomycin solution. To estimate the contribution of oxidative phosphorylation, divide the total ATP level by the ratio calculated between the plateau generated after the addition of oligomycin and the plateau generated after the addition of 2DG. To compute the amount of ATP produced by glycolysis, divide the total ATP level by the rate of oxidative phosphorylation (see Fig. 16.2B).

2.11. Notes

Unlike fluorescent reporters, luciferase has the advantage that no external light source is required to reveal its presence; thus, there is no background light originating from the sample, and it is possible to avoid photobleaching, phototoxicity, and autofluorescence (Welsh & Kay, 2005). However, this enzyme has its own disadvantages. For instance, luciferase becomes extensively inactivated during the coupling reaction. Importantly, the ATP concentration and light emission are not proportional. Furthermore, this assay is highly dependent on ATP and oxygen; thus, it is necessary to define the endogenous ATP levels to accurately determine the kinetics of ATP production. Several strategies have been developed to overcome these problems (Wibom, Lundin, & Hultman, 1990). Adding aliquots of an ATP standard during a continuous luminescence recording ensures a quantification of the

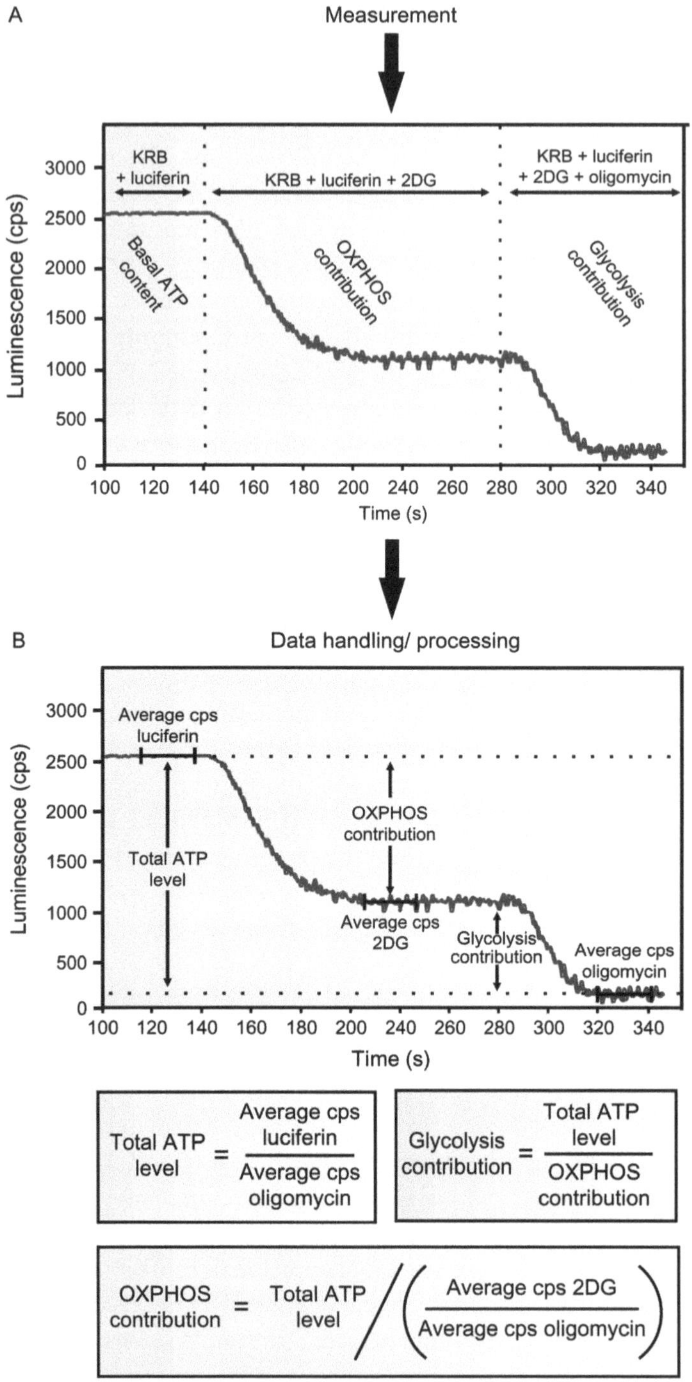

Figure 16.2 *Assessment of the OXPHOS and glycolytic contribution to ATP content.* (A) A luminescence trace for determining glycolytic and OXPHOS ATP in the presence of 2DG and oligomycin. (B) The data were processed to determine the contribution of glycolysis and OXPHOS to ATP production. The luminescence is expressed in relative light units. cps, counts per seconds; 2DG, 2-deoxyglucose.

ATP concentration (Lemasters & Hackenbrock, 1976). Alternatively, it is possible to perform measurements with mutant luciferases that improve the enzyme's stability or increase its catalytic efficiency (Fujii et al., 2007; Hirokawa, Kajiyama, & Murakami, 2002).

Nevertheless, the major drawback of this method is that luciferin is an amphipathic molecule that, due to its carboxyl group, is charged at physiological pH, thus preventing easy passage across cell membranes. It is possible to increase the rate of luciferin entry into cells by manipulating the buffer conditions. For example, improved luciferin uptake into cells has been demonstrated using a buffer with a low pH (Craig, Simmonds, Watmore, McCapra, & White, 1991). Other ways to enhance luciferin entry in mammalian cells include the employment of uncharged luciferin derivatives (Gould & Subramani, 1988) or membrane-permeable luciferin esters (Harwood, Mofford, Reddy, & Miller, 2011).

3. MONITORING INTRACELLULAR ATP USING ALTERNATIVE ENZYMATIC METHODS

The main advantage of the luciferin–luciferase assay is the very rapid detection of ATP levels, but this technique requires a luminometer. Should a luminometer not be available, it is possible to use alternative procedures that take advantage of standard enzymatic methods. Two alternative enzymatic methods for ATP measurement are proposed below.

The first method is based on the glycerokinase/glycerolphosphate oxidase/horseradish peroxidase-coupled assay. Briefly, the assay is composed of a three-step enzymatic reaction:

1. ATP (from a measured sample) + glycerol + glycerokinase → glycerol-3-phosphate + ADP
2. Glycerol-3-phosphate + O_2 + glycerolphosphate oxidase → phosphodioxyacetone + H_2O_2
3. H_2O_2 + TMPD (TetraMethylPhenyleneDiamine) + horseradish peroxidase → H_2O + TMPD*

   *The absorbance of oxidized TMPD is measured spectrophotometrically at 610 nm.

This method was originally proposed for the detection of triglycerides rather than ATP.

The second method is based on the hexokinase/glucose-6-phosphate dehydrogenase-coupled assay. This method is the most common and was originally used for glucose detection. After some modification, it was also used for hexokinase detection in subfractionated mitochondrial protein complexes, as described by Beutner, Ruck, Riede, Welte, and Brdiczka (1996) or by Wieckowski, Brdiczka, and Wojtczak (2000) and Wieckowski et al. (2001). Briefly, the assay is composed of a two-step enzymatic reaction:

1. ATP + glucose + hexokinase → glucose-6-phosphate + ADP
2. Glucose-6-phosphate + NADP + glucose-6-phosphate dehydrogenase → 6-phosphogluconate + NADPH

   The absorbance of reduced NADPH can be measured spectrophotometrically at 340 nm or fluorimetrically at an excitation/emission of 340/450 nm, respectively.

Below, we include an accurate description of this tool and provide a precise standardization of the method for quantifying the amount of ATP.

3.1. Reagent setup

1. Reaction buffer: 100 m*M* TEA (triethanolamine), 16 m*M* $MgSO_4$ (magnesium sulfate) and 10 m*M* EDTA (ethylenediaminetetraacetic acid disodium salt dehydrate); pH 7.6
2. Stock solutions: glucose (1 M stock), NAD (1 M stock), ATP (0.75 M stock), hexokinase (0.5 U/ml stock), and glucose-6-phosphate dehydrogenase (200 U/ml stock)

3.2. Sample preparation

1. Resuspend the cells in a small amount of PBS (depending on the pellet size).
2. Add TCA (trichloroacetic acid) to a final concentration of 2%. The TCA should contain 0.001% of xylenol blue dye as a pH indicator. A red color indicates an acidic pH.
3. After the TCA extraction, the sample should be diluted to 0.1% (concentration of TCA).
4. Neutralize the sample using the Tris–acetate buffer, pH 7.7. The color of the neutralized sample should change from red to yellow. If necessary, add more of the Tris–acetate buffer, pH 7.7.
5. Keep the sample on ice and use it for ATP measurements or store it at −80 °C.

An alternative method has been described by Manfredi, Yang, Gajewski, and Mattiazzi (2002).

1. Resuspend the cells in a small amount of PBS (depending on the pellet size).
2. Add 10 μl of ice-cold 0.4 *M* perchloric acid per milligram of protein.
3. Keep the suspension on ice for 30 min, then centrifuge it at 14,000 rpm for 10 min at 4 °C.
4. Collect the supernatant and neutralize it with 10 μl of 4 *M* K_2CO_3 per 100 μl of supernatant.
5. Keep the suspension on ice for 10 min and then at −80 °C for 1–2 h to precipitate the perchlorate.
6. Centrifuge the sample again at 14,000 rpm for 10 min at 4 °C.
7. Keep the collected supernatant on ice and use it for ATP measurements. Alternatively, store it at −80 °C.

3.3. Measurements

1. Dispense 2.5 ml of the reaction buffer into the cuvette and add the following:
   - 10 μl of 1 *M* glucose
   - 10 μl of 1 *M* NAD^+
   - 20 μl of glucose-6-phosphate dehydrogenase (200 U/ml stock)
   - 25–50 μl of the sample
2. Record the fluorescence trace. Begin the reaction with the addition of 20 μl of hexokinase (0.5 U/ml stock) and wait for the plateau. If the change in fluorescence is small, add more sample. To ensure that the amount of hexokinase in the reaction was not limiting, it is possible to add an additional hexokinase to the cuvette. No significant change in the fluorescence should be observed (see Fig. 16.3).

3.4. Standardization of the method

After reaching the plateau, add small volumes of the ATP stock solution (diluted to the proper concentration, if necessary). Measure the increase in fluorescence (see Fig. 16.3). It is important to perform this standardization on the same day of the experiment (e.g., for each ATP stock and sample buffer).

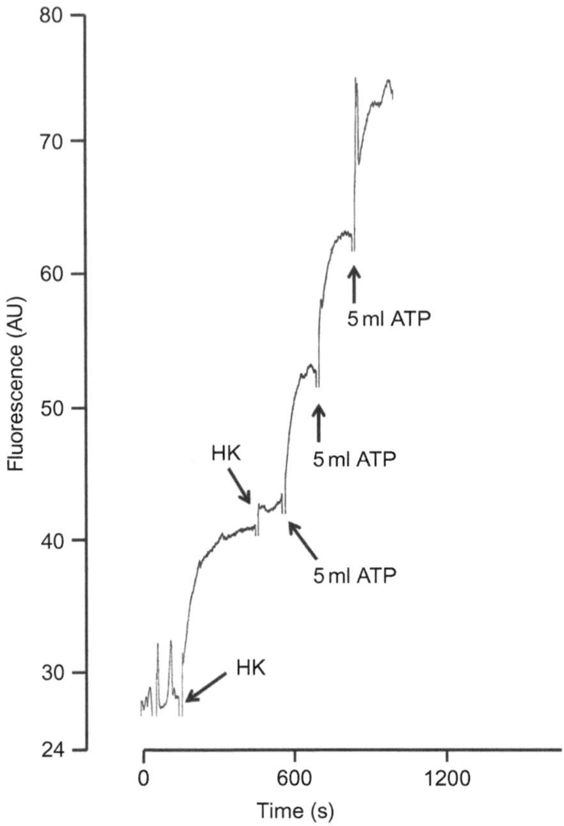

Figure 16.3 *Representative trace of an ATP measurement in Zajdela tumor cell lysates.* The data were obtained from 50 μl of Zajdela tumor cell lysates. HK, hexokinase; ml, milliliter; AU, arbitrary unit.

3.5. Notes

The protocol presented above (in Section 3) is useful for measuring the total cellular ATP level. It is very difficult to distinguish between the cytosolic and mitochondrial ATP levels using this method. These difficulties are based on the practical impossibility of separating the cytosolic fraction without damaging the mitochondria (during mild homogenization or plasma membrane permeabilization using digitonin). Mitochondrial damage will cause the release of ATP from broken mitochondria and contaminate the cytosolic ATP pool. Recently, Soccio et al. described a fast method for isolating cytosol-enriched fractions in which mitochondrial integrity is preserved by using an isotonic grinding medium (Soccio, Laus, Trono, & Pastore,

2013). This fluorimetric method to assay ATP content is practically identical to the one presented in the section above (Section 3).

4. SUMMARY

Because ATP regulates many aspects of cellular life, detecting the levels of this molecule is fundamentally important in many biological fields. Importantly, the alteration of ATP storage is found in several human disorders, including cancer. Currently, the measurement of ATP levels is also used to evaluate tumor growth and to determine tumor chemosensitivity (Bradbury, Simmons, Slater, & Crouch, 2000; Sevin et al., 1988). Thus, an efficient measurement of intracellular ATP levels will allow the assessment of proliferation and the effective energetic state of a tumor cell. Furthermore, it will be possible to determine whether a specific pharmacological treatment is able to block tumor growth and to reverse the malignant phenotype.

In this chapter, we have described luciferase/luciferin-based techniques that guarantee a rapid and highly accurate determination of ATP levels in different compartments, particularly in the cytosol (the site of glycolysis) and the mitochondria (the site of OXPHOS). However, in certain cases, the use of firefly luciferase may not always be possible, and alternative methods are necessary. We have presented two efficient techniques to measure intracellular ATP without luciferase: one based on the glycerokinase/glycerolphosphate oxidase/horseradish peroxidase-coupled assay and the other based on the hexokinase/glucose-6-phosphate dehydrogenase-coupled assay (for the latter, we also provided a detailed description).

ACKNOWLEDGMENTS

This study was supported by the Italian Association for Cancer Research (AIRC); Telethon (GGP11139B); local funds from the University of Ferrara; the Italian Ministry of Education, University and Research (COFIN, FIRB, and Futuro in Ricerca); and an Italian Ministry of Health grant to Paolo Pinton. S. P. was supported by a FISM (Fondazione Italiana Sclerosi Multipla) research fellowship (2012/B/11). M. R. W. was supported by the Polish National Science Center (UMO-2011/11/M/NZ3/02128), Polish Ministry of Science and Higher Education Grant W100/HFSC/2011 and BIO-IMAGing in Research Innovation and Education (FP7-REGPOT-2010-1). A. K. W. was supported by the POIG.02.01.00-14-059/09 project titled "Multi-profile modernization of scientific and research infrastructure of the Children's Memorial Health Institute" that was submitted within the frameworks of the competition for Measure 2.1 "Development of high potential research centers", Priority II "R&D Infrastructure" within the frameworks of the Innovative Economy Operational Programme, 2007–2013.

REFERENCES

Ainscow, E. K., & Rutter, G. A. (2002). Glucose-stimulated oscillations in free cytosolic ATP concentration imaged in single islet beta-cells: Evidence for a Ca^{2+}-dependent mechanism. *Diabetes*, *51*(Suppl. 1), S162–S170.

Bell, C. J., Manfredi, G., Griffiths, E. J., & Rutter, G. A. (2007). Luciferase expression for ATP imaging: Application to cardiac myocytes. *Methods in Cell Biology*, *80*, 341–352.

Beutner, G., Ruck, A., Riede, B., Welte, W., & Brdiczka, D. (1996). Complexes between kinases, mitochondrial porin and adenylate translocator in rat brain resemble the permeability transition pore. *FEBS Letters*, *396*(2–3), 189–195.

Bonora, M., Bononi, A., De Marchi, E., Giorgi, C., Lebiedzinska, M., Marchi, S., et al. (2013). Role of the c subunit of the FO ATP synthase in mitochondrial permeability transition. *Cell Cycle*, *12*(4), 674–683.

Bonora, M., Giorgi, C., Bononi, A., Marchi, S., Patergnani, S., Rimessi, A., et al. (2013). Subcellular calcium measurements in mammalian cells using jellyfish photoprotein aequorin-based probes. *Nature Protocols*, *8*(11), 2105–2118.

Bonora, M., Patergnani, S., Rimessi, A., De Marchi, E., Suski, J. M., Bononi, A., et al. (2012). ATP synthesis and storage. *Purinergic Signal*, *8*(3), 343–357.

Bowers, K. C., Allshire, A. P., & Cobbold, P. H. (1992). Bioluminescent measurement in single cardiomyocytes of sudden cytosolic ATP depletion coincident with rigor. *Journal of Molecular and Cellular Cardiology*, *24*(3), 213–218.

Bradbury, D. A., Simmons, T. D., Slater, K. J., & Crouch, S. P. (2000). Measurement of the ADP:ATP ratio in human leukaemic cell lines can be used as an indicator of cell viability, necrosis and apoptosis. *Journal of Immunological Methods*, *240*(1–2), 79–92.

Brini, M., Pinton, P., King, M. P., Davidson, M., Schon, E. A., & Rizzuto, R. (1999). A calcium signaling defect in the pathogenesis of a mitochondrial DNA inherited oxidative phosphorylation deficiency. *Nature Medicine*, *5*(8), 951–954.

Chiesa, A., Rapizzi, E., Tosello, V., Pinton, P., de Virgilio, M., Fogarty, K. E., et al. (2001). Recombinant aequorin and green fluorescent protein as valuable tools in the study of cell signalling. *The Biochemical Journal*, *355*(Pt. 1), 1–12.

Chinopoulos, C., Vajda, S., Csanady, L., Mandi, M., Mathe, K., & Adam-Vizi, V. (2009). A novel kinetic assay of mitochondrial ATP-ADP exchange rate mediated by the ANT. *Biophysical Journal*, *96*(6), 2490–2504.

Cobbold, P. H., Cuthbertson, K. S., Goyns, M. H., & Rice, V. (1983). Aequorin measurements of free calcium in single mammalian cells. *Journal of Cell Science*, *61*, 123–136.

Craig, F. F., Simmonds, A. C., Watmore, D., McCapra, F., & White, M. R. (1991). Membrane-permeable luciferin esters for assay of firefly luciferase in live intact cells. *The Biochemical Journal*, *276*(Pt. 3), 637–641.

DeLuca, M., & McElroy, W. D. (1974). Kinetics of the firefly luciferase catalyzed reactions. *Biochemistry*, *13*(5), 921–925.

de Wet, J. R., Wood, K. V., DeLuca, M., Helinski, D. R., & Subramani, S. (1987). Firefly luciferase gene: Structure and expression in mammalian cells. *Molecular and Cellular Biology*, 7(2), 725–737.

Duee, E. D., & Vignais, P. V. (1969). Kinetics and specificity of the adenine nucleotide translocation in rat liver mitochondria. *The Journal of Biological Chemistry*, *244*(14), 3920–3931.

Fujii, H., Noda, K., Asami, Y., Kuroda, A., Sakata, M., & Tokida, A. (2007). Increase in bioluminescence intensity of firefly luciferase using genetic modification. *Analytical Biochemistry*, *366*(2), 131–136.

Gatenby, R. A., & Gillies, R. J. (2004). Why do cancers have high aerobic glycolysis? *Nature Reviews Cancer*, *4*(11), 891–899.

Giorgi, C., Baldassari, F., Bononi, A., Bonora, M., De Marchi, E., Marchi, S., et al. (2012). Mitochondrial Ca(2+) and apoptosis. *Cell Calcium*, *52*(1), 36–43.

Giorgi, C., Ito, K., Lin, H. K., Santangelo, C., Wieckowski, M. R., Lebiedzinska, M., et al. (2010). PML regulates apoptosis at endoplasmic reticulum by modulating calcium release. *Science, 330*(6008), 1247–1251.

Giorgi, C., Romagnoli, A., Pinton, P., & Rizzuto, R. (2008). Ca2+ signaling, mitochondria and cell death. *Current Molecular Medicine, 8*(2), 119–130.

Gould, S. J., & Subramani, S. (1988). Firefly luciferase as a tool in molecular and cell biology. *Analytical Biochemistry, 175*(1), 5–13.

Harwood, K. R., Mofford, D. M., Reddy, G. R., & Miller, S. C. (2011). Identification of mutant firefly luciferases that efficiently utilize aminoluciferins. *Chemistry & Biology, 18*(12), 1649–1657.

Hirokawa, K., Kajiyama, N., & Murakami, S. (2002). Improved practical usefulness of firefly luciferase by gene chimerization and random mutagenesis. *Biochimica et Biophysica Acta, 1597*(2), 271–279.

Houtkooper, R. H., Pirinen, E., & Auwerx, J. (2012). Sirtuins as regulators of metabolism and healthspan. *Nature Reviews Molecular Cell Biology, 13*(4), 225–238.

Jouaville, L. S., Pinton, P., Bastianutto, C., Rutter, G. A., & Rizzuto, R. (1999). Regulation of mitochondrial ATP synthesis by calcium: Evidence for a long-term metabolic priming. *Proceedings of the National Academy of Sciences of the United States of America, 96*(24), 13807–13812.

Kawamata, H., Starkov, A. A., Manfredi, G., & Chinopoulos, C. (2010). A kinetic assay of mitochondrial ADP-ATP exchange rate in permeabilized cells. *Analytical Biochemistry, 407*(1), 52–57.

Kennedy, H. J., Pouli, A. E., Ainscow, E. K., Jouaville, L. S., Rizzuto, R., & Rutter, G. A. (1999). Glucose generates sub-plasma membrane ATP microdomains in single islet beta-cells. Potential role for strategically located mitochondria. *The Journal of Biological Chemistry, 274*(19), 13281–13291.

Lee, M. S., Park, W. S., Kim, Y. H., Ahn, W. G., Kwon, S. H., & Her, S. (2012). Intracellular ATP assay of live cells using PTD-conjugated luciferase. *Sensors (Basel, Switzerland), 12*(11), 15628–15637.

Lemasters, J. J., & Hackenbrock, C. R. (1976). Continuous measurement and rapid kinetics of ATP synthesis in rat liver mitochondria, mitoplasts and inner membrane vesicles determined by firefly-luciferase luminescence. *European Journal of Biochemistry, 67*(1), 1–10.

Liemburg-Apers, D. C., Imamura, H., Forkink, M., Nooteboom, M., Swarts, H. G., Brock, R., et al. (2011). Quantitative glucose and ATP sensing in mammalian cells. *Pharmaceutical Research, 28*(11), 2745–2757.

Manfredi, G., Yang, L., Gajewski, C. D., & Mattiazzi, M. (2002). Measurements of ATP in mammalian cells. *Methods, 26*(4), 317–326.

Manoli, I., Alesci, S., Blackman, M. R., Su, Y. A., Rennert, O. M., & Chrousos, G. P. (2007). Mitochondria as key components of the stress response. *Trends in Endocrinology and Metabolism, 18*(5), 190–198.

Marchi, S., Patergnani, S., & Pinton, P. (2014). The endoplasmic reticulum-mitochondria connection: One touch, multiple functions. *Biochimica et Biophysica Acta, 1837*(4), 461–469.

Marchi, S., & Pinton, P. (2014). The mitochondrial calcium uniporter complex: Molecular components, structure and physiopathological implications. *The Journal of Physiology, 592*(Pt 5), 829–839.

Patergnani, S., Marchi, S., Rimessi, A., Bonora, M., Giorgi, C., Mehta, K. D., et al. (2013). PRKCB/protein kinase C, beta and the mitochondrial axis as key regulators of autophagy. *Autophagy, 9*(9), 1367–1385.

Pollak, M. (2013). Targeting oxidative phosphorylation: Why, when, and how. *Cancer Cell, 23*(3), 263–264.

Porcelli, A. M., Pinton, P., Ainscow, E. K., Chiesa, A., Rugolo, M., Rutter, G. A., et al. (2001). Targeting of reporter molecules to mitochondria to measure calcium, ATP, and pH. *Methods in Cell Biology*, *65*, 353–380.

Rimessi, A., Giorgi, C., Pinton, P., & Rizzuto, R. (2008). The versatility of mitochondrial calcium signals: From stimulation of cell metabolism to induction of cell death. *Biochimica et Biophysica Acta*, *1777*(7–8), 808–816.

Rimessi, A., Marchi, S., Fotino, C., Romagnoli, A., Huebner, K., Croce, C. M., et al. (2009). Intramitochondrial calcium regulation by the FHIT gene product sensitizes to apoptosis. *Proceedings of the National Academy of Sciences of the United States of America*, *106*(31), 12753–12758.

Rodriguez-Enriquez, S., Carreno-Fuentes, L., Gallardo-Perez, J. C., Saavedra, E., Quezada, H., Vega, A., et al. (2010). Oxidative phosphorylation is impaired by prolonged hypoxia in breast and possibly in cervix carcinoma. *The International Journal of Biochemistry & Cell Biology*, *42*(10), 1744–1751.

Scatena, R. (2012). Mitochondria and cancer: A growing role in apoptosis, cancer cell metabolism and dedifferentiation. *Advances in Experimental Medicine and Biology*, *942*, 287–308.

Sevin, B. U., Peng, Z. L., Perras, J. P., Ganjei, P., Penalver, M., & Averette, H. E. (1988). Application of an ATP-bioluminescence assay in human tumor chemosensitivity testing. *Gynecologic Oncology*, *31*(1), 191–204.

Shackelford, D. B., & Shaw, R. J. (2009). The LKB1-AMPK pathway: Metabolism and growth control in tumour suppression. *Nature Reviews Cancer*, *9*(8), 563–575.

Shapiro, A. B., Hajec, L., & Gao, N. (2011). High-throughput, homogeneous, fluorescence intensity-based measurement of adenosine diphosphate and other ribonucleoside diphosphates with nanomolar sensitivity. *Analytical Biochemistry*, *415*(2), 190–196.

Soccio, M., Laus, M., Trono, D., & Pastore, D. (2013). A new simple fluorimetric method to assay cytosolic ATP content: Application to durum wheat seedlings to assess modulation of mitochondrial potassium channel and uncoupling protein activity under hyperosmotic stress. *Biologia*, *68*(3), 421–432.

Voronina, S. G., Barrow, S. L., Simpson, A. W., Gerasimenko, O. V., da Silva Xavier, G., Rutter, G. A., et al. (2010). Dynamic changes in cytosolic and mitochondrial ATP levels in pancreatic acinar cells. *Gastroenterology*, *138*(5), 1976–1987.

Webb, S. E., Rogers, K. L., Karplus, E., & Miller, A. L. (2010). The use of aequorins to record and visualize Ca(2+) dynamics: From subcellular microdomains to whole organisms. *Methods in Cell Biology*, *99*, 263–300.

Welsh, D. K., & Kay, S. A. (2005). Bioluminescence imaging in living organisms. *Current Opinion in Biotechnology*, *16*(1), 73–78.

Wibom, R., Lundin, A., & Hultman, E. (1990). A sensitive method for measuring ATP-formation in rat muscle mitochondria. *Scandinavian Journal of Clinical and Laboratory Investigation*, *50*(2), 143–152.

Wieckowski, M. R., Brdiczka, D., & Wojtczak, L. (2000). Long-chain fatty acids promote opening of the reconstituted mitochondrial permeability transition pore. *FEBS Letters*, *484*(2), 61–64.

Wieckowski, M. R., Vyssokikh, M., Dymkowska, D., Antonsson, B., Brdiczka, D., & Wojtczak, L. (2001). Oligomeric C-terminal truncated Bax preferentially releases cytochrome c but not adenylate kinase from mitochondria, outer membrane vesicles and proteoliposomes. *FEBS Letters*, *505*(3), 453–459.

Zamaraeva, M. V., Sabirov, R. Z., Manabe, K., & Okada, Y. (2007). Ca(2+)-dependent glycolysis activation mediates apoptotic ATP elevation in HeLa cells. *Biochemical and Biophysical Research Communications*, *363*(3), 687–693.

CHAPTER SEVENTEEN

Measurement of ADP–ATP Exchange in Relation to Mitochondrial Transmembrane Potential and Oxygen Consumption

Christos Chinopoulos[*,1], **Gergely Kiss**[*], **Hibiki Kawamata**[†], **Anatoly A. Starkov**[†]

[*]Department of Medical Biochemistry, Semmelweis University, Budapest, Hungary
[†]Brain and Mind Research Institute, Weill Medical College of Cornell University, New York, USA
[1]Corresponding author: e-mail address: chinopoulos.christos@eok.sote.hu

Contents

Abstract

We have previously described a fluorometric method to measure ADP–ATP exchange rates in mitochondria of permeabilized cells, in which several enzymes that consume substantial amounts of ATP and other competing reactions interconverting adenine

Methods in Enzymology, Volume 542
ISSN 0076-6879
http://dx.doi.org/10.1016/B978-0-12-416618-9.00017-0

nucleotides are present. This method relies on recording changes in free extramitochondrial Mg^{2+} with the Mg^{2+}-sensitive fluorescent indicator Magnesium Green (MgGr)™, exploiting the differential affinity of ADP and ATP for Mg^{2+}. In particular, cells are permeabilized with digitonin in the presence of BeF_3^- and Na_3VO_4, inhibiting all ATP- and ADP-utilizing reactions but mitochondrial exchange of ATP with ADP catalyzed by the adenine nucleotide translocase. The rate of ATP appearing in the medium upon the addition of ADP to energized mitochondria is then calculated from the rate of change in free extramitochondrial Mg^{2+} using standard binding equations. Here, we describe a variant of this method involving an improved calibration step. This step minimizes errors that may be introduced during the conversion of the MgGr™ signal into free extramitochondrial $[Mg^{2+}]$ and ATP. Furthermore, we describe an approach for combining this methodology with the measurement of mitochondrial membrane potential and oxygen consumption in the same sample. The method described herein is useful for the study of malignant cells, which are known to thrive in hypoxic environments and to harbor mitochondria with profound functional alterations.

ABBREVIATIONS

ANT adenine nucleotide translocase
AP_5A P^1,P^5-di(adenosine-5′) pentaphosphate
cATR carboxyatractyloside
EDTA ethylenediaminetetraacetic acid
EGTA ethylene glycol-bis(2-aminoethylether)-*N*,*N*,*N*′,*N*′-tetraacetic acid
MgGr Magnesium Green
SF 6847 tyrphostin 9, RG-50872, Malonaben, 3,5-di-*tert*-butyl-4-hydroxybenzylidene-malononitrile, 2,6-di-*t*-butyl-4-(2′,2′-dicyanovinyl)phenol

1. INTRODUCTION

Adenine nucleotide translocase (ANT) exchanges ATP for ADP across the inner mitochondrial membrane (Klingenberg, 1980; Pebay-Peyroula & Brandolin, 2004). Because ANT transports adenine nucleotides only in the Mg^{2+}-free state (Klingenberg, 1980; Kramer, 1980), and Mg^{2+} has differential affinity for ADP and ATP, we are able to measure ANT activity using the membrane-impermeable Mg^{2+}-sensitive fluorescent indicator "Magnesium Green (MgGr)." In this assay, the rate of change in free extramitochondrial $[Mg^{2+}]$ in the experimental medium is measured following the addition of ADP to mitochondria. The changes in free extramitochondrial $[Mg^{2+}]$ exhibit complete sensitivity to submicromolar amounts of the ANT inhibitor, carboxyatractyloside (cATR;

Chinopoulos et al., 2009; Metelkin, Demin, Kovacs, & Chinopoulos, 2009). The rate of change in free [Mg^{2+}] is converted to rate of ATP exported from mitochondria using standard binding equations (Chinopoulos et al., 2009).The ATP–ADP exchange rate mediated by the ANT from isolated mitochondria has been validated in Chinopoulos et al. (2009), especially in view of the contribution of the ATP-Mg^{2+}/P_i carrier (Aprille, 1993) and a homolog of the Mrs2 protein originally described in yeast that mediates an electrophoretic uptake of Mg^{2+} in mitochondria (Kolisek et al., 2003).

In isolated mitochondria, besides ANT, adenylate kinase and creatine kinase in the intermembrane space interconvert adenine nucleotides, but the former is effectively inhibited by P^1,P^5-di(adenosine-5′) pentaphosphate (AP_5A; Lienhard & Secemski, 1973), and the latter by excluding creatine or its phosphate derivatives in the medium (Chinopoulos et al., 2009). In permeabilized cells, however, numerous additional enzymes that interconvert adenine nucleotides exist. These include phosphorylases, phosphatases and kinases, as well as the Na^+/K^+ ATPase, the plasmalemmal, and endoplasmic Ca^{2+} ATPase, and in contractile cells the myosin ATPase. These reactions hamper the use of the binding equations that convert free [Mg^{2+}] to ANT-dependent ADP–ATP exchange rate (Chinopoulos et al., 2009). Thus, in permeabilized cells, all competing adenine nucleotide interconverting reactions need to be inhibited. This is accomplished by the use of BeF_3^- and vanadium compounds, which effectively inhibit ADP- and/or ATP-utilizing reactions (Baukrowitz, Hwang, Nairn, & Gadsby, 1994; Cantley et al., 1977; Davies & Hol, 2004; Gordon, 1991; Mukherjee et al., 2004; Robinson, Davis, & Steinberg, 1986; Werber, Peyser, & Muhlrad, 1992). In Kawamata, Starkov, Manfredi, and Chinopoulos (2010), we used this approach and demonstrated ANT-dependent ADP–ATP exchange rate in permeabilized cells, using BeF_3^- and Na_3VO_4. This method was also applied in determining ANT dysfunction in relation to mitochondrial membrane potential in myotubes expressing ANT1-harboring mutations linked to autosomal-dominant progressive external opthalmoplegia (Kawamata, Tiranti, Magrane, Chinopoulos, & Manfredi, 2011). In these studies, the conversion of dye emission signal to [Mg^{2+}], and subsequently to ATP, was calibrated by obtaining the maximal fluorescence signal with excess [Mg^{2+}] and the minimal fluorescence by the addition of the cation chelator, ethylenediaminetetraacetic acid (EDTA). However, because MgGr is a fluorimetric dye with single excitation and emission, it is subjected to the potential pitfalls of nonratiometric dyes related to variations in dye concentration and/or bleaching. In practice, this method introduced variability in the measurement of ADP–ATP exchange rate and a better calibration was required. Below we describe a step-by-step measurement of ANT-dependent adenine nucleotide exchange with a modified calibration method, allowing for an error-free conversion of MgGr signal to free extramitochondrial [Mg^{2+}]. Furthermore, we present the methods to correlate ADP–ATP exchange rate to mitochondrial membrane potential and oxygen consumption. These could become particularly informative when studying bioenergetic parameters of cancer cell mitochondria that exhibit decreased or complete loss of electron flux associated with impaired respiration and ATP synthesis (Kwong, Henning, Starkov, & Manfredi, 2007).

2. METHODOLOGY

2.1. Reagents

Standard laboratory chemicals, AP_5A, safranin O, and digitonin from Sigma (St. Louis, MO, USA). MgGr $5K^+$ salt from Life Technologies (Carlsbad, CA, USA). cATR from Calbiochem (San Diego, CA, USA). Tyrphostin 9, RG-50872, Malonaben, 3,5-di-*tert*-butyl-4-hydroxybenzylidenemalononitrile, 2,6-di-*t*-butyl-4-(2′,2′-dicyanovinyl)phenol (SF 6847) from Biomol (catalog number EI-215; BIOMOL GmbH, Hamburg, Germany).

2.2. Stock reagents and buffers

25 m*M* Na_3VO_4, pH 8.7 (see below how to prepare)
0.2 *M* $BeSO_4$ (see below how to prepare)
0.5 *M* NaF (see below how to prepare)
25 m*M* Ap_5A, dissolved in distilled water
1 m*M* cATR, dissolved in distilled water
5 m*M* Oligomycin (if a mixture of oligomycins A, B, and C are used, that is, item # O4876 from Sigma-Aldrich, assume MW = 800), dissolved in ethanol
0.2 *M* EDTA
1 m*M* Safranin O, dissolved in distilled water
1.1 m*M* MgGr $5K^+$ salt, dissolved in distilled water
1 *M* Glutamate, dissolved in distilled water, pH 7 with KOH

0.5 *M* Malate, dissolved in distilled water, pH 7 with KOH
0.2 *M* ADP, K^+ salt (see below how to prepare)
0.2 *M* ATP K^+ salt (see below how to prepare)
2.5 m*M* Digitonin, dissolved in DMSO (see below)
1 *M* $MgCl_2$
1 m*M* SF 6847, dissolved in ethanol. Do not substitute with FCCP, CCCP, or 2,4-dinitrophenol (see below).

Buffer A
8 m*M* KCl
110 m*M* K-gluconate
10 m*M* NaCl
10 m*M* Hepes
10 m*M* KH_2PO_4
0.005 m*M* EGTA
10 m*M* Mannitol
1 m*M* $MgCl_2$
0.5 mg ml^{-1} Bovine serum albumin (fatty-acid-free)
pH 7.2

Buffer B
Buffer A without $MgCl_2$, but containing 25 μ*M* AP_5A, 5 m*M* NaF, 0.2 m*M* $BeSO_4$, 30 μ*M* Na_3VO_4, 5 μ*M* EDTA (to chelate nominal amount of Mg^{2+} in the buffer) and 50 μ*M* digitonin.

Buffer C
Same as buffer A, but without $MgCl_2$, and containing 4 μ*M* cATR, 2 μ*M* oligomycin, and 5 μ*M* EDTA.

2.3. Preparation of reagents

ADP and ATP are purchased as a K^+ salt of the highest purity available and titrated to pH 6.9 with KOH to make stocks of 0.2 *M*. Concentration of ADP and ATP stock solutions is corrected by measuring absorbance at 260 nm using an extinction coefficient $\varepsilon M = 15{,}400\ M^{-1}\ cm^{-1}$.

All other reagents should be dissolved in distilled water, except SF 6847 and oligomycin which should be dissolved in ethanol, and digitonin which should be dissolved in DMSO.

2.4. Preparation of sodium orthovanadate (Na_3VO_4) and beryllium trifluoride (BeF_3^-)

A 25 m*M* Na_3VO_4 solution is prepared in distilled water (>17 mΩ resistance). The pH of the solution is adjusted to 8.7 with HCl, upon which it turns yellow. This solution is boiled until it turns colorless and cooled to room temperature. The pH is reassessed and readjusted to pH 8.7 with HCl, upon which the solution turns yellow again. This cycle of boiling until colorless and readjusting the pH is repeated until the solution remains colorless at pH 8.7. Finally, the solution is brought up to the initial volume with distilled water and stored in aliquots at −80 °C. This treatment removes all decavanadate ions present in the Na_3VO_4 solution, which induces mitochondrial membrane depolarization and inhibition of oxygen consumption (Aureliano & Crans, 2009). Orthovanadate inhibits the oxidation of only disrupted mammalian mitochondria (Byczkowski, Zychlinski, & Tluczkiewicz, 1979). Likewise, fluoroberyllium nucleoside diphosphate complexes inhibit only the exposed F_1F_0-ATPase (Issartel, Dupuis, Lunardi, & Vignais, 1991). $BeSO_4$ and NaF are prepared as aqueous stock solutions of 0.2 *M* and 0.5 *M*, respectively, and kept at +4 °C for several years. BeF_3^- is formed immediately in solution upon mixing of $BeSO_4$ and NaF, provided that NaF is in excess in the working solution.

Vanadate, beryllium, and fluoride salts are highly toxic to tissues and to the environment, and thus require proper handling and disposal. The combination of orthovanadate and BeF_3^- will inhibit kinases, mutases, phosphatases, and ATPases (Climent, Bartrons, Pons, & Carreras, 1981; Ray, Moore, & Rao, 1990). However, some kinases, such as pyruvate kinase, will remain uninhibited (Lord & Reed, 1990). In this respect, upon permeabilization of the cells one has to totally separate pyruvate kinase from its substrate, phosphoenol pyruvate, that is, there must be no glucose present in the medium prior to permeabilization, and a few minutes lag time must be allowed prior to ADP–ATP exchange rate measurements in order for the remaining reactions by kinases to "die-out."

2.5. Digitonin stock

Our digitonin powder stocks were purchased as "approximately 50% estimated by thin layer chromatography"; therefore, we cannot be certain of the exact concentration of digitonin present in the measuring chamber. Optimum digitonin amount added to the chamber is determined by measuring oxygen consumption in nonpermeabilized cells using succinate as respiratory substrate, as detailed in Kawamata et al. (2010) and is the one conferring the highest respiratory control ratio. The entire study should be performed using the same digitonin stock solution.

2.6. Equipment

Oroboros Oxygraph-2k (Oroboros Instruments, Innsbruck, Austria) equipped with an LED exhibiting a wavelength maximum of 465 ± 25 nm (current for light intensity adjusted to 2 mA, i.e., level '4') and either an <505-nm shortpass excitation filter (dye-based, filter set "safranin") or an <495-nm shortpass excitation filter (dye-based, filter set "MgG/CaG"). Emitted light is detected by a photodiode (range of sensitivity: 350–700 nm), through either an >560-nm longpass emission filter (dye-based) for safranin O or an >525-nm longpass emission filter for MgGr. A cuvette-based fluorimeter equipped with the capability of stirring and thermoregulation would also be suitable (e.g., Hitachi F-7000 spectrofluorimeter; Hitachi High Technologies, Maidenhead, UK).

2.7. Culturing and harvesting of the cells

The amount of cells required varies from one type to another. In cell types that contain extremely high amounts of mitochondria, such as in differentiated myotubes from mouse C_2C_{12} myoblasts, plating 140,000 myocytes for differentiation (by exchanging horse for fetal bovine serum at the time when myocytes are over 90% confluent) is sufficient for the assay. For other cells, such as human fibroblasts, 1–2 million cells are required. Below, we present results obtained from HEK293 cells (~400,000 cells per assay). Regardless of the cell type, cells are washed once in phosphate-buffered saline and harvested with 0.1 ml of 0.25% trypsin–EDTA, inactivated by 0.9 ml calf serum, followed by centrifugation at 1100 *g* for 2 min. Next, cells are washed once in buffer A without disturbing the pellet. After the wash, cells are resuspended in 0.2 ml of buffer B. This buffer contains the same base composition as buffer A described above, but includes substrates (glutamate and malate) as well as inhibitors. See steps below for specific concentrations of substrates and inhibitors to include. The rationale for using this particular buffer composition is elaborated in Chinopoulos et al. (2009). Glutamate and malate as mitochondrial substrates were chosen on the basis that they support mitochondrial substrate-level phosphorylation, and as such, they contribute to greater ATP efflux rates (Chinopoulos et al., 2010; Kiss et al., 2013).

2.8. K_d determination of ATP and ADP for Mg^{2+}

First, one should measure the apparent K_d values of ADP for Mg^{2+} and ATP for Mg^{2+} for the pertaining conditions (media, temperature, ionic strength, type and amount of cells, etc.), steps 1–7.

1. Cells resuspended in 2 ml of buffer C are added to a chamber of an Oroboros Oxygraph-2k. The presence of cATR and oligomycin in the buffer is only required for the K_d determination. MgGr fluorescence is recorded by the O2k-Fluorescence LED2-Module at a 1-Hz acquisition rate. Experiments are performed at 37 °C. Digitonin (see above regarding remarks about its concentration) and MgGr $5K^+$ salt (MgGr; 1 μ*M*) are subsequently added to the chamber.
2. MgGr fluorescence signal is recorded upon stepwise additions of 0.1 m*M* $MgCl_2$ for a total of 10 additions (about 50 s of recording time per addition). This is shown in Fig. 17.1A, on the *left* side of the dashed line.
3. Continuing in *same cells in the chamber*, add 0.25 m*M* ADP in subsequent steps for a total of 19 steps (about 50 s of recording time per addition). This is shown in Fig. 17.1A, on the *right* side of the dashed line.
4. Likewise, in a new cell preparation, repeat steps 1–3, but with 11 additions of 0.2 m*M* ATP instead of ADP, also shown in Fig. 17.1A.
5. Convert the MgGr signal of *both left and right part* of Fig. 17.1A to free $[Mg^{2+}]$. To do this, plot the steady states of MgGr after each addition of $MgCl_2$ (as seen in the *left part* of Fig. 17.1A) on the *y*-axis as a function of $[Mg^{2+}]$ (*x*-axis) and apply the following fit equation (you can use either trace): $f = y_0 + a(1 - \exp(-b \cdot x))$. This is an exponential rise to maximum equation with a three-parameter function, where y_0, a, and b are coefficients. The results are shown in Fig. 17.1B.
6. Calibrate the *right part* of Fig. 17.1A, using the coefficients y_0, a, and b determined from step 5, shown in Fig. 17.1B, by solving the following exponential function for x: $x = (-1/b)\ln(1 - ((f - y_0)/a))$. After calibration of MgGr signals obtained in Fig. 17.1A, it should look like Fig. 17.1C. If calibration was performed correctly, the left parts of both traces should be almost identical.
7. Next, from the calibrated *right part* of Fig. 17.1C, the K_d of ADP for Mg^{2+} and the K_d of ATP for Mg^{2+} is estimated *by fitting the following equation with the least squares method to the data points*:

$$[Mg^{2+}]_f = 0.5 \cdot \left([Mg^{2+}]_t - K_d - [L]_t + \sqrt{\left([Mg^{2+}]_t - K_d - [L]_t\right)^2 + 4K_d[Mg^{2+}]_t} \right)$$

where $[Mg^{2+}]_t$ is the total $[Mg^{2+}]$ (1 m*M*), $[L]_t$ is the total concentration of added ADP (or ATP). The least square fitting method can be

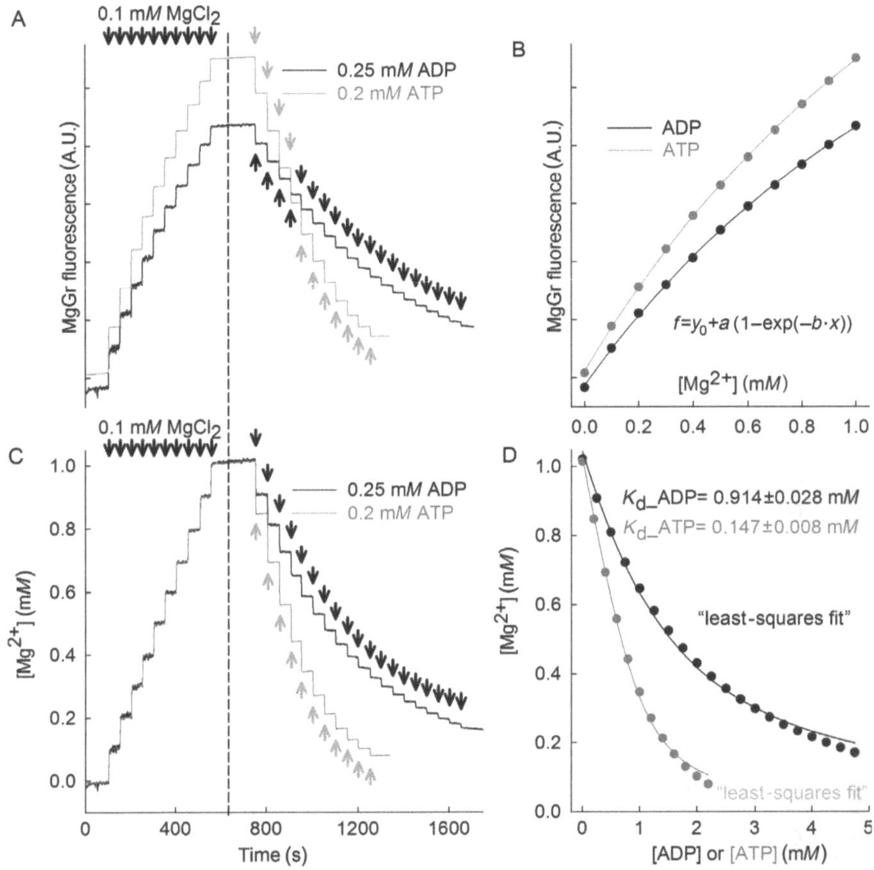

Figure 17.1 *Magnesium Green fluorescence calibration and estimation of* K_d *of ATP and ADP for* Mg^{2+}. (A) Reconstructed time recordings of MgGr raw fluorescence traces in permeabilized HEK293 cells as a function of extramitochondrial [Mg^{2+}] (left part of the traces), and as a function of extramitochondrial ADP or ATP (right parts of the traces). (B) MgGr fluorescence changes are dependent on extramitochondrial [Mg^{2+}]. (C) Calibrated time recordings of extramitochondrial [Mg^{2+}] (left part of the traces), and as a function of extramitochondrial ADP and ATP (right parts of the traces) are shown. (D) Calibrated extramitochondrial Mg^{2+} plots as a function of ADP or ATP are shown, from which we estimated K_d of ADP and ATP for Mg^{2+} using the least squares method to fit the data.

performed by a customized Excel tool or other freely available software. The fitted curves are shown in Fig. 17.1D for both ADP and ATP. The K_d determined are K_d_ADP $= 0.914 \pm 0.028$ m*M*, and K_d_ATP $= 0.147 \pm 0.008$ m*M*. Since K_d values are now known *for the pertaining experimental conditions*, ADP–ATP exchange rates can be calculated from the MgGr recordings.

2.9. $[Mg^{2+}]_{free}$ determination from MgGr fluorescence in permeabilized cells and conversion to ADP–ATP exchange rate

1. Add 1.1 μ*M* of MgGr in 1.8 ml of buffer B. Record MgGr fluorescence for a few minutes and allow the signal to stabilize. In the meantime, harvest cells by trypsinization as elaborated above. Cells are washed once with buffer A and then resuspended in 0.2 ml of buffer B.
2. Upon addition of cells to the measuring chamber, wait for 3 min and resume recording. After 2 min, add a known amount of ADP; we suggest 2 m*M* ADP. When ADP is added, the fluorescence signal will drop, as shown in Fig. 17.2A. Subsequently, the uncoupler SF 6847 is added to the chamber in 10 n*M* increments (~100 s between each interval); this will be used later on to correlate the ATP efflux rates with membrane potential changes, see below. It is important to use SF 6847 and not the more widely used FCCP, CCCP, or 2,4-dinitrophenol, because SF 6847 is the only uncoupler that does not quench MgGr fluorescence up to a concentration of 1 μ*M* (SF 6847 is also appropriate to use with Calcium Green, a Ca^{2+}-sensitive dye; this is not unexpected as both Calcium Green and MgGr are fluorescein derivatives).
3. Figure 17.2B shows the free $[Mg^{2+}]$ calibrated from MgGr raw fluorescence. One may notice that initial total $[Mg^{2+}]$ was measured to be 1.1486 m*M*, that is, not exactly 1 m*M*, which was the actual amount added into the chamber. After the addition of 2 m*M* ADP, free $[Mg^{2+}]$ was measured to attain the value of0.5553 m*M*, that is, not the exact value of 0.3958 m*M*, which is what would be expected. The value of 0.3958 m*M* was derived from the equation below:

$$[Mg^{2+}]_f = 0.5 \cdot \left([Mg^{2+}]_t - K_d - [L]_t + \sqrt{\left([Mg^{2+}]_t - K_d - [L]_t\right)^2 + 4K_d[Mg^{2+}]_t} \right)$$

where $[Mg^{2+}]_t$ is the total $[Mg^{2+}]$ (1 m*M*), $[L]_t$ is the total concentration of added ADP (or ATP), and K_d is the fitted value of K_{ADP} or K_{ATP}, respectively. These errors stem from the fact that MgGr is not a ratiometric dye, and thus it is subject to the pitfalls elaborated above. It is therefore necessary to adjust the whole trace shown in Fig. 17.2B so that the value just prior to the addition of 2 m*M* ADP is 1 m*M* $[Mg^{2+}]$ (the total), and the first value of $[Mg^{2+}]$ after the addition of 2 m*M* ADP is 0.3958 m*M*. From Fig. 17.1B, it is apparent that the relation between

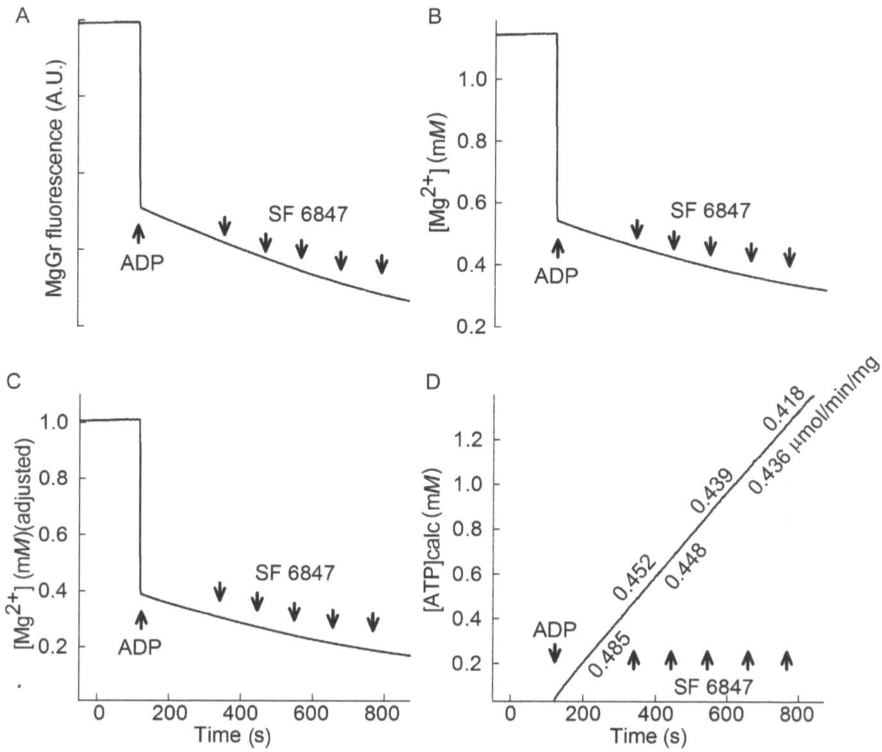

Figure 17.2 *Determination of extramitochondrial [Mg^{2+}] and conversion to ATP.* (A) Reconstructed time recording of MgGr raw fluorescence in permeabilized HEK293 cells upon addition of 2 m*M* ADP (where indicated), followed by incremental 10 n*M* additions of the uncoupler SF 6847. (B) Calibrated time recording of extramitochondrial [Mg^{2+}] obtained from panel (A). (C) Corrected calibrated trace of panel (B), as described in the text. (D) Calculated amount of ATP appearing in the medium converted from panel (C). The rate of ATP appearing in the medium is indicated in µmol min^{-1} mg $protein^{-1}$.

free [Mg^{2+}] and MgGr fluorescence signal in the 0–0.5 m*M* range is fairly linear, see also Chinopoulos et al. (2009). Thus, one may apply a series of simple arithmetic operations to correct the errors: the difference between the measured total [Mg^{2+}] (1.1486 m*M*) and the measured first value of [Mg^{2+}] after the addition of 2 m*M* ADP (0.5553 m*M*) is 0.5933 m*M*, while the difference between the actual total [Mg^{2+}] (1 m*M*) and the expected first value of [Mg^{2+}] after the addition of 2 m*M* ADP (0.3958 m*M*) is 0.6042 m*M*; therefore, the measured changes in free [Mg^{2+}] are underestimated by a factor of 0.6042/0.5553 = 1.01837. Thus, the whole trace of Fig. 17.2B is multiplied by 1.01837. This would generate a trace in which the value of

$[Mg^{2+}]$ just prior to the addition of 2 m*M* ADP is 1.169 m*M* (the estimated total), and the first value of $[Mg^{2+}]$ after the addition of 2 m*M* ADP is 0.5655 m*M* (not shown). By subtracting the difference of 1.169–1 = 0.169 m*M* from this whole trace, the final trace should look like the one depicted in Fig. 17.2C. Note that in the trace shown in Fig. 17.2C, the value just prior to the addition of 2 m*M* ADP is 1 m*M* $[Mg^{2+}]$ (the total), and the first value of $[Mg^{2+}]$ after the addition of 2 m*M* ADP is 0.3958 m*M*. The trace depicting the estimated $[Mg^{2+}]$ has therefore been corrected, and the conversion to ATP appearing in the medium can now be applied, as shown in the next step.

4. We convert the corrected $[Mg^{2+}]$ values to ATP appearing in the medium, using the equation below:

$$[\mathrm{ATP}]_t = \frac{\left(\frac{[\mathrm{Mg}^{2+}]_t}{[\mathrm{Mg}^{2+}]_f} - 1 - \frac{[\mathrm{ADP}]_t(t=0) + [\mathrm{ATP}]_t(t=0)}{K_{\mathrm{ADP}} + [\mathrm{Mg}^{2+}]_f}\right)}{\left(\frac{1}{K_{\mathrm{ATP}} + [\mathrm{Mg}^{2+}]_f} - \frac{1}{K_{\mathrm{ADP}} + [\mathrm{Mg}^{2+}]_f}\right)}$$

The above equation is available for download as an executable file at http://www.tinyurl.com/ant-calculator. The rate of ATP appearing in the medium can be calculated by making a linear regression for the ATP values as a function of time, as shown in Fig. 17.2D.

2.10. Mitochondrial membrane potential ($\Delta\Psi m$) determination in *in situ* mitochondria of permeabilized cells

$\Delta\Psi m$ is estimated using fluorescence quenching of the cationic dye safranin O due to its accumulation inside energized mitochondria (Akerman & Wikstrom, 1976), also taking into account the considerations discussed in Perevoshchikova, Sorochkina, Zorov, and Antonenko (2009) and Figueira, Melo, Vercesi, and Castilho (2012).

1. Cells are treated exactly as described for free $[Mg^{2+}]$ determination, except that MgGr is replaced by 5 μ*M* safranin O.
2. Fluorescence is recorded in an Oroboros Oxygraph-2k at a 1-Hz acquisition rate, using the 495-nm excitation and 585-nm emission wavelengths. Experiments are performed at 37 °C. After the baseline signal has stabilized, 2 m*M* ADP is added to the chamber and fluorescence is allowed to stabilize. The raw signal of safranin O fluorescence is shown in Fig. 17.3A.

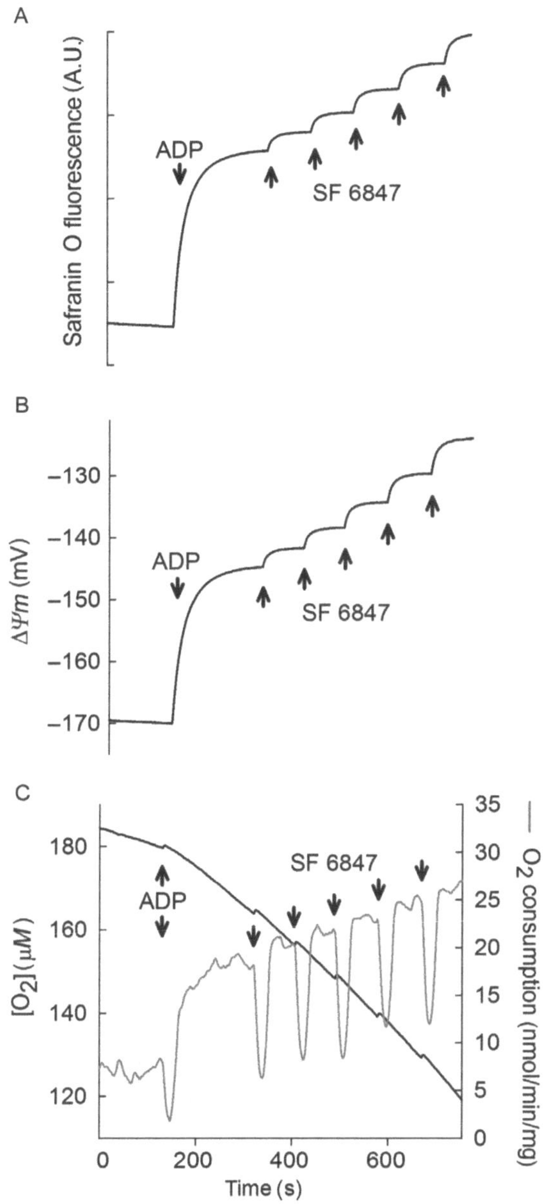

Figure 17.3 *Mitochondrial membrane potential and oxygen consumption determination in permeabilized cells.* (A) Reconstructed time recording of safranin O raw fluorescence in permeabilized HEK293 cells upon addition of 2 m*M* ADP (where indicated), followed by incremental 10 n*M* additions of the uncoupler SF 6847. (B) Calibrated time recording of $\Delta\Psi m$ obtained from panel (A). (C) Reconstructed time recording of oxygen concentration in the medium (black trace) and oxygen flux (gray trace) recorded simultaneously with either MgGr signal (panel A) or safranin O signal (panel A).

3. After the addition of ADP, $\Delta\Psi m$ is further depolarized stepwise by the addition of 10 n*M* SF 6847 until the signal after depolarization becomes stable (~100 s). Safranin O fluorescence is converted to mV (Fig. 17.3B) using the Nernst equation and assuming a matrix $[K^+] = 120$ m*M*, which was determined by a voltage-fluorescence calibration curve of safranin O fluorescence in the presence of 2 n*M* valinomycin and increasing $[K^+]$ (0.2–120 m*M*; Akerman & Wikstrom, 1976). An important note is that cancer cells express very variable levels of IF1 (Sanchez-Arago et al., 2013), which inhibits the ATP hydrolytic function of the ATPase, thus hindering reversal rates. Therefore, it is not recommended to study ATP *influx* rates in mitochondria from cancer cells, that is, at very depolarized values. The method is suitable for studying cancer mitochondria during ATP synthesis.

2.11. Mitochondrial respiration

Oxygen consumption is simultaneously measured with safranin O reflecting $\Delta\Psi m$ or MgGr fluorescence reflecting ADP–ATP exchange rates in the same samples polarographically using the Oxygraph-2k. Oxygen concentration (black trace of Fig. 17.3C) and oxygen flux (gray trace of Fig. 17.3C expressed as pmol s^{-1} mg^{-1}; negative time derivative of oxygen concentration, divided by tissue mass per volume and corrected for instrumental background oxygen flux arising from oxygen consumption of the oxygen sensor and back-diffusion into the chamber) were recorded using DatLab software (Oroboros Instruments).

From the above measured parameters, one is able to estimate the rate of ATP appearing in the medium as a function of $\Delta\Psi m$ or oxygen consumption rate.

3. CONCLUSIONS AND COMMENTS

Using this technique, we have previously addressed ANT-dependent adenine nucleotide exchange in disease conditions (Kawamata et al., 2011). By generating ADP–ATP exchange rate/$\Delta\Psi m$ profile for myotubes expressing ANT1 with pathogenic mutations, we found that the mutations caused a reduced ADP-induced depolarization of membrane potential. This suggested that mutant ANT1-expressing cells require more $\Delta\Psi m$ to generate the same amount of ATP. In the same cells, ATP synthesis measured by a luciferase-based method did not show significant differences between mutant and wild-type ANT1-expressing cells. Thus, this is a sensitive method to determine ATP production from mitochondria. Furthermore, with the presently described methodology, one is provided the capacity to measure ADP–ATP exchange rate in a kinetic mode together with oxygen consumption, in addition to recording changes in $\Delta\Psi m$. This may be particularly informative when studying bioenergetic parameters of cancer cell mitochondria, as these are known to exhibit decreased or complete loss of electron flux, which leads to impaired respiration and ATP synthesis (Kwong et al., 2007).

ACKNOWLEDGMENTS

This work was supported by the National Institute of Health/National Institute of Neurological Disease and Stroke Grant F31 NS054554 to H. K., PO AG014930 grant by NIH/NIA to A. A. S, the Országos Tudományos Kutatási Alapprogram Grants NNF 78905, NNF2 85658, K 100918, and the MTA-SE Lendület Neurobiochemistry Research Division Grant 95003 to C. C.

REFERENCES

Akerman, K. E., & Wikstrom, M. K. (1976). Safranine as a probe of the mitochondrial membrane potential. *FEBS Letters*, *68*, 191–197.

Aprille, J. R. (1993). Mechanism and regulation of the mitochondrial ATP-Mg/P(i) carrier. *Journal of Bioenergetics and Biomembranes*, *25*, 473–481.

Aureliano, M., & Crans, D. C. (2009). Decavanadate (V10 O28 6-) and oxovanadates: Oxometalates with many biological activities. *Journal of Inorganic Biochemistry*, *103*, 536–546.

Baukrowitz, T., Hwang, T. C., Nairn, A. C., & Gadsby, D. C. (1994). Coupling of CFTR Cl- channel gating to an ATP hydrolysis cycle. *Neuron*, *12*, 473–482.

Byczkowski, J., Zychlinski, L., & Tluczkiewicz, J. (1979). Interaction of vanadate with respiratory chain of rat liver and wheat seedling mitochondria. *The International Journal of Biochemistry*, *10*, 1007–1011.

Cantley, L. C., Jr., Josephson, L., Warner, R., Yanagisawa, M., Lechene, C., & Guidotti, G. (1977). Vanadate is a potent (Na, K)-ATPase inhibitor found in ATP derived from muscle. *The Journal of Biological Chemistry*, *252*, 7421–7423.

Chinopoulos, C., Gerencser, A. A., Mandi, M., Mathe, K., Torocsik, B., Doczi, J., et al. (2010). Forward operation of adenine nucleotide translocase during F0F1-ATPase reversal: critical role of matrix substrate-level phosphorylation. *FASEB Journal*, *24*, 2405–2416.

Chinopoulos, C., Vajda, S., Csanady, L., Mandi, M., Mathe, K., & Adam-Vizi, V. (2009). A novel kinetic assay of mitochondrial ATP-ADP exchange rate mediated by the ANT. *Biophysical Journal*, *96*, 2490–2504.

Climent, F., Bartrons, R., Pons, G., & Carreras, J. (1981). Effect of vanadate on phosphoryl transfer enzymes involved in glucose metabolism. *Biochemical and Biophysical Research Communications*, *101*, 570–576.

Davies, D. R., & Hol, W. G. (2004). The power of vanadate in crystallographic investigations of phosphoryl transfer enzymes. *FEBS Letters*, *577*, 315–321.

Figueira, T. R., Melo, D. R., Vercesi, A. E., & Castilho, R. F. (2012). Safranine as a fluorescent probe for the evaluation of mitochondrial membrane potential in isolated organelles and permeabilized cells. *Methods in Molecular Biology*, *810*, 103–117.

Gordon, J. A. (1991). Use of vanadate as protein-phosphotyrosine phosphatase inhibitor. *Methods in Enzymology*, *201*, 477–482.

Issartel, J. P., Dupuis, A., Lunardi, J., & Vignais, P. V. (1991). Fluoroaluminum and fluoroberyllium nucleoside diphosphate complexes as probes of the enzymatic mechanism of the mitochondrial F1-ATPase. *Biochemistry*, *30*, 4726–4733.

Kawamata, H., Starkov, A. A., Manfredi, G., & Chinopoulos, C. (2010). A kinetic assay of mitochondrial ADP-ATP exchange rate in permeabilized cells. *Analytical Biochemistry*, *407*, 52–57.

Kawamata, H., Tiranti, V., Magrane, J., Chinopoulos, C., & Manfredi, G. (2011). adPEO mutations in ANT1 impair ADP-ATP translocation in muscle mitochondria. *Human Molecular Genetics*, *20*, 2964–2974.

Kiss, G., Konrad, C., Doczi, J., Starkov, A. A., Kawamata, H., Manfredi, G., et al. (2013). The negative impact of alpha-ketoglutarate dehydrogenase complex deficiency on matrix substrate-level phosphorylation. *The FASEB Journal*, *27*, 2392–2406.

Klingenberg, M. (1980). The ADP-ATP translocation in mitochondria, a membrane potential controlled transport. *Journal of Membrane Biology*, *56*, 97–105.

Kolisek, M., Zsurka, G., Samaj, J., Weghuber, J., Schweyen, R. J., & Schweigel, M. (2003). Mrs2p is an essential component of the major electrophoretic Mg2+ influx system in mitochondria. *The EMBO Journal*, *22*, 1235–1244.

Kramer, R. (1980). Influence of divalent cations on the reconstituted ADP, ATP exchange. *Biochimica et Biophysica Acta*, *592*, 615–620.

Kwong, J. Q., Henning, M. S., Starkov, A. A., & Manfredi, G. (2007). The mitochondrial respiratory chain is a modulator of apoptosis. *The Journal of Cell Biology*, *179*, 1163–1177.

Lienhard, G. E., & Secemski, I. I. (1973). P 1, P 5 -Di(adenosine-5')pentaphosphate, a potent multisubstrate inhibitor of adenylate kinase. *The Journal of Biological Chemistry*, *248*, 1121–1123.

Lord, K. A., & Reed, G. H. (1990). Vanadyl(IV) complexes with pyruvate kinase: Activation of the enzyme and electron paramagnetic resonance properties of ternary complexes with the protein. *Archives of Biochemistry and Biophysics*, *281*, 124–131.

Metelkin, E., Demin, O., Kovacs, Z., & Chinopoulos, C. (2009). Modeling of ATP-ADP steady-state exchange rate mediated by the adenine nucleotide translocase in isolated mitochondria. *The FEBS Journal*, *276*, 6942–6955.

Mukherjee, B., Patra, B., Mahapatra, S., Banerjee, P., Tiwari, A., & Chatterjee, M. (2004). Vanadium—An element of atypical biological significance. *Toxicology Letters*, *150*, 135–143.

Pebay-Peyroula, E., & Brandolin, G. (2004). Nucleotide exchange in mitochondria: Insight at a molecular level. *Current Opinion in Structural Biology*, *14*, 420–425.

Perevoshchikova, I. V., Sorochkina, A. I., Zorov, D. B., & Antonenko, Y. N. (2009). Safranine O as a fluorescent probe for mitochondrial membrane potential studied on the single particle level and in suspension. *Biochemistry (Mosc)*, *74*, 663–671.

Ray, B. D., Moore, J. M., & Rao, B. D. (1990). ^{31}P NMR studies of enzyme-bound substrate complexes of yeast 3-phosphoglycerate kinase: III. Two ADP binding sites and their Mg(II) affinity; effects of vanadate and arsenate on enzymic complexes with ADP and 3-P-glycerate. *Journal of Inorganic Biochemistry*, *40*, 47–57.

Robinson, J. D., Davis, R. L., & Steinberg, M. (1986). Fluoride and beryllium interact with the (Na+K)-dependent ATPase as analogs of phosphate. *Journal of Bioenergetics and Biomembranes*, *18*, 521–531.

Sanchez-Arago, M., Formentini, L., Martinez-Reyes, I., Garcia-Bermudez, J., Santacatterina, F., Sanchez-Cenizo, L., et al. (2013). Expression, regulation and clinical relevance of the ATPase inhibitory factor 1 in human cancers. *Oncogenesis*, *2*, e46.

Werber, M. M., Peyser, Y. M., & Muhlrad, A. (1992). Characterization of stable beryllium fluoride, aluminum fluoride, and vanadate containing myosin subfragment 1-nucleotide complexes. *Biochemistry*, *31*, 7190–7197.

CHAPTER EIGHTEEN

Real-Time Assessment of the Metabolic Profile of Living Cells with Genetically Encoded NADH Sensors

Yuzheng Zhao*, Yi Yang*,[1], Joseph Loscalzo†,[1]
*Synthetic Biology and Biotechnology Laboratory, State Key Laboratory of Bioreactor Engineering, School of Pharmacy, East China University of Science and Technology, Shanghai, China
†Cardiovascular Division, Department of Medicine, Brigham and Women's Hospital, Harvard Medical School, Boston, Massachusetts, USA
[1]Corresponding authors: e-mail addresses: yiyang@ecust.edu.cn; jloscalzo@partners.org

Contents

Abstract

Redox metabolism plays a critical role in multiple pathophysiological settings, including oncogenesis and tumor progression. Until recently, however, our knowledge of key redox processes in living systems was limited by the lack of an adequate methodology

Methods in Enzymology, Volume 542
ISSN 0076-6879
http://dx.doi.org/10.1016/B978-0-12-416618-9.00018-2

to monitor redox potential. Nicotinamide adenine dinucleotide, in its reduced (NADH) and oxidized (NAD^+) forms, is perhaps the most important small molecule in the redox metabolism of mammalian cells. We have previously developed a series of genetically encoded fluorescent sensors allowing for the quantification of intracellular NADH. Here, we present experimental components and considerations that are required to perform a standardized quantification of intracellular NADH based on these probes. Moreover, we present the initial calibration experiments necessary to obtain reliable data from this approach, we detail a protocol to measure intracellular NADH levels in steady-state kinetic experiments, and we provide consideration on the processing of data. Among various applications, this technique is suitable for the study of redox alterations in malignant cells.

1. INTRODUCTION

1.1. Cellular redox systems

Oxidation–reduction reactions exist in all cells and are critical for cell homeostasis and signaling, including energy metabolism, gene expression, cell cycle regulation, immune response, cell growth, and cell apoptosis (Balaban, Nemoto, & Finkel, 2005; Delaunay, Pflieger, Barrault, Vinh, & Toledano, 2002; Jones, 2008; Menon & Goswami, 2007; Ying, 2008). The intracellular redox states are mainly determined by pyridine nucleotide redox systems and thiol/disulfide redox systems. The pyridine nucleotide redox system comprises reduced and oxidized nicotinamide adenine dinucleotide ($NADH/NAD^+$) and reduced and oxidized nicotinamide adenine dinucleotide phosphate ($NADPH/NADP^+$). $NADH/NAD^+$ are the most important cofactors involved in energy metabolism, receiving electrons during the oxidation of various nutrients and then oxidized via mitochondrial respiration. In addition, this redox couple also function in circadian rhythms, apoptosis, aging, development, and carcinogenesis (Chen et al., 2009; Dumollard, Ward, Carroll, & Duchen, 2007; Lin, Ford, Haigis, Liszt, & Guarente, 2004; Nakahata, Sahar, Astarita, Kaluzova, & Sassone-Corsi, 2009; Ramsey et al., 2009; Zhang et al., 2006). In contrast, NADPH provides the reducing equivalents for biosynthetic reactions of anabolic pathways and antioxidant defense against oxidative stress, and is also responsible for generating free radicals, such as superoxide and nitric oxide. The thiol/disulfide redox system comprises the glutathione redox system and the thioredoxin redox system. Both reduced glutathione and thioredoxin are regenerated by reductases that consume NADPH (Fig. 18.1).

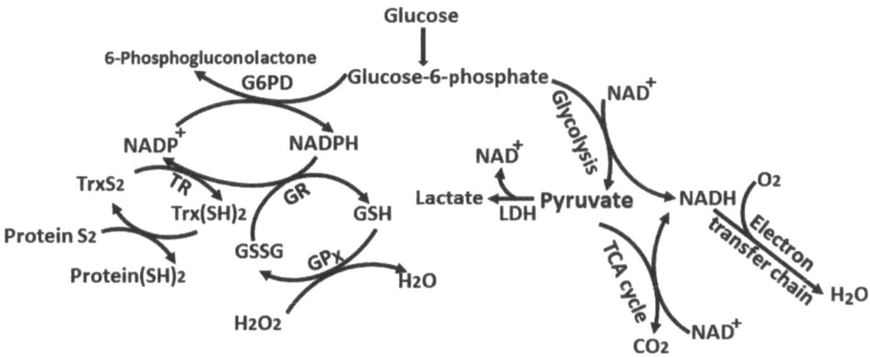

Figure 18.1 *Simplified scheme for intracellular redox systems.* NADH produced during glycolysis and the tricarboxylic acid (TCA) cycle is consumed by the electron transfer chain. NADPH generated by the catalysis of glucose-6-phosphate dehydrogenase (G6PD) is utilized to reduce oxidized glutathione (GSSG) by glutathione reductase (GR) or oxidized thioredoxin (Trx) by thioredoxin reductase (TR).

1.2. Genetically encoded fluorescent sensors for NADH

For many years, researchers relied on weak autofluorescence to measure the NADH levels in living cells (Kasischke, Vishwasrao, Fisher, Zipfel, & Webb, 2004; Mayevsky & Rogatsky, 2007; Patterson, Knobel, Arkhammar, Thastrup, & Piston, 2000; Pellerin & Magistretti, 2004). This method has the drawbacks of a weak signal, low sensitivity and specificity, and cell injury resulting from the ultraviolet irradiation required to induce the autofluorescent signal. To probe dynamic cellular events in living cells with high spatial and temporal resolution, researchers have developed a diverse set of genetically encoded biosensors, which generally involve the fusion of a fluorescent protein with a specific sensing protein domain. In bacteria, the Rex family proteins are especially sensitive to NADH (Brekasis & Paget, 2003; Gyan, Shiohira, Sato, Takeuchi, & Sato, 2006; Sickmier et al., 2005; Wang et al., 2008). Crystallographic studies have shown that NADH binding induces dramatic conformational changes in the Rex dimer, which shifts from an open to a closed form (McLaughlin et al., 2010). By fusing the Rex protein and circularly permuted fluorescent proteins (cpFPs), our group and Yellen's group independently developed genetically encoded NADH sensors that allow specific monitoring of dynamic changes in NADH levels as affected by NADH transport, glucose metabolism, mitochondrial electron transport, and redox environment (Hung, Albeck, Tantama, & Yellen, 2011; Zhao et al., 2011). The Frex sensors (Zhao et al., 2011; Fig. 18.2A) specifically detect NADH levels with large dynamic range

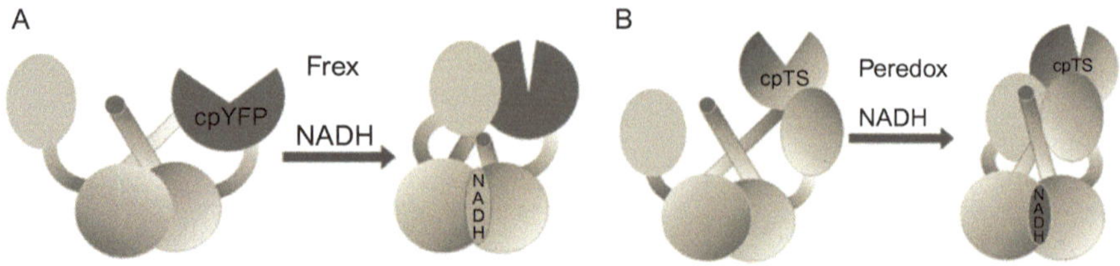

Figure 18.2 *Schematic representations of Frex and Peredox.* (A) Design of Frex, which is a fusion protein of a complete Rex monomer, cpYFP, and the NADH-binding domain of a second Rex molecule. (B) Design of Peredox, which is a fusion protein of two complete Rex monomers connected by a circular permuted T-Sapphire molecule. *From Zhao et al. Cell Metab 2011;14(4):555-566.*

and different affinities; however, their fluorescence is pH sensitive. Peredox sensors (Hung et al., 2011; Fig. 18.2B) are much more resistant to local pH and partially reflect the more physiologically relevant NADH:NAD$^+$ ratio; however, their dynamic range of fluorescence is much smaller than that of Frex sensors (150% versus 800% upon NADH binding). The affinity of Peredox for NADH is quite high, which makes these sensors largely saturated under normal physiological conditions (Hung et al., 2011) and, therefore, difficult to detect redox fluctuations.

Genetically encoded sensors have also been used to detect disulfides (Cannon & James Remington, 2009; Dooley et al., 2004; Hanson et al., 2004; Meyer & Dick, 2010; Ostergaard, Henriksen, Hansen, & Winther, 2001), oxidized glutathione (Bjornberg, Ostergaard, & Winther, 2006; Gutscher et al., 2008), and hydrogen peroxide (Belousov et al., 2006). Here, we present a detailed protocol for live cell imaging and real-time measurement of the intracellular NADH level using the Frex series of genetically encoded sensors.

2. EXPERIMENTAL COMPONENTS AND CONSIDERATIONS

2.1. Expression and purification of recombinant Frex sensors

Sufficient quantities of purified Frex sensors are desirable to ensure consistent and reproducible results from *in vitro* titration. Although there are several options for protein purification, we recommend immobilized metal affinity chromatography and Ni^{2+} resin for the purification of His-tagged proteins. Protein purity is assessed by SDS-PAGE analysis. The aliquots of purified Frex sensors are stored at −20 °C until experimental use. The steps for the sensor protein preparation are as follows:

1. *Escherichia coli* BL21 (DE3) pLys cells carrying the pRSETb-Frex expression plasmid are grown in 10 ml LB media containing 34 μg/ml chloramphenicol and 100 μg/ml ampicillin. Cells are incubated with shaking at 37 °C until OD 600 nm reaches 0.6–0.8.
2. Add IPTG from a 100 m*M* stock to a final concentration of 0.1 m*M* and incubate with shaking at 18 °C overnight.
3. Harvest the cells by centrifugation at 4000 × *g* for 30 min at 4 °C.
4. Resuspend the cell pellets in 1 ml 50 m*M* potassium phosphate buffer (Kpi buffer), pH 7.4, containing 0.5 *M* sodium chloride and 20 m*M* imidazole (Buffer A), and lyse cells by ultrasonication.
5. Purify His6-tagged proteins using His MultiTrp 96-well filter plates. First, wash the plate with 2 column volumes of wash buffer (Buffer A contain 50 m*M* imidazole). Second, elute the protein from the resin using 50 m*M* Kpi buffer, pH 7.4, containing 0.5 *M* sodium chloride and 300 m*M* imidazole (Buffer B).
6. Desalt and exchange the protein preparations into 100 m*M* potassium phosphate buffer (pH 7.4) before assay.
7. Measure the protein concentration by the Lowry assay and store the aliquotted samples at −20 °C.

2.2. Frex sensor expression in mammalian cells

Frex sensors were expressed in different subcellular compartments by tagging these proteins with organelle-specific signal peptides and observing the cells 24–30 h post-transfection, as follows:

The coding sequences of Frex were subcloned into pcDNA3.1 Hygro (+) (Invitrogen) upstream of a Kozak sequence for mammalian expression. For nuclear targeting of Frex, the threefold nuclear localization signal (3 × NLS) DPKKKRKVDPKKKRKVDPKKKRKV was added to the C-terminus. For mitochondrial targeting, the mitochondrial localization signal MRKMLAAVSRVLSGASQKPASRVLVASRNFANDATF was inserted at the N-terminus. 293FT cells (Invitrogen, USA) were maintained in DMEM (high glucose) supplemented with 10% FBS, 0.1 m*M* MEM with nonessential amino acids (Invitrogen), 6 m*M* L-glutamine, and 1 m*M* sodium pyruvate (Invitrogen) at 37 °C in a humidified atmosphere of 95% air and 5% CO_2. Cells were plated in antibiotic-free high glucose-DMEM supplemented with 10% FBS 16 h before transfection. We typically used 0.8 μg endotoxin-free plasmids with 3.2 μl Lipofectamine 2000 (Invitrogen) for each well of a 12-well plate according to the manufacturer's protocol.

2.3. Fluorescent microplate reading

For high-throughput detection of intracellular NADH levels, a Synergy 2 Multi-Mode Microplate Reader (Biotek) was used for measuring the fluorescence of suspended cells expressing NADH sensors in 96-well or 384-well black flat-bottom microplates. In general, 293FT cells were harvested 24–48 h after transfection, counted, washed, and suspended in phosphate-buffered saline preheated at 37 °C, and aliquots of cells were incubated at 37 °C with compounds of interest during the measurement. Dual-excitation ratios were obtained using a Synergy MX Multi-Mode Microplate Reader (Biotek) at 410 and 500 nm excitation, and 528 nm emission for both excitation wavelengths. Filters used were 528 BP 20 nm for emission, and 410 BP 20 nm or 485 BP 20 nm for excitation.

2.4. Laser scanning confocal microscopy

For real-time imaging of intracellular NADH, we utilized the Zeiss Laser Scanning Confocal Microscopy (LSCM) system on a Zeiss Axio Observer Z1 inverted microscope equipped with a Plan Apo 63 × 1.4 NA oil immersion objective. 293FT cells expressing the NADH sensor were maintained at 37 °C in a humidified atmosphere using a CO_2 incubator (PECON). We excited NADH sensors sequentially line by line with the 405 nm and 488 nm laser line and with the emission detection set to 500–550 nm. We strongly recommend that microscopes provided with "line mode" scanning ability be used to measure the fluorescence ratio of each pixel accurately by minimizing subcellular particle movements and the effects of vibration. Scanning was performed using the "line mode," 1024 × 1024 format, 12 bit depth, 2 × line average, and 3.0 pinhole.

2.5. Automatic wide field fluorescence microscopy

To assess Frex and C3L194K sensor occupancy by NADH in mitochondria, a high-performance fluorescence microscopy system equipped with a Nikon Eclipse Ti-E automatic microscope, Plan Apo VC 60 × 1.2 NA water immersion objective, Evolve 512 EMCCD (Photometrics), and the highly stable Lambda XL light source (Sutter Instruments) was used for ratiometric fluorescence imaging of living cells and for calibration with purified recombinant sensor proteins. 410 BP 20 nm and 480 BP 30 nm band-pass excitation filters, and a 535/40 emission filter were switched by a Lambda 10-XL filter wheel (Sutter Instruments) for dual-excitation ratio imaging. Images were captured using a 512 × 512 format, 16 bit depth, and 100 ms exposure for both channels.

3. INITIAL CHARACTERIZATION EXPERIMENTS

Initial experiments are necessary to determine the fluorescent properties of Frex sensors for each preparation (K_d for NADH and dynamic range), sensor concentration, instrument parameters, and pH fluctuation for NADH detection. This approach will ensure that both valid and relevant results are achieved in the steady-state and kinetics experiments.

3.1. *In vitro* titration of Frex sensor by NADH

The Frex sensor had an apparent K_d for NADH and NADH analogs of 3.7 μ*M* and ≫400 μ*M*, respectively (Zhao et al., 2011). For titration experiments *in vitro*, when removing samples from −20 °C storage, we recommend that the aliquot of Frex sensors be rapidly defrosted by hand and then placed immediately on ice. Use one aliquot per set of experiments as multiple freeze-thaw cycles may denature and inactivate Frex sensors. Ideally, all experiments should be conducted with Frex sensors from the same purification preparation. A sample protocol is as follows:

1. Prepare the NADH stock solution by dissolving 7.09 mg NADH ml^{-1} in 100 m*M* Kpi buffer, pH 7.4. Dilute 20 μl of the stock solution with 980 μl of 100 m*M* Kpi buffer to prepare the maximal standard concentration (200 μ*M*) and then make serial dilutions of NADH to concentration of: 200, 100, 50, 20, 10, 5, 2, 1, 0.5, 0.2, and 0.1 μ*M*.
2. Add 50 μl of different concentrations of NADH to 96-well black flat-bottom microplates.
3. Dilute Frex protein into 100 m*M* Kpi buffer (pH 7.4) with 0.1% bovine serum albumin to a final concentration of 1 μ*M*, and then add 50 μl of Frex protein to 96-well black flat-bottom microplates using a dispenser.
4. Mix immediately and then measure the dual-excitation ratios of the Frex sensor using a synergy 2 Multi-Mode Microplate Reader with 420 and 485 nm excitation and 528 nm emission for both excitation wavelengths.
5. Fluorescence values are background-corrected by subtracting the signal intensity of different concentrations of NADH solution.
6. *Note*: For calibration of mitochondrial NADH levels, prepare NADH stock solutions in 100 m*M* Kpi buffer, pH 8.0. Dilute Frex and C3L194K protein into 100 m*M* KPi buffer with 0.1% bovine serum

albumin (pH 8.0) to final concentrations of 0.1 and 0.3 μ*M*, respectively. For *in vitro* titration of the Frex sensor and the C3L194K sensor, a series of NADH solutions with concentration (0–200 μ*M*) and (0–1 m*M*), respectively, were used as above.

3.2. Correction of pH effects with cpYFP

As with many other genetically encoded sensors based on cpYFP, Frex sensors are sensitive to pH. The fluorescence of the cpYFP moiety excited at 485 nm increased significantly with an increase of pH, while the fluorescence excited at 420 nm was much less affected (Fig. 18.3A). Fortunately, the Frex sensor is one of the most responsive genetic encoded sensors currently available, exhibiting an eightfold increase of fluorescence with excitation at 490 nm upon NADH binding, whereas there is less than a twofold increase in fluorescence with excitation at 490 nm when the pH increases from 7.0 to 7.4, the range of physiological pH in the cytosol. Thus, pH variations complicate modestly the quantification of free intracellular NADH levels. In order to identify pH fluctuations during the study of biological processes or effects of specific treatments, we suggest that one always measure the fluorescence of cpYFP alone in cells, or purified cpYFP in solution, before or in parallel with measurements in Frex-expressing cells. Under many circumstances, however, the pH variation in the cytosol is too small to induce significant changes in Frex fluorescence (Zhao et al., 2011). pH effects were not obvious in the studies of other cpYFP-based sensors (Belousov et al., 2006; Berg, Hung, & Yellen, 2009; Nakai, Ohkura, & Imoto, 2001). In contrast, the effects of pH variations are more obvious in studies of mitochondrial function, which heavily depend on electron transfer and respiration. Under such circumstances, the Frex fluorescence signal should be corrected by cpYFP fluorescence measured in parallel experiments, owing to the similar pH dependency of Frex and cpYFP (Fig. 18.3B). For example, we found that there was a biphasic increase of mitochondrial-targeted Frex fluorescence during the 1 h time course upon glucose supplementation in glucose-starved cells (Fig. 18.3C). It turned out that the slow phase was an artifact caused by the increase in pH in the mitochondrial matrix as shown by the cpYFP fluorescence. To correct this artifact, Frex-Mit fluorescence was normalized by cpYFP-Mit fluorescence. This corrected Frex-Mit fluorescence reached its maximum over 10 min after glucose supplementation, consistent with reports in the literature using other cells (Eto et al., 1999).

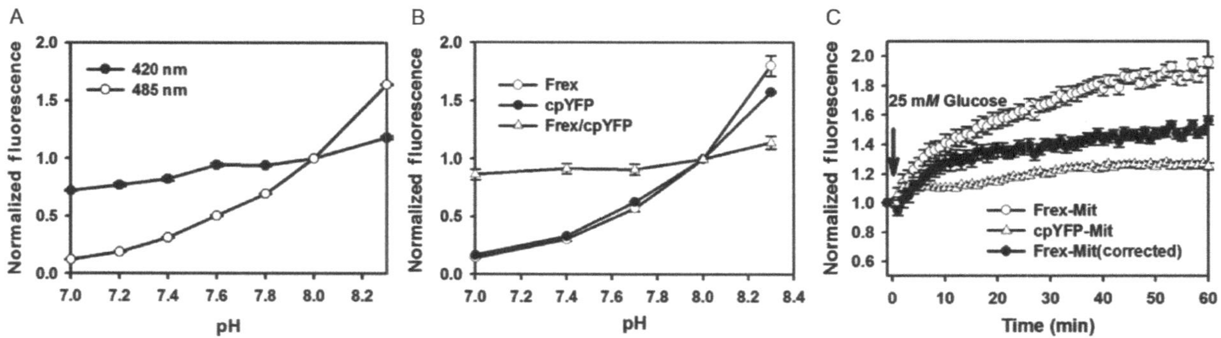

Figure 18.3 *pH effect correction of Frex fluorescence.* (A) pH dependency of cpYFP fluorescence excited at 420 and 485 nm. (B) pH dependency of Frex and cpYFP fluorescence excited at 485 nm. (C) Kinetics of fluorescence response of Frex-Mit and cpYFP-Mit after glucose supplementation. The solid symbol represents the fluorescence response of Frex-Mit corrected for pH effects. *From Zhao et al. Cell Metab 2011;14(4): 555–566.*

In fluorescence microscopy studies of live cells when cells are imaged with an on-line CO_2 incubation system, significant changes in pH may occur when the cells are moved from or placed into a CO_2 incubator, as the pH of cell culture medium is typically buffered by bicarbonate salts and is highly dependent on the dissolved CO_2 level. Such pH variations may be monitored using cpYFP-expressing cells. The following precautions should be taken under these conditions to control pH:

1. Adjust the pH of different stock solutions to pH 7.4 to maintain the pH of the buffers and medium for the cells.
2. Add 100 m*M* HEPES (pH 7.4) to phenol red-free complete medium to increase the pH buffering capacity of the medium.
3. Keep cells in the incubator for 10 min before imaging to equilibrate the dissolved CO_2.

4. STEADY-STATE AND KINETICS EXPERIMENTS

After determination of the critical experimental parameters and precautions from the initial characterization experiments, a full steady-state kinetics experiment can be conducted.

4.1. Quantification of NADH in subcellular compartments

Measuring NADH concentrations in living cells is vital for understanding fluctuations in metabolic states. For many years, researchers relied on endogenous NAD(P)H fluorescence for measuring mitochondrial NADH levels (Mayevsky, 2009). The total NAD^+–NADH pool in the matrix varies from 300 μ*M* to 3 m*M* depending on the cell type (Joubert, Fales, Wen, Combs, & Balaban, 2004; Yang et al., 2007). Previous reports showed that the NAD^+/NADH ratio varies from 2 to 16 in the mitochondrial matrix (Henley & Laughrey, 1970; Kasimova et al., 2006; Williamson, Lund, & Krebs, 1967); however, the ratio of free/bound NADH varies significantly from almost 0:1 (Wakita, Nishimura, & Tamura, 1995), 1.5:1 (Blinova et al., 2005), to 1:4 (Yu & Heikal, 2009) as measured by time-resolved fluorescence, fluorescence anisotropy, and fluorescence spectral decomposition analysis. These data may reflect different metabolic conditions in the mitochondria of different organisms or the limitations of the techniques used. Unfortunately, the traditional NAD(P)H fluorescence method is awkward and challenging for measuring free NADH levels in the cytosol or nucleus owing either to very low concentrations of NADH in these organelles or to the NADPH pool bearing the majority of cytosolic reduced pyridine dinucleotide (Jones, 1981; Shigemori et al., 1996).

NADH sensors are intrinsically ratiometric with two excitation wavelengths (Fig. 18.4A) and allow quantification of free intracellular NADH concentrations in different subcellular compartments (Fig. 18.4B). A standard protocol is as follows:

For the digitonin permeabilization assay

1. Prepare digitonin stock solution by dissolving 30 mg digitonin in 1 ml PBS.
2. For digitonin permeabilization assay of Frex-expressing cells, resuspend cells with PBS buffer (pH 7.4) containing 0.001% digitonin.
3. For digitonin permeabilization assay of Frex-Mit-expressing cells, resuspend cells with PBS and then lyse them with 100 m*M* KPi buffer (pH 8.0) containing 0.3% digitonin.
4. Measure the fluorescence (excited at 485 nm) of cells or cell lysates in the presence or absence of 100 μ*M* NADH.

For ratiometric fluorescence imaging

1. Measure the ratiometric fluorescence of purified recombinant sensor proteins (0.1 μ*M* Frex or 0.3 μ*M* C3L194K) with or without 100 μ*M* NADH in living cells using a high-performance fluorescence microscopy system equipped with a stable light source and high-performance EMCCD camera.
2. Assess Frex and C3L194K sensor occupancy by NADH in the mitochondria and then calculate mitochondrial NADH concentration.

For the FrexH fluorescence assay

1. Measure the ratiometric fluorescence of purified recombinant FrexH sensor proteins (0.1 μ*M* FrexH) with or without 10 μ*M* NADH and FrexH-expressing cells using a fluorescence plate reader.
2. Assess the FrexH sensor occupancy by NADH in the cytosol and then calculate cytosolic NADH concentration.

Considering the highly specific affinity of sensors toward NADH compared to other pyridine and adenine nucleotides (Zhao et al., 2011), Frex (K_d ~3.7 μ*M* at pH 7.4) and FrexH (K_d ~40 n*M* at pH 7.4) sensors are recommended for NADH detection in the cytosol and nucleus, where the free NADH concentration is extremely low. By contrast, Frex (K_d ~11 μ*M* at pH 8.0) and C3L194K (K_d ~50 μ*M* at pH 8.0) sensors are recommended for NADH detection in the mitochondria, where the free NADH concentration is comparatively high. We measured subcellular NADH levels in 293FT cells through Frex and FrexH using a microplate reader and found

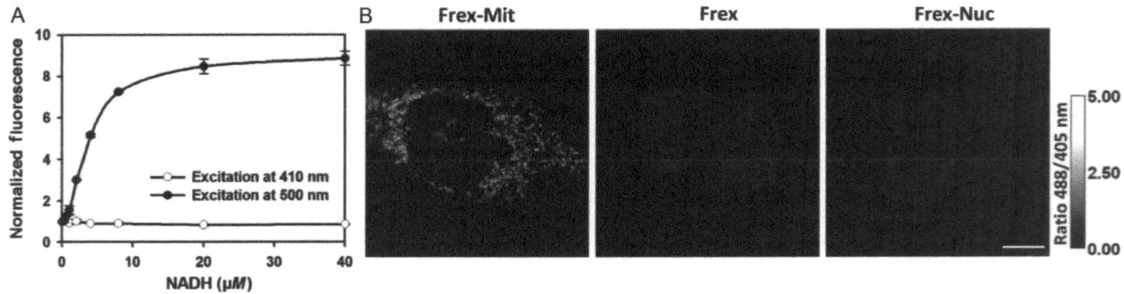

Figure 18.4 *Subcellular distribution of NADH in mammalian cells.* (A) Fluorescence intensities with excitation at 410 or 500 nm normalized to the initial value; emission at 528 nm. (B) Ratiometric fluorescence images of cells expressing Frex in mitochondria, the nucleus, and cytosol. The pixel-by-pixel ratios of the 488 nm excitation image to the 405 nm excitation image of the same cell were used to gray the images. Scale bar, 10 μm. *From Zhao, et al. Cell Metab 2011;14(4):555-566.*

extremely low cytosolic NADH levels ranging from 120 n*M* to 130 n*M* (Table 18.1), in agreement with previous estimations in the nucleus (Zhang et al., 2002). Ratiometric fluorescence imaging and digitonin permeabilization assays showed that the Frex-Mit sensors were ∼70–75% saturated by NADH in the mitochondrial matrix of 293FT cells, suggesting that the mitochondrial free NADH concentration is 33 ± 9 and 26 ± 7 μ*M*, respectively (Table 18.1). The free NADH concentration in the mitochondrial matrix was also determined to be 27 ± 5 μ*M* using a low-affinity sensor, C3L194K (Table 18.1).

Exogenous NADH-binding protein expression, that is, Frex sensor expression, may contribute to the buffering systems for NADH. However, the fluorescence of Frex expressed in the cytosol responds immediately to exogenous NADH, lactate, and glucose, suggesting that such temporal buffering effects of the Frex sensor are minimal. Cells have several "buffering" systems for NADH, including several NADH-binding proteins in the cytosol and mitochondria. Furthermore, NADH and NAD^+ are buffered by coupled reaction pairs, pyruvate and lactate via lactate dehydrogenase in the cytosol and oxoglutarate and glutamate via β-hydroxybutyrate dehydrogenase in the mitochondria. The concentrations of these metabolites are in the millimolar range. The Frex protein concentration expressed in the cytosol ranges from 4 to 13 μ*M*, depending on cell type and transfection efficiency. Frex protein concentration expressed in the mitochondria was estimated to be within 10–30 μ*M*, although only 70% was bound with NADH. We estimated that the Frex-Mit-bound NADH is less than 12%

Table 18.1 Subcellular NADH concentrations in 293FT cells

Sensors	K_d for NADH (μ*M*)	Subcellular location	Detection methods	Sensor occupancy (%)	[NADH] (μ*M*)	Reference (μ*M*)
FrexH-Cyt	0.04	Cytosol	Microplate	∼76	∼0.13	∼0.13[a]
Frex-Cyt	3.7	Cytosol	Microplate	∼3.3	∼0.12	
Frex-Mit	11[b]	Mitochondria	Imaging	∼70–75	$\sim 33 \pm 9$	∼60–75[c]
Frex-Mit	11[b]	Mitochondria	Microplate	∼70–75	$\sim 26 \pm 7$	
C3L194K	50[b]	Mitochondria	Imaging	∼35	$\sim 27 \pm 5$	

[a]The concentration of free NADH in the nucleus is ∼130 n*M* (Zhang, Piston, & Goodman, 2002).
[b]Measured at pH 8.0.
[c]The concentration of NADH in mitochondria of intact pancreatic islets at resting glucose levels (4–5 m*M*) is ∼60–75 μ*M* (Patterson et al., 2000).

of the total mitochondrial NADH in resting cells. According to these calculations, Frex sensors are unlikely to have a major impact on the total NADH concentration and the metabolic status of the cells in which they are expressed (Zhao & Yang, 2012).

4.2. Dynamic monitoring of NADH in single cells

Frex sensors are brighter and more sensitive than endogenous NAD(P)H fluorescence. Thus, they are more suitable for dynamic monitoring of NADH in single cells, particularly in the cytosol and nucleus where the endogenous NAD(P)H fluorescence is extremely weak and derived mainly from NADPH. For single cell imaging, a sample protocol is as follows:

1. Plate 293 FT cells or cells of interest onto a 35-mm glass-bottom dish with phenol red-free fresh growth medium containing 10% FBS, 0.1 m*M* MEM with nonessential amino acids, 6 m*M* L-glutamine, and 1 m*M* sodium pyruvate.
2. Transfect the cells with 0.8 μg per dish of cDNA (Frex, Frex-Nuc, or Frex-Mit) using Lipofectamine 2000 according to the manufactures's instruction, and incubate the cells in a humidified atmosphere at 37 °C under 5% CO_2 for 24–30 h.
3. Replace the medium with fresh growth medium containing 100 m*M* HEPES (pH 7.4), and keep cells in a humidified atmosphere using a CO_2 incubator (PeCon) for 10 min to equilibrate the dissolved CO_2 before imaging.
4. For single cell imaging, choose moderately bright cells in which the fluorescence is well distributed in the cytosol (Frex), in the nucleus (Frex-Nuc), or in the mitochondria (Frex-Mit), and use the Zeiss 710 laser scanning confocal microscopy (LSCM) system on a Zeiss Axio Observer Z1 inverted microscope equipped with a Plan Apo 63 × 1.4 NA oil immersion objective for green fluorescence of the indicator (Frex).
5. Switch on the lasers, controller, scanning head, and computer. *Note*: The lasers need 30 min to warm up and stabilize.
6. Start the scanning process using the "line mode" and excite the indicator (Frex) sequentially line by line with the 405 and 488 nm laser line and with emission detection set to 500–550 nm.
7. For fluorescence images, adjust the 405 and 488 nm laser intensity to the minimum level that allows easy identification of single cells, adjust the size of the pinhole for acceptable depth of the image, and adjust the

sensitivity of the photomultiplier tube (PMT) for optimal signal-to-noise ratio.

8. Acquire images every 15 s for 5–10 min.
9. During the image acquisition, add fresh growth medium containing reagents of interest.
10. Determine the image ratio by dividing the image intensities acquired with excitation at 488 nm by those acquired at 405 nm.

4.3. High-throughput measurement of NADH by microplate reader

Frex sensors are highly specific, have a large dynamic range of fluorescence upon NADH binding, and can be targeted to different subcellular compartments. These advantages make Frex compatible with high-throughput measurement of NADH by microplate reader. A sample protocol is as follows:

1. Plate 293 FT cells into 12-well plates with fresh growth medium containing 10% FBS, 0.1 m*M* MEM with nonessential amino acids, 6 m*M* L-glutamine, and 1 m*M* sodium pyruvate.
2. Transfect the cells with 0.8 μg per dish of cDNA (Frex-Mit or Frex-cpYFP) using 3.2 μl Lipofectamine 2000 according to the manufacturer's instruction, and incubate the cells in a humidified atmosphere at 37 °C under 5% CO_2 for 24–30 h.
3. Harvest 293FT cells expressing Frex-Mit and cpYFP-Mit from 12-well plates by trypsinization, and then centrifuge at 100 × *g* for 5 min.
4. Carefully remove the supernatant with a 1-ml pipette tip or an aspirator.
5. Gently add PBS preheated to 37 °C along the wall of 1.5 ml Eppendorf tube, and carefully remove the supernatant.
6. Maintain the cells in PBS without glucose for 1 h to allow consumption of intracellular glucose.
7. Prepare glucose stock solution by dissolving 90 mg glucose ml^{-1} in PBS with (500 m*M* glucose), dilute 100 μl of the stock solution with 900 μl of PBS to make the maximal standard concentration (50 m*M*), and then make serial dilutions for a series of glucose solutions with concentration: 50, 20, 10, 5, 2, 1, 0.5, 0.2, 0.1, 0.05, and 0.02 m*M*.
8. Add 50 μl of different concentrations of glucose to 96-well black flat-bottom microplates.
9. Count cells with a hemacytometer, and then add 50 μl of cell suspensions to 96-well black flat-bottom microplates using a dispenser.

10. Mix immediately and then measure the dual-excitation ratios of the Frex sensor using a synergy 2 Multi-Mode Microplate Reader with 420 and 485 nm excitation and 528 nm emission for both excitation wavelengths.
11. Correct the fluorescence values by subtracting the intensity of 293FT cell samples not expressing Frex.
12. Correct the Frex-Mit fluorescence signal by cpYFP fluorescence measured in parallel experiments to eliminate the effects of pH variation.

5. DATA HANDLING/PROCESSING

For microplate reader assays, samples containing equal numbers of Frex-expressing cells or untransfected (control) cells were both measured. Fluorescence values were background-corrected by subtracting the intensity of 293FT cell samples not expressing Frex. Unlike endogenous NAD(P)H fluorescence, it is possible that the free NADH concentration in different cells and different organelles can be quantified and compared after calibration of Frex fluorescence in live cells. For the calibration of cytosolic NADH levels, Frex-expressing cells were resuspended in PBS buffer, pH 7.4, containing 0.001% digitonin. The fluorescence (excited at 485 nm) of cells or cell lysates in the presence or absence of 100 μ*M* NADH was then measured.

For imaging, raw data were exported to Image J software as 12-bit TIF for analysis. The pixel-by-pixel ratio of the 488 nm excitation image by the 405 nm excitation image of the same cell were used to pseudocolor the images in HSB gray space. Simply, the RGB value (255, 0, 255) represents the lowest ratio, and the red (255, 0, 0) represents the highest ratio, while the gray brightness is proportional to the fluorescence signals in both channels. In order to assess Frex sensor occupancy by NADH in the mitochondrial matrix, a calibration standard of sensor protein with or without NADH should be obtained using the same microscopy setting. For the calibration of mitochondrial NADH level, Frex protein was diluted into 100 m*M* KPi buffer with 0.1% bovine serum albumin (pH 8.0) to a final concentration of 0.1 μ*M*.

6. SUMMARY

This chapter presents methods for intracellular NADH detection using genetically encoded fluorescence sensors. This protocol is adaptable to determine NADH levels using Frex sensors by a variety of instruments (microplate reader, fluorescence spectrofluorimeter, flow cytometers, laser scanning confocal microscope, or fluorescence microscope). Compared to endogenous NAD(P)H fluorescence, Frex sensors are highly specific, manifest a large change in fluorescence upon NADH binding, and can be targeted to different subcellular compartments. These advantages allow for real-time imaging of cytosolic and nuclear NADH concentrations, which are very useful for better understanding of cellular redox metabolism and signaling involving NADH.

ACKNOWLEDGMENTS

We thank Stephanie Tribuna for expert secretarial assistance. Work supported by the 973 Program (2013CB531200), National Natural Science Foundation of China (Grants 31225008, 31170815, and 31071260), Key Project of Shanghai Innovation of Science and Technology Plan (NO 12JC1402900), Dawn Program of Shanghai Education Commission (Grant 11SG31), Doctoral Fund of Ministry of Education of China, and by NIH Grants HL 061795, HL 108630, HL 070819, and HL 048743 (to J.L.).

REFERENCES

Balaban, R. S., Nemoto, S., & Finkel, T. (2005). Mitochondria, oxidants, and aging. *Cell, 120*, 483–495.

Belousov, V. V., Fradkov, A. F., Lukyanov, K. A., Staroverov, D. B., Shakhbazov, K. S., Terskikh, A. V., et al. (2006). Genetically encoded fluorescent indicator for intracellular hydrogen peroxide. *Nature Methods, 3*, 281–286.

Berg, J., Hung, Y. P., & Yellen, G. (2009). A genetically encoded fluorescent reporter of ATP:ADP ratio. *Nature Methods, 6*, 161–166.

Bjornberg, O., Ostergaard, H., & Winther, J. R. (2006). Mechanistic insight provided by glutaredoxin within a fusion to redox-sensitive yellow fluorescent protein. *Biochemistry, 45*, 2362–2371.

Blinova, K., Carroll, S., Bose, S., Smirnov, A. V., Harvey, J. J., Knutson, J. R., et al. (2005). Distribution of mitochondrial NADH fluorescence lifetimes: Steady-state kinetics of matrix NADH interactions. *Biochemistry, 44*, 2585–2594.

Brekasis, D., & Paget, M. S. (2003). A novel sensor of NADH/NAD+ redox poise in *Streptomyces coelicolor* A3(2). *EMBO Journal, 22*, 4856–4865.

Cannon, M. B., & James Remington, S. (2009). Redox-sensitive green fluorescent protein: Probes for dynamic intracellular redox responses. A review. *Methods in Molecular Biology, 476*, 50–64.

Chen, S., Whetstine, J. R., Ghosh, S., Hanover, J. A., Gali, R. R., Grosu, P., et al. (2009). The conserved NAD(H)-dependent corepressor CTBP-1 regulates *Caenorhabditis elegans* life span. *Proceedings of the National Academy of Sciences of the United States of America, 106*, 1496–1501.

Delaunay, A., Pflieger, D., Barrault, M. B., Vinh, J., & Toledano, M. B. (2002). A thiol peroxidase is an H2O2 receptor and redox-transducer in gene activation. *Cell, 111*, 471–481.

Dooley, C. T., Dore, T. M., Hanson, G. T., Jackson, W. C., Remington, S. J., & Tsien, R. Y. (2004). Imaging dynamic redox changes in mammalian cells with green fluorescent protein indicators. *Journal of Biological Chemistry, 279*, 22284–22293.

Dumollard, R., Ward, Z., Carroll, J., & Duchen, M. R. (2007). Regulation of redox metabolism in the mouse oocyte and embryo. *Development, 134*, 455–465.

Eto, K., Tsubamoto, Y., Terauchi, Y., Sugiyama, T., Kishimoto, T., Takahashi, N., et al. (1999). Role of NADH shuttle system in glucose-induced activation of mitochondrial metabolism and insulin secretion. *Science, 283*, 981–985.

Gutscher, M., Pauleau, A. L., Marty, L., Brach, T., Wabnitz, G. H., Samstag, Y., et al. (2008). Real-time imaging of the intracellular glutathione redox potential. *Nature Methods, 5*, 553–559.

Gyan, S., Shiohira, Y., Sato, I., Takeuchi, M., & Sato, T. (2006). Regulatory loop between redox sensing of the NADH/NAD(+) ratio by Rex (YdiH) and oxidation of NADH by NADH dehydrogenase Ndh in *Bacillus subtilis*. *Journal of Bacteriology, 188*, 7062–7071.

Hanson, G. T., Aggeler, R., Oglesbee, D., Cannon, M., Capaldi, R. A., Tsien, R. Y., et al. (2004). Investigating mitochondrial redox potential with redox-sensitive green fluorescent protein indicators. *Journal of Biological Chemistry, 279*, 13044–13053.

Henley, K. S., & Laughrey, E. G. (1970). The redox state of the mitochondrial NAD system in cirrhosis of the liver and in chronic quantitative undernutrition in the rat. *Biochimica et Biophysica Acta, 201*, 9–12.

Hung, Y. P., Albeck, J. G., Tantama, M., & Yellen, G. (2011). Imaging cytosolic NADH-NAD(+) redox state with a genetically encoded fluorescent biosensor. *Cell Metabolism, 14*, 545–554.

Jones, D. P. (1981). Determination of pyridine dinucleotides in cell extracts by high-performance liquid chromatography. *Journal of Chromatography, 225*, 446–449.

Jones, D. P. (2008). Radical-free biology of oxidative stress. *American Journal of Physiology Cell Physiology, 295*, C849–C868.

Joubert, F., Fales, H. M., Wen, H., Combs, C. A., & Balaban, R. S. (2004). NADH enzyme-dependent fluorescence recovery after photobleaching (ED-FRAP): Applications to enzyme and mitochondrial reaction kinetics, *in vitro*. *Biophysical Journal, 86*, 629–645.

Kasimova, M. R., Grigiene, J., Krab, K., Hagedorn, P. H., Flyvbjerg, H., Andersen, P. E., et al. (2006). The free NADH concentration is kept constant in plant mitochondria under different metabolic conditions. *Plant Cell, 18*, 688–698.

Kasischke, K. A., Vishwasrao, H. D., Fisher, P. J., Zipfel, W. R., & Webb, W. W. (2004). Neural activity triggers neuronal oxidative metabolism followed by astrocytic glycolysis. *Science, 305*, 99–103.

Lin, S. J., Ford, E., Haigis, M., Liszt, G., & Guarente, L. (2004). Calorie restriction extends yeast life span by lowering the level of NADH. *Genes & Development, 18*, 12–16.

Mayevsky, A. (2009). Mitochondrial function and energy metabolism in cancer cells: Past overview and future perspectives. *Mitochondrion, 9*, 165–179.

Mayevsky, A., & Rogatsky, G. G. (2007). Mitochondrial function *in vivo* evaluated by NADH fluorescence: From animal models to human studies. *American Journal of Physiology Cell Physiology, 292*, C615–C640.

McLaughlin, K. J., Strain-Damerell, C. M., Xie, K., Brekasis, D., Soares, A. S., Paget, M. S., et al. (2010). Structural basis for NADH/NAD+ redox sensing by a Rex family repressor. *Molecular Cell, 38*, 563–575.

Menon, S. G., & Goswami, P. C. (2007). A redox cycle within the cell cycle: Ring in the old with the new. *Oncogene, 26*, 1101–1109.

Meyer, A. J., & Dick, T. P. (2010). Fluorescent protein-based redox probes. *Antioxidants & Redox Signaling, 13*, 621–650.

Nakahata, Y., Sahar, S., Astarita, G., Kaluzova, M., & Sassone-Corsi, P. (2009). Circadian control of the NAD+salvage pathway by CLOCK-SIRT1. *Science, 324*, 654–657.

Nakai, J., Ohkura, M., & Imoto, K. (2001). A high signal-to-noise Ca(2+) probe composed of a single green fluorescent protein. *Nature Biotechnology, 19*, 137–141.

Ostergaard, H., Henriksen, A., Hansen, F. G., & Winther, J. R. (2001). Shedding light on disulfide bond formation: Engineering a redox switch in green fluorescent protein. *EMBO Journal, 20*, 5853–5862.
Patterson, G. H., Knobel, S. M., Arkhammar, P., Thastrup, O., & Piston, D. W. (2000). Separation of the glucose-stimulated cytoplasmic and mitochondrial NAD(P)H responses in pancreatic islet beta cells. *Proceedings of the National Academy of Sciences of the United States of America, 97*, 5203–5207.
Pellerin, L., & Magistretti, P. J. (2004). Neuroscience. Let there be (NADH) light. *Science, 305*, 50–52.
Ramsey, K. M., Yoshino, J., Brace, C. S., Abrassart, D., Kobayashi, Y., Marcheva, B., et al. (2009). Circadian clock feedback cycle through NAMPT-mediated NAD + biosynthesis. *Science, 324*, 651–654.
Shigemori, K., Ishizaki, T., Matsukawa, S., Sakai, A., Nakai, T., & Miyabo, S. (1996). Adenine nucleotides via activation of ATP-sensitive K + channels modulate hypoxic response in rat pulmonary artery. *American Journal of Physiology, 270*, L803–L809.
Sickmier, E. A., Brekasis, D., Paranawithana, S., Bonanno, J. B., Paget, M. S., Burley, S. K., et al. (2005). X-ray structure of a Rex-family repressor/NADH complex insights into the mechanism of redox sensing. *Structure, 13*, 43–54.
Wakita, M., Nishimura, G., & Tamura, M. (1995). Some characteristics of the fluorescence lifetime of reduced pyridine nucleotides in isolated mitochondria, isolated hepatocytes, and perfused rat liver *in situ*. *Journal of Biochemistry, 118*, 1151–1160.
Wang, E., Bauer, M. C., Rogstam, A., Linse, S., Logan, D. T., & von Wachenfeldt, C. (2008). Structure and functional properties of the *Bacillus subtilis* transcriptional repressor Rex. *Molecular Microbiology, 69*, 466–478.
Williamson, D. H., Lund, P., & Krebs, H. A. (1967). The redox state of free nicotinamide-adenine dinucleotide in the cytoplasm and mitochondria of rat liver. *Biochemical Journal, 103*, 514–527.
Yang, H., Yang, T., Baur, J. A., Perez, E., Matsui, T., Carmona, J. J., et al. (2007). Nutrient-sensitive mitochondrial NAD + levels dictate cell survival. *Cell, 130*, 1095–1107.
Ying, W. (2008). NAD +/NADH and NADP +/NADPH in cellular functions and cell death: Regulation and biological consequences. *Antioxidants & Redox Signaling, 10*, 179–206.
Yu, Q., & Heikal, A. A. (2009). Two-photon autofluorescence dynamics imaging reveals sensitivity of intracellular NADH concentration and conformation to cell physiology at the single-cell level. *Journal of Photochemistry and Photobiology B, 95*, 46–57.
Zhang, Q., Piston, D. W., & Goodman, R. H. (2002). Regulation of corepressor function by nuclear NADH. *Science, 295*, 1895–1897.
Zhang, Q., Wang, S. Y., Nottke, A. C., Rocheleau, J. V., Piston, D. W., & Goodman, R. H. (2006). Redox sensor CtBP mediates hypoxia-induced tumor cell migration. *Proceedings of the National Academy of Sciences of the United States of America, 103*, 9029–9033.
Zhao, Y., Jin, J., Hu, Q., Zhou, H. M., Yi, J., Yu, Z., et al. (2011). Genetically encoded fluorescent sensors for intracellular NADH detection. *Cell Metabolism, 14*, 555–566.
Zhao, Y., & Yang, Y. (2012). Frex and FrexH: Indicators of metabolic states in living cells. *Bioengineered Bugs, 3*, 181–188.

CHAPTER NINETEEN

^{13}C Isotope-Assisted Methods for Quantifying Glutamine Metabolism in Cancer Cells

Jie Zhang[1], Woo Suk Ahn[1], Paulo A. Gameiro[1], Mark A. Keibler, Zhe Zhang, Gregory Stephanopoulos[2]

Department of Chemical Engineering, Massachusetts Institute of Technology, Cambridge, Massachusetts, USA

[1]These authors contributed equally to this paper

[2]Corresponding author: e-mail address: gregstep@mit.edu

Contents

Abstract

Glutamine has recently emerged as a key substrate to support cancer cell proliferation, and the quantification of its metabolic flux is essential to understand the mechanisms by which this amino acid participates in the metabolic rewiring that sustains the survival and growth of neoplastic cells. Glutamine metabolism involves two major routes, glutaminolysis and reductive carboxylation, both of which begin with the deamination of glutamine to glutamate and the conversion of glutamate into α-ketoglutarate. In glutaminolysis, α-ketoglutarate is oxidized via the tricarboxylic acid cycle and decarboxylated to pyruvate. In reductive carboxylation, α-ketoglutarate is reductively converted into isocitrate, which is isomerized to citrate to supply acetyl-CoA for *de novo* lipogenesis. Here, we describe methods to quantify the metabolic flux of glutamine through these two routes, as well as the contribution of glutamine to lipid synthesis. Examples of how these methods can be applied to study metabolic pathways of oncological relevance are provided.

Methods in Enzymology, Volume 542
ISSN 0076-6879
http://dx.doi.org/10.1016/B978-0-12-416618-9.00019-4

1. INTRODUCTION

In recent years, glutamine has emerged as a central precursor in the metabolism of cancer cells. Not only does glutamine, a nonessential amino acid, serve as the major mechanism of nitrogen transport into cells, but it also supplements glucose as a substantial carbon source via anaplerosis into the tricarboxylic acid (TCA) cycle (Daye & Wellen, 2012; DeBerardinis & Cheng, 2010). Given the necessity of transformed cells to perform elevated macromolecular biosynthesis to continue their growth and invasion within the body, targeting glutamine metabolism represents a promising opportunity for disrupting tumor proliferation (Vander Heiden, 2011).

As has been the case with glucose, mutated genes and malfunctioned signaling pathways in cancers have been found to influence the regulation of glutamine metabolism, including K-Ras (Gaglio et al., 2011; Son et al., 2013), p53 (Hu et al., 2010; Suzuki et al., 2010), and mTOR (Csibi et al., 2013). Most strikingly, c-Myc has been found to elicit "addiction" to the amino acid by inducing the expression of genes involved in glutamine metabolism, such as the glutamine transporter ASCT2 and glutaminase (GLS) (Gao et al., 2009; Wise et al., 2008).

Once taken up by the cell, glutamine is directed toward protein synthesis or deaminated, typically by GLS; nonproteinogenic glutamate is then converted to α-ketoglutarate via either glutamate dehydrogenase or transamination. After reaching this step, glutamine-derived α-ketoglutarate can be further metabolized along the TCA cycle through two different routes: The first, glutaminolysis, traditionally refers to oxidation of this α-ketoglutarate to malate and subsequent decarboxylation to pyruvate by malic enzyme (ME) or further oxidation to oxaloacetate by malate dehydrogenase. This progression contributes to ATP production through generation of substrates for oxidation in aerobic respiration and enables redox control from NADPH production through ME, formation of precursors for macromolecular biosynthesis such as alanine and pyruvate, or excretion of carbon as lactate by lactate dehydrogenase in Fig. 19.1 (DeBerardinis & Cheng, 2010). The second major route of glutamine metabolism, RC, has been shown to dominate in cell lines under hypoxic stress or disrupted mitochondrial functioning; in these situations, glutamine-derived α-ketoglutarate has been found to preferentially undergo reductive metabolism through the

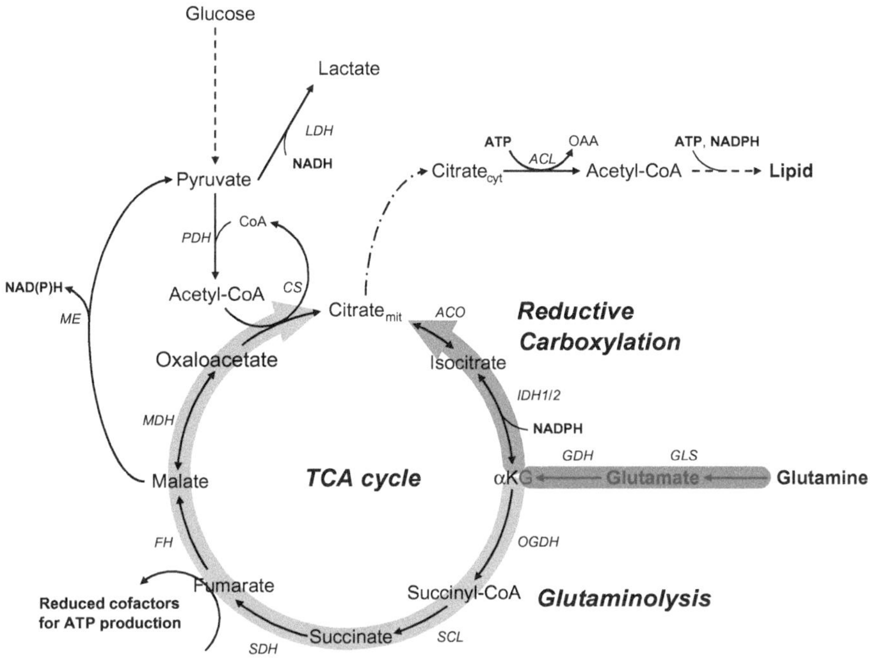

Figure 19.1 *Glutamine metabolism in mammalian cells.* The figure shows the two major routes for glutamine metabolism in mammalian cells, glutaminolysis and reductive carboxylation. In either pathway, glutamine is first deaminated to glutamate, which is then converted to α-ketoglutarate via glutamate dehydrogenase (GDH) or transamination. In glutaminolysis, α-ketoglutarate is oxidized to malate via the TCA cycle and subsequently decarboxylated to pyruvate via malic enzyme (ME) or oxidized to oxaloacetate via malate dehydrogenase (MDH). In reductive carboxylation, α-ketoglutarate is reductively converted via isocitrate dehydrogenase (IDH) to isocitrate, which is then isomerized to citrate. Abbreviations: LDH, lactate dehydrogenase; PDH, pyruvate dehydrogenase; CS, citrate synthase; ACL, ATP citrate lyase; ACO, aconitase; OGDH, oxoglutarate dehydrogenase; SDH, succinate dehydrogenase; FH, fumarase; MDH, malate dehydrogenase; GLS, glutaminase.

TCA cycle to isocitrate and then citrate, where it can then be converted to acetyl-CoA for lipid synthesis (Metallo et al., 2012; Mullen et al., 2012; Wise et al., 2011). Induction of this pathway has been shown to be controlled by mass action via conditions that perturb the citrate-to-α-ketoglutarate ratio, such as stabilization of the HIF-2α oncogene and/or oxidative energetic stress, and its activity has been demonstrated both *in vitro* and *in vivo*; targeting glutamine metabolism via GLS inhibition holds promise as a potential therapeutic strategy under these conditions, especially in combination with other anticancer drugs (Fendt et al., 2013a, 2013b; Gameiro et al., 2013).

In recognizing the significance of glutamine anaplerosis and its potentially divergent fates toward meeting the demands of either energy and combating oxidative stress or synthesizing macromolecules, it is necessary to have a means of quantifying these fates experimentally. Stable isotope labeling provides a direct readout of intracellular metabolism, and it can be combined with the known stoichiometry of biochemical pathways to estimate the activity of corresponding enzyme fluxes (Keibler, Fendt, & Stephanopoulos, 2012). We describe here how stable isotope tracers can be used to assess the use of glutamine by cancer cells for survival and proliferation.

2. METHODS

The typical flow of a stable isotopic tracer-assisted study in cancer cells is schemed as in Fig. 19.2. Depending on the goals of each experiment, various factors need to be considered in each of these steps. In this section, we describe and discuss some of the common considerations to be taken during labeling experiments using cultured cancer cells. Although we limit our discussion to *in vitro* cell culture systems, many principles, such as the selection of tracers and analyses of intracellular metabolites, can be extended to *in vivo* animal models in the study of cancer metabolism as well (Bier et al., 1977; Gameiro et al., 2013; Maher et al., 2012; Yuneva et al., 2012). (Despite these parallels, the complexity of whole-body metabolism and the logistical difficulties of live animal experiments present challenges that make such *in vivo* studies beyond the scope of this chapter.)

2.1. Design of experiment

Experimental design is critical for obtaining useful metabolic information from the cell culture system. In particular, in isotopic labeling studies, the choice of tracer determines the range of possible labeling metabolite patterns and therefore strongly influences the observability as well as accuracy of the estimated intracellular fluxes. A wide collection of stable isotopes is now available, and some investigations have been conducted to assess the effectiveness of different ^{13}C isotopic tracers to determine specific fluxes in mammalian cells, as well as the optimal label or combination of labels that should be used for a particular purpose (Metallo, Walther, & Stephanopoulos, 2009;

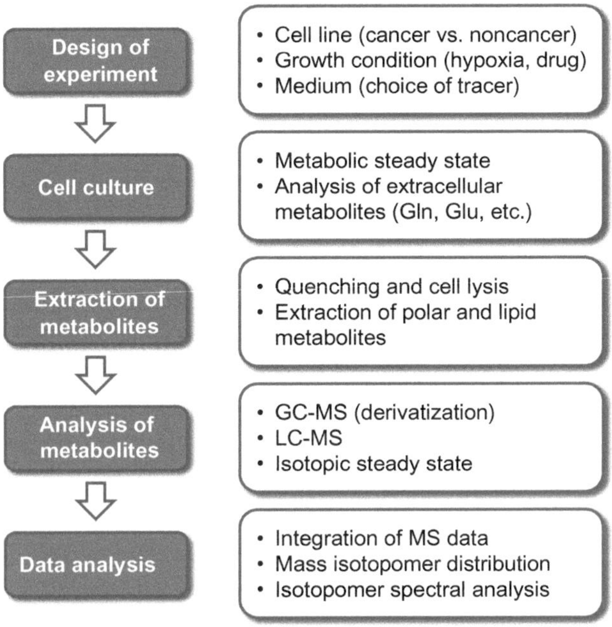

Figure 19.2 *Typical flow of isotopic tracer experiments to study cancer metabolism.* The scheme summarizes the main steps in an isotopic tracer-assisted study of cancer metabolism and common factors need to be considered, which are discussed in this section.

Walther, Metallo, Zhang, & Stephanopoulos, 2012). Here, we present typical ^{13}C-labeled glutamine tracers that are used to trace the catabolism of glutamine carbons in central carbon metabolism. Generally, [U-$^{13}C_5$]glutamine is a good tracer to effectively evaluate the total contribution of glutamine in the TCA cycle and lipogenesis (Yoo, Stephanopoulos, & Kelleher, 2004). Other partially labeled glutamine tracers such as [1-^{13}C] and [5-^{13}C]glutamine can also be useful to evaluate the fraction of glutamine that is metabolized through RC (Gameiro et al., 2013). Examples of mass isotopomer distribution (MID) analysis procedures that can be performed using these tracers are provided in Section 3.1.

2.2. Cell culture using stable isotopic tracers

Cell culture is usually performed with commercially available media formulations containing glucose, amino acids, vitamins, and inorganic salts, such as Dulbecco's modified Eagle medium (DMEM) and RPMI-1640. In stable isotope tracer experiments, specialized basal media without glucose, glutamine, and/or any other substrate of interest are used, allowing the use of specific ^{13}C-labeled tracer (typically glucose or glutamine). As is common with cancer cell lines, the culture medium is supplemented with 5–10% fetal bovine serum (FBS), which is complex and contains many unspecified molecules that may dilute ^{13}C-labeling of intracellular metabolites and interfere in cellular metabolism. Thus, in tracer experiments, normal FBS is replaced by dialyzed FBS, in which unspecified small molecules (typical cut-off, 10,000 Da) were minimized to avoid those unwanted effects.

The labeling patterns of intracellular metabolites in a stable isotope experiment are strongly influenced by many factors, including the growth conditions, medium, and particularly the metabolic status of the cells. To obtain reproducibility and comparability, it is of utmost importance not only to carry out the experiments in a consistent manner but also to ensure cells have reached metabolic steady state before quenching and extracting metabolites. Metabolic steady state is the condition in which both extracellular and intracellular metabolic fluxes, as well as intracellular metabolite concentrations, do not change with time. Generally, a constant specific growth rate (either stationary or exponential growth phase) is required for metabolic steady state, although these conditions alone do not guarantee it. Further validations could be performed to ensure that metabolic steady state is established (Ahn & Antoniewicz, 2013; Deshpande, Yang, & Heinzle, 2009).

Similar to but distinct from the concept of metabolic steady state is the isotopic steady state where metabolites reach a steady state in which the isotopic labeling pattern does not change with time. Since the incorporation of ^{13}C atoms into unlabeled metabolites is determined by the biochemical reactions of the metabolic network, it is only possible to reach isotopic steady state after a metabolic steady state has established. In addition, the metabolite pool size has a major impact on how fast it becomes enriched in isotopic label. Owing to various dynamics of isotope incorporation, different metabolites may reach isotopic steady state in distinct timescales. For example, it has been reported that using [U-$^{13}C_6$]glutamine tracer allowed metabolites in TCA cycle to reach isotopic steady state within 3 h and using [1,2-$^{13}C_2$] glucose tracer the glycolytic metabolites could reach isotopic steady state within 1.5 h at exponential and stationary phase of cell growth (Ahn & Antoniewicz, 2013).

An example protocol for an isotopic tracer experiment is given below, using the human lung adenocarcinoma epithelial A549 cell line (ATCC, CCL-185). Parameters in the following experimental procedure, such as initial cell density and incubation time with labeled substrate, have been determined from previous studies (Metallo et al., 2012), and preliminary experiments to ensure metabolic and isotopic steady state should be performed if different cell lines are used.

1. Day 0, seed 200,000 cells/well in a 6-well plate (or a 35-mm culture dish) in 2 mL DMEM containing 25 m*M* glucose, 4 m*M* glutamine, 10% (v/v) FBS, and 100 U/mL penicillin/streptomycin. Incubate the cells in a humidified incubator controlled at 5% CO_2 and 37 °C. Allow at least 6-h culture to attach cells to the plate.
2. Day 1, remove the growth media completely by aspiration and wash two times with 2 mL phosphate-buffered saline (PBS) (room temperature). Add 2 mL DMEM containing 25 m*M* glucose, 4 m*M* [U-$^{13}C_5$]glutamine, 10% (v/v) dialyzed FBS, and 100 U/mL penicillin/streptomycin. Depending on the experimental design, other tracers can be used.
3. Day 2 (24 h after replacing with tracer medium), the cells are ready for the extraction of metabolites.

2.3. Measurement of extracellular glutamine and glutamate

Different cancer cell lines, depending on the oncogenes activated or the tumor suppressors inactivated, have distinct patterns of glutamine metabolism. For example, c-Myc induces the overexpression of GLS in some cancer cells and, as a consequence, increases the glutamine uptake of those cells (Gao et al., 2009; Wise et al., 2008). Measuring consumption of glutamine provides a readout of its total use by cells. In some cases, the glutamine consumption rate is also required as a parameter for metabolic flux analysis (MFA), a methodology that can estimate intracellular metabolic fluxes based on stable isotopes (e.g., ^{13}C-isotopic tracers) and metabolic models (Stephanopoulos, 1999). However, it is beyond the scope of this chapter to describe MFA in great detail, and for further reading the readers are referred to Zamboni (2011).

The concentration of glutamine and glutamate can be measured by numerous ways, for example, enzymatic assay (Pye, Stonier, & McGale, 1978), high-performance liquid chromatography (Tcherkas, Kartsova, & Krasnova, 2001), or mass spectrometry-based analytical methods such as GC- or LC-MS (Darmaun, Manary, & Matthews, 1985; Qu et al., 2002). Here, we use the YSI 7100 Bioanalytical System (YSI Life Science) which applies immobilized enzymes to catalyze the corresponding chemical reactions to measure glutamine and glutamate. Assuming that the cells are in exponential growth phase and the culture has reached its metabolic steady state, that is, all intra- and extracellular fluxes are constant, the specific growth rate μ, specific uptake rate of glutamine q_{Gln} (it has a negative value because glutamine concentration in the medium decreases), and specific production rate of ammonium $q_{\mathrm{NH_3}}$ can be determined by solving the following equations as described previously (Glacken, Adema, & Sinskey, 1988).

$$\frac{\mathrm{d}X}{\mathrm{d}t} = \mu X \tag{19.1}$$

$$\frac{\mathrm{d}[\mathrm{Gln}]}{\mathrm{d}t} = q_{\mathrm{Gln}} X - k[\mathrm{Gln}] \tag{19.2}$$

$$\frac{\mathrm{d}[\mathrm{NH_3}]}{\mathrm{d}t} = k[\mathrm{Gln}] + q_{\mathrm{NH_3}} X \tag{19.3}$$

where k is the glutamine decomposition rate constant, which was determined to be approximately 0.0020 h^{-1} under the typical culture condition (Ahn & Antoniewicz, 2011; Ozturk & Palsson, 1990).

If measurements are available at multiple time points, it is possible to validate whether the cells are in exponential growth phase by comparing the specific growth rates or glutamine consumption rates over different periods of time. However, a constant specific growth rate does not necessarily ensure the establishment of metabolic steady state; thus, this validation only provides a partial check of the requirements.

2.4. Extraction of intracellular metabolites

While extracellular metabolite measurements only provide information on those metabolites that are consumed or secreted, intracellular metabolite concentrations contain much richer information on the state of cellular metabolism. However, analyzing for intracellular metabolites poses some extra challenges relatively to measuring extracellular metabolites. First, most intracellular metabolites cannot cross the cell membrane so that their analysis requires that the cells are broken to release the metabolites. Second, metabolites vary in the level of polarity; therefore, it is necessary to apply both polar and nonpolar solvents to extract metabolites once they are released after cell lysis. Third, in the process of cell lysis and metabolite extraction, cells are exposed to extremely harsh conditions and the metabolic steady state may be dramatically perturbed; therefore, an effective quenching procedure is of paramount importance to rapidly stop all enzymatic activities so that concentration of metabolites is not significantly altered. As such, quenching and extraction methods fall into two main categories dictated by whether the cell culture is in suspension or adherent. Of the various methods that have been developed to address these challenges, we describe one method that is simple and feasible to most biological laboratories. An example protocol for extracting metabolites from a 6-well plate is given below, and it can be modified accordingly. For other methods, the readers are referred to references (Lorenz, Burant, & Kennedy, 2011; Sellick, Hansen, Stephens, Goodacre, & Dickson, 2011).

Example protocol for extraction of metabolites:

1. After taking the 6-well plate from 37 °C incubator, quickly remove the medium by aspiration or store the spent medium at −20 °C freezer until further analysis.
2. Wash the cells with 2 mL of saline solution and remove quickly by aspiration.

   Note: Saline rather than PBS is used to avoid the interference of phosphate with the gas chromatography–mass spectrometry (GC-MS) analysis.
3. Place the plate on ice and quickly apply 400 μL of −20 °C methanol directly to the cells to quench metabolism.
4. Add 200 μL of ice-cold distilled water and briefly mix by gentle shaking.
5. Scrape the cells off using a cell scraper.
6. Transfer the methanol–water cell suspension into a microcentrifuge tube containing 400 μL of chloroform.
7. Vortex the microcentrifuge tubes vigorously at 4 °C for 10 min.
8. Separate the two phases by centrifugation at >14,000 × *g* for at least 10 min, preferably at 4 °C. Upper phase (methanol and water) contains polar metabolites and lower phase (chloroform) contains nonpolar metabolites (i.e., lipids). Cell debris, proteins, and nucleic acids are in the middle layer.
9. Transfer the metabolite fraction of interest into a clean microcentrifuge tube without disturbing the middle layer.
10. Dry the metabolite extract in a freeze dryer or a vacuum centrifuge. Alternatively, samples can be dried on an evaporator under air or nitrogen gas flow.

2.5. GC-MS analysis of intracellular metabolites

GC-MS is widely used to analyze intracellular metabolites labeled by ^{13}C isotopic tracers (Wittmann, 2002). For GC-MS analysis, the metabolites are chemically derivatized for better volatility in GC separation and analyzed by electron impact ionization in MS. The generated mass spectrum from GC-MS allows us to decipher cellular metabolism (Ahn & Antoniewicz, 2013). A broad spectrum of reagents is available for derivatization (Wittmann, 2007). Here, we provide example protocols of derivatization and GC program for both polar metabolites and fatty acids.

2.5.1 Derivatization of polar metabolites

1. Completely dry samples (in 1.5-mL centrifuge tubes) under air or nitrogen gas flow.
   Note: The presence of water will hinder the derivatization reaction!
2. Add 20 μL of 2 wt% methoxyamine hydrochloride in pyridine (MOX reagent; Thermo Scientific) to the tube and vortex for 5 s.
3. Incubate at 37 °C for 90 min on a heating block.
4. Add 25 μL of *N*-methyl-*N*-(*tert*-butyldimethylsilyl)-trifluoroacetamide + 1% *tert*-butyldimethylchlorosilane (Thermo Scientific).
5. Incubate at 60 °C for 1 h on a heating block.
6. Centrifuge at >14,000 × *g* for 10 min and transfer the supernatant into GC vials with inserts.

2.5.2 Derivatization of fatty acids

1. Add 500 μL of 2% sulfuric acid in methanol to dried fatty acid samples. Incubate samples at 60 °C for at least 2 h (no more than 8 h).
2. Add 600 μL hexane and 150–200 μL saturated NaCl to samples. Vortex (at room temperature) for at least 10 min.
3. Spin samples down for 1 min at >10,000 rpm.
4. Transfer 350–400 μL of upper (hexane) phase into a new microcentrifuge tube without disturbing the lower phase.
   Note: The lower phase contains sulfuric acid which will destroy the GC column!
5. Dry samples under air flow for at least 10 min. Add 40–50 μL hexane and resuspend the samples by vortexing.
6. Centrifuge at >10,000 × *g* for 1 min. Transfer the samples into GC vials with glass inserts.

2.5.3 GC-MS analysis

The GC-MS system that we use consists of an Agilent 6890N GC equipped with DB-35ms (30 m × 0.25 mm i.d. × 0.25 μm; Agilent J&W Scientific) capillary column and 5975B Inert XL MS system under electron ionization at 70 eV and quadrupole mass analyzer. The MS source and quadrupole were held at 230 and 150 °C, respectively. Helium was used as carrier gas at a flow rate of 1 mL/min. Here, we provide GC programs for analyzing polar metabolites and fatty acids (Table 19.1). These programs can also be optimized for different sample types or derivatization methods.

The resulting MID is obtained by integration of ion chromatograms from each set of metabolite mass fragments (Antoniewicz, Kelleher, & Stephanopoulos, 2007a). The MID is characterized as the fractional abundance of mass isotopomers defined by M0, M1 to Mn. Here, M is the base mass of an ion fragment while the following number from 0 to *n* (active carbon number) indicates the mass shift from M (Table 19.2). To account for the contribution in the labeling of a metabolite that is derived from only the ^{13}C atoms from ^{13}C isotopic tracers, MIDs need to be corrected for natural isotope abundances (Fernandez, Des Rosiers, Previs, David, & Brunengraber, 1996). This is an important step in the correct interpretation of isotopic labeling data. Intracellular metabolites can be identified by

Table 19.1 GC programs for polar metabolite and fatty acid samples

Polar metabolites	**Fatty acids**
Hold at 70 °C for 2 min	Hold at 180 °C for 2 min
Increase to 140 °C at 3 °C/min	Increase to 200 °C at 8 °C/min
Increase to 150 °C at 1 °C/min	Hold at 200 °C for 8 min
Increase to 280 °C at 3 °C/min	Increase to 280 °C at 10 °C/min
Hold at 280 °C for 6.33 min	Hold at 280 °C for 7.5 min

Table 19.2 Mass fragments of intracellular metabolites for GC-MS analysis

Metabolite	**Retention time (min)**	**Mass (*m/z*)**	**Carbon atoms**	**Fragment formula**
Pyruvate	8.8	174	1–2–3	$C_6H_{12}O_3NSi$
Succinate	24.0	289	1–2–3–4	$C_{12}H_{25}O_4Si_2$
Fumarate	24.5	287	1–2–3–4	$C_{12}H_{23}O_4Si_2$
Malate	32.5	419	1–2–3–4	$C_{18}H_{39}O_5Si_3$
Aspartate	33.7	418	1–2–3–4	$C_{18}H_{40}O_4NSi_3$
Citrate	44.0	459	1–2–3–4–5–6	$C_{20}H_{39}O_6Si_3$
Palmitate	8.5	270	1 to 16	$C_{17}H_{34}O_2$

searching against libraries of the metabolite's retention time (using the specific GC program) and its characteristic fragmentation pattern (mass spectrum at full range, e.g., *m/z* 100–600).

3. APPLICATIONS

3.1. Choice of tracers and MID analysis

Various glutamine tracers are available, and the choice depends on which specific pathway or reaction needs to be monitored. Uniformly ^{13}C-labeled glutamine ([U-$^{13}C_5$]glutamine), [1-^{13}C]glutamine, and [5-^{13}C]glutamine are good isotopic tracers to analyze the major pathways of glutamine metabolism in mammalian cells.

To trace glutamine catabolism in the TCA cycle, we can use [U-$^{13}C_5$] glutamine and [1-^{13}C]glutamine. The [U-$^{13}C_5$]glutamine tracer transfers *four* ^{13}C atoms to TCA cycle intermediates through oxidation (forward TCA cycle), generating M4 malate, fumarate, and aspartate; whereas it incorporates *five* ^{13}C atoms through RC of α-ketoglutarate, generating M5 citrate, M3 oxaloacetate, M3 aspartate, M3 malate, and M3 fumarate. Acetyl-CoA molecules originating from [U-$^{13}C_5$]glutamine are fully labeled. Therefore, [U-$^{13}C_5$]glutamine can be used to trace the overall contribution of glutamine (via RC *and* glutaminolysis) to lipid synthesis (Fig. 19.3, taken from Gameiro, 2013). Of note, since the glutaminolytic pathway generates M3 pyruvate from [U-$^{13}C_5$]glutamine (via mitochondrial ME), M5 citrate can in principle be formed via condensation of glutaminolysis-derived acetyl-CoA (M2 AcCoA) with M3 oxaloacetate (not depicted in Fig. 19.3). Thus, the M5 citrate from [U-$^{13}C_5$]glutamine can be an ambiguous readout of RC activity in cells that heavily rely on complete glutaminolysis (glutamine-to-pyruvate conversion).

To specifically trace the contribution of RC to the TCA cycle, we can use [1-^{13}C]glutamine, in which ^{13}C-labeled carbon is lost during oxidation of α-ketoglutarate but is retained on citrate and downstream metabolites through RC activity (Fig. 19.4, taken from Gameiro, 2013). To trace the contribution of RC to lipid synthesis, the [5-^{13}C]glutamine tracer should be used, as the [1-^{13}C]glutamine-derived isotopic label cannot be incorporated in acetyl-CoA through RC. [5-^{13}C]glutamine transfers one ^{13}C atom to acetyl-CoA and fatty acids through RC *only*. Although the [5-^{13}C] glutamine-derived isotopic label is transferred to TCA cycle metabolites oxidatively, it cannot be incorporated in the acetyl-CoA carbon skeleton of TCA cycle metabolites (see the acetyl-CoA moiety highlighted by dashed

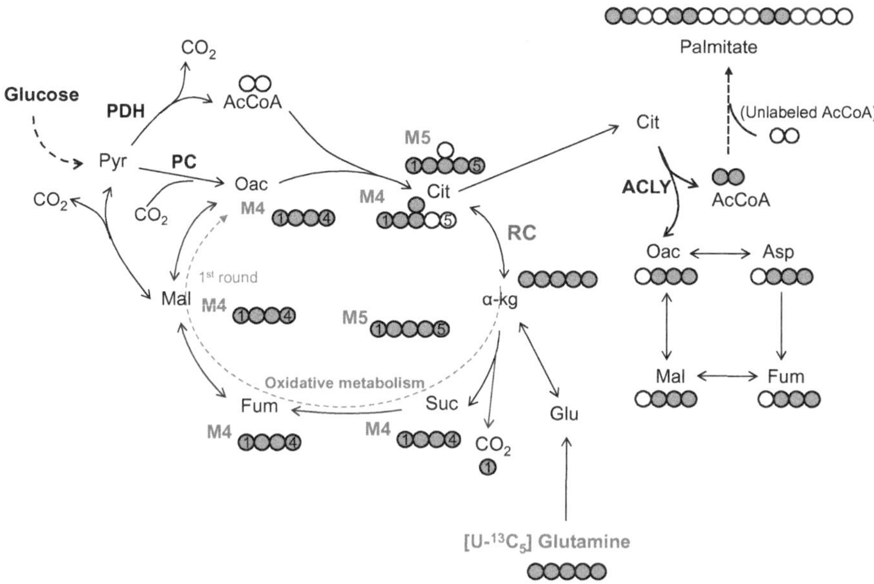

Figure 19.3 *Carbon atom transition map for [U-$^{13}C_5$]glutamine.* The map illustrates the fate of [U-$^{13}C_5$]glutamine in the TCA cycle and palmitate (^{12}C atoms are represented by empty circles and ^{13}C-atoms are represented by filled circles). Mass isotopomers generated by RC include M5 citrate (Cit), M3 oxaloacetate (Oac), M3 aspartate (Asp), M3 malate (Mal), and M3 fumarate (Fum). Mass isotopomers generated by oxidative metabolism include M4 Cit, M4 succinate (Suc), M4 Fum, and M4 Oac. Both RC and glutaminolysis (not shown) generate fully labeled acetyl-CoA (M2 AcCoA). For simplicity, labeling patterns arising from molecular symmetry and cellular compartments are not shown. The first round of the TCA cycle is illustrated, and positional labeling is depicted for relevant metabolites.

circles in Fig. 19.4). Consequently, [5-^{13}C]glutamine is specific to trace the RC-to-lipid flux.

It should also be noted that oxidation of [U-$^{13}C_5$]- or [1-^{13}C]glutamine via TCA cycle can yield ^{13}C-labeled CO_2, which can be further incorporated into some metabolites (e.g., malate) through a carboxylation reaction. Although in a cell culture system this reincorporation of ^{13}C can be considerably diluted by unlabeled bicarbonate in the media and the high level of unlabeled CO_2 maintained in the incubator, it may become necessary to account for this process if *in vivo* labeling experiments are performed.

3.2. Isotopomer spectral analysis

The principle of isotopomer spectral analysis (ISA) was initially developed by Kelleher and colleagues to quantify *de novo* synthesis of fatty acids using

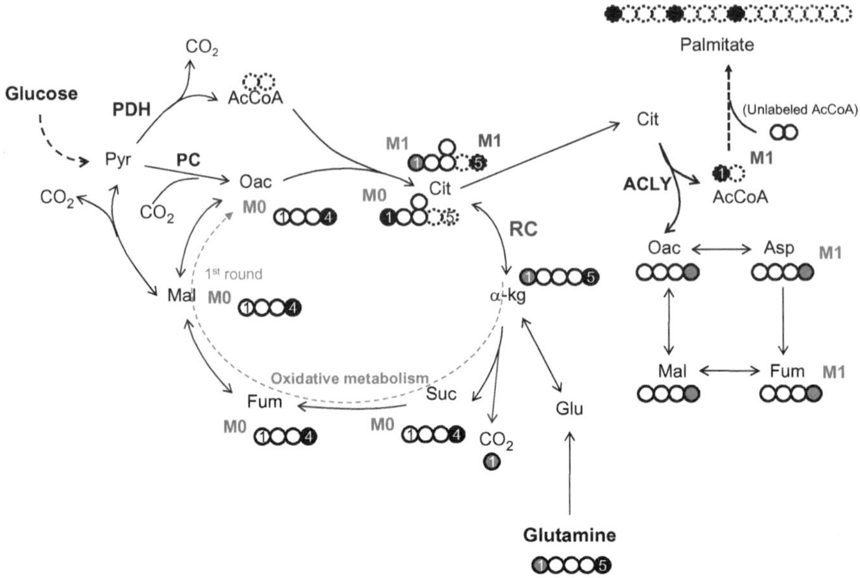

Figure 19.4 *Carbon atom transition map for [1-^{13}C]glutamine and [5-^{13}C]glutamine.* The map illustrates the fate of [1-^{13}C] and [5-^{13}C]glutamine used to trace the *reductive* TCA cycle (carbon atoms are represented by circles). The [1-^{13}C]glutamine-derived isotopic label (gray circle) is lost during oxidation of α-ketoglutarate, but it is retained during the reductive TCA cycle and transferred to citrate and downstream metabolites. [5-^{13}C] glutamine-derived isotopic label (black circle) is incorporated into acetyl-CoA and fatty acids through RC *only*; isotopic label from [5-^{13}C]glutamine cannot be transferred to fatty acids through oxidative TCA cycle. Metabolites containing the acetyl-CoA carbon skeleton are highlighted by dashed circles. For simplicity, labeling patterns arising from molecular symmetry and cellular compartments are not shown. Positional labeling is depicted for relevant metabolites only.

^{13}C-labeled acetate tracers (Kelleher & Masterson, 1992; Kharroubi, Masterson, Aldaghlas, Kennedy, & Kelleher, 1992). ISA provides a general framework to measure the contribution of different sources to the precursor pool (e.g., acetyl-CoA) of a given product (e.g., palmitate); it requires the biochemistry to follow the polymerization principle, with monomeric subunits (the precursors) condensing into a polymer (the product). There are two dimensionless parameters that are determined by the ISA approach: D, which is the fractional contribution of ^{13}C-enriched atoms of the tracer versus endogenous pathways $(1-D)$ as sources of the precursor (lipogenic acetyl-CoA); $g(t)$ indicates the fraction of newly synthesized product in the sample (e.g., palmitate) (Fig. 19.5, taken from Gameiro, 2013). ISA assumes that the precursor mixture rapidly achieves metabolic and isotopic steady state during the incubation with the tracer. A key component of the ISA

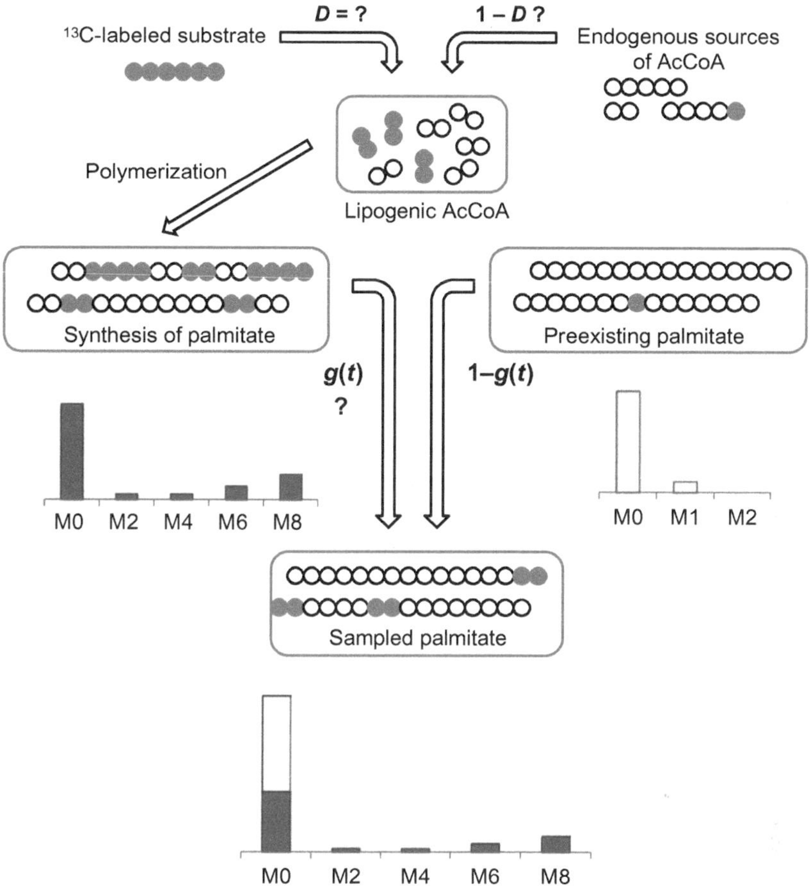

Figure 19.5 *The isotopomer spectral analysis (ISA) method.* ^{13}C-labeled glucose tracer is used to trace the glucose-to-lipid flux. The glucose-derived ^{13}C atoms and endogenous sources (e.g., acetate and glutamine) contribute to the pool of lipogenic acetyl-CoA. The *D* parameter indicates the unknown fractional contribution (or relative flux) of the glucose tracer to lipogenic acetyl-CoA. During the incubation with the tracer, palmitate is synthesized from eight molecules of acetyl-CoA. The newly synthesized palmitate mixes with an existing pool of palmitate (prior to tracer addition) and the fraction of *de novo* palmitate synthesis after a period of time, *t* is *g*(*t*). Each bar graph represents the palmitate MID. Preexisting palmitate has a significant M1 enrichment due to natural abundance. The parameters *D* and *g*(*t*) are estimated by fitting the predicted to the measured MID in the sampled palmitate.

model is that the sampled product is not required to attain isotopic steady state. The ISA model uses equations to calculate the probability of the appearance of each mass isotopomer (M0, M1, M2, etc.) in the polymer molecule, based on test values for *D* and *g*(*t*) (Kelleher & Masterson, 1992; Yoo et al., 2004). These probabilities are compared with MID of the polymer measured by GC-MS. A fitting procedure to minimize differences of measured and simulated values outputs a best estimate for D and $g(t)$. Here, we performed ISA using the elementary metabolite unit framework-based software *Metran* (Antoniewicz, Kelleher, & Stephanopoulos, 2007b; Young, Walther, Antoniewicz, Yoo, & Stephanopoulos, 2008) on measured palmitate and estimated the percent contribution to lipogenic acetyl-CoA (D value) and the percent *de novo* lipogenesis ($g(t)$ value), together with their associated confidence intervals. The confidence intervals of D and $g(t)$ values can be estimated based on measurement errors of the palmitate MID and the metabolic model of ISA (Fig. 19.5) (Antoniewicz, Kelleher, & Stephanopoulos, 2006). Since the precursor acetyl-CoA can come from both glucose-derived pyruvate and glutamine-derived α-ketoglutarate (Fig. 19.4), it is possible to use [U-$^{13}C_6$]glucose and [U-$^{13}C_5$]glutamine to determine their relative contribution to the *de novo* lipogenesis (Gameiro et al., 2013).

For calculation of the absolute *de novo* lipogenic flux (in picomoles of palmitate per cell per hour), the quantity of newly synthesized palmitate can be determined by multiplying the percentage of *de novo* lipogenesis ($g(t)$ value) by the total cellular palmitate. Similarly, the absolute flux of a given substrate (e.g., glutamine) to palmitate can be calculated by multiplying the percent contribution to lipogenic acetyl-CoA (the D value) by the amount of newly synthesized palmitate and dividing by the integral viable cell density over the course of the experiment.

3.3. Hypoxia case study

A recent example from our lab demonstrates the application of MID analysis and ISA to demonstrate the role of reductive glutamine metabolism in cancer cell *de novo* lipogenesis under hypoxia (Metallo et al., 2012). After culturing A549 lung adenocarcinoma cells in 1% O_2, we found that, while consumption of glutamine increased relative to 21% O_2 culture conditions, secretion of glutamate into the media did not change (Fig. 19.6A), suggesting an additional biosynthetic role for glutamine activated under hypoxia. To investigate, we first examined the potential fates for glutamine under normoxic conditions, namely oxidative and reductive TCA cycle metabolism. In several cancer cell lines tested, the specific contribution of RC to lipogenesis was estimated by ISA using [5-^{13}C]glutamine, which showed that RC, in contrast to glutaminolysis, accounted for the majority of total lipogenic flux from glutamine (determined by ISA using [U-$^{13}C_5$]

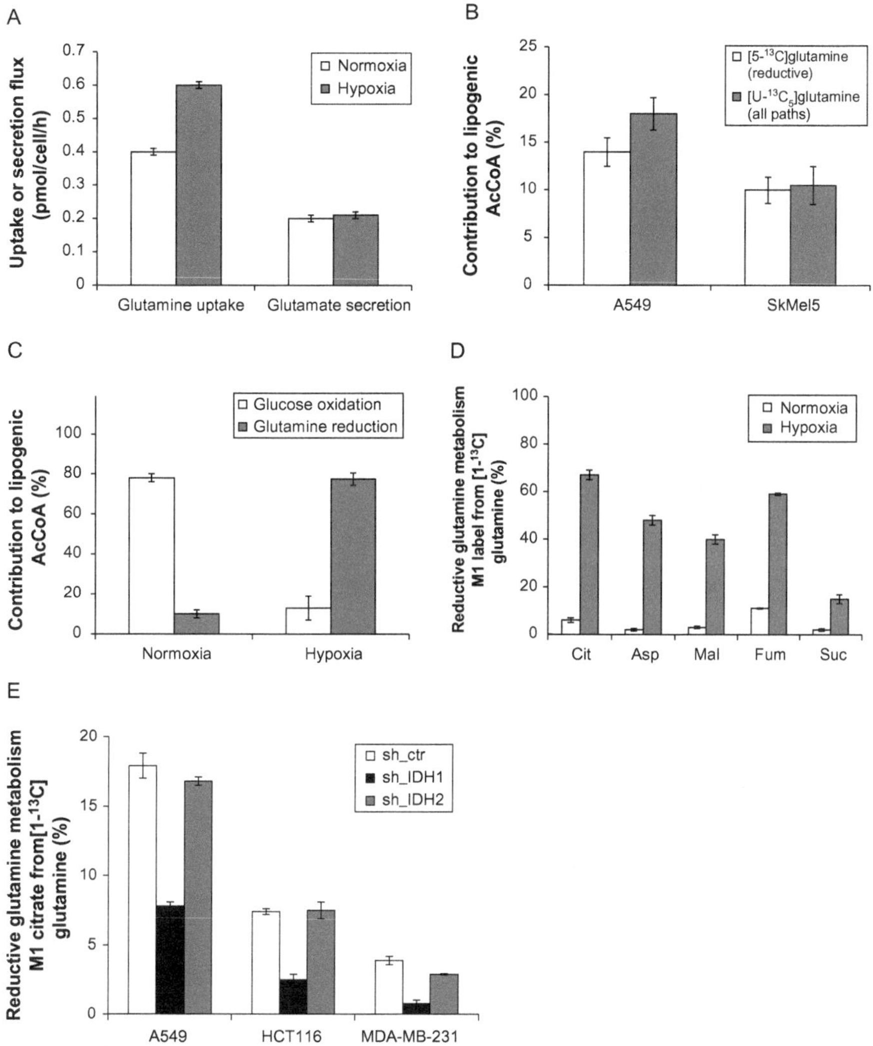

Figure 19.6 *Hypoxia case study data.* Representative data based on authors' previous study on RC and hypoxia. (A) Cell-specific uptake and secretion rates of glutamine and glutamate, respectively, in A549 cells under normoxic and hypoxic conditions. (B) Fractional contribution of total glutamine and RC-metabolized glutamine to lipogenic acetyl-CoA for multiple cancer cell lines, as determined by ISA. (C) Fractional contribution of glucose oxidation and glutamine-derived αKG reduction toward lipid synthesis in A549 cells under normoxic and hypoxic conditions. (D) Fraction of TCA cycle metabolites containing M1 label when cultured in [1-^{13}C]glutamine in A549 cells under normoxic and hypoxic conditions. (E) Small-hairpin RNA-induced knockdown of *IDH1* but not *IDH2* expression reduces citrate M1 labeling in various cancer cell lines when cultured in [1-^{13}C]glutamine.

glutamine) (Fig. 19.6B). We further observed an increase of RC flux in A549 cells under hypoxia, as indicated by a significantly higher contribution of glutamine-derived acetyl-CoA (estimated using [5-^{13}C]glutamine) as compared to the total contribution of glucose-derived acetyl-CoA (estimated using [U-$^{13}C_6$]glucose) to lipid synthesis (Fig. 19.6C), as well as a substantial increase in ^{13}C label from [1-^{13}C]glutamine retained in citrate and downstream metabolites (Fig. 19.6D). Knockdown of the isocitrate dehydrogenase (IDH) isoform encoded by *IDH1* (cytosolic, $NADP^+$), but not *IDH2* (mitochondrial, $NADP^+$), reduced ^{13}C labeling of TCA cycle intermediates in [1-^{13}C]glutamine-cultured cells (Fig. 19.6E), demonstrating that IDH1 was responsible for the RC flux under these conditions.

4. SUMMARY

Glutamine has been well known as a central precursor for protein and nucleotide synthesis in proliferating cells. However, in an anaplerotic pathway upregulated in many cancer cells, it can also be converted to α-ketoglutarate and incorporated in the TCA cycle, where it can serve as a supplementary carbon. Strikingly, under conditions of hypoxia or defective mitochondrial function, glutamine can become the major source of lipogenic acetyl-CoA through reductive carboxylation. Given the fact that glutamine has become increasingly important in understanding the metabolism of cancer cells, tools to quantify glutamine metabolism are urgently needed.

Here, we described the method of using stable isotope tracers to study the fate of glutamine and its important role for cancer cell proliferation and survival. First, we described the methods for tracer experiments, measurement of extracellular metabolites, extraction of intracellular metabolites, and GC-MS analysis. Then, we discussed optimal tracer selection for determining glutamine contribution to the TCA cycle via glutaminolysis and RC through MID analysis. Finally, we described the framework of isotopomer spectral analysis to assess the fractional contribution of glutamine to lipid synthesis.

Glutamine appears to be essential for cancer cells beyond the need as a nitrogen source but also as a key metabolite that participates in central carbon metabolism as well as energy production. However, the mechanisms by which glutamine metabolism is regulated are not fully understood. In this chapter, we describe stable isotope-assisted methods that can be applied to investigate the regulation of glutamine catabolism and how glutamine promotes cancer cell proliferation and survival.

ACKNOWLEDGMENTS

Research on cancer metabolism in Stephanopoulos Lab is funded by NIH grants 1R01DK075850-01 and 1R01CA160458-01A1. J. Z. is supported by a fellowship from Luxembourg Centre for Systems Biomedicine, University of Luxembourg. M. A. K. is funded by the David H. Koch Graduate Fellowship Fund and the Ludwig Fund for Cancer Research.

REFERENCES

Ahn, W. S., & Antoniewicz, M. R. (2011). Metabolic flux analysis of CHO cells at growth and non-growth phases using isotopic tracers and mass spectrometry. *Metabolic Engineering, 13*(5), 598–609.

Ahn, W. S., & Antoniewicz, M. R. (2013). Parallel labeling experiments with [1,2-(13)C] glucose and [U-(13)C]glutamine provide new insights into CHO cell metabolism. *Metabolic Engineering, 15*, 34–47.

Antoniewicz, M. R., Kelleher, J. K., & Stephanopoulos, G. (2006). Determination of confidence intervals of metabolic fluxes estimated from stable isotope measurements. *Metabolic Engineering, 8*(4), 324–337.

Antoniewicz, M. R., Kelleher, J. K., & Stephanopoulos, G. (2007a). Accurate assessment of amino acid mass isotopomer distributions for metabolic flux analysis. *Analytical Chemistry, 79*(19), 7554–7559.

Antoniewicz, M. R., Kelleher, J. K., & Stephanopoulos, G. (2007b). Elementary metabolite units (EMU): A novel framework for modeling isotopic distributions. *Metabolic Engineering, 9*(1), 68–86.

Bier, D. M., Arnold, K. J., Sherman, W. R., Holland, W. H., Holmes, W. F., & Kipnis, D. M. (1977). In-vivo measurement of glucose and alanine metabolism with stable isotopic tracers. *Diabetes, 26*(11), 1005–1015.

Csibi, A., Fendt, S. M., Li, C., Poulogiannis, G., Choo, A. Y., Chapski, D. J., et al. (2013). The mTORC1 pathway stimulates glutamine metabolism and cell proliferation by repressing SIRT4. *Cell, 153*(4), 840–854.

Darmaun, D., Manary, M. J., & Matthews, D. E. (1985). A method for measuring both glutamine and glutamate levels and stable isotopic enrichments. *Analytical Biochemistry, 147*(1), 92–102.

Daye, D., & Wellen, K. E. (2012). Metabolic reprogramming in cancer: Unraveling the role of glutamine in tumorigenesis. *Seminars in Cell & Developmental Biology, 23*(4), 362–369.

DeBerardinis, R. J., & Cheng, T. (2010). Q's next: The diverse functions of glutamine in metabolism, cell biology and cancer. *Oncogene, 29*(3), 313–324.

Deshpande, R., Yang, T. H., & Heinzle, E. (2009). Towards a metabolic and isotopic steady state in CHO batch cultures for reliable isotope-based metabolic profiling. *Biotechnology Journal, 4*(2), 247–263.

Fendt, S. M., Bell, E. L., Keibler, M. A., Davidson, S. M., Wirth, G. J., Fiske, B., et al. (2013). Metformin decreases glucose oxidation and increases the dependency of prostate cancer cells on reductive glutamine metabolism. *Cancer Research, 73*(14), 4429–4438.

Fendt, S. M., Bell, E. L., Keibler, M. A., Olenchock, B. A., Mayers, J. R., Wasylenko, T. M., et al. (2013). Reductive glutamine metabolism is a function of the alpha-ketoglutarate to citrate ratio in cells. *Nature Communications, 4*, 2236.

Fernandez, C. A., Des Rosiers, C., Previs, S. F., David, F., & Brunengraber, H. (1996). Correction of 13C mass isotopomer distributions for natural stable isotope abundance. *Journal of Mass Spectrometry*, *31*(3), 255–262.

Gaglio, D., Metallo, C. M., Gameiro, P. A., Hiller, K., Danna, L. S., Balestrieri, C., et al. (2011). Oncogenic K-Ras decouples glucose and glutamine metabolism to support cancer cell growth. *Molecular Systems Biology*, 7, 523.

Gameiro, P. A. (2013). *On the reprogramming of the Krebs cycle in hypoxic and VHL-deficient cancer cells*. Ph.D. Thesis, Coimbra, Portugal: University of Coimbra.

Gameiro, P. A., Yang, J., Metelo, A. M., Perez-Carro, R., Baker, R., Wang, Z., et al. (2013). In vivo HIF-mediated reductive carboxylation is regulated by citrate levels and sensitizes VHL-deficient cells to glutamine deprivation. *Cell Metabolism*, *17*(3), 372–385.

Gao, P., Tchernyshyov, I., Chang, T. C., Lee, Y. S., Kita, K., Ochi, T., et al. (2009). c-Myc suppression of miR-23a/b enhances mitochondrial glutaminase expression and glutamine metabolism. *Nature*, *458*(7239), 762–765.

Glacken, M. W., Adema, E., & Sinskey, A. J. (1988). Mathematical descriptions of hybridoma culture kinetics: I. Initial metabolic rates. *Biotechnology and Bioengineering*, *32*(4), 491–506.

Hu, W., Zhang, C., Wu, R., Sun, Y., Levine, A., & Feng, Z. (2010). Glutaminase 2, a novel p53 target gene regulating energy metabolism and antioxidant function. *Proceedings of the National Academy of Sciences of the United States of America*, *107*(16), 7455–7460.

Keibler, M. A., Fendt, S. M., & Stephanopoulos, G. (2012). Expanding the concepts and tools of metabolic engineering to elucidate cancer metabolism. *Biotechnology Progress*, *28*(6), 1409–1418.

Kelleher, J. K., & Masterson, T. M. (1992). Model equations for condensation biosynthesis using stable isotopes and radioisotopes. *The American Journal of Physiology*, *262*(1 Pt. 1), E118–E125.

Kharroubi, A. T., Masterson, T. M., Aldaghlas, T. A., Kennedy, K. A., & Kelleher, J. K. (1992). Isotopomer spectral analysis of triglyceride fatty acid synthesis in 3T3-L1 cells. *The American Journal of Physiology*, *263*(4 Pt 1), E667–E675.

Lorenz, M. A., Burant, C. F., & Kennedy, R. T. (2011). Reducing time and increasing sensitivity in sample preparation for adherent mammalian cell metabolomics. *Analytical Chemistry*, *83*(9), 3406–3414.

Maher, E. A., Marin-Valencia, I., Bachoo, R. M., Mashimo, T., Raisanen, J., Hatanpaa, K. J., et al. (2012). Metabolism of [U-13 C]glucose in human brain tumors in vivo. *NMR in Biomedicine*, *25*(11), 1234–1244.

Metallo, C. M., Gameiro, P. A., Bell, E. L., Mattaini, K. R., Yang, J., Hiller, K., et al. (2012). Reductive glutamine metabolism by IDH1 mediates lipogenesis under hypoxia. *Nature*, *481*(7381), 380–384.

Metallo, C. M., Walther, J. L., & Stephanopoulos, G. (2009). Evaluation of 13C isotopic tracers for metabolic flux analysis in mammalian cells. *Journal of Biotechnology*, *144*(3), 167–174.

Mullen, A. R., Wheaton, W. W., Jin, E. S., Chen, P. H., Sullivan, L. B., Cheng, T., et al. (2012). Reductive carboxylation supports growth in tumour cells with defective mitochondria. *Nature*, *481*(7381), 385–388.

Ozturk, S. S., & Palsson, B. O. (1990). Chemical decomposition of glutamine in cell culture media: Effect of media type, pH, and serum concentration. *Biotechnology Progress*, *6*(2), 121–128.

Pye, I. F., Stonier, C., & McGale, H. F. (1978). Double-enzymatic assay for determination of glutamine and glutamic acid in cerebrospinal fluid and plasma. *Analytical Chemistry*, *50*(7), 951–953.

Qu, J., Chen, W., Luo, G., Wang, Y., Xiao, S., Ling, Z., et al. (2002). Rapid determination of underivatized pyroglutamic acid, glutamic acid, glutamine and other relevant amino acids in fermentation media by LC-MS-MS. *The Analyst*, *127*(1), 66–69.

Sellick, C. A., Hansen, R., Stephens, G. M., Goodacre, R., & Dickson, A. J. (2011). Metabolite extraction from suspension-cultured mammalian cells for global metabolite profiling. *Nature Protocols*, *6*(8), 1241–1249.

Son, J., Lyssiotis, C. A., Ying, H., Wang, X., Hua, S., Ligorio, M., et al. (2013). Glutamine supports pancreatic cancer growth through a KRAS-regulated metabolic pathway. *Nature*, *496*(7443), 101–105.

Stephanopoulos, G. (1999). Metabolic fluxes and metabolic engineering. *Metabolic Engineering*, *1*(1), 1–11.

Suzuki, S., Tanaka, T., Poyurovsky, M. V., Nagano, H., Mayama, T., Ohkubo, S., et al. (2010). Phosphate-activated glutaminase (GLS2), a p53-inducible regulator of glutamine metabolism and reactive oxygen species. *Proceedings of the National Academy of Sciences of the United States of America*, *107*(16), 7461–7466.

Tcherkas, Y. V., Kartsova, L. A., & Krasnova, I. N. (2001). Analysis of amino acids in human serum by isocratic reversed-phase high-performance liquid chromatography with electrochemical detection. *Journal of Chromatography. A*, *913*(1–2), 303–308.

Vander Heiden, M. G. (2011). Targeting cancer metabolism: A therapeutic window opens. *Nature Reviews Drug Discovery*, *10*(9), 671–684.

Walther, J. L., Metallo, C. M., Zhang, J., & Stephanopoulos, G. (2012). Optimization of 13C isotopic tracers for metabolic flux analysis in mammalian cells. *Metabolic Engineering*, *14*(2), 162–171.

Wise, D. R., DeBerardinis, R. J., Mancuso, A., Sayed, N., Zhang, X. Y., Pfeiffer, H. K., et al. (2008). Myc regulates a transcriptional program that stimulates mitochondrial glutaminolysis and leads to glutamine addiction. *Proceedings of the National Academy of Sciences of the United States of America*, *105*(48), 18782–18787.

Wise, D. R., Ward, P. S., Shay, J. E., Cross, J. R., Gruber, J. J., Sachdeva, U. M., et al. (2011). Hypoxia promotes isocitrate dehydrogenase-dependent carboxylation of alpha-ketoglutarate to citrate to support cell growth and viability. *Proceedings of the National Academy of Sciences of the United States of America*, *108*(49), 19611–19616.

Wittmann, C. (2002). Metabolic flux analysis using mass spectrometry. *Advances in Biochemical Engineering/Biotechnology*, *74*, 39–64.

Wittmann, C. (2007). Fluxome analysis using GC-MS. *Microbial Cell Factories*, *6*, 6.

Yoo, H., Stephanopoulos, G., & Kelleher, J. K. (2004). Quantifying carbon sources for de novo lipogenesis in wild-type and IRS-1 knockout brown adipocytes. *Journal of Lipid Research*, *45*(7), 1324–1332.

Young, J. D., Walther, J. L., Antoniewicz, M. R., Yoo, H., & Stephanopoulos, G. (2008). An elementary metabolite unit (EMU) based method of isotopically nonstationary flux analysis. *Biotechnology and Bioengineering*, *99*(3), 686–699.

Yuneva, M. O., Fan, T. W., Allen, T. D., Higashi, R. M., Ferraris, D. V., Tsukamoto, T., et al. (2012). The metabolic profile of tumors depends on both the responsible genetic lesion and tissue type. *Cell Metabolism*, *15*(2), 157–170.

Zamboni, N. (2011). 13C metabolic flux analysis in complex systems. *Current Opinion in Biotechnology*, *22*(1), 103–108.

CHAPTER TWENTY

Measurement of Fatty Acid Oxidation Rates in Animal Tissues and Cell Lines

Frank K. Huynh[*], Michelle F. Green[*], Timothy R. Koves[*,†,‡], Matthew D. Hirschey[*,†,‡,§,1]

[*]Duke Molecular Physiology Institute, Duke University Medical Center, Durham, North Carolina, USA
[†]Sarah W. Stedman Nutrition and Metabolism Center, Duke University Medical Center, Durham, North Carolina, USA
[‡]Department of Medicine, Duke University Medical Center, Durham, North Carolina, USA
[§]Department of Pharmacology and Cancer Biology, Duke University Medical Center, Durham, North Carolina, USA
[1]Corresponding author: e-mail address: matthew.hirschey@duke.edu

Contents

Abstract

While much oncological research has focused on metabolic shifts in glucose and amino acid oxidation, recent evidence suggests that fatty acid oxidation (FAO) may also play an important role in the metabolic reprogramming of cancer cells. Here, we present a simple method for measuring FAO rates using radiolabeled palmitate, common laboratory reagents, and standard supplies. This protocol is broadly applicable for measuring FAO rates in cultured cancer cells as well as in both malignant and nontransformed animal tissues.

Methods in Enzymology, Volume 542
ISSN 0076-6879
http://dx.doi.org/10.1016/B978-0-12-416618-9.00020-0

1. INTRODUCTION

In order to sustain rapid cellular proliferation and growth, cancer cells undergo a complex metabolic rearrangement characterized by suppressing several catabolic pathways involved in energy production and activating anabolic processes including nucleotide and lipid biosynthesis (DeBerardinis et al., 2007; Frezza & Gottlieb, 2009; Lunt & Vander Heiden, 2011). Compared to studies on glucose or amino acid oxidation, remarkably few studies have addressed changes in lipid oxidation during oncogenesis or for cancer cell survival.

Lipids are oxidized primarily in the mitochondria in a multistep pathway called β-oxidation (or fatty acid oxidation, FAO) (McGarry & Foster, 1980). This process begins by importing long-chain fatty acyl-CoAs into mitochondria, followed by a four-step reaction. In the first step, acyl-CoA dehydrogenase oxidizes the long-chain acyl-CoA. In the second step, enoyl-CoA hydratase hydrates the acyl-chain forming hydroxy-acyl-CoA. In the third step, hydroxy-acyl-CoA dehydrogenase oxidizes the substrate forming a second keto-group. In the fourth step, thiolase cleaves acetyl-CoA and adds a free CoA to the new substrate, generating an acyl-CoA shortened by two carbons. This shortened acyl-CoA continues through the four-step reaction until the entire chain is oxidized into acetyl-CoA (Fig. 20.1). A major result of metabolic shifts toward FAO is increased ATP production, and this is often associated with stress responses and survival.

Evidence is emerging for an important role for FAO in cancer (for a review, see Carracedo, Cantley, & Pandolfi, 2013). One recent study found that cells that undergo loss of attachment from solid tumors have less glucose uptake and oxidation, which results in decreased ATP and NADPH and increased reactive oxygen species (ROS) (Schafer et al., 2009). In this study, elevated ROS inhibited FAO, while antioxidant administration counteracted ROS and reactivated FAO, increased ATP levels, and prevented cell death by anoikis, demonstrating a crucial role for antioxidants and FAO in cancer cell survival. Another study identified a novel peroxisome proliferator-activated receptor-dependent mechanism by which the promyelocytic leukemia protein regulates FAO (Ito et al., 2012). In a final example, an atypical carnitine palmitoyltransferase 1 (CPT1) isoform C was identified as a potential oncogene, outside of its normal role in the brain (Zaugg et al., 2011). CPT1 converts long-chain fatty acyl-CoAs to fatty acylcarnitines to translocate lipids into the mitochondria, where acyl-CoAs

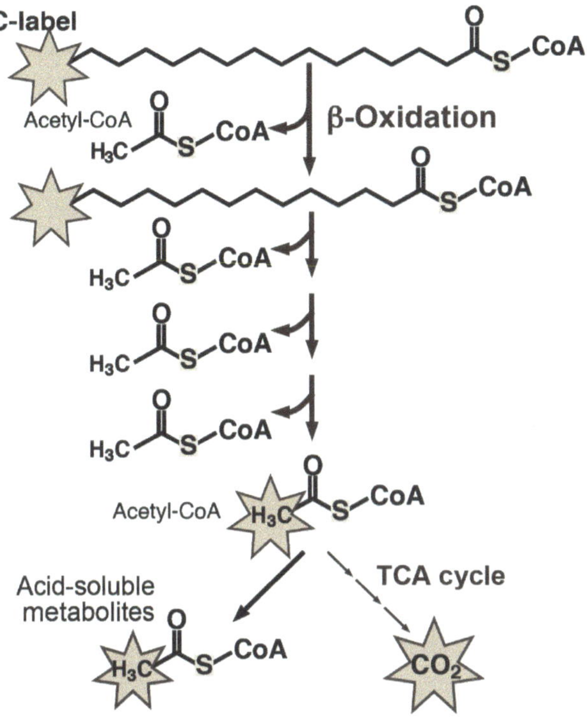

Figure 20.1 *Biochemical overview of the assay.* ^{14}C-palmitate is oxidized via β-oxidation to generate acetyl-CoA, which can then enter the tricarboxylic acid (TCA) cycle or be incorporated into acid-soluble metabolites.

are reformed to undergo FAO. In these examples, FAO is often associated with cancer cell survival, similar to postmitotic, metabolic cell types.

Thus, a picture is emerging for FAO to be an important metabolic pathway in cancer. Here, we describe detailed methods to measure mitochondrial FAO, which were adapted from previous studies (Kim, Hickner, Cortright, Dohm, & Houmard, 2000; Mannaerts, Debeer, Thomas, & De Schepper, 1979; Shindo, Osumi, & Hashimoto, 1978; Van Veldhoven et al., 1991). These methods use common supplies and reagents to measure FAO rates in either whole tissue homogenates (Fig. 20.2) or cultured cells (Fig. 20.3).

2. PROTOCOL OVERVIEW

2.1. Biochemical principle of the assay

^{14}C-labeled palmitate is oxidized via β-oxidation to generate acetyl-CoA, which can then enter the tricarboxylic acid (TCA) cycle and be oxidized

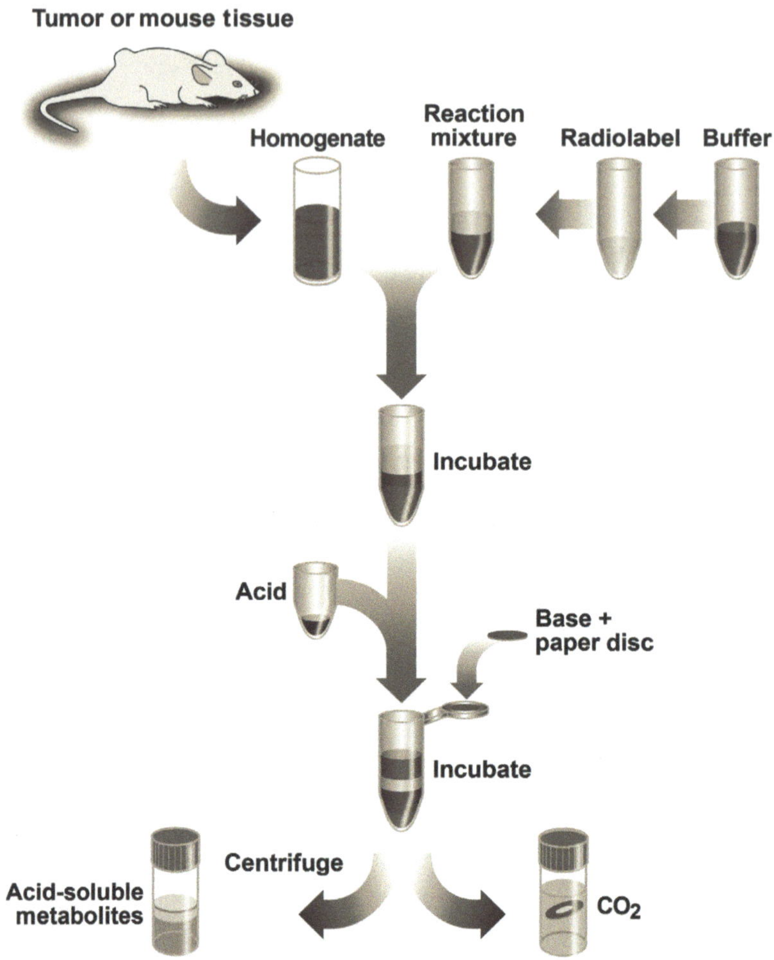

Figure 20.2 *Schematic of the protocol for tissue homogenates.* The tissue or tumor of interest is homogenized with a Dounce homogenizer to lyse cells, while keeping the mitochondria intact. The tissue homogenate is then incubated with a reaction mixture containing radiolabeled palmitate. A strong acid is added to stop the reaction. Unoxidized bound palmitate will precipitate out of solution upon addition of acid. Acetyl-CoA, acetylcarnitines, ketone bodies, and tricarboxylic acid (TCA) cycle intermediates (acid-soluble metabolites) will remain soluble after the addition of the strong acid. Acetyl-CoA that enters the TCA cycle can be oxidized to CO_2 and this will be captured by the strong base in the Whatman paper disc. The amount of ^{14}C in the acid-soluble metabolite fraction and the paper disc CO_2 fraction can each be measured by scintillation counting. (See the color plate.)

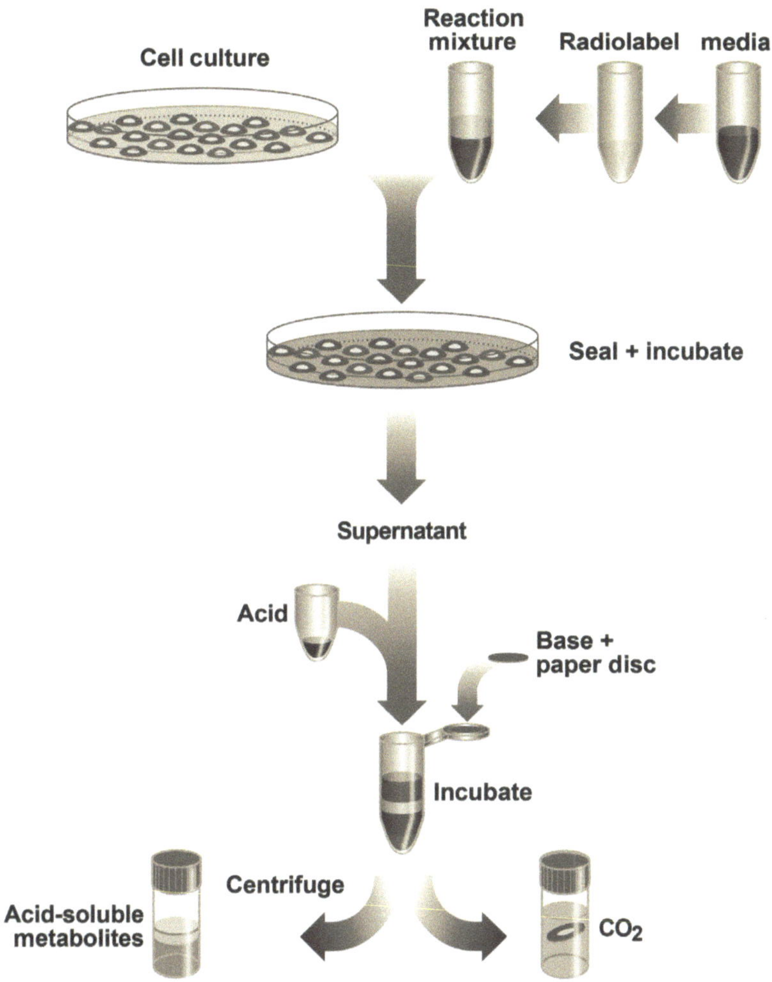

Figure 20.3 *Schematic of the protocol for cultured cells.* In a similar manner to tissue oxidation, the cells of interest are incubated in media containing ^{14}C-palmitate. The supernatant/cells are then added to a strong acid to stop the reaction and the amount of ^{14}C in the acid-soluble metabolite fraction and the paper disc CO_2 fraction can each be measured by scintillation counting. (See the color plate.)

to CO_2 (Fig. 20.1). Figure 20.1 shows an example where the ^{14}C label is at position C-16 (IUPAC numbering), which allows for the measurement of complete oxidation of palmitate. Depending on the process of interest (i.e., complete oxidation or incomplete oxidation), palmitate with the ^{14}C label at different positions can be used. Maximal signal will likely be obtained with the radiolabel at position C-1, and this reagent is more readily available. This protocol has been optimized for use with radiolabeled palmitate, but other radiolabeled fatty acids can also be used.

The radiolabel can have several fates. ^{14}C-palmitate that does not get oxidized to fatty acyl-CoAs shorter than ~6 carbons in length will precipitate out of solution upon addition of perchloric acid. Incompletely oxidized acid-soluble metabolites containing the ^{14}C radiolabel can include palmitoyl-carnitine, acetyl-carnitine, acetyl-CoA, ketone bodies (in the liver), gluconeogenic intermediates (depending on cell type), cataplerotic TCA cycle intermediates, and fatty acyl-CoAs that are shorter than 6 carbons in length (if using palmitate labeled at the C-16 position). Radiolabeled acetyl-CoA can also enter the TCA cycle and be oxidized to $^{14}CO_2$.

2.2. Overview of the protocol

Figure 20.2 provides an overview of the protocol for measuring FAO in animal tissues. Similarly, Fig. 20.3 shows the protocol adapted for cultured cells.

3. EQUIPMENT

Mouse dissection tools (forceps and scissors)
Bell jar or other anesthesia apparatus
Isoflurane or other mouse anaesthetic
2-mL Dounce homogenizer (Kimble Kontes, 8853000002)
Water bath or heating block
1.5-mL Eppendorf tubes
Whatman filter paper
Scissors
Scintillation counter, fluid, and vials
24-Well culture plates

4. MATERIALS

1 *M* sucrose
1 *M* Tris–HCl, pH 7.4
50 m*M* EDTA, pH 8.0
Radiolabeled fatty acid (e.g., 1-^{14}C-palmitate)
Nonradiolabeled fatty acid (same fatty acid as above)
30% fatty acid-free bovine serum albumin (BSA) (Sigma, A9205)
1 *M* KH_2PO_4

1 *M* KCl
1 *M* $MgCl_2$
50 m*M* L-carnitine
20 m*M* malate
10 m*M* coenzyme A
100 m*M* ATP
1 *M* dithiothreitol (DTT)
1 *M* perchloric acid or other strong acid
1 *M* sodium hydroxide or other strong base
Milli-Q or other purified water (ddH_2O)
Dulbecco's modified Eagle's medium (DMEM)

4.1. STE buffer

0.25 *M* sucrose
10 m*M* Tris–HCl
1 m*M* EDTA
pH to 7.4

4.2. Preparation of 7% BSA/5 m*M* cold palmitate mixture

BSA is required to act as a carrier for palmitate to ensure lipid solubility in an aqueous solution. Cold palmitate is included in the FAO reaction in order to increase the total rate of oxidation and signal measured in the assay. In order to prepare a solution of 7% BSA/5 m*M* palmitate (~5:1 molar ratio), do the following:

1. Prepare a 7.5% BSA solution by diluting a 30% stock (e.g., fatty acid-free BSA; Sigma, A9205) in ddH_2O. Warm to 42 °C in a water bath. Prepare 20 mL for the 7% BSA/5 m*M* palmitate solution and 20 mL for a 7% BSA only control solution.
2. Weigh 27.8 mg sodium palmitate (Sigma, P9767) and put in a 50-mL conical tube. Add 1.3 mL of ddH_2O and keep cap closed but loosened.
3. Place the tube in a boiling water bath until the fatty acids are dissolved (a few minutes).
4. Cool the palmitate until it can be held in a bare hand, but the fatty acids are still dissolved (~70 °C). If the palmitate precipitates, then warm again.
5. Immediately add 18.7 mL of the 7.5% BSA. Place in a 42 °C water bath for 30 min. For the 7% BSA only control, add 18.7 mL of the 7.5% BSA to 1.3 mL of ddH_2O.

6. If particles form and are visible to the naked eye, sonicate for 5 min. Increase the temperature to 47 °C if particles are still present, but do not go above 50 °C to avoid denaturing the BSA.
7. Store unused aliquots at −80 °C for no more than 6 months. Avoid freeze–thaw cycles.

Tip: If performing this assay for the first time, it may be useful to vary the amount of cold palmitate per reaction to find the ideal concentration of total palmitate per reaction for your sample, tissue of choice, and experimental conditions. To do this, use the 7% BSA control solution to adjust the concentration of cold palmitate. Typical concentrations of cold palmitate for tissue homogenates and cells are 0.1–0.5 m*M* and 0.3–1.0 m*M*, respectively.

4.3. Solubilization of radiolabeled palmitate into the aqueous phase

The radiolabeled palmitate is typically stored as a liquid in ethanol, which is not compatible with the FAO assay. In order to remove the ethanol and resolubilize the palmitate in BSA, perform the following:

1. Calculate the amount of radiolabeled palmitate required. Each sample should be assayed in triplicate with 0.4 μCi per reaction for animal tissues and 0.2 μCi per well for cell lines.

   Tip: When optimizing this assay, up to 1–2 μCi labeled palmitate can be used to increase the signal.

   Tip: Be sure to include enough for a triplicate blank sample as well as a triplicate measurement of the total radioactivity input per reaction.
2. Dry down the radiolabeled palmitate in an Eppendorf tube under nitrogen gas. The ethanol will evaporate and leave a film of radiolabeled palmitate at the bottom of the tube.
3. For tissue samples, resolubilize the radiolabeled palmitate in the appropriate amount of 7% BSA/5 m*M* palmitate warmed to 37 °C to get a solution of 7% BSA/5 m*M* palmitate/0.01 μCi/μL ^{14}C-palmitate. For cultured cells, use the 7% BSA control solution to adjust the 7% BSA/5 m*M* palmitate solution to a concentration of 7% BSA/2.5 m*M* palmitate. Resolubilize the radiolabeled palmitate in the appropriate amount of 7% BSA/2.5 m*M* palmitate that when added to the required amount of DMEM will give a final media concentration of 0.3% BSA/100 μ*M* palmitate/0.4 μCi/mL ^{14}C-palmitate. 500 μL of media will be required for each well of a 24-well plate.
4. Heat the radioactive BSA/palmitate solution at 37 °C for at least 30 min with frequent vortexing to ensure complete solubilization. Alternatively,

the solution can be heated at 37 °C overnight in a thermomixer to aid solubilization.

Tip: The radioactive BSA/palmitate solution can be prepared the day before and stored at 37 °C overnight.

5. PROTOCOL

5.1. FAO in animal tissues

1. Prepare the oxidation reaction mixture according to Table 20.1 and keep at 37 °C for 1–2 h.
2. Prechill the required amount of sucrose–Tris–EDTA (STE) buffer at 4 °C.
3. Prechill the 2 mL Dounce homogenizer on ice.
4. Anesthetize the mouse (e.g., isoflurane) according to institutional animal safety protocols.
5. Quickly remove the tissue(s) of interest and euthanize the mouse.

   Tip: For liver, about 200 mg of liver should be enough. For heart tissue, use the whole heart. A tumor xenograft could also be used. The appropriate amount to collect for other tissues should be determined empirically. More tissue homogenate is always better to increase the signal in this assay.
6. Rinse the tissue several times in ice-cold STE buffer to remove any blood.

   Tip: For more collagenous tissues, such as heart tissue, use a razor blade and mince the tissue in ice-cold STE buffer which will facilitate homogenization by the Dounce homogenizer.
7. Place the tissue directly into the 2-mL Dounce homogenizer containing approximately 5 volumes of chilled STE buffer.

   Tip: For 200 mg of liver, use 1 mL of STE buffer. For one regular-sized mouse heart, use 0.8 mL of STE buffer.
8. Homogenize liver tissue using 5 strokes of the loose-fitting A pestle of the Dounce homogenizer (5 down and 5 up-strokes).

   Tip: For tough, firm tissues such as the heart, use the A pestle and perform 1 firm down stroke right to the bottom of the homogenizer. For the up-strokes, twist the A pestle while gently pulling up in order to avoid forming a strong vacuum pressure that will damage the mitochondria. The first complete up-stroke can take 1–2 min to perform, but each successive up-stroke will become easier. Repeat this process up to 25 times for heart tissue to ensure thorough homogenization.

Table 20.1 Oxidation reaction mixture for animal tissues

Component	Concentration of stock	Volume of stock per 400 μL reaction (μL)	Final concentration per reaction mixture
Sucrose[a]	1 *M*	40	100 m*M*
Tris–HCl, pH 7.4	1 *M*	4	10 m*M*
KH_2PO_4	1 *M*	2	5 m*M*
EDTA, pH 8.0	50 m*M*	1.6	0.2 m*M*
KCl	1 *M*	32	80 m*M*
$MgCl_2$	1 *M*	0.4	1 m*M*
L-Carnitine	50 m*M*	16	2 m*M*
Malate	20 m*M*	2	0.1 m*M*
Coenzyme A	10 m*M*	2	0.05 m*M*
ddH_2O	–	221.6	–
pH to 8.0			
ATP[b]	100 m*M*	8	2 m*M*
DTT	1 *M*	0.4	1 m*M*
BSA:palmitate solution	7% BSA/5 m*M* palmitate/0.01 μCi/μL ^{14}C-palmitate	40	0.7% BSA/500 μ*M* palmitate/0.4 μCi ^{14}C-palmitate
Tissue homogenate	–	30	30 μL/rxn

[a]Components highlighted in dark grey (blue in online) can be combined ahead of time and stored at 4 °C for up to 6 months. Heat the solution to 37 °C before adding the components highlighted in light grey (green in online).
[b]Components highlighted in light grey (green in online) should be added fresh. After the addition of these reagents, incubate the reaction mixture at 37 °C for 1–2 h to ensure complete solubilization of the palmitate.

9. Pour the homogenate from the Dounce homogenizer into a prechilled Eppendorf tube and place on ice.
10. Continue until all tissues of interest are homogenized.

    Tip: No more than 30 min should pass between the time the tissue is collected and when the FAO reaction is started. Waiting longer than 30 min will result in a poor assay signal. Working quickly will ensure the integrity of the mitochondria and strengthen the signal in the assay. It may be useful to have more than one person and more than one Dounce homogenizer to expedite tissue processing.

Tip: If harvesting tissues from several animals, the whole tissue is more stable than the homogenate. Leave tissues intact in ice-cold STE buffer until all tissues are harvested and then process homogenates as quickly as possible.

11. Centrifuge the crude homogenates at 420 × *g* for 10 min at 4 °C.
12. Decant the supernatant containing crude mitochondria and place in fresh, chilled Eppendorf tubes. Keep on ice.
13. Resuspend the pellet in 400 μL STE buffer and centrifuge again at 420 × *g* for 10 min at 4 °C. Decant the supernatant and pool with the previous supernatant. (This step is optional but may release additional mitochondria trapped in the low-speed pellet.)
14. Discard the pellet.
15. Take a small aliquot of homogenate and set aside for a BCA assay to measure protein concentration.

    Tip: A 1:10–1:50 dilution of each tissue homogenate should be sufficient for the BCA assay.
16. Add 30 μL of tissue homogenate to fresh, chilled Eppendorf tubes and keep on ice.

    Tip: Each tissue sample should be assayed in triplicate. Be sure to include a triplicate blank sample containing 30 μL of STE buffer *in lieu* of tissue homogenate.
17. Immediately add 370 μL of the oxidation reaction mixture.
18. Incubate at 37 °C for 30 min.
19. While the reaction is incubating, prepare acidification vials with a $^{14}CO_2$ trap. First, cut out small round discs of Whatman filter paper that are slightly larger than the inside of a cap of an Eppendorf tube. Use a cap to make an imprint on the filter paper, which acts as a cutting guide. Stuff the filter paper into the inside of the cap of a fresh Eppendorf tube, ensuring that it does not easily fall out.
20. Add 200 μL of 1 *M* perchloric acid into the tube.

    Tip: Perform the acid and base steps in a fume hood.
21. To the same tube containing acid, add 20 μL of 1 *M* NaOH or other strong base to the paper disc in the cap.

    Tip: Make sure that the paper disc will not fall out when the cap is closed. Use the large bore of a pipette tip to ensure the filter paper is firmly placed. The caps should be open when the strong base is applied to the filter paper resting in the tube top (Fig. 20.2).
22. When the reaction is done incubating, transfer the entire reaction mixture into the tubes containing perchloric acid and the paper disc.

23. Quickly close the cap. Work quickly but gently to prevent the paper disc from falling out.
24. Incubate for 1 h at room temperature.
    Tip: Gently shaking on an orbital shaker can increase the signal, but also risks the paper disc from dislodging from the cap.
    Tip: Incubating for more than 1 h does not result in greater signal and often causes the paper disc to fall into the acid solution. If this happens, the sample is not salvageable.
25. Carefully open the reaction vial in a fume hood and transfer the paper disc to a scintillation vial.
    Tip: A pipette tip or small disposable tweezers can be used to transfer the paper disc to the scintillation vial.
26. Centrifuge the remaining acid solution at maximum speed ($\geq 14{,}000 \times g$) for 10 min at 4 °C.
27. Transfer 400 μL of the supernatant to a scintillation vial.
28. Additionally, add 3× 370 μL of nonincubated reaction mixture to three scintillation vials to get a triplicate measurement of the amount of radioactivity input into each reaction.
    Tip: This will be used to calculate the specific activity, which will allow for converting counts per minute into moles of palmitate.
29. Add 4 mL of scintillation fluid to all scintillation vials and measure the average counts per minute over 3 min with a standard scintillation counter.

5.2. FAO in cell lines

1. Plate cells in 24-well plates and grow until confluent.
   Tip: This step needs to be optimized depending on the cell type. Growing cells have very different metabolic needs than confluent cells. For example, MEFs become senescent when they reach confluency and therefore should be used at a stage immediately prior to confluence (e.g., approximately 85%; equal to 50,000 cells/well plated 48 h before the assay).
2. Add the BSA/palmitate mixture prepared in Section 4.3 to the required amount of prewarmed DMEM to obtain DMEM containing 0.3% BSA/100 μ*M* palmitate/0.4 μCi/mL ^{14}C-palmitate.
3. Add 1 m*M* carnitine to the media.
   Tip: Depending on the radiolabeled substrate used and the cell type, serum-free media may help to increase the signal in the assay.

4. Rinse cells with PBS two times to remove residual media.
5. Add 500 μL of media containing radiolabeled palmitate to each well.
6. Seal the plate with parafilm and incubate at 37 °C for 3 h.

   Tip: Because the plate is sealed with the addition of HEPES, incubation can be done in either a CO_2 or standard incubator.
7. During the incubation, set up the acidification vials and $^{14}CO_2$ trap as described in Steps 19–21 of Section 5.1.
8. At the end of the incubation, add 400 μL of the media into the acidification vial to stop the reaction and quickly close the tube.

   Tip: Cells could be scraped from the plate and placed in acidification media to determine the amount of acid-soluble metabolites remaining in the cells. Alternatively, cells can be used for protein determination or other metabolite measurements.
9. Measure the radioactivity in the captured $^{14}CO_2$ and acid-soluble metabolites as outlined in Steps 23–29 of Section 5.1.
10. In order to normalize to the amount of protein in each well, collect the cells and perform a BCA assay.

   Tip: Make sure to take special care when measuring protein concentration in cells containing radiolabeled palmitate. Alternatively, a separate replicate plate of cells that does not get ^{14}C-palmitate added could be used to estimate protein concentrations per well.

5.3. Alternative method of $^{14}CO_2$ capture

If the above carbon capture method is not successful, this alternative method of $^{14}CO_2$ capture has also been used with success.

1. Cut ~5 mm off of the bottom tip of a 200-μL pipette tip.
2. Cut the lid off of a 0.25-mL microdialysis tube (Fisher, 02-681-230) and press firmly into the top of the 200 μL tip.
3. Add 200 μL of 1 *M* NaOH to the microdialysis tube.
4. At the end of the reaction incubation, transfer 400 μL of either the tissue reaction mixture or the cell culture media into a 5-mL disposable culture tube (Simport, T400-3).
5. Gently place the tip/tube assembly into the culture tube and cover with a rubber septum (Sigma, Z124567).
6. Using a 23-gauge 1.5-in. needle and syringe, inject 100 μL of 1 *M* perchloric acid through the rubber gasket into the reaction mixture while avoiding introducing acid into the NaOH in the trap tube.
7. Place the culture tubes upright in a rack and shake on an orbital shaker for 1 h.

8. Transfer the 200 μL of NaOH containing trapped $^{14}CO_2$ to a scintillation vial.
9. Add 4 mL of scintillation fluid to the scintillation vials and measure the average counts per minute over 3 min with a scintillation counter.
10. Transfer the reaction mixture containing acid-soluble metabolites to an Eppendorf tube and allow to settle at 4 °C overnight. Centrifuge at 10,000 × *g* for 10 min at room temperature.
11. Transfer 150 μL of the reaction mixture to a scintillation vial.
12. Add 4 mL of scintillation fluid to the scintillation vials and measure the average counts per minute over 3 min with a scintillation counter.

Tip: To further increase the signal obtained from either animal tissues or cell culture, perform the assay on isolated mitochondria instead of whole tissue lysates or cultured cells. A detailed method for isolating intact mitochondria has been previously published (Frezza, Cipolat, & Scorrano, 2007).

6. DATA HANDLING AND CALCULATIONS

Determine the specific activity of the reaction mixture by dividing the disintegrations per minute (DPM) of the input by the number of nanomoles of palmitate (cold + hot) per reaction. The specific activity can be used to convert DPM to nanomoles for the other samples.

$$\text{Specific activity} = \frac{\text{DPM of input}}{\text{total nmol of palmitate}}$$

To determine the rate of conversion of ^{14}C-palmitate to $^{14}CO_2$, divide the nanomoles of $^{14}CO_2$ by the amount of time the reaction was allowed to run before addition of perchloric acid. The rate of conversion can also be normalized to the amount of protein or milligrams of tissue per reaction. A similar calculation can be done to determine the rate of conversion of ^{14}C-palmitate to acid-soluble metabolites.

ACKNOWLEDGMENTS

We would like to acknowledge the members of the Hirschey and Muoio Labs for thoughtful feedback. We would also like to acknowledge the funding support to the Hirschey Lab from the American Heart Association grants 12SDG8840004 and 12IRG9010008, The Edward Mallinckrodt, Jr. Foundation, The Ellison Medical Foundation, and the National Institutes of Health grants AA022146 and AG045351. M. F. G. is supported by a Post-doctoral Fellowship from the NIH/NCI (CA059365-19). F. K. H. is supported by an American Diabetes Association/Canadian Diabetes Association Joint Post-doctoral Fellowship (PF-3-13-4342-FH).

REFERENCES

Carracedo, A., Cantley, L. C., & Pandolfi, P. P. (2013). Cancer metabolism: Fatty acid oxidation in the limelight. *Nature Reviews. Cancer*, *13*, 227–232.

DeBerardinis, R. J., Mancuso, A., Daikhin, E., Nissim, I., Yudkoff, M., Wehrli, S., et al. (2007). Beyond aerobic glycolysis: Transformed cells can engage in glutamine metabolism that exceeds the requirement for protein and nucleotide synthesis. *Proceedings of the National Academy of Sciences of the United States of America*, *104*, 19345–19350.

Frezza, C., Cipolat, S., & Scorrano, L. (2007). Organelle isolation: Functional mitochondria from mouse liver, muscle and cultured fibroblasts. *Nature Protocols*, *2*, 287–295.

Frezza, C., & Gottlieb, E. (2009). Mitochondria in cancer: Not just innocent bystanders. *Seminars in Cancer Biology*, *19*, 4–11.

Ito, K., Carracedo, A., Weiss, D., Arai, F., Ala, U., Avigan, D. E., et al. (2012). A PML-PPAR-delta pathway for fatty acid oxidation regulates hematopoietic stem cell maintenance. *Nature Medicine*, *18*, 1350–1358.

Kim, J. Y., Hickner, R. C., Cortright, R. L., Dohm, G. L., & Houmard, J. A. (2000). Lipid oxidation is reduced in obese human skeletal muscle. *American Journal of Physiology. Endocrinology and Metabolism*, *279*, E1039–E1044.

Lunt, S. Y., & Vander Heiden, M. G. (2011). Aerobic glycolysis: Meeting the metabolic requirements of cell proliferation. *Annual Review of Cell and Developmental Biology*, *27*, 441–464.

Mannaerts, G. P., Debeer, L. J., Thomas, J., & De Schepper, P. J. (1979). Mitochondrial and peroxisomal fatty acid oxidation in liver homogenates and isolated hepatocytes from control and clofibrate-treated rats. *The Journal of Biological Chemistry*, *254*, 4585–4595.

McGarry, J. D., & Foster, D. W. (1980). Regulation of hepatic fatty acid oxidation and ketone body production. *Annual Review of Biochemistry*, *49*, 395–420.

Schafer, Z. T., Grassian, A. R., Song, L., Jiang, Z., Gerhart-Hines, Z., Irie, H. Y., et al. (2009). Antioxidant and oncogene rescue of metabolic defects caused by loss of matrix attachment. *Nature*, *461*, 109–113.

Shindo, Y., Osumi, T., & Hashimoto, T. (1978). Effects of administration of di-(2-ethylhexyl)phthalate on rat liver mitochondria. *Biochemical Pharmacology*, *27*, 2683–2688.

Van Veldhoven, P. P., Vanhove, G., Vanhoutte, F., Dacremont, G., Parmentier, G., Eyssen, H. J., et al. (1991). Identification and purification of a peroxisomal branched chain fatty acyl-CoA oxidase. *The Journal of Biological Chemistry*, *266*, 24676–24683.

Zaugg, K., Yao, Y., Reilly, P. T., Kannan, K., Kiarash, R., Mason, J., et al. (2011). Carnitine palmitoyltransferase 1C promotes cell survival and tumor growth under conditions of metabolic stress. *Genes & Development*, *25*, 1041–1051.

CHAPTER TWENTY-ONE

Methods to Assess Lipid Accumulation in Cancer Cells

Jørgen Sikkeland*, Yang Jin*, Fahri Saatcioglu*,†,1
*Department of Biosciences, University of Oslo, Oslo, Norway
†Institute for Cancer Genetics and Informatics, Oslo University Hospital, Oslo, Norway
1Corresponding author: e-mail address: fahris@ibv.uio.no

Contents

Methods in Enzymology, Volume 542
ISSN 0076-6879
http://dx.doi.org/10.1016/B978-0-12-416618-9.00021-2

Abstract

Oncogenesis and tumor progression are associated with significant alterations in cellular metabolism. One metabolic pathway that is commonly deregulated in malignant cells is *de novo* lipogenesis. Lipogenesis is indeed highly upregulated in several types of cancer, a phenomenon that is linked to tumor progression and poor prognosis. Steroid hormones play an essential role in the growth of a variety of cancers and have been shown to increase the expression and activity of several lipogenic factors, including fatty acid synthase and sterol regulatory element-binding proteins. Such an altered gene expression profile promotes lipid biogenesis and may result in the accumulation of neutral lipids, which become visible as cytoplasmic lipid droplets. By using breast and prostate cancer cells exposed to steroid hormones as a model, here we describe methods for the direct qualitative and quantitative assessment of neutral lipid accumulation in malignant cells.

1. INTRODUCTION

Alterations in metabolic pathways to meet increased demand for energy and components for growth and proliferation have now been recognized as a hallmark for cancer (Cairns, Harris, & Mak, 2011; Hanahan & Weinberg, 2011; Tennant, Duran, & Gottlieb, 2010). The best-recognized metabolic change in cancer cells is the escalated utilization of glucose for glycolysis compared with normal cells (the "Warburg effect"; Cairns et al., 2011; Tennant et al., 2010). This increase in energy production fuels elevated *de novo* protein, DNA, and fatty acid (FA) synthesis (Menendez & Lupu, 2007).

High proliferation in cancer cells necessitates *de novo* synthesis of FAs for a steady supply of lipids for membrane production, alternate routes of energy production, and palmitoylation of signaling proteins. *De novo* lipogenesis is highly active in embryogenesis, but in adult cells the biosynthesis of FA by the key enzyme fatty acid synthase (FASN) is mainly confined to cells with active lipid metabolism and/or hormone sensitivity (e.g., cells from liver, adipose tissue, prostate, and mammary glands; Kusakabe et al., 2000). Many human cancers express high levels of FASN and display upregulated lipogenesis that is linked to poor prognosis, such as in colon, breast, and prostate cancers (Kuhajda, 2006; Swinnen, Brusselmans, & Verhoeven, 2006). In earlier *in vitro* studies, FASN upregulation in breast cancer cells was observed upon progestin treatment (Chalbos, Chambon, Ailhaud, & Rochefort, 1987). Later studies documented that testosterone also induced FASN expression along with accumulation of triglyceride (TG) synthesis in both breast and prostate cancer cells (Chambon, Rochefort, Vial, & Chalbos, 1989; Swinnen, Esquenet, Goossens, Heyns, & Verhoeven, 1997; Swinnen, Van Veldhoven, Esquenet, Heyns, & Verhoeven, 1996). Increase in lipid droplet formation was later also described in other cancer types, including colon adenocarcinoma (Accioly et al., 2008), invasive squamous cervical carcinoma (Than et al., 2003), and human brain tumors (Opstad, Bell, Griffiths, & Howe, 2008). Characterization of lipogenesis and its potential role in cancer is still in progress. Nevertheless, emerging evidence suggest that proteins associated with lipid droplets, their formation, or lipid droplets themselves may affect cancer progression, and could be useful as prognostic markers or therapeutic targets in cancer treatment (Bozza & Viola, 2010; Mashima, Seimiya, & Tsuruo, 2009; Zhao, Butler, & Tan, 2013). This underscores the importance of lipogenesis analysis in cancer cells.

Here, we provide a detailed protocol to directly detect and quantify accumulation of neutral lipids *in vitro*, exemplified by staining of human breast and prostate cancer cells after treatment with steroid hormones, using three different lipid-specific dyes: Oil Red O, HCS LipidTOX Red, and AdipoRed. In addition, protocols are provided for concomitant assessment of changes in lipogenic gene expression.

2. MATERIALS

2.1. Cell culture and steroid treatment

1. T47D and LNCaP (hormone-sensitive breast and prostate cancer cell lines, respectively) cells were obtained from the American Type Culture Collection and maintained in medium supplemented with 10% fetal bovine serum (FBS). Cells were propagated in 100-mm plastic dishes at 37 °C in a humidified 5% CO_2 + 95% air incubator.
2. Regular medium: RPMI 1640 with L-glutamine (Lonza), supplemented with 50 U/ml penicillin/streptomycin (Lonza).
3. Fetal bovine serum.
4. 1× Phosphate-buffered saline (PBS), pH 7.4: Mix 150 m*M* sodium chloride (NaCl), 6.7 m*M* disodium hydrogen phosphate dihydrate, 1.85 m*M* sodium dihydrogen phosphate monohydrate in distilled water, adjust to pH 7.4, and sterilize by autoclaving.
5. Trypsin EDTA (Lonza).

6. Synthetic androgen R1881 (Dupont-NEN) and progesterone (Sigma), for both 1000 × stock (10^{-5} *M*) dissolved in ethanol (see *Note 1*).
7. Activated charcoal (Sigma)-treated FBS (CT-FBS).

2.2. RNA isolation

1. Trizol (Invitrogen).
2. Chloroform.
3. Isopropyl alcohol.
4. Diethylpyrocarbonate (DEPC)-treated water.
5. 70% ethanol in DEPC-treated water.
6. NanoDrop (Thermo Scientific) or equivalent spectrophotometer.

2.3. cDNA synthesis

1. Superscript II Reverse Transcriptase (RT) and 5 × RT buffer (Invitrogen).
2. 0.1 *M* 1,4-Dithiothreitol (DTT).
3. Poly(dT) oligonucleotides dissolved in sterilized distilled H_2O (sdH_2O) to a final concentration of 0.5 mg/ml.
4. dNTP Mix (10 m*M* each).
5. RNasin (Promega).

2.4. Quantitative PCR amplification

1. Lightcycler 480 SYBR green I Master mix (Roche Diagnostics).
2. Lightcycler 480 Instrument (Roche Diagnostics).
3. Lightcycler 480 96-well plates (Roche Diagnostics).
4. Oligonucleotide primers are dissolved in sdH_2O to a stock concentration of 10 μ*M*. Primer sequences are: *GAPDH*: forward 5′-GGC CTC CAA GGA GTA AGA CC-3′, reverse 5′-AGG GGT CTA CAT GGC AAC TG-3′. *FASN*: forward 5′-GCA CCT CTC AGG CAT CGA C-3′, reverse 5′-CTG TGG TCC CAC TTG ATG AG-3′.

2.5. Fixing cells

1. 4% (w/v) Paraformaldehyde (PFA): Make fresh. Prepare 1 ml per well (six-well plate). Dissolve PFA in PBS at 70 °C with a few drops of 1 *M* NaOH. When all is dissolved, cool down to room temperature and neutralize solution with 1 *M* HCl to pH 7.4 (see *Note 2*).
2. 1 × PBS, pH 7.4.

2.6. Mounting of cover slips

1. Mowiol mounting medium:
   - Mix 2.4 g poly(vinyl alcohol; Mowiol) with 6 g glycerol in a 50-ml tube.
   - Add 6 ml sdH_2O and rotate at room temperature for 2 h.
   - Add 12 ml of 200 m*M* Tris–HCl, pH 8.5, and incubate at 50 °C until the Mowiol is dissolved.
   - Clarify by centrifugation at 1900 RCF for 20 min at room temperature.
   - Add 1,4-diazabicyclo[2.2.2]octane (DABCO) from a 20% (w/v) stock (dissolved in sdH_2O water) to a final concentration of 0.2% (w/v).
   - Aliquot and store at −20 °C.
2. Tissue paper.
3. Glass slides.

2.7. Oil Red O staining

1. Oil Red O solution stock: Dissolve 0.5 g Oil Red O (Sigma) in 200 ml isopropyl alcohol at 56 °C for 1 h. Cool down before further use. Store at room temperature.
2. Oil Red O working solution: Make fresh the day of use. Mix four parts distilled water with six parts of stock solution. Let solution rest for at least 15 min. Precipitate will form. Filter solution through a filter paper, for example, 3MM Chr Whatman paper. Keep at room temperature until use.
3. 1 × PBS, pH 7.4.
4. 60% (v/v) isopropyl alcohol in distilled water.

2.8. Oil Red O extraction

1. Isopropyl alcohol.
2. Spectrophotometer or colorimeter.

2.9. HCS LipidTOX Red staining

1. LipidTOX Red staining solution: Dilute HCS LipidTOX Red (Invitrogen) 1:200 in PBS. Prepare 40 μl per cover slip; allow 10% extra for pipetting errors.
2. 0.2% Triton X-100 in PBS (v/v).
3. Incubation chamber: Prepare a dark chamber big enough to store all cover slips. On base of the chamber, place a parafilm layer and secure

it with tape. In the inner sides of the chamber, place moist tissue paper (see *Note 3*).
4. 1 × PBS, pH 7.4.
5. 4′,6-Diamidino-2-phenylindole (DAPI) staining solution: Prepare a stock dissolving DAPI in sdH_2O at room temperature at 1 mg/ml, aliquot in small volumes (e.g., 200 μl) to prevent unnecessary freeze/thawing and light exposure, and store at −80 °C. DAPI working solution is a 1:1000 dilution of the stock in PBS. Prepare 1 ml per well (six-well plate) +10% to allow for pipetting errors.
6. Any light-proof cover, for example, aluminum foil.
7. Tissue paper.

2.10. AdipoRed assay

1. 1 × PBS, pH 7.4.
2. AdipoRed reagent (Lonza) diluted 3:100 in PBS. Prepare 2 ml per well (six-well plate); allow 10% for pipetting errors. Make fresh just before use.
3. Fluorometric plate reader.

3. METHODS

3.1. Cell culture

Use the same culture conditions for LNCaP and T47D cells.

1. Grow sufficient amount of cells. Wash with PBS, trypsinize, and determine the cell number using a hemocytometer.
2. Plate out 1×10^5 cells into each well of a six-well plate in regular growth medium. If the cells are to be used for qualitative imaging, add cover slips to the bottom of each well before plating the cells (see *Note 4*).
3. After 1 day, change medium (regular growth medium supplemented with 5% CT-FBS) for steroid starvation.
4. Two days later, add hormone (final concentrations: 10^{-8} *M* for both R1881 and progesterone) and culture the cells for 96 h before collection/further manipulation (see *Note 5*).

3.2. RNA isolation

This and the following two sections (Sections 3.3 and 3.4) explain in detail a method for isolating RNA from cultured cells, converting it to cDNA, and quantitatively measuring gene expression of interest (e.g., *FASN*).

1. Wash cells once with PBS, and add 0.5 ml Trizol to each well (six-well plate). Let the cells dissolve for 5 min and collect the solution in a 1.5-ml tube.
2. Add 0.1 ml chloroform and mix by vortexing for 15 s.
3. After incubating 2–3 min at room temperature, centrifuge tubes for 15 min at 12,000 RCF at 4 °C.
4. Transfer the phenol-free transparent top layer, which contains the RNA, to a new tube and mix with 0.25 ml 100% isopropyl alcohol.
5. After 10 min of incubation at room temperature, centrifuge the samples for 10 min at 12,000 RCF at 4 °C.
6. Wash the RNA pellets, which should be visible, with 70% ethanol and air-dry for approximately 5 min.
7. Dissolve the pellets in DEPC-treated water and determine the total RNA concentration by spectrophotometric analysis at 260 nm.

3.3. cDNA synthesis

1. Mix 2 μg of total RNA, 0.5 μl poly(dT) primers, and 0.5 μl dNTP mix in PCR tubes to a final volume of 5 μl in DEPC-treated water.
2. Incubate samples at 65 °C for 5 min and cool down on ice.
3. Prepare a master mix containing 0.25 μl RT (50 units), 2 μl 5× RT buffer, 0.25 μl RNasin, 1 μl of 0.1 *M* DTT, and 1.5 μl DEPC-treated water per sample and add to the tubes from above followed by 50 min incubation at 42 °C. A control sample containing no RT should be included to test for genomic DNA contamination in the RNA samples.
4. Inactivate the enzymatic activity by 15 min incubation at 70 °C. Store at −20 °C until use.

3.4. Quantitative PCR amplification

1. Make a cDNA pool using 2 μl nondiluted cDNA from each sample and make serial dilutions of 1:1, 1:10, 1:100, and 1:1000 with sdH_2O.
2. Dilute the remaining cDNA 1:10 with sdH_2O.
3. Make a mixture of 5 μl SYBR green I Master mix, 0.5 μl forward primer, 0.5 μl reverse primer, and 4 μl diluted cDNA (or 4 μl sdH_2O as a negative control) and add to a 96-well plate. Load the plate in the PCR machine and run with the following program:

Denaturation	95 °C for 5 min
Amplification (45 cycles)	95 °C for 10 s
	62 °C for 20 s
	72 °C for 30 s, single acquisition
Melting curve analysis	95 °C for 5 s
	65 °C for 1 min
	Continuous acquisition until 97 °C
Cooling	40 °C for 30 s

4. To confirm exclusion of nonspecific PCR by-products, perform a melting curve analysis for each reaction. The change in fluorescence of SYBR Green dye in every cycle gives the crossing point (CP) values, defined as the points at which the fluorescence increased appreciably above background. Use the CP values for each sample to calculate the relative CP value to that of the control group, while also correcting for primer-specific reaction efficiency with an internal standard curve. Normalize the relative value to the expression of one or several reference genes (in this experiment, *GAPDH* was used) to compensate for variations in the quantity or quality of the starting mRNA as well as for differences in RT efficiency. Figure 21.1 shows the relative mRNA expression of *FASN* in LNCaP (A) and T47D (B) cells in the absence or presence of 10^{-8} *M* R1881 or progesterone.

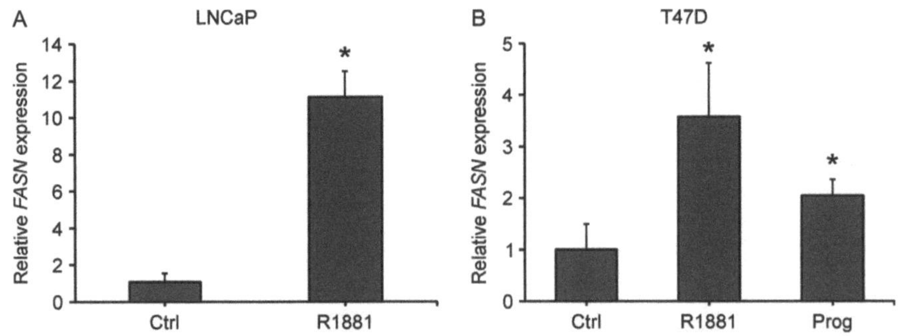

Figure 21.1 *Steroid treatment increases mRNA expression of* FASN *in LNCaP (A) and T47D (B) cells.* Cells were cultured in medium containing 5% CT-FBS for 48 h and treated with 10^{-8} *M* R1881, 10^{-8} *M* progesterone (Prog), or vehicle (Ctrl) for 96 h upon which RNA was isolated. Relative mRNA levels were determined by qPCR and normalized to the expression of the reference gene *GAPDH*. $^{*}p < 0.05$.

3.5. Fixing cells

There are several agents and procedures for cell fixation. This section describes a method that works well for the subsequent cell staining alternatives found in Sections 3.7 and 3.9.

1. Wash cells very carefully with PBS as they are easy to detach from the plate.
2. Add 1 ml of 4% PFA per well and incubate for 20 min at room temperature on a level surface with no shaking or tilting.
3. Discard the PFA in a hazardous waste chamber and wash cells twice for 5 min in PBS (see *Note 6*).

3.6. Mounting of cover slips

1. Add a 40 μl drop of mounting solution in the middle of a glass slide (see *Note 7*).
2. Remove the cover slip from the six-well plate using tweezers and dry the edges by blotting with tissue paper. Then place the cover slip on the mounting solution (side with cells facing down) taking care not to trap bubbles.
3. Incubate 2 h at room temperature in the dark to let the mounting solution solidify.
4. Wipe off any precipitate on the cover slip with a moist tissue paper.
5. Keep the slides in the dark at 4 °C.

3.7. Oil Red O staining

Oil Red O is a good and cost-effective staining agent for both quantitative and qualitative measurement of lipid droplet formation. It is, however, less efficient for high throughput analyses.

1. Fix cells and remove PBS (see Section 3.5).
2. Wash cells with 60% isopropyl alcohol.
3. Add 1.5 ml of Oil Red O working solution per well and incubate for 15–30 min. Observe under microscope until cells are properly stained. Make sure to stop the staining when/if precipitate starts to form. Wash cells with 60% isopropyl alcohol (see *Note 8*).
4. Wash cells with 1 ml PBS and leave in PBS until further use.
5. Mount cover slips on glass slides (see Section 3.6) or use the stained cells for quantification by Oil Red O extraction (see Section 3.8). The Oil Red O-stained cells can be photographed directly, or higher magnification images of cells on the mounted cover slips can be obtained using any

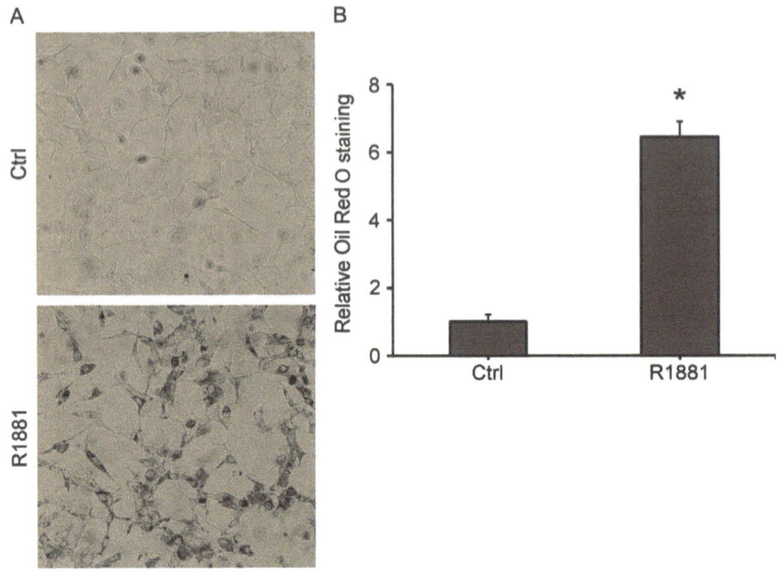

Figure 21.2 *Increased lipid accumulation in LNCaP cells in response to androgen treatment detected by Oil Red O.* LNCaP cells were cultured in medium containing 5% CT-FBS for 48 h and treated with 10^{-8} *M* R1881 or vehicle (Ctrl) for 96 h, fixed with PFA, and stained with Oil Red O. (A) Representative bright-field images of stained cells. (B) Quantified Oil Red O staining after isopropyl alcohol extraction. $^{*}p < 0.05$. (See the color plate.)

microscope equipped with an RGB camera. Figure 21.2A shows a representative image of Oil Red O-stained LNCaP cells that were either left untreated or treated with R1881 (10^{-8} *M*) for 96 h. Figure 21.4A shows a representative image of Oil Red O-stained T47D cells that were treated for 96 h with R1881 (10^{-8} *M*), progesterone (10^{-8} *M*), or left untreated.

3.8. Oil Red O extraction and quantification

1. Remove PBS from stained cells.
2. Add 1 ml of isopropyl alcohol and incubate on a tilting board for 5 min.
3. Transfer the isopropyl alcohol solution containing the extracted Oil Red O solution to a 1.5-ml tube and spin down at 2300 RCF for 1 min at room temperature. Transfer supernatant to a new tube.
4. Quantify Oil Red O levels by spectrophotometric or colorimetric analysis at 518 nm (see *Note 9*). Use isopropyl alcohol as blank control.

Figure 21.2B shows the relative absorbance of Oil Red O in LNCaP cells that were either left untreated or treated with R1881 (10^{-8} *M*) for 96 h.

3.9. HCS LipidTOX Red staining

This section describes cell staining with HCS LipidTOX Red (Invitrogen). Alternative colors of the same reagent and similar reagents from other sources are available and should work just as well. HCS LipidTOX Red can also be applied to live cell imaging.

1. Fix cells (see Section 3.5) and remove the PBS.
2. Keep the cells in 0.2% Triton X-100 on a tilting board for 15 min for permeabilization (see *Note 10*).
3. Wash cells 2 × 5 min with PBS.
4. Place 40 μl of LipidTOX Red staining solution on the parafilm in the incubation chamber (see *Note 11*).
5. Remove the cover slip from the well using tweezers and dry by blotting the edges with tissue paper.
6. Place the cover slip on the drop of LipidTOX Red staining solution (cells facing down) being careful not to generate bubbles. Close the box and incubate at room temperature for 1 h. Minimize the light exposure to the cells from this point on.
7. Bring the cover slip out of the chamber and back to the six-well plate (cells facing up).
8. Wash cells 5 min in PBS.
9. Add 1 ml of DAPI staining solution per well and stain for 5 min in the dark.
10. Wash cells 3 × 5 min in PBS with a cover.
11. Mount the cover slips on glass slides (see Section 3.6).
12. Use a fluorescence microscope to observe the stained cells. HCS LipidTOX Red has absorption and emission wavelengths that are similar to Alexa Fluor 594, of 590 and 617 nm, respectively. DAPI has absorption and emission maximum at 358 and 491 nm, respectively. Figure 21.3A shows a representative overlay of DAPI and HCS LipidTOX Red-stained LNCaP cells that were either left untreated or treated with R1881 (10^{-8} *M*) for 96 h. Figure 21.5A shows a representative overlay of DAPI and HCS LipidTOX Red-stained T47D cells that were treated for 96 h with R1881 (10^{-8} *M*), progesterone (10^{-8} *M*), or left untreated.

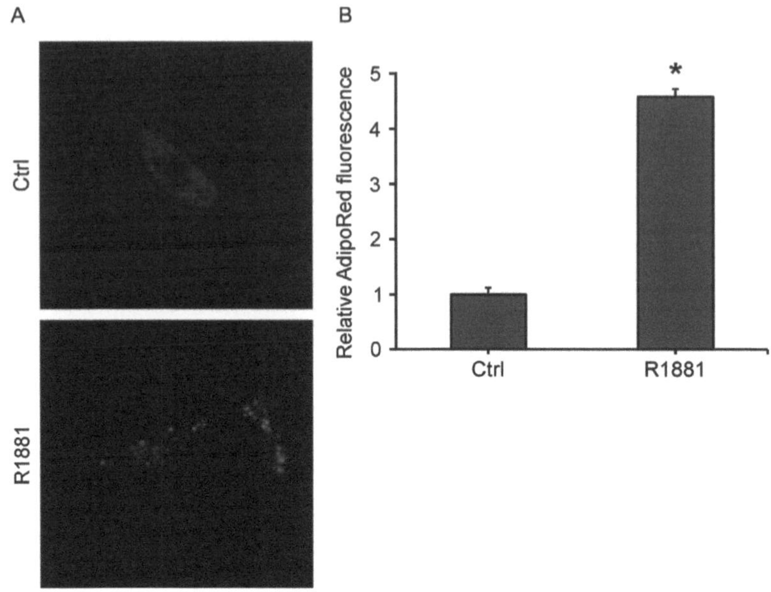

Figure 21.3 *Increased lipid accumulation in LNCaP cells in response to androgen treatment detected by HCS LipidTOX Red and AdipoRed.* LNCaP cells were cultured in medium containing 5% CT-FBS for 48 h and treated with 10^{-8} *M* R1881 or vehicle (Ctrl) for 96 h. (A) Representative fluorescence confocal images of cells stained with HCS LipidTOX Red (red) and DAPI (blue). (B) Relative fluorimetric detection using the AdipoRed assay. $^{*}p < 0.05$. (See the color plate.)

3.10. AdipoRed assay

The AdipoRed kit used here is designed for the detection of neutral lipids in adipocytes, but is sensitive enough to detect small lipid droplets formed in viable T47D and LNCaP cells.

1. Wash cells carefully with PBS.
2. Add 2 ml of diluted AdipoRed reagent mix per well (six-well plate) and incubate for 20–30 min at room temperature (see *Note 12*).
3. Quantify AdipoRed staining of neutral lipids using a fluorimeter with excitation at 485 nm and emission at 572 or 535 nm. Figure 21.3B shows the AdipoRed staining in LNCaP cells that were either left untreated or treated with R1881 (10^{-8} *M*) for 96 h.

3.11. Image quantification

The images (e.g., *.tif) of cells stained with Oil Red O or HCS LipidTOX Red can be used for quantification as an alternative to quantification using Oil Red O extraction or AdipoRed assay. There are many different methods and softwares available for this purpose. Here, we describe two easy approaches to image quantification using the freely available software ImageJ (download at http://rsbweb.nih.gov/ij).

1. Determination of Oil Red O-stained area:
   - Open the image in ImageJ.
   - Define image area covered by cells with *free hand selection*, lock (Ctrl+T) and measure the area (Ctrl+M).
   - Define the area covered by Oil Red O by adjusting the saturation threshold of the image (Shift+T). Use the same adjustment settings for all images.
   - Convert the image to binary and measure the remaining area within the previously defined cell area.
   - The Oil Red O area: Cell area ratio can be used for comparison of culture conditions.

   Figure 21.4B shows the relative Oil Red O-stained area of T47D cells left in the absence or presence of 10^{-8} *M* R1881 or progesterone for 96 h.

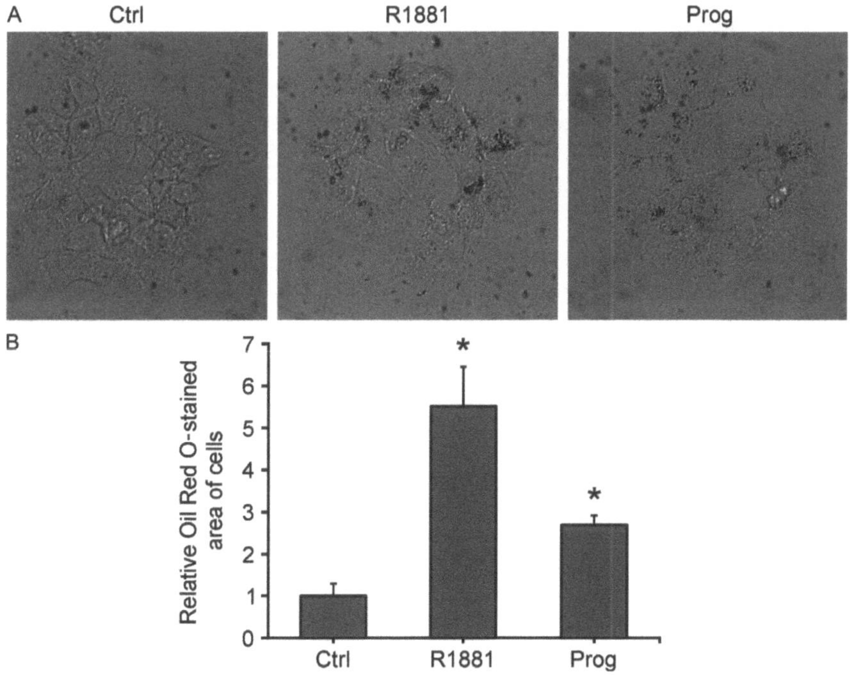

Figure 21.4 *Increased lipid accumulation in T47D cells in response to steroid treatment detected by Oil Red O.* T47D cells were cultured in medium containing 5% CT-FBS for 48 h and treated with 10^{-8} *M* R1881, 10^{-8} *M* progesterone (Prog), or vehicle (Ctrl) for 96 h, fixed with PFA, and stained with Oil Red O. (A) Representative bright-field images of stained cells. (B) Relative Oil Red O-stained area of the cells in (A) quantified using ImageJ. $^{*}p<0.05$. (See the color plate.)

2. Determination of lipid droplets detected by HCS LipidTOX Red:
- Open the image in ImageJ. The HCS LipidTOX Red signal should be in a separate channel or image.
- Exclude background signals by adjusting the signal threshold (Shift + T). Use the same adjustment settings for all images.
- Measure the number of lipid droplets using the *Analyze Particles* option. Use the same parameters for all images (see *Note 13*).
- Normalize the number of detected lipid droplet particles per cell nucleus.

Figure 21.5B shows the relative number of lipid droplets per nucleus in T47D cells left in the absence or presence of 10^{-8} *M* R1881 or progesterone for 96 h.

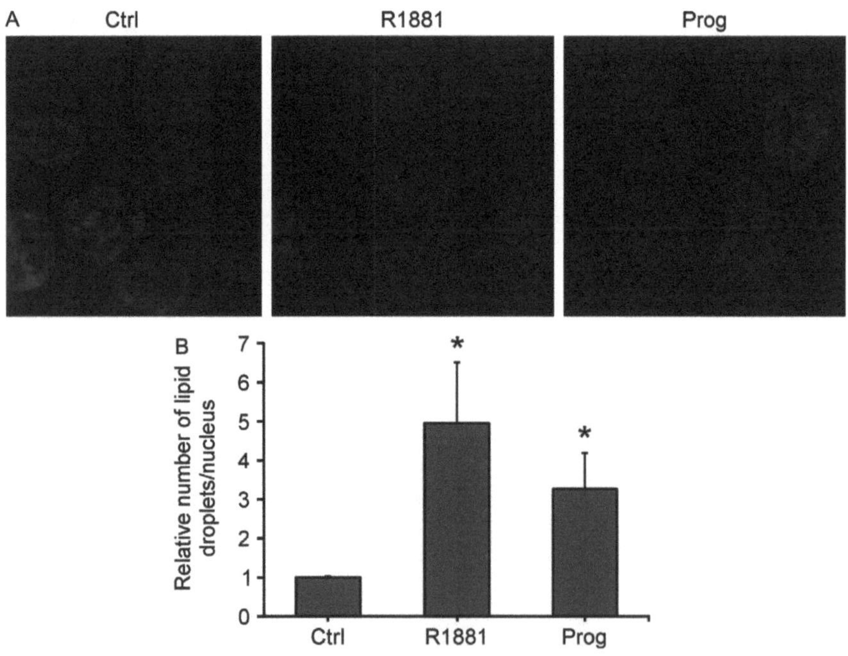

Figure 21.5 *Increased lipid accumulation in T47D cells in response to steroid treatment detected by HCS LipidTOX Red.* T47D cells were cultured in medium containing 5% CT-FBS for 48 h and treated with 10^{-8} *M* R1881, 10^{-8} *M* progesterone (Prog), or vehicle (Ctrl) for 96 h. (A) Representative fluorescence confocal images of cells stained with HCS LipidTOX Red (red) and DAPI (blue). (B) Relative number of lipid droplets per nucleus from (A) quantified using ImageJ. $^{*}p < 0.05$. (See the color plate.)

4. NOTES

1. T47D and LNCaP cells are treated with R1881 and progesterone to stimulate lipid accumulation. In previous studies, other hormones have been used as well, such as anabolic steroids dihydrotestosterone, testosterone, and mibolerone (Chambon et al., 1989; Swinnen et al., 1996).
2. PFA is a possible carcinogen. One should work with it in a ventilation hood and discard it safely.
3. How the chamber looks and what it is made of is not important. The main point is that the sample is protected from light and that the chamber is relatively airtight so that the air inside will stay humid during an hour of incubation at room temperature. We typically use a $30 \times 30 \times 3$ cm plastic chamber wrapped with aluminum foil. This can easily contain 24 cover slips on a sheet of parafilm at once. The humid air inside the chamber is maintained by placing the moist tissue paper next to the inside walls.
4. When using 12- or 24-well plates, LNCaP cells may clump at the center of the well and may influence the outcome of the experiment. We therefore recommend using six-well plates. This may not be an issue for other cell types (e.g., T47D).
5. In our hands, treatment with 10^{-8} *M* R1881 for 96 h is best for consistent induction of lipid accumulation in LNCaP cells, but the effect has been successfully detected with concentrations as low as 10^{-10} *M* R1881 (Swinnen et al., 1996). For direct measurement of enzyme expression/activity involved in the process of lipid accumulation, cells can be harvested earlier than the 96-h treatment. For example, significant *FASN* expression is detected in LNCaP cells treated with 10^{-8} *M* R1881 after only 16 h (Swinnen et al., 1997).
6. After the cells are fixed with PFA and properly washed, they can be stored for a few days in 2.5 ml PBS per well at 4 °C before further steps. Seal the plate with parafilm.
7. Before applying the mounting solution, it is useful to cut the end of the pipette tip. This makes it easier to avoid bubbles, as it is a very viscous solution.
8. Be sure to remove all Oil Red O precipitate and excess working solution when washing with 60% isopropyl alcohol and PBS. The precipitate not only impairs the clarity in final images taken, but poorly washed cells can of course give erroneous quantification results.

9. 518 nm is the absorption maximum of Oil Red O in the visible spectrum, but the peak is broad so proximal wavelengths can be used if necessary.
10. If DAPI stain is not needed along with HCS LipidTOX Red staining, then it is not necessary to permeabilize the cells.
11. If staining is used together with immunofluorescence, apply HCS LipidTOX Red simultaneously with secondary antibody.
12. Less than half of the reagent used in manufacturers' protocol is sufficient. The assay is designed for TG detection mainly in adipocytes and generally requires a longer incubation time with LNCaP and T47D cells.
13. Dense populations of lipid droplets may be hard to separate. Convert image to binary and use the *Watershed* and *Fill Holes* commands.

ACKNOWLEDGMENTS

This work was supported by grants from the Norwegian Cancer Society and the Norwegian Research Council to F. S.

REFERENCES

Accioly, M. T., Pacheco, P., Maya-Monteiro, C. M., Carrossini, N., Robbs, B. K., Oliveira, S. S., et al. (2008). Lipid bodies are reservoirs of cyclooxygenase-2 and sites of prostaglandin-E2 synthesis in colon cancer cells. *Cancer Research*, *68*(6), 1732–1740.

Bozza, P. T., & Viola, J. P. (2010). Lipid droplets in inflammation and cancer. *Prostaglandins, Leukotrienes, and Essential Fatty Acids*, *82*(4–6), 243–250.

Cairns, R. A., Harris, I. S., & Mak, T. W. (2011). Regulation of cancer cell metabolism. *Nature Reviews. Cancer*, *11*(2), 85–95.

Chalbos, D., Chambon, M., Ailhaud, G., & Rochefort, H. (1987). Fatty acid synthetase and its mRNA are induced by progestins in breast cancer cells. *The Journal of Biological Chemistry*, *262*(21), 9923–9926.

Chambon, M., Rochefort, H., Vial, H. J., & Chalbos, D. (1989). Progestins and androgens stimulate lipid accumulation in T47D breast cancer cells via their own receptors. *Journal of Steroid Biochemistry*, *33*(5), 915–922.

Hanahan, D., & Weinberg, R. A. (2011). Hallmarks of cancer: The next generation. *Cell*, *144*(5), 646–674.

Kuhajda, F. P. (2006). Fatty acid synthase and cancer: New application of an old pathway. *Cancer Research*, *66*(12), 5977–5980.

Kusakabe, T., Maeda, M., Hoshi, N., Sugino, T., Watanabe, K., Fukuda, T., et al. (2000). Fatty acid synthase is expressed mainly in adult hormone-sensitive cells or cells with high lipid metabolism and in proliferating fetal cells. *The Journal of Histochemistry and Cytochemistry*, *48*(5), 613–622.

Mashima, T., Seimiya, H., & Tsuruo, T. (2009). De novo fatty-acid synthesis and related pathways as molecular targets for cancer therapy. *British Journal of Cancer*, *100*(9), 1369–1372.

Menendez, J. A., & Lupu, R. (2007). Fatty acid synthase and the lipogenic phenotype in cancer pathogenesis. *Nature Reviews. Cancer*, *7*(10), 763–777.

Opstad, K. S., Bell, B. A., Griffiths, J. R., & Howe, F. A. (2008). An investigation of human brain tumour lipids by high-resolution magic angle spinning ^{1}H MRS and histological analysis. *NMR in Biomedicine*, *21*(7), 677–685.
Swinnen, J. V., Brusselmans, K., & Verhoeven, G. (2006). Increased lipogenesis in cancer cells: New players, novel targets. *Current Opinion in Clinical Nutrition and Metabolic Care*, *9*(4), 358–365.
Swinnen, J. V., Esquenet, M., Goossens, K., Heyns, W., & Verhoeven, G. (1997). Androgens stimulate fatty acid synthase in the human prostate cancer cell line LNCaP. *Cancer Research*, *57*(6), 1086–1090.
Swinnen, J. V., Van Veldhoven, P. P., Esquenet, M., Heyns, W., & Verhoeven, G. (1996). Androgens markedly stimulate the accumulation of neutral lipids in the human prostatic adenocarcinoma cell line LNCaP. *Endocrinology*, *137*(10), 4468–4474.
Tennant, D. A., Duran, R. V., & Gottlieb, E. (2010). Targeting metabolic transformation for cancer therapy. *Nature Reviews. Cancer*, *10*(4), 267–277.
Than, N. G., Sumegi, B., Bellyei, S., Berki, T., Szekeres, G., Janaky, T., et al. (2003). Lipid droplet and milk lipid globule membrane associated placental protein 17b (PP17b) is involved in apoptotic and differentiation processes of human epithelial cervical carcinoma cells. *European Journal of Biochemistry*, *270*(6), 1176–1188.
Zhao, Y., Butler, E. B., & Tan, M. (2013). Targeting cellular metabolism to improve cancer therapeutics. *Cell Death & Disease*, *4*, e532.

CHAPTER TWENTY-TWO

Analysis of Hypoxia-Induced Metabolic Reprogramming

Chendong Yang[*,1], **Lei Jiang**[*,1], **Huafeng Zhang**[†], **Larissa A. Shimoda**[‡], **Ralph J. DeBerardinis**[*], **Gregg L. Semenza**[§,¶,||,2]

[*]Children's Medical Center Research Institute, University of Texas Southwestern Medical Center, Dallas, Texas, USA
[†]School of Life Science, University of Science and Technology of China, Hefei, PR China
[‡]Division of Pulmonary and Critical Care Medicine, Johns Hopkins University School of Medicine, Baltimore, Maryland, USA
[§]Vascular Program, Institute for Cell Engineering, Johns Hopkins University School of Medicine, Baltimore, Maryland, USA
[¶]Departments of Pediatrics, Medicine, Oncology, Radiation Oncology, and Biological Chemistry, Johns Hopkins University School of Medicine, Baltimore, Maryland, USA
[||]McKusick-Nathans Institute of Genetic Medicine, Johns Hopkins University School of Medicine, Baltimore, Maryland, USA
[1]These authors contributed equally.
[2]Corresponding author: e-mail address: gsemenza@jhmi.edu

Contents

Methods in Enzymology, Volume 542
ISSN 0076-6879
http://dx.doi.org/10.1016/B978-0-12-416618-9.00022-4

Abstract

Hypoxia is a common finding in advanced human tumors and is often associated with metastatic dissemination and poor prognosis. Cancer cells adapt to hypoxia by utilizing physiological adaptation pathways that promote a switch from oxidative to glycolytic metabolism. This promotes the conversion of glucose into lactate while limiting its transformation into acetyl coenzyme A (acetyl-CoA). The uptake of glucose and the glycolytic flux are increased under hypoxic conditions, mostly owing to the upregulation of genes encoding glucose transporters and glycolytic enzymes, a process that depends on hypoxia-inducible factor 1 (HIF-1). The reduced delivery of acetyl-CoA to the tricarboxylic acid cycle leads to a switch from glucose to glutamine as the major substrate for fatty acid synthesis in hypoxic cells. In addition, hypoxia induces (1) the HIF-1-dependent expression of BCL2/adenovirus E1B 19-kDa interacting protein 3 (BNIP3) and BNIP3-like (BNIP3L), which trigger mitochondrial autophagy, thereby decreasing the oxidative metabolism of both fatty acids and glucose, and (2) the expression of the sodium–hydrogen exchanger NHE1, which maintains an alkaline intracellular pH. Here, we present a compendium of methods to study hypoxia-induced metabolic alterations.

1. INTRODUCTION

Intratumoral hypoxia represents a major driving force in cancer progression. Primary tumors in the cervix and breast have significantly decreased oxygenation (median partial pressure of oxygen (p_{O_2}) values = 8 and 10 mmHg, respectively) compared with normal tissue in the same organ (p_{O_2} = 42 and 65 mmHg, respectively), and cancers with p_{O_2} values below 10 mmHg are at significantly increased risk of metastasis and patient mortality (Vaupel, Mayer, & Hockel, 2004). In order to survive, cancer cells must adapt to reduced O_2 availability by reprogramming their metabolism. Many of the critical metabolic responses to hypoxia are regulated by hypoxia-inducible factor 1 (HIF-1), which coordinately activates the transcription of genes encoding glycolytic enzymes and other key mediators of metabolic reprogramming (Iyer et al., 1998; Seagroves et al., 2001; Semenza, 2011). This chapter will review the biochemical and molecular mechanisms underlying metabolic adaptations to hypoxia and present methods for analyzing the metabolism of hypoxic cancer cells.

2. EXPOSURE OF CELLS TO HYPOXIC CULTURE CONDITIONS

Cancer cells are typically cultured in a standard tissue culture incubator containing 5% CO_2 and 95% air and are thus exposed to an ambient environment of 20% O_2 (air contains 21% O_2 and $0.95 \times 21\% = 20\%$). To subject cells to hypoxic conditions, tissue culture plates are placed in a modular incubator chamber (Billups-Rothenberg Inc.) that is flushed with a gas mixture containing 1% O_2, 5% CO_2, and 94% N_2. Flushing the chamber at 2 psi for 5 min is sufficient to lower the ambient O_2 concentration to 1%. However, it should be recognized that the diffusion of gas through the tissue culture medium occurs much more slowly and, for many responses to hypoxia, may actually represent the rate-limiting step. Minimizing the amount of tissue culture media will facilitate more rapid equilibration. It is also advisable to remove the lids of the tissue culture plates for incubation times of <24 h.

One percent O_2 is chosen because it is low enough to provide a strong hypoxic stimulus (a 20-fold reduction in O_2 concentration compared with standard culture conditions), but 1% O_2 is not low enough to directly cause significant death of most cell types. 1% O_2 corresponds to a p_{O_2} of 7 mmHg at sea level, which is comparable to the median observed in breast and cervical cancers as described in the preceding text and thus represents a reasonable model of intratumoral hypoxia.

For many cancer lines, extended hypoxic exposure results in cell death that is not a direct effect of hypoxia but is secondary to acidosis. Thus, for hypoxic exposures of more than 24 h, it is recommended that the tissue culture media be supplemented with 25 m*M* HEPES in order to buffer the H^+ ions produced by glycolysis, provided of course that the measurement of pH changes is not a goal of the experiment.

3. ALTERATIONS IN GLUCOSE UPTAKE AND METABOLISM

3.1. Molecular mechanisms

The classic intracellular adaptation to hypoxia is the switch from oxidative to glycolytic metabolism. Under normoxic conditions, glucose is metabolized via the Embden–Meyerhof pathway to pyruvate, which is then converted to acetyl coenzyme A (acetyl-CoA) through the activity of pyruvate dehydrogenase (PDH) for entry into the tricarboxylic acid (TCA) cycle. Under hypoxic conditions, HIF-1 activates the transcription of the *PDK1* gene encoding PDH kinase, which phosphorylates and inactivates the catalytic subunit of PDH, thereby inhibiting the conversion of pyruvate to acetyl-CoA (Kim, Tchernyshyov, Semenza, & Dang, 2006; Papandreou, Cairns, Fontana, Lim, & Denko, 2006). There are three other genes encoding PDK isoforms in the human genome, and of these, *PDK3* is also regulated by HIF-1 (Lu, Lin, Chen, Lai, & Tsai, 2008). In concert with the transactivation of *PDK1*, HIF-1 also activates transcription of the *LDHA* gene encoding lactate dehydrogenase A, which catalyzes the conversion of pyruvate to lactate (Semenza et al., 1996). Thus, the combined effect of PDK1 and LDHA is to decrease flux from pyruvate to acetyl-CoA and increase flux from pyruvate to lactate.

Glycolysis is inherently less efficient than oxidative phosphorylation, with a net synthesis of 2 and 36 mol of ATP, respectively, per mole of glucose. To compensate for the reduced efficiency, HIF-1 activates transcription of the *SLC2A1* and *SLC2A3* genes encoding the glucose transporters GLUT1 and GLUT3, respectively, to increase delivery of substrate for glycolysis (Ebert, Firth, & Ratcliffe, 1995; Iyer et al., 1998). In addition, HIF-1 coordinately activates transcription of the genes encoding glycolytic enzymes in hypoxic cells (Iyer et al., 1998). As a result of these transcriptional responses, flux through the glycolytic pathway is significantly increased under conditions of chronic hypoxia, thereby compensating for the reduced efficiency of glycolysis as a means of generating ATP.

Although oxidative metabolism is reduced under hypoxic conditions, it is not eliminated and HIF-1 increases the efficiency of electron transfer from complex IV (cytochrome *c* oxidase) to O_2 under hypoxic conditions by orchestrating an isoform switch in which the regulatory subunit COX4-1 is replaced by COX4-2. To accomplish this, HIF-1 activates transcription of the *LON* gene, which encodes a mitochondrial protease that is required for the degradation of COX4-1, and the *COX4I2* gene, which encodes COX4-2 (Fukuda et al., 2007).

3.2. Measurement of glucose uptake

Methods for measurement of glucose uptake are described in TeSlaa and Teitell (2014).

3.3. Analysis of glycolysis

A number of experimental approaches exist to measure the rate of glycolysis, which is defined here as the rate at which glucose in the culture medium is converted to lactate and then secreted into the extracellular space. These approaches range in complexity from simple quantitation of lactate abundance using enzymatic assays or automated analyzers to much more detailed inspection of stable isotope transfer from glucose to lactate using mass spectrometry (MS) or nuclear magnetic resonance (NMR). Several of these approaches are described in the succeeding text. Choosing the most appropriate approach for a specific experiment requires consideration of cost, access to instrumentation, demand for sensitivity, and the importance of quantitative precision, along with other factors. However, even the simplest assays require careful preparation of reagents and handling of the cells to maximize the chances of obtaining reproducible results. The sections in the succeeding text describe many of these issues. One additional glycolytic assay that warrants mention here but is not described further in this chapter involves the transfer of ^{3}H from [5-^{3}H]-glucose to water. This assay has the advantage of the extremely high sensitivity afforded by radioisotopes, but it does not directly enable the user to measure the abundance of lactate. Descriptions of this assay can be found in the literature (Liang, Buettger, Berner, & Matschinsky, 1997).

3.3.1 Culture medium for glycolytic assays

As concentrations of glucose, lactate, growth factors, and other components may influence glycolytic rate, it is important to carefully define the culture medium used in assays to measure glycolysis and other metabolic pathways. One issue to consider is whether or not to use dialyzed serum when measuring the glycolytic rate. Physiological concentrations of glucose and lactate in adult serum are approximately 5 and 0.5–2 m*M*, respectively. However, fetal calf serum often contains extremely high lactate levels, sometimes exceeding 30 m*M*. Thus, even if fetal calf serum is diluted to 10%, supraphysiological lactate levels will result. There are no major technical hurdles to measuring glycolytic rates in the presence of a large preexisting lactate pool. However, the investigator must be aware that this lactate is present in the medium at time 0, or the resulting glycolytic rate will be overestimated. A similar consideration is whether an exogenous source of pyruvate is included in the culture medium.

Our practice has been to prepare medium from a base powder (e.g., Dulbecco's Modified Eagle's medium (DMEM); Sigma, D5030), lacking glucose, glutamine, and pyruvate, and then to supplement with 10% dialyzed fetal calf serum. Glucose and glutamine can be added later to the desired concentration, including ^{13}C-labeled nutrients if indicated for isotope tracing experiments. Glutamine is best added on the day of the assay to minimize the spontaneous release of ammonia prior to the experiment.

Details of medium preparation

- Use DMEM powder (Sigma, D5030) lacking glucose, glutamine, pyruvate, sodium bicarbonate, and phenol red.
- Add water to bring volume to 90% of the desired total.

- Add sodium bicarbonate to the final concentration of 42.5 m*M*.
- Optional: add HEPES to the final concentration of 25 m*M*.
- If desired, add 10 U/mL penicillin and 10 μg/mL streptomycin.
- Supplement with dialyzed 10% fetal calf serum.
- Adjust pH to 7.4 with sodium hydroxide and add water to obtain the final volume.
- Phenol red may be added if desired at 0.015 g/L. Phenol red may limit the linear range of some colorimetric enzyme assays, and leaving it out does not impact the health of short-term cultures.
- On the day of the experiment, add glucose and glutamine to the desired final concentration. For a standard 6–8 h metabolic assay, we use 10 m*M* glucose and 4 m*M* glutamine. Few if any cell lines will exhaust either of these nutrients over such a short time frame.
- Note that some cell lines prefer medium formulations other than DMEM. Many of these, for example, Roswell Park Memorial Institute 1640, are sold as glucose-free and/or glutamine-free. Many cell lines can also easily be adapted to growth in DMEM for short-term metabolic assays.

3.3.2 Cell culture

Various assays to measure glucose and lactate require different volumes of culture medium, and this determines how many cells are needed for each experiment. Automated analyzers may require up to 0.6 mL of culture medium. Our practice has been to plate cells into 60-mm dishes, culture them until they approach 80% confluence, and begin the assay at that time. Details of cell culture and medium harvest

- The specific plating protocol should be optimized for each cell line, but plating 7.5×10^5 to 1×10^6 cells works well for murine embryonic fibroblasts and most cancer cell lines. It is important to allow the cells to adhere to the dish and resume proliferation in order to obtain reproducible metabolic rates. Most cell lines will be 80% confluent within 18–36 h after plating.
- At time zero, rinse the cells twice with prewarmed phosphate-buffered saline (PBS).
- Overlay with 2 mL of medium. This volume is large enough to keep the cells covered for several hours but is small enough that the glucose and lactate concentrations will change appreciably over this time. With larger volumes, changes in concentration may be too small to measure reliably.

- Save and freeze an aliquot of the starting medium to analyze at the end of the assay.
- If glutamine utilization or ammonia secretion is to be measured along with glucose and lactate, set up a parallel dish with no cells, and culture alongside the dishes with cells. Medium from this dish will be used to control for the small amounts of spontaneous glutamine degradation and ammoniagenesis.
- Culture the cells for 6–8 h.
- At the end of the culture period, use a pipette to transfer the medium from the dish to a conical tube. Centrifuge for 5 min at $1000 \times g$ to remove debris and then aliquot into Eppendorf tubes. Analyze immediately or store at −80 °C.
- In order to normalize the apparent metabolic rates, it is necessary to know the number of cells or the protein content of each dish. There are several ways to do this. Cells can be counted after removing the medium at the end of the culture. Keep in mind that if you are comparing cell lines with different rates of proliferation, there can be substantial drift in cell number from time zero, even over just 6–8 h. This problem is magnified over longer cultures.
- Alternatively, if substantial size differences exist between cells or conditions, it is preferable to normalize according to the protein content. After removing the culture medium, rinse the cells briefly with PBS, add 0.5 mL of 0.1 N NaOH (or RIPA lysis buffer), scrape into an Eppendorf tube, and determine protein abundance from each dish.

3.3.3 Determination of glucose and lactate content in culture medium

Glucose and lactate can be measured using automated analyzers or various enzymatic assays, many of which exist as commercial kits. We have had success with an automated electrochemical analyzer (BioProfile Basic-4 analyzer, NOVA). This instrument analyzes glucose, lactate, glutamine, and glutamate from medium aliquots of ~0.6 mL. Analysis of each sample takes about 90 s. Commercial kits for glucose and lactate typically use just a few microliters and can be performed in 96-well microtiter plates. These systems are preferable when higher-throughput applications or smaller cultures are involved.

Do not report utilization rates in terms of the change in medium concentration, because this value depends on the volume of the medium. Rather, rates should be reported in terms of molar quantities consumed or secreted per unit of time.

Details for rate calculation

- Most of the assays described in the preceding text report glucose and lactate as concentrations in mmol/L, and therefore, these values should first be converted to the total abundance of each molecule in the volume of medium used in the culture (e.g., 2 mL).
- Then, determine the change in abundance between the medium samples from the start and end of the assay. For glutamine, a separate spontaneous degradation rate is calculated using the dish with no cells, and this rate is subtracted from the dishes containing cells. These values are the total molar amount of metabolite used or secreted over the course of the assay.
- Convert these values to rates by dividing by the number of hours of culture (usually 6–8) the biological unit of choice (e.g., million cells and mg protein).
- If performed properly, this assay should produce no more than 10% variance among biological replicates processed in parallel.
- Note that using this method to calculate a rate implies that the rate does not change over the course of the assay (i.e., that glucose consumption, lactate secretion, etc., are essentially linear). This is usually a fair assumption in short cultures such as those described here. With cultures of 18 h or longer, it is not uncommon for a nutrient to become limiting in the latter stages of the experiment. This causes the rate calculated by comparing starting and ending concentrations to underestimate the rates that occurred when nutrient availability was sufficient.
- An example data set generated with an automated analyzer is shown in Fig. 22.1. In this experiment, rates were compared between cells treated with vehicle and those with an inhibitor against the serine/threonine kinase Akt, which stimulates glycolysis in many cell lines (Yang et al., 2009). Note that the rate of lactate secretion is almost twice as high as the rate of glucose consumption, as is typical of cancer cell lines (DeBerardinis et al., 2007). The rate of glutamine utilization was approximately one-fifth of the rate of glucose utilization, which is also typical. The Akt inhibitor (Akti) reduced both glucose consumption and lactate secretion without causing a large effect on glutamine metabolism.

3.3.4 Measurement of conversion of [1,6-^{13}C]glucose to extracellular [3-^{13}C]lactate

Many cancer cell lines have such high glycolytic rates that it is practically impossible to introduce ^{13}C-labeled glucose to a culture and sample the medium quickly enough to avoid observing ^{13}C-lactate. The small amounts

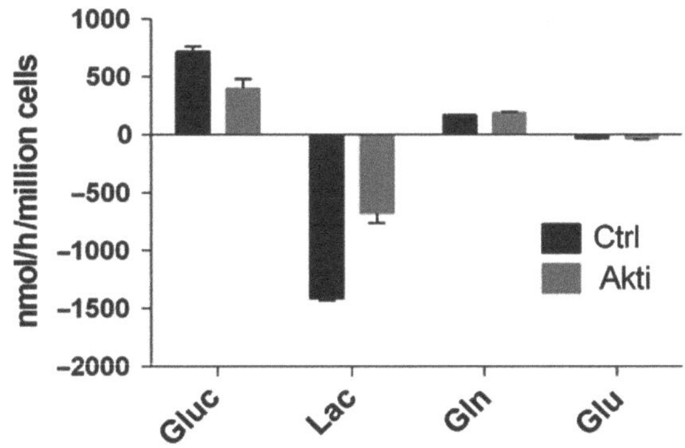

Figure 22.1 *Analysis of metabolite utilization and secretion by cultured cancer cells.* SF188 glioblastoma cells were analyzed as described in the text, in the presence and absence of an Akt inhibitor (Akti). Culture medium was analyzed for the abundance of glucose (Gluc), lactate (Lac), glutamine (Gln), and glutamate (Glu) using an automated electrochemical analyzer. Positive values indicate consumption and negative values indicate secretion. Data are the average and standard deviation of three independent cultures.

of lactate released over seconds to minutes are too small to be detected reliably by the assays described in the preceding text. However, with access to MS, it becomes straightforward to obtain very sensitive and precise measurements of the rate at which glucose is converted to lactate over time periods as short as a few minutes. We have used this approach in cancer cell lines (Harrison et al., 2012) and small populations of primary cells including hematopoietic stem cells (Simsek et al., 2010). For established cell lines, cells should be plated as described in the preceding text and cultured until they reach 80% confluence. Medium is prepared as in the preceding text and then supplemented with [1,6-^{13}C]glucose. Labeling glucose in this fashion produces lactate containing one additional mass unit from ^{13}C, and this can be used to calculate both the fractional enrichment of the extracellular lactate pool and the total abundance of ^{13}C-lactate.

Details of procedure

- Rinse the cells with warm PBS and aspirate.
- Overlay with 2 mL of medium supplemented with [1,6-^{13}C]glucose, prewarmed to 37 °C.
- Harvest aliquots of medium (20 μL) at several time points, starting at 30 s and extending over 60 min. Save an aliquot of the time zero medium for comparison.
- Keep aliquots on ice until all are collected.

- Add an internal standard of Na-[U-^{13}C]lactate to each aliquot and vortex. We use 10.7 nmol of Na-[U-^{13}C]lactate per aliquot, but this can be adjusted if necessary so that the amount of this uniformly labeled lactate is relatively similar to the amount of lactate produced by the cells.
- Extract lactate by sequential addition of 1 mL each of methanol, chloroform, and water, vortexing vigorously after each. Centrifuge the mixture at 1000 × *g* for 10 min and transfer the upper (aqueous) phase to a screw-topped glass tube (16 × 100 mm; Fisher 14-959-35AA).
- Evaporate the aqueous phase to dryness under blown air or nitrogen gas at 42 °C or using a speed-vac.
- Add 100 μL of a derivatizing agent (e.g., the trimethylsilyl donor Tri-Sil; Thermo-Fisher) and derivatize at 42 °C for 30 min.
- Transfer the derivatized sample to a screw-topped autoinjector vial (Agilent, #5182-0716) fitted with a glass sample insert (Agilent, #5181-3377).
- Prepare a set of five standards of increasing fractional enrichment (0%, 25%, 50%, 75%, and 100% ^{13}C) by mixing unlabeled 4 m*M* Na-lactate with 4 m*M* Na-[3-^{13}C]lactate (10 μL of Na-lactate and no Na-[3-^{13}C]lactate for 0%, 7.5 μL of Na-lactate and 2.5 μL of Na-[3-^{13}C]lactate for 25%, etc., bringing the total volume of each tube to 20 μL with water). Process these samples using the same extraction, evaporation, and derivatization methods as for the medium samples.
- Inject 3 μL of each standard and sample for analysis by gas chromatography/mass spectrometry (GC/MS). We use an Agilent 6890 gas chromatography system networked to an Agilent 5973 mass selective detector. Monitor the abundance of fragment ions corresponding to unlabeled ($m+0$) through fully labeled ($m+3$) derivatized lactate. For TMS-derivatized lactate, these correspond to m/z 219–222.

Details of data analysis

- Generate a standard curve by plotting the known fractional enrichment of each standard on the y-axis against the empirical enrichment calculated by the formula ($m/z\ 220_{\text{area}}/(m/z\ 219_{\text{area}} + m/z\ 220_{\text{area}})$) on the x-axis. This should produce a straight line with an R^2 value very close to 1.0. Calculate the linear equation of the trend line ($y = mx + b$).
- Calculate the fractional enrichment of [3-^{13}C]lactate at each time point using the empirical enrichment to solve for y in the linear equation. An example of a time course of lactate enrichment is given in Fig. 22.2A. Note that the fractional enrichment is initially 0% and that all subsequent

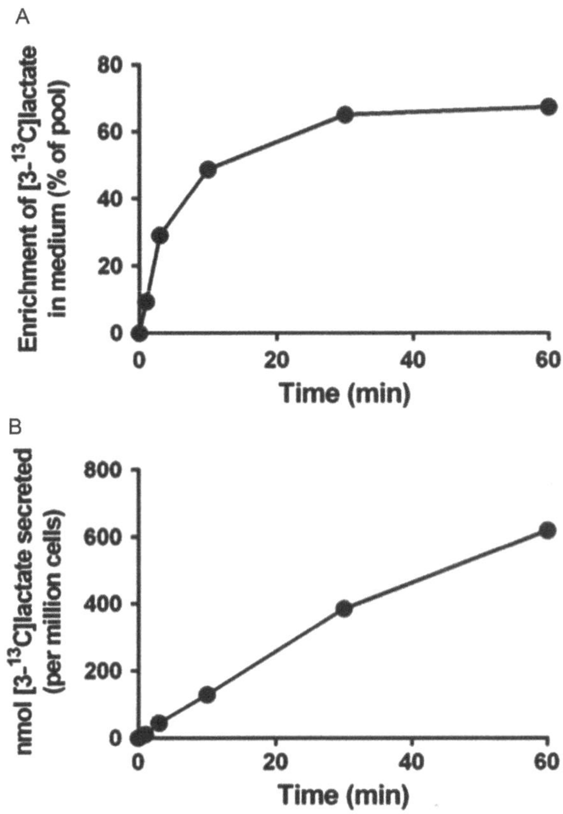

Figure 22.2 *Quantitation of lactate secretion rate.* (A) Fractional enrichment of lactate secreted by SF188 glioblastoma cells cultured with [1,6-^{13}C]glucose and unlabeled glutamine. (B) Total abundance of [3-^{13}C]lactate in the same cultures.

time points contain labeled lactate. Note also that empirical determination of fractional enrichment using a standard curve accounts for natural abundance of $m+1$ lactate (i.e., lactate containing an additional mass unit because of the natural abundance of ^{13}C, ^{29}Si, etc.), as opposed to ^{13}C that originated on [1,6-^{13}C]glucose. The fractional enrichment rises rapidly and begins to reach a plateau within an hour, often within 15 min or less.

- Calculate the total abundance of lactate in the sample by normalizing the combined abundance of unlabeled ($m+0$) and singly labeled ($m+1$) lactate against the internal standard ($m+3$). Knowing the abundance of the internal standard (e.g., 10.7 nmol per 20 μL aliquot) allows the abundance of lactate in the sample to be quantified. For example, if the combined abundance of unlabeled and singly labeled lactate is twice the abundance of the standard, then the 20 μL aliquot of medium contained 21.4 nmol of lactate (10.7 nmol × 2).

- Determine the amount of [3-^{13}C]lactate in each sample of medium by multiplying the fractional enrichment by the total abundance at each time point. The initial lactate abundance is related to the abundance of unlabeled lactate in the time zero medium and should be very low when dialyzed serum is used. Total [3-^{13}C]lactate should rise linearly over the first hour. This allows the [3-^{13}C]lactate secretion rate to be calculated easily using the slope of the line or by dividing the abundance of [3-^{13}C]lactate at the final time point by the duration of the culture. For example, in Fig. 22.2B, the rate is 10.5 nmol/10^6 cells/min.
- If desired, one can also account for the small, progressive loss of volume due to repeated sampling of the culture medium. This would involve correcting the volume of medium at each time point by subtracting the amount that had been withdrawn. In most experiments, this correction will not substantially change the calculated rate. For example, in Fig. 22.2, the total loss of volume over the experiment was only 100 μL, amounting to 5% of a 2 mL starting volume.
- This assay can also be used to calculate rates of turnover for intracellular lactate (Harrison et al., 2012).

3.4. Using ^{13}C to trace intracellular metabolic fates of glucose

Glucose supplies many metabolic pathways beyond glycolysis, and stable isotope tracing (i.e., ^{13}C) is very effective at determining the fates of glucose carbon within the cell. MS and NMR are widely used for these purposes. In both approaches, the choice of the specific form of ^{13}C-labeled glucose determines the types of conclusions that may be drawn from the data. This issue has been reviewed in detail (Walther, Metallo, Zhang, & Stephanopoulos, 2012). Straightforward examples include the ability of [1,2-^{13}C]glucose to analyze the contribution of the oxidative pentose phosphate pathway to lactate formation and [3,4-^{13}C]glucose to detect pyruvate carboxylase-mediated formation of oxaloacetate for use in the TCA cycle (Cheng et al., 2011; Marin-Valencia, Cho, et al., 2012). Transfer of ^{13}C can be tracked into a large number of intracellular metabolites from diverse pathways, including biosynthetic precursors, lipids, and nucleotides. We discuss the example of using ^{13}C-glucose to analyze the TCA cycle, although the general approach can be applied to analyze many other pathways as well.

Note that a similar experimental setup can also be used to trace the metabolic fates of pyruvate, glutamine, acetate, fatty acids, and many other nutrients consumed by cancer cells.

Preparation of culture medium for ^{13}C tracer assays is similar to the approach described in the preceding text for glycolysis assays. Our practice is to prepare medium from powder and to supplement the labeled nutrient prior to the assay. The duration of the labeling period depends on the experimental question; a key consideration is whether dynamic or steady-state labeling data are desired. Dynamic labeling of TCA cycle intermediates can be detected within a few minutes, whereas steady-state labeling is only established after several hours of culture. Tracking the residual fraction of an unlabeled target metabolite (e.g., citrate) by MS is an effective way to determine when the system is beginning to approach isotopic steady state. Many cancer cell lines reach a plateau of labeling after 6–8 h. Consideration must also be given to the concentrations of glucose and glutamine, the two major nutrients catabolized in the TCA cycle. For most applications, 10–15 m*M* glucose and 2–4 m*M* glutamine will suffice for assays of this duration.

As with the assays described in the preceding text, adherent cells should be plated into a 60-mm dish and cultured such that they will reach 80% confluence on the morning of the assay. For some cell lines, using a larger (100-mm) dish will significantly improve metabolite abundance and signal in MS analysis. Although the procedure in the succeeding text was optimized for adherent cells, the same principles apply to cells grown in suspension. For suspension cells, metabolites can be extracted after rapidly pelleting the cells to remove the culture medium.

Details of procedure

- Rinse the cells once with prewarmed PBS.
- Overlay each culture with 2 mL of medium supplemented with ^{13}C-glucose.
- Culture the cells for the desired duration (e.g., 6 h) and then place the dishes on an ice tray.
- If the experiment calls for analysis of the culture medium as well, carefully remove it with a pipette and transfer it to a conical tube on ice. Set aside until the cells have been lysed and flash-frozen.
- Rinse the cells once with ice-cold normal saline to remove unlabeled metabolites in the medium and metabolites that have been secreted from the cells. Avoiding PBS at this stage reduces the appearance of a massive inorganic phosphate peak on GC/MS analysis.

- Immediately add 0.8 mL of a cold 1:1 mixture of methanol/water, scrape up the cells, and transfer to an Eppendorf tube.
- Add a standard compound to control for subsequent derivatization and MS. A labeled form of a target compound can be used for quantitation, or an organic acid not typically found in cellular extracts can be added. Our practice is to preload the Eppendorf tubes with 50 nmol of sodium-2-oxobutyrate.
- Drop the Eppendorf tube into liquid nitrogen to flash-freeze the cells and quench metabolism.
- Subject the cells to three rapid freeze–thaw cycles by transferring to a 37 °C heat block or water bath and then dropping back into liquid nitrogen.
- Centrifuge at maximum speed on a tabletop microfuge at 4 °C to pellet cellular debris. The pellet contains most of the cellular protein and can be used for normalization purposes if desired.
- Transfer the supernatant to a screw-topped glass tube, and evaporate completely using blown air or a speed-vac.
- Add 100 μL of a derivatizing agent (e.g., the trimethylsilyl donor Tri-Sil; Thermo-Fisher) and derivatize at 42 °C for 30 min.
- Transfer the derivatized sample to a screw-topped autoinjector vial fitted with a glass sample insert.
- Inject 3 μL of each sample for analysis by GC/MS. Monitor abundance of informative fragment ions from metabolites of interest. Retention times and mass fragmentation signatures of all metabolites should be validated using pure standards.

Details of data analysis

- Relative metabolite abundance across samples can be estimated by summing the total ion current peak and normalizing against the internal standard and protein content.
- Choose a fragment ion containing all carbons from the parent molecule. (For a description of useful fragment ions for TMS-derivatized TCA cycle intermediates and other molecules, see Cheng et al., 2011.)
- Use commercial software (e.g., MSDChem; Agilent) to determine the mass isotopomer distribution of target metabolites. This involves extracting the area under the curve for each relevant mass isotopomer (i.e., $m+0$, $m+1$, . . ., $m+n$), where n = the maximum number of ^{13}C nuclei that could have been transferred from ^{13}C-glucose. For practical purposes, n is usually equal to the number of carbon atoms in the

target metabolite. Thus, for a 6-carbon metabolite like citrate, ions corresponding to $m+0$ through $m+6$, at a minimum, should be analyzed.

- The raw signal for each mass isotopomer is then corrected mathematically to account for natural abundance isotopes. Descriptions of this process can be found in the literature (Fernandez, Des Rosiers, Previs, David, & Brunengraber, 1996). These corrections are essential because 1.1% of all carbon in nature is ^{13}C, such that an "unlabeled" population of citrate molecules would contain in excess of 6% $m+1$. Using silylated derivatizing agents adds considerably to the natural abundance isotopes, as nearly 4.7% of Si is naturally ^{29}Si ($m+1$).
- An example data set from cells cultured in [U-^{13}C]glucose is shown in Fig. 22.3. Note the anticipated utilization of glucose carbon (Fig. 22.3A) and the observed isotope distribution (Fig. 22.3B). Approximately, 80% of the lactate pool is labeled, as demonstrated by the unlabeled fraction ($m+0$) of approximately 20%. Most of the labeled lactate is in the form of $m+3$, as would be expected if glycolysis is active. The majority of the citrate pool is also labeled. The appearance of abundant $m+2$ is best explained by the activity of PDH, which converts fully labeled pyruvate ($m+3$) to fully labeled acetyl-CoA ($m+2$). This acetyl-CoA enters the TCA cycle, delivering two ^{13}C nuclei to citrate. Higher-order labeling in citrate is related to numerous factors, including the processing of citrate $m+2$ through successive rounds of the TCA cycle, and the contribution of alternative pathways including pyruvate carboxylase (Cheng et al., 2011). In malate, note that although $m+2$ is the most abundant labeled form, the overall enrichment is low relative to citrate. This is explained by a large contribution of unlabeled carbon from glutamine at the level of α-ketoglutarate. This could easily be validated by culturing a parallel dish with ^{13}C-glutamine and unlabeled glucose.
- If desired, any of a number of computational approaches can be used to estimate metabolic flux quantitatively from ^{13}C enrichment data.

3.5. Use of NMR spectroscopy

As an alternative to MS, NMR can be used to track the fates of glucose carbon. Although it is far less sensitive than MS and requires significantly larger samples for analysis, NMR offers the advantage of positional ^{13}C specificity, enabling much more confidence about the metabolic activity that produces a particular labeling pattern. Although NMR is now less widely used than MS

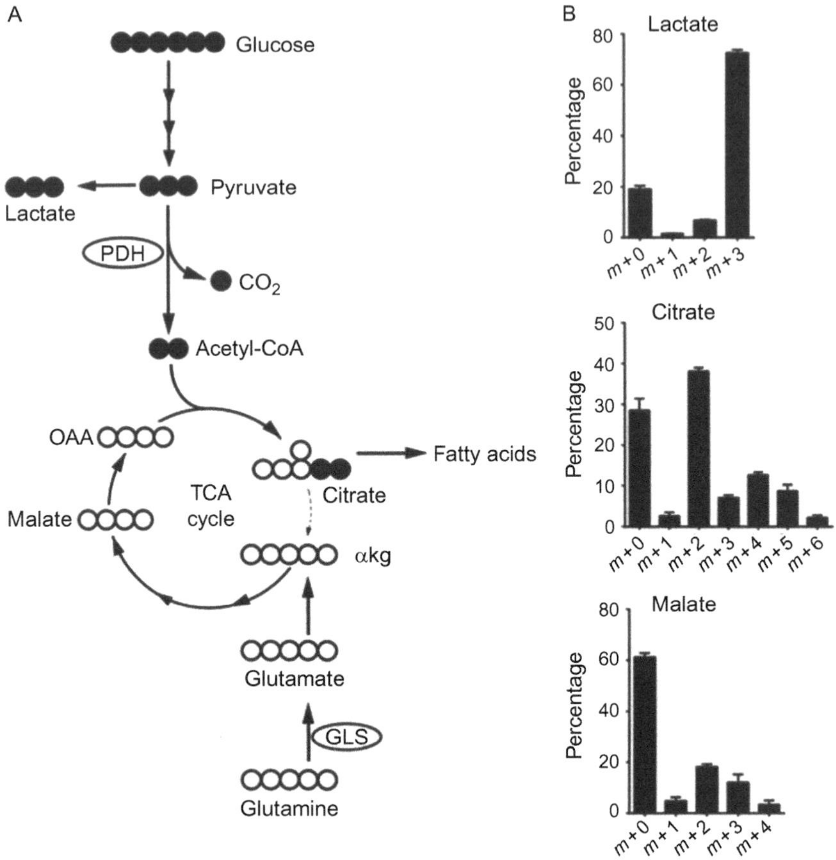

Figure 22.3 *Analysis of glucose metabolism by mass spectrometry.* (A) Transfer of ^{13}C from [U-^{13}C]glucose to various intracellular metabolites. Black symbols indicate ^{13}C and white symbols indicate ^{12}C (unlabeled). (B) Isotopomer distribution for lactate, citrate, and malate in Hep-G2 hepatoma cancer cells cultured in [U-^{13}C]glucose and unlabeled glutamine. Data are the average and standard deviation of three independent cultures. OAA, oxaloacetate; αkg, α-ketoglutarate; PDH, pyruvate dehydrogenase; GLS, glutaminase.

for cell-based metabolic research, we include a description of a method to obtain high-quality NMR spectra amenable to metabolic flux models.

Details of preparation of samples for ^{13}C NMR

- Prepare labeling medium as described in the preceding text.
- Culture cells in 150-mm dishes until they reach 80% confluence. Use four to eight dishes per condition.
- On the morning of the assay, rinse the cells with prewarmed PBS.
- Overlay each culture with 15 mL of medium supplemented with ^{13}C-glucose.

- Culture the cells for the desired duration (e.g., 6 h) and then place the dishes on an ice tray.
- Rinse the cells briefly with ice-cold PBS or saline.
- Add 2 mL of ice-cold 4% perchloric acid per dish, and scrape cellular material into a 50-mL conical tube. Combine material from all dishes for each condition in the same conical tube.
- Centrifuge at $1000 \times g$ for 5 min to remove debris.
- Transfer the supernatant to a new conical tube and neutralize with 8 *M* potassium hydroxide. Potassium perchlorate will precipitate as a salt.
- Centrifuge the neutralized solution at $3000 \times g$ for 10 min and transfer the supernatant to a new conical tube.
- Evaporate the sample by lyophilization.
- Reconstitute lyophilized metabolites in deuterium oxide (D_2O) and transfer to a 3-mm NMR tube.
- Acquire proton-decoupled ^{13}C NMR spectra. We have had success with a Varian ANOVA 14.1T spectrometer equipped with a 3-mm broadband probe. Conditions for acquisition: pulse flip angle 45°, repetition time 1.5 s, spectral width 35 kHz, number of data points 104,986, and number of scans 7000–30,000. Free induction decays are then zero filled to 131,072 points and apodized with exponential multiplication.
- Use commercial software (e.g., Advanced Chemistry Development ACD/Labs SpecManager) to analyze spectra and extract relevant peak areas.
- Example spectra from culture with a cancer cell line are provided in Fig. 22.4. The top spectrum was generated from cells cultured in [1,6-^{13}C]glucose, and the bottom spectrum was generated from cells cultured under identical conditions with [U-^{13}C]glucose. Note that the same carbons are labeled in both experiments but that the quality of the peaks differs. This phenomenon is related to spin-coupling, in which adjacent ^{13}C nuclei cause the NMR signal to be split from a single peak into multiplets. Thus, the prominent singlets in the [1,6-^{13}C]glucose experiment are caused by nonadjacent labeling, whereas those prominent multiplets in the [U-^{13}C]glucose experiment are caused by the transfer of adjacent ^{13}C nuclei to glycolytic and TCA cycle intermediates. As with MS, choosing the form of labeled glucose depends on the goals of the experiment. One very useful application of [U-^{13}C]glucose is *in vivo* infusions of mice and humans (Maher et al., 2012; Marin-Valencia, Yang, et al., 2012). In these experiments, enrichment in the glucose pool can be fairly low, making it difficult to determine whether ^{13}C singlets are related to metabolism or natural abundance ^{13}C. In

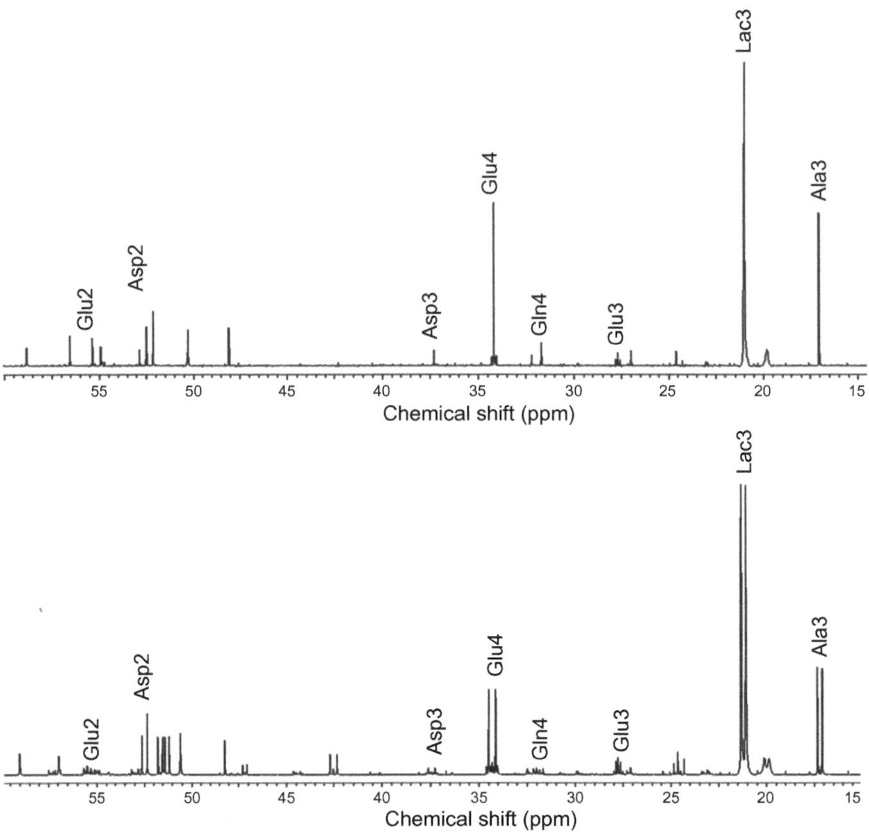

Figure 22.4 *Analysis of glucose metabolism by NMR spectroscopy.* Top: ^{13}C NMR spectrum from SF188 glioblastoma cells cultured in [1,6-^{13}C]glucose and unlabeled glutamine. Bottom: ^{13}C NMR spectrum from SF188 glioblastoma cells cultured in [U-^{13}C] glucose and unlabeled glutamine. Each labeled peak is a carbon position in a metabolite supplied by glucose metabolism (e.g., Lac3 is carbon 3 of lactate). Glu, glutamate; Asp, aspartate; Gln, glutamine; Lac, lactate; Ala, alanine.

contrast, prominent multiplets are produced by metabolism of [U-^{13}C] glucose and do not appear from natural abundance ^{13}C.

4. INDUCTION OF MITOCHONDRIA-SELECTIVE AUTOPHAGY

4.1. Molecular mechanisms

The induction of PDK1 and LDHA expression shunts pyruvate away from the mitochondria, thereby reducing flux of acetyl-CoA through the TCA cycle and reducing the production of NADH and the subsequent consumption of O_2 as the final electron acceptor in the electron transport chain. The efficiency of electron transport under hypoxic conditions is reduced, leading to increased mitochondrial ROS production (Guzy & Schumacker, 2006). In knockout mouse embryo fibroblasts (MEFs) that lack HIF-1 activity, failure to induce PDK1 results in a dramatic increase in the levels of reactive oxygen species (ROS) that ultimately results in cell death under conditions of chronic hypoxia (Kim et al., 2006). Thus, the switch from oxidative to glycolytic metabolism represents an adaptive response to prevent oxidant excess and energy deficiency.

Although PDK1 and LDHA effectively reduce the generation of acetyl-CoA derived from glucose, they have no effect on the generation of acetyl-CoA derived from other substrates, such as fatty acids. One global strategy to prevent oxidative metabolism is to reduce mitochondrial mass through the process of autophagy. In hypoxic cells, HIF-1 activates transcription of the *BNIP3* and *BNIP3L* genes, which encode mitochondrial proteins that trigger mitochondria-selective autophagy in hypoxic cells (Bellot et al., 2009; Zhang et al., 2008). Failure to induce *BNIP3* expression in HIF-1-deficient or BNIP3-deficient MEFs results in increased ROS production under conditions of chronic hypoxia and ROS-induced cell death (Zhang et al., 2008). In HIF-1-deficient MEFs, induction of both *PDK1* and *BNIP3* expressions is lost, and forced expression of either PDK1 or BNIP3 rescues the cells from death due to the accumulation of excess ROS under conditions of chronic hypoxia (Kim et al., 2006; Zhang et al., 2008).

4.2. Methods for detecting autophagy

Autophagy is defined as the process of sequestrating cytoplasmic proteins or even entire organelles into the lysosome for degradation. During autophagy, the fusion of membrane forms a closed double-membrane structures called autophagosomes, which fuse with lysosomes to form autolysosomes in which the sequestered components are degraded by lysosomal hydrolases (Levine & Klionsky, 2004). In mammalian cells, microtubule-associated protein 1A/1B light chain 3 (LC3) is a homologue of the yeast autophagy protein Atg8. A 30-kDa precursor (pro-LC3) is synthesized and processed by cleavage in the C-terminal region to form LC3-I, which is a soluble protein that is distributed ubiquitously throughout mammalian cells. When autophagy is stimulated, LC3-I is conjugated to phosphatidylethanolamine (PE) to form LC3-II, with the PE group mediating the binding of LC3-II to lipid in the membranes of autophagosomes and, to a lesser extent, autolysosomes (Kabeya et al., 2004; Tanida et al., 2004). Thus, LC3-II accumulation is a marker of autophagy, and the detection of LC3 by immunoblot or immunofluorescence assay has become a reliable method for monitoring autophagy. Consensus guidelines for the use and interpretation of assays for monitoring autophagy have been published (Klionsky et al., 2012) and should be consulted by anyone performing studies in this field.

4.2.1 Detection of LC3-I and LC3-II by immunoblot assay

Cells are lysed in radioimmunoprecipitation assay (RIPA) buffer, which has strong denaturing capabilities that are particularly useful for disruption of membrane structures and which consists of 50 m*M* Tris–HCl (pH 7.4), 150 m*M* NaCl, 1% Triton X-100, 0.5% sodium deoxycholate, 0.1% SDS, and 5 m*M* EDTA. Immediately prior to lysing cells, the buffer is brought to 2 m*M* PMSF, 2 μg/mL sodium orthovanadate, and 1 × protease inhibitor cocktail (Roche). Cells are lysed in this complete RIPA buffer on ice for 1 h followed by protein quantification. Aliquots containing equal amounts of protein are fractionated by 12% SDS-PAGE, followed by conventional membrane transfer and blocking with 5% fat-free milk. The membrane is incubated with LC3 antibodies (Novus Biologicals), followed by conventional membrane wash and horseradish peroxidase (HRP)-conjugated secondary antibody incubation. The antibody signal is detected using enhanced chemiluminescence substrate (PerkinElmer). Blots were stripped and reprobed with a polyclonal antibody against β-actin to confirm equal protein loading. In the immunoblot analysis, LC3 appears as two bands at 18 and 16 kDa, which correspond to LC3-I and LC3-II, respectively. The ratio of LC3-II/LC3-I is often used as one measure of autophagy.

4.2.2 Analysis of cells expressing GFP–LC3

As LC3-II is accumulated in autophagosomes during autophagy, it can also be imaged as a marker of autophagosomes in living cells. For this purpose, cDNA encoding LC3 is ligated to *Bgl*II- and *Kpn*I-digested pEGFP-C1 (Clontech) expression vector encoding green fluorescent protein (GFP) to generate a vector that encodes a GFP–LC3 fusion protein. The control vector pEGFP-C1 or the recombinant vector expressing GFP–LC3 is transfected into living cells on chamber slides. After overnight incubation, cells can be subjected to different stress conditions to induce autophagy and examined by fluorescence microscopy. Cells expressing GFP will manifest a diffuse fluorescence throughout the cell, whereas cells expressing GFP–LC3 will show punctate staining (representing the concentration of

GFP–LC3 in autophagosomes) under conditions in which autophagy is induced. Both the number and fluorescence intensity of the puncta, as well as the percentage of cells with punctate staining, can be quantified as measures of autophagy.

4.2.3 Caveats

Consistent results from multiple assays are required to ensure that autophagy is actually being analyzed and additional assays are available, with the gold standard being electron microscopy. False-positive results may be obtained by LC3 immunoblot if LC3-II is generated by a stimulus that does not induce autophagy, in which case, LC-II will be detected by immunoblot but will not result in localization to autophagosomes. False-negative results may be obtained if LC3-II is rapidly turned over in autophagosomes (Tanida, Minematsu-Ikeguchi, Ueno, & Kominami, 2005). Alternatively, if fusion of autophagosomes with lysosomes is blocked, autophagosomes will accumulate. Several drugs block autophagolysosome formation or lysosome protease activity, including bafilomycin, chloroquine, E64d, and pepstatin A. The reader is encouraged to consult recent guidelines for monitoring autophagy (Klionsky et al., 2012).

4.2.4 Analysis of mitochondria-selective autophagy

Whereas many stimuli, such as starvation, induce nonselective macroautophagy, in which autophagosomes contain a variety of cellular organelles including endoplasmic reticulum (ER), Golgi apparatus, and mitochondria, hypoxia induces a selective form of autophagy in which mitochondria but not other cellular organelles are degraded. To distinguish between these processes, cells can be stained with 10 n*M* nonyl acridine orange (NAO), which binds to mitochondrial membranes irrespective of mitochondrial membrane potential and is thus a measure of mitochondrial mass, or 1 μ*M* ER-Tracker dye (BODIPY® FL glibenclamide; Molecular Probes), which binds to ER, at 37 °C for 15 min in PBS containing 5% fetal bovine serum (FBS), followed by rinsing with PBS. Stained cells are then filtered and analyzed in a FACScan flow cytometer (BD Bioscience). Conditions that induce mitochondria-selective autophagy will result in reduced NAO staining but no change in staining with ER-Tracker (Zhang et al., 2008).

Changes in the ratio of mitochondrial to nuclear DNA also provide a measure of mitochondria-selective autophagy. The amount of mitochondrial DNA relative to nuclear DNA is determined by quantitative real-time

PCR using primers for *Nd2* (NADH dehydrogenase subunit 2; mitochondrial genome) and *Nme1* (nuclear genome). Relative *Nd2* copy number is calculated based on the threshold cycle (C_t) as $2^{-\Delta(\Delta C_t)}$, where $\Delta C_t = C_{t_{Nd2}} - C_{t_{Nme1}}$ and $\Delta(\Delta C_t) = \Delta C_{t_{sample}} - \Delta C_{t_{control}}$ (Zhang et al., 2008).

4.3. Analysis of ROS

Most ROS are generated during the process of oxidative phosphorylation in the mitochondria. The production of ROS by the mitochondrial respiratory chain is central to mitochondrial oxidative damage and redox signaling. ROS include, predominantly, superoxide anion ($O_2^{\cdot -}$), hydrogen peroxide (H_2O_2), and hydroxyl radical ($\cdot OH$). ROS are continually produced during metabolic processes and have, in general, a very short half-life. ROS can serve as both intracellular and intercellular messengers. Excess ROS can lead to cellular toxicity by damaging DNA, lipids, and proteins. Thus, both ROS production and destruction are tightly regulated. ROS levels can be measured in isolated mitochondria, cultured cells, and tissue samples.

4.3.1 Molecular probes used to measure ROS

2′,7′-Dichlorodihydrofluorescein diacetate (H_2DCFDA) is probably the most commonly used probe for measuring cellular H_2O_2 (McLennan & Degli Esposti, 2000). The cell-permeable H_2DCFDA diffuses into cells and is deacetylated by cellular esterases to form 2′,7′-dichlorodihydrofluorescein (H_2DCF). In the presence of ROS, predominantly H_2O_2, H_2DCF is rapidly oxidized to 2′,7′-dichlorofluorescein (DCF), which is highly fluorescent, with excitation and emission wavelengths of 498 and 522 nm, respectively. The H_2DCFDA assay can provide reliable measurements of H_2O_2 levels in cells and mitochondrial preparations. Originally, oxidation of H_2DCF to DCF was thought to be specific for H_2O_2, but recent evidence has shown that other ROS such as hydroxyl radical, hydroperoxides, and peroxynitrite can also oxidize H_2DCF, but with greatly reduced sensitivity as compared with that of H_2O_2.

Amplex Red (10-acetyl-3,7-dihydroxyphenoxazine) is another probe to detect the release of H_2O_2 from isolated mitochondria, live cells, and tissues. In combination with HRP, Amplex Red reacts with H_2O_2 in a 1:1 stoichiometry to produce a strong red fluorescent product (excitation/emission wavelength ~568/581 nm) with great stability and less background. Its improved version, Amplex UltraRed, offers brighter fluorescence and enhanced sensitivity on a per mole basis in the presence of HRP.

Dihydroethidium is designed for highly selective detection of superoxide in the mitochondria of live cells. It is the most sensitive and most frequently used probe for measuring cellular superoxide (Zamzami et al., 1995). Dihydroethidium exhibits blue fluorescence; however, once this probe is oxidized by superoxide to 2-hydroxyethidium, a strong red fluorescence emits with excitation peak at ~400 nm and emission detection at ~590 nm. This probe provides optimal discrimination of superoxide from other ROS species. However, one major limit of this probe is indirect interference from changes in mitochondrial membrane potential (Budd, Castilho, & Nicholls, 1997).

MitoSOX Red is a derivative of dihydroethidium. The cationic triphenylphosphonium substituent of MitoSOX Red is responsible for the electrophoretically driven uptake of the probe in actively respiring mitochondria. MitoSOX Red indicator provides more specific mitochondrial localization as compared with dihydroethidium. In some cases, researchers have used dihydroethidium and MitoSOX Red to provide discrete indications of cytosolic and mitochondrial superoxide production, respectively.

Flow cytometry have been widely used to measure the fluorescence produced by these probes in live cells. Trypsinized cells are washed with PBS followed by staining with 1 μM H_2DCFDA (Molecular Probes) at 37 °C for 15 min in PBS containing 5% FBS. Stained cells are washed with PBS, filtered, and analyzed immediately with a FACScan flow cytometer (BD Bioscience) with excitation at 485–495 nm and emission at 520–530 nm. Equivalent numbers of nonstained cells are used as a blank. All gain and amplifier settings are held constant for the duration of the experiment.

Multiple methods of measurement with an increasing variety of probes that can detect cellular ROS are available today, each with their own strengths and limitations. It should be noted that there are no probes that are absolutely specific for any particular ROS. In addition, the quantitative analysis of ROS within cells is difficult because many factors affect ROS production; for example, trypsinization may increase ROS, whereas the presence of free radical-quenching agents such as spermine and albumin may reduce ROS. Thus, as in the case of autophagy, the use of several independent methods is required to properly evaluate ROS in live cells.

4.4. Measurement of cellular O_2 consumption

The O_2 consumption rate is used as a key parameter to study mitochondrial function and oxidative metabolism. Currently, three methods are commonly used to measure O_2 consumption *in vitro*: the Clark O_2 electrode, electron paramagnetic resonance (EPR) oximetry, and the MitoXpress fluorescent assay. Each method has its advantages and disadvantages. The EPR method uses a paramagnetic material in the presence of O_2-consuming cells in a closed system (Diepart, Jordan, & Gallez, 2009). Although EPR oximetry is sensitive and not limited by the cell numbers, a key limitation is that it is technically not possible to add additional compounds during the experiment. The MitoXpress assay is based on a phosphorescent O_2-sensitive probe whose excited state can be quenched by O_2 (Hynes, Hill, & Papkovsky, 2006). The MitoXpress fluorescent method is a cheap, simple, and direct way of measuring cellular O_2 consumption; however, as with EPR, it does not provide the opportunity of adding additional compounds during the measurement. In addition, the MitoXpress method does not provide the actual p_{O_2}, and it is not possible to measure the absolute quantity of O_2 consumed.

The Clark-type O_2 electrode (Clark, Wolf, Granger, & Taylor, 1953) has long been the main technique for measuring p_{O_2} and O_2 consumption, and it is still widely used for a variety of teaching and research applications. It is a reproducible and sensitive technique for studying enzymatic and chemical reactions that consume O_2. The Clark-type O_2 electrode method can measure O_2 consumption in isolated mitochondria or in intact living cells. Besides, a unique advantage of this method is that it provides the opportunity to add compounds during the experiment, which is particularly important in studying the mitochondrial function in which additional substrates (e.g., glutamate, succinate, and ADP) are often added during the experiment run.

The Clark-type O_2 electrode consists of an anode and a cathode, which is in contact with an electrolyte solution and covered by a semi-permeable membrane. O_2 diffuses through the membrane to the cathode, where the O_2 is reduced. The current produced by the electrode is proportional to the O_2 tension in the solution. The Strathkelvin 782 Oxytherm electrode unit (Strathkelvin Instruments, North Lanarkshire, Scotland) utilizes Clark-type polarographic oxygen probes immersed in magnetically stirred sample chambers and produces p_{O_2} evolution curves in 2–15 min. This two-channel instrument can measure two independent samples simultaneously. To maintain a stable chamber temperature, an additional surrounding chamber is filled with water and connected with a water circulation system. All of the measurements are carried out at 37 °C. The respiration chamber volume can be varied from 50 μL to 3 mL. Air-saturated water and zero oxygen calibrating solution are used to calibrate the electrode to the high point and zero value of oxygen concentration, respectively.

Cells are trypsinized and suspended at $0.5–1 \times 10^7$ cells/mL in prewarmed DMEM supplemented with 10% FBS. For each set of experiments, equal numbers of cells in the same volume are placed into the chamber, in which the Clark-type microelectrode will monitor the dissolved O_2 concentration in the sealed chamber over time. O_2 can be measured in units of % saturation of water or partial pressure (mmHg) or kPa in which concentration of O_2 at air saturation in the units is specified (μmol/L/Torr or μmol/L/kPa). It is important to ensure that no air bubbles remain attached to the tip of the electrode holder when it is inserted into the chamber. During the measurement, small quantities of inhibitor or substrate may be added to the chamber by injecting it through a fine stainless needle connected to a syringe pipette. When doing this, care should be taken that there is no air bubble at the end of the needle when it enters the chamber. The data are exported to a computerized chart recorder. O_2 consumption rates are assessed by determining the rate of increase in the signal for each sample by linear regression, using instrument software. These values are then corrected with respect to blanks (baseline slopes in the absence of cells) and normalized with respect to control sample intensity to give the normalized O_2 consumption rates.

There are other O_2 measuring systems, such as chambers that measure single-cell O_2 consumption rates (Molter et al., 2008) and NMR spectroscopy (Pilatus et al., 2001). The XF Analyzer (Seahorse Bioscience) measures carbon dioxide and O_2 simultaneously in real time, and detailed methods for the use of this system are described in TeSlaa and Teitell (2014).

5. MAINTENANCE OF INTRACELLULAR pH

5.1. Molecular mechanisms

The increased production of lactic acid in hypoxic cells has the potential to increase intracellular $[H^+]$. However, hypoxia also induces the expression of multiple sodium–hydrogen exchangers and carbonic anhydrases, which maintain an alkaline intracellular pH (pH_i) and an acidic extracellular pH (pH_e) (Rey, Luo, Shimoda, & Semenza, 2011; Shimoda, Fallon, Pisarcik, Wang, & Semenza, 2006; Swietach, Vaughan-Jones, & Harris, 2007). Lactate is transported out of hypoxic cells by the monocarboxylate transporter

MCT4, the expression of which is also induced by hypoxia (Ullah, Davies, & Halestrap, 2006). A method for profiling pH_e and pH_i in tumors by [^{31}P] NMR was recently reported (Lutz, Le Fur, Chiche, Pouyssegur, & Cozzone, 2013). In the succeeding text, we describe methods for measuring pH_i and sodium–hydrogen exchanger activity in cultured cells.

5.2. Measurement of pH_i

The development of pH-sensitive fluorescent dyes has allowed for the quick, reliable, and reproducible measurement of pH_i. The most widely used pH-sensitive dye is 2′,7′-bis-(2-carboxyethyl)-5(and-6)-carboxyfluorescein (BCECF; Rink, Tsien, & Pozzan, 1982). BCECF has a pK_a of 7.0, which closely matches the normal pH_i range, and is available in a cell-permeant acetoxymethyl ester form (BCECF AM; Molecular Probes). Once in the cytoplasm, nonspecific intracellular esterases hydrolyze the AM group, converting BCECF AM into the fluorescent form that is retained in the cytoplasm. Since cleavage only occurs in healthy cells, cytosolic retention of the fluorescent product provides a measure of cell viability.

To measure pH_i, cells are grown on glass coverslips (25 mm; Fisher Scientific) that can be inserted into a polycarbonate, laminar flow cell chamber (Warner Instruments, model RC-21BR) and perfused with the desired extracellular solutions (see the succeeding text). Coverslips should be coated with the appropriate matrix (i.e., gelatin and collagen) for the cell type under study prior to seeding. Stock solutions of BCECF AM (5 m*M*) are made with Pluronic F-127 DMSO (Molecular Probes) on the day of the experiment, and adherent cells are incubated with 5 μ*M* BCECF AM for 30–60 min at 37 °C under 95% air/5% CO_2. BCECF AM should be made in limited quantities as needed but can be stored in a freezer desiccator at ≤20 °C for up to 6 months. After loading, the edge of the coverslip is dried with a tissue, and coverslips are mounted in the flow chamber using vacuum grease along the outside edges to create a water-tight seal. Special care should be taken to ensure that excess vacuum grease is removed from the inside of the chamber and does not block the inflow and outflow ports. A second glass coverslip is mounted in the same manner on the top of the chamber. The chamber is then filled with extracellular solution, checked for leaks, placed in a heated platform (Warner Instruments, model PH-2) on a stage adapter (Warner Instruments), and perfused at a rate of 0.5–1.0 mL/min for 15 min at 37 °C to remove extracellular dye and allow complete de-esterification of cytosolic dye. Extracellular solutions are heated and gassed (as appropriate) in water-jacketed 50-mL glass reservoirs (Radnoti) connected via tubing to a multi-input manifold (MPP-5; 5-to-1 manifold), allowing for rapid changes between solutions. Inflow perfusate is heated prior to entry into the cell chamber by an in-line heater (Warner Instruments, SH-27B) connected to a dual-output temperature controller (Warner Instruments, TC-324B) that is also used to set the temperature for the heated platform. Fluid withdrawal, and thus flow rate, within the chamber is maintained by a variable flow minipump (Fisher Scientific) on the outflow tubing.

Most cell types express multiple exchangers that are responsible for control of pH_i, including Na^+/H^+ exchangers and Cl^-/HCO_3^- exchangers (Na^+ dependent and Na^+ independent). To measure baseline pH_i under conditions that approximate those *in vivo*, with all exchangers active, cells are perfused with a modified Krebs–Ringer bicarbonate solution containing (in m*M*) 118.3 NaCl, 4.7 KCl, 1.2 $MgSO_4$, 25 $NaHCO_3$, 2.5 $CaCl_2$, 1.1 glucose, and 1.2 KH_2PO_4 and gassed with 16% O_2/5% CO_2 to maintain solution pH at 7.4. The contribution of the Na^+/H^+ exchanger to maintenance of baseline pH_i can be isolated by perfusing cells with HEPES-buffered saline solution (HBSS) containing (in m*M*) 130 NaCl, 5 KCl, 1 $MgCl_2$, 1.5 $CaCl_2$, 10 glucose, and 20 HEPES with solution pH adjusted to 7.4 with NaOH, where the addition of HEPES and removal of bicarbonate/CO_2 eliminate contributions from the Cl^-/HCO_3^- exchangers.

Ratiometric measurement of fluorescence from BCECF can be performed on any number of commercially available imaging workstations. We have used a workstation from Intracellular Imaging Inc. consisting of a Nikon TSE 100 Eclipse inverted microscope with epifluorescence attachments. Light from a xenon arc lamp is filtered by interference filters (Omega Optical) at 490 nm (pH sensitive) and 440 nm (pH insensitive) and focused onto the cells under examination via a 20 × fluorescence objective (Super Fluor 20; Nikon). Light emitted from the cells is filtered at 530 nm and detected by a digital imaging camera (Pixelfly 200XS; PCO-TECH Inc.). Images can be captured at any rate desired by the user; we found that 5 images/min provides suitable resolution to observe changes in pH_i in most cells. An electronic shutter (Sutter Instrument Co.) is used to minimize photobleaching of dye between image captures. All protocols are executed and data collected online with InCyte software (Intracellular Imaging Inc.). At the end of each experiment, pH_i is estimated from a two-point *in situ* calibration. Fluorescence is monitored as cells are perfused first with a solution containing (in m*M*) 105 KCl, 1 $MgCl_2$, 1.5 $CaCl_2$, 10 glucose, 20 HEPES–Tris, and 0.01 nigericin, an ionophore that allows pH_i to equilibrate to external pH, with solution pH adjusted to 6.5 with 5 *M* KOH. After 10 min, the solution is switched to one where pH is adjusted to 7.5 with KOH. During data analysis, the ratio of 490:440 nm fluorescence is converted to pH_i from a simple linear regression using the two-point calibration values. Intracellular H^+ ion concentration ($[H^+]_i$) is determined from pH_i using the formula $pH_i = -\log([H^+]_i)$. For most cell types, pH_i will average around 7.2 in HCO_3^--containing extracellular solution and between 6.6 and 6.9 with HBSS.

5.3. Measurement of sodium–hydrogen exchanger activity

The activity of Na^+/H^+ exchangers can be measured using the ammonium pulse technique (Fig. 22.5). BCECF-loaded cells are perfused with solution 1 containing (in m*M*) 130 NaCl, 5 KCl, 1 $MgCl_2$, 1.5 $CaCl_2$, 10 glucose, and 20 HEPES with pH adjusted to 7.4 with NaOH at 37 °C. Baseline pH_i is measured for 2 min before cells are exposed to NH_4Cl (ammonium pulse) by perfusing for 3 min with solution 2 containing (in m*M*) 110 NaCl, 20 NH_4Cl, 5 KCl, 1 $MgCl_2$, 1.5 $CaCl_2$, 10 glucose, and 20 HEPES at a pH of 7.4 using NaOH. The ammonium pulse causes intracellular alkalinization due to influx of NH_3 and buffering of intracellular H^+. Washout of NH_4Cl in the absence of extracellular Na^+ using a Na^+- and NH_4^+-free

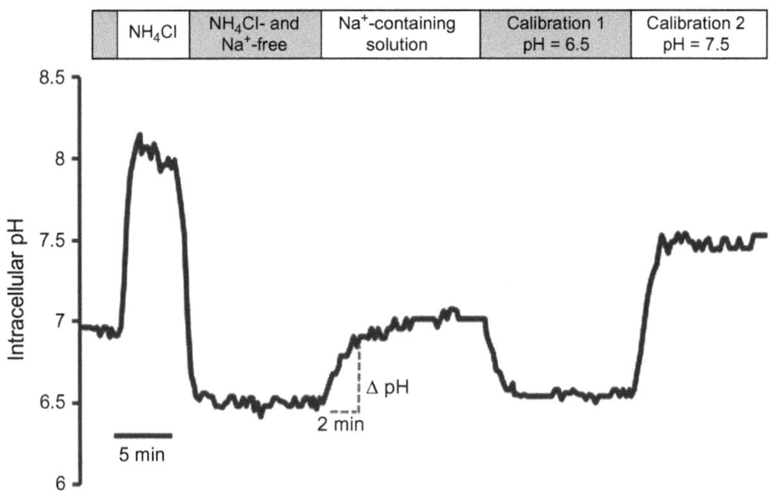

Figure 22.5 *Measurement of Na^+/H^+ exchange activity in BCECF-loaded cells.* Cells are initially perfused with Na^+-containing solution to establish a baseline followed by solution containing NH_4Cl (ammonium pulse). Perfusate is then switched to an NH_4Cl- and Na^+-free solution to induce acidosis. After 10 min, extracellular Na^+ is reintroduced, and Na^+/H^+ exchanger activity is estimated from the rise of pH over the first 2 min. Finally, an *in situ* calibration is performed.

solution (solution 3) containing (in m*M*) 130 choline chloride, 5 KCl, 1 $MgCl_2$, 1.5 $CaCl_2$, 10 glucose, and 20 HEPES at a pH of 7.4 using KOH for 10 min results in intracellular acidification due to rapid diffusion and washout of NH_3, leaving behind H^+ ions. The external solution is then switched back to solution 1 for 10 min. Readdition of extracellular Na^+ allows activation of Na^+/H^+ exchange and recovery from acidification to basal levels. The rate of Na^+-dependent recovery from intracellular acidification, usually calculated over the first 2 min and expressed as pH units/min, corresponds to Na^+/H^+ exchanger activity.

ACKNOWLEDGMENTS

G. L. S. is an American Cancer Society Research Professor and the C. Michael Armstrong Professor at the Johns Hopkins University School of Medicine and is supported by grants from the American Cancer Society, Congressionally Directed Medical Research Programs, National Cancer Institute and State of Maryland Department of Health and Mental Hygiene. R. J. D. is supported by NIH Grant R01-CA157996.

REFERENCES

Bellot, G., Garcia-Medina, R., Gounon, P., Chiche, J., Roux, D., Pouyssegur, J., et al. (2009). Hypoxia-induced autophagy is mediated through hypoxia-inducible factor induction of BNIP3 and BNIP3L via their BH3 domains. *Molecular and Cellular Biology, 29*(10), 2570–2581.

Budd, S. L., Castilho, R. F., & Nicholls, D. G. (1997). Mitochondrial membrane potential and hydroethidine-monitored superoxide generation in cultured cerebellar granule cells. *FEBS Letters, 415*(1), 21–24.

Cheng, T., Sudderth, J., Yang, C., Mullen, A. R., Jin, E. S., Mates, J. M., et al. (2011). Pyruvate carboxylase is required for glutamine-independent growth of tumor cells. *Proceedings of the National Academy of Sciences of the United States of America, 108*(21), 8674–8679.

Clark, L. C., Jr., Wolf, R., Granger, D., & Taylor, Z. (1953). Continuous recording of blood oxygen tensions by polarography. *Journal of Applied Physiology, 6*(3), 189–193.

DeBerardinis, R. J., Mancuso, A., Daikhin, E., Nissim, I., Yudkoff, M., Wehrli, S., et al. (2007). Beyond aerobic glycolysis: Transformed cells can engage in glutamine metabolism that exceeds the requirement for protein and nucleotide synthesis. *Proceedings of the National Academy of Sciences of the United States of America, 104*(49), 19345–19350.

Diepart, C., Jordan, B. F., & Gallez, B. (2009). A new EPR oximetry protocol to estimate the tissue oxygen consumption in vivo. *Radiation Research, 172*(2), 220–225.

Ebert, B. L., Firth, J. D., & Ratcliffe, P. J. (1995). Hypoxia and mitochondrial inhibitors regulate expression of glucose transporter-1 via distinct cis-acting sequences. *The Journal of Biological Chemistry, 270*(49), 29083–29089.

Fernandez, C. A., Des Rosiers, C., Previs, S. F., David, F., & Brunengraber, H. (1996). Correction of ^{13}C mass isotopomer distributions for natural stable isotope abundance. *Journal of Mass Spectrometry, 31*(3), 255–262.

Fukuda, R., Zhang, H., Kim, J., Shimoda, L., Dang, C. V., & Semenza, G. L. (2007). HIF-1 regulates cytochrome oxidase subunits to optimize efficiency of respiration in hypoxic cells. *Cell, 129*(1), 111–122.

Guzy, R. D., & Schumacker, P. T. (2006). Oxygen sensing by mitochondria at complex III: The paradox of increased reactive oxygen species during hypoxia. *Experimental Physiology, 91*(5), 807–819.

Harrison, C., Yang, C., Jindal, A., DeBerardinis, R. J., Hooshyar, M. A., Merritt, M., et al. (2012). Comparison of kinetic models for analysis of pyruvate-to-lactate exchange by hyperpolarized ^{13}C NMR. *NMR in Biomedicine, 25*(11), 1286–1294.

Hynes, J., Hill, H. R., & Papkovsky, D. B. (2006). The use of a fluorescence-based oxygen uptake assay in the analysis of cytotoxicity. *Toxicology In Vitro, 20*(5), 785–792.

Iyer, N. V., Kotch, L. E., Agani, F., Leung, S. W., Laughner, E., Wenger, R. H., et al. (1998). Cellular and developmental control of O_2 homeostasis by hypoxia-inducible factor 1α. *Genes and Development, 12*(2), 149–162.

Kabeya, Y., Mizushima, N., Yamamoto, A., Oshitani-Okamoto, S., Ohsumi, Y., et al. (2004). LC3, GABARAP and GATE16 localize to autophagosomal membrane depending on form-II formation. *Journal of Cell Science, 117*(13), 2805–2812.

Kim, J. W., Tchernyshyov, I., Semenza, G. L., & Dang, C. V. (2006). HIF-1-mediated expression of pyruvate dehydrogenase kinase: A metabolic switch required for cellular adaptation to hypoxia. *Cell Metabolism, 3*(3), 177–185.

Klionsky, D. J., Abdalla, F. C., Abeliovich, H., Abraham, R. T., Acevedo-Arozena, A., Adeli, K., et al. (2012). Guidelines for the use and interpretation of assays for monitoring autophagy. *Autophagy, 8*(4), 445–544.

Levine, B., & Klionsky, D. J. (2004). Development by self-digestion: Molecular mechanisms and biological functions of autophagy. *Developmental Cell, 6*(4), 463–477.

Liang, Y., Buettger, C., Berner, D. K., & Matschinsky, F. M. (1997). Chronic effect of fatty acids on insulin release is not through the alteration of glucose metabolism in a pancreatic beta-cell line (beta HC9). *Diabetologia, 40*(9), 1018–1027.

Lu, C. W., Lin, S. C., Chen, K. F., Lai, Y. Y., & Tsai, S. J. (2008). Induction of pyruvate dehydrogenase kinase-3 by hypoxia-inducible factor-1 promotes metabolic switch and drug resistance. *The Journal of Biological Chemistry, 283*(42), 28106–28114.

Lutz, N. W., Le Fur, Y., Chiche, J., Pouyssegur, J., & Cozzone, P. J. (2013). Quantitative in vivo characterization of intracellular and extracellular pH profiles in heterogeneous tumors: A novel method enabling multiparametric pH analysis. *Cancer Research, 73*(15), 4616–4628.

Maher, E. A., Marin-Valencia, I., Bachoo, R. M., Mashimo, T., Raisanen, J., Hatanpaa, K. J., et al. (2012). Metabolism of [U-^{13}C]glucose in human brain tumors in vivo. *NMR in Biomedicine, 25*(11), 1234–1244.

Marin-Valencia, I., Cho, S. K., Rakheja, D., Hatanpaa, K. J., Kapur, P., Mashimo, T., et al. (2012). Glucose metabolism via the pentose phosphate pathway, glycolysis and Krebs cycle in an orthotopic mouse model of human brain tumors. *NMR in Biomedicine, 25*(10), 1177–1186.

Marin-Valencia, I., Yang, C., Mashimo, T., Cho, S., Baek, H., Yang, X. L., et al. (2012). Analysis of tumor metabolism reveals mitochondrial glucose oxidation in genetically diverse human glioblastomas in the mouse brain in vivo. *Cell Metabolism, 15*(6), 827–837.

McLennan, H. R., & Degli Esposti, M. (2000). The contribution of mitochondrial respiratory complexes to the production of reactive oxygen species. *Journal of Bioenergetics and Biomembranes, 32*(2), 153–162.

Molter, T. W., McQuaide, S. C., Holl, M. R., Meldrum, D. R., Dragavon, J. M., Anderson, J. B., et al. (2008). A new approach for measuring single-cell oxygen consumption rates. *IEEE Transactions on Automation Science and Engineering, 5*(1), 32–42.

Papandreou, I., Cairns, R. A., Fontana, L., Lim, A. L., & Denko, N. C. (2006). HIF-1 mediates adaptation to hypoxia by actively downregulating mitochondrial oxygen consumption. *Cell Metabolism, 3*(3), 187–197.

Pilatus, U., Aboagye, E., Artemov, D., Mori, N., Ackerstaff, E., & Bhujwalla, Z. M. (2001). Real-time measurements of cellular oxygen consumption, pH, and energy metabolism using nuclear magnetic resonance spectroscopy. *Magnetic Resonance in Medicine*, *45*(5), 749–755.

Rey, S., Luo, W., Shimoda, L. A., & Semenza, G. L. (2011). Metabolic reprogramming by HIF-1 promotes the survival of bone marrow-derived angiogenic cells in ischemic tissue. *Blood*, *117*(18), 4988–4998.

Rink, T. J., Tsien, R. Y., & Pozzan, T. (1982). Cytoplasmic pH and free Mg^{2+} in lymphocytes. *Journal of Cell Biology*, *95*(1), 189–196.

Seagroves, T. N., Ryan, H. E., Lu, H., Wouters, B. G., Knapp, M., Thibault, P., et al. (2001). Transcription factor HIF-1 is a necessary mediator of the Pasteur effect in mammalian cells. *Molecular and Cellular Biology*, *21*(10), 3436–3444.

Semenza, G. L. (2011). Regulation of metabolism by hypoxia-inducible factor 1. *Cold Spring Harbor Symposium on Quantitative Biology*, *76*, 347–353.

Semenza, G. L., Jiang, B.-H., Leung, S. W., Passantino, R., Concordet, J.-P., Maire, P., et al. (1996). Hypoxia response elements in the aldolase A, enolase 1, and lactate dehydrogenase A gene promoters contain essential binding sites for hypoxia-inducible factor 1. *The Journal of Biological Chemistry*, *271*(51), 32529–32537.

Shimoda, L. A., Fallon, M., Pisarcik, S., Wang, J., & Semenza, G. L. (2006). HIF-1 regulates hypoxic induction of NHE1 expression and alkalinization of intracellular pH in pulmonary arterial myocytes. *American Journal of Physiology—Lung Cellular and Molecular Physiology*, *291*(5), L941–L949.

Simsek, T., Kocabas, F., Zheng, J., DeBerardinis, R. J., Mahmoud, A. I., Olsen, E. N., et al. (2010). The distinct metabolic profile of hematopoietic stem cells reflects their location in a hypoxic niche. *Cell Stem Cell*, 7(3), 380–390.

Swietach, P., Vaughan-Jones, R. D., & Harris, A. L. (2007). Regulation of tumor pH and the role of carbonic anhydrase 9. *Cancer Metastasis Reviews*, *26*(2), 299–310.

Tanida, I., Minematsu-Ikeguchi, N., Ueno, T., & Kominami, E. (2005). Lysosomal turnover, but not a cellular level, of endogenous LC3 is a marker for autophagy. *Autophagy*, *1*(2), 84–91.

Tanida, I., Sou, Y. S., Ezaki, J., Minematsu-Ikeguchi, N., Ueno, T., & Kominami, E. (2004). HsAtg4B/HsApg4B/autophagin-1 cleaves the carboxyl termini of three human Atg8 homologues and delipidates microtubule-associated protein light chain 3- and GABAA receptor-associated protein-phospholipid conjugates. *The Journal of Biological Chemistry*, *279*(35), 36268–36276.

TeSlaa, T., & Teitell, T. A. (2014). Conventional techniques to measure glycolytic flux. *Methods in Enzymology*, *542*.

Ullah, M. S., Davies, A. J., & Halestrap, A. P. (2006). The plasma membrane lactate transporter MCT4, but not MCT1, is up-regulated by hypoxia through a HIF-1α-dependent mechanism. *The Journal of Biological Chemistry*, *281*(14), 9030–9037.

Vaupel, P., Mayer, A., & Hockel, M. (2004). Tumor hypoxia and malignant progression. *Methods in Enzymology*, *381*, 335–354.

Walther, J. L., Metallo, C. M., Zhang, J., & Stephanopoulos, G. (2012). Optimization of ^{13}C isotopic tracers for metabolic flux analysis in mammalian cells. *Metabolic Engineering*, *14*(2), 162–171.

Yang, C., Sudderth, J., Dang, T., Bachoo, R. M., McDonald, J. G., & DeBerardinis, R. J. (2009). Glioblastoma cells require glutamate dehydrogenase to survive impairments of glucose metabolism or Akt signaling. *Cancer Research*, *69*(20), 7986–7993.

Zamzami, N., Marchetti, P., Castedo, M., Decaudin, D., Macho, A., Hirsch, T., et al. (1995). Sequential reduction of mitochondrial transmembrane potential and generation of reactive oxygen species in early programmed cell death. *Journal of Experimental Medicine*, *182*(2), 367–377.

Zhang, H., Bosch-Marce, M., Shimoda, L. A., Tan, Y. S., Baek, J. H., Wesley, J. B., et al. (2008). Mitochondrial autophagy is an HIF-1-dependent adaptive metabolic response to hypoxia. *The Journal of Biological Chemistry*, *283*(16), 10892–10903.

AUTHOR INDEX

Note: Page numbers followed by "*f*" indicate figures and "*np*" indicate footnotes.

C

D

E

F

G

K

M

N

O

P

Q

R

S

T

U

V

W

X

Y

Z

SUBJECT INDEX

Note: Page numbers followed by "*f*" indicate figures and "*t*" indicate tables.

D

E

H

S

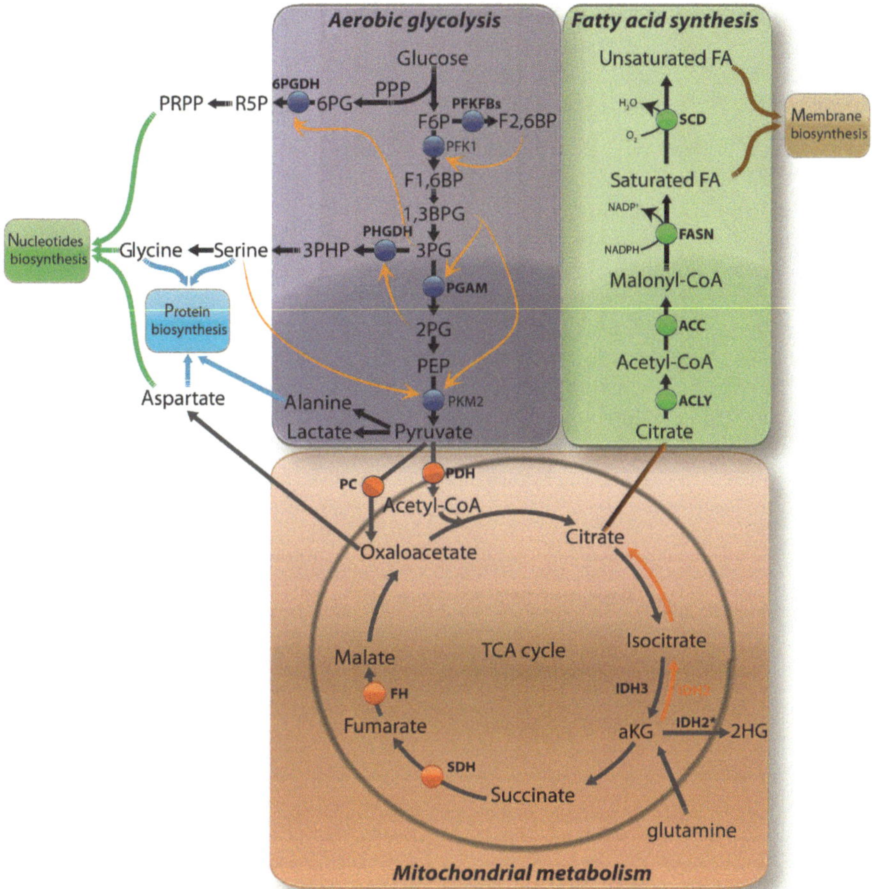

Marco Sciacovelli *et al.*, Figure 1.2 *The metabolic alterations of cancer cells.* Three major components of the metabolic transformation of cancer cells are depicted: aerobic glycolysis, mitochondrial metabolism, and lipid synthesis. Glucose branching to biosynthetic processes is regulated by various glycolytic enzymes, indicated as blue circles. The activity of these enzymes is subjected to allosteric control and posttranslational modification, indicated by the orange arrows. Mitochondrial metabolism is fueled by glucose and glutamine. PDH (red circle) is a mitochondrial gatekeeper and its activity is tightly regulated by genetic and environmental cues. When PDH is inactive, glutamine becomes the major carbon source for TCA cycle intermediates and anabolic precursors such as oxaloacetate and citrate. Both oxidative (gray arrows) and reductive (red arrows) metabolism of glutamine is used by cancer cells. PC could also provide glucose-derived carbons to the mitochondria, when PDH is inactive. Fatty acid biosynthesis requires acetyl-CoA obtained from citrate, which can be synthesized from glucose or glutamine. Several regulated steps (green circles) lead to the synthesis of palmitate, which is further converted into unsaturated lipids by the oxygen-dependent enzyme SCD. Dotted arrows indicate multiple enzymatic steps. Abbreviations: PPP, pentose phosphate

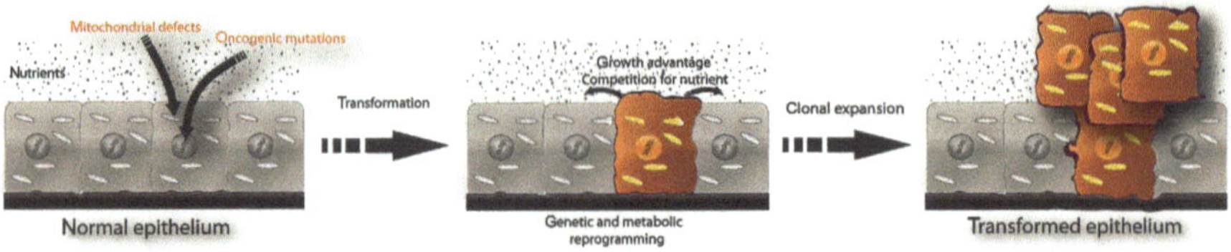

Marco Sciacovelli ***et al.*****, Figure 1.3** *Metabolic evolution of cancer cells.* We propose a general model of tumor formation and evolution: cancer transformation occurs upon mutations of key nuclear or mitochondrial enzymes. These mutations lead to a profound genetic and metabolic reprogramming that supports proliferation. Cancer-relevant mutations increase the fitness of cancer cells in the context of competition for nutrients and grant cancer cells a selective growth advantage. Only the mutations that support this feature will be selected in the process of clonal expansion. We propose that promotion of anabolism is a general characteristic of the most common oncogenic lesions.

---

pathway; PFK1, phosphofructokinase 1, PFKFBs, 6-phosphofructo-2-kinase/fructose-2,6-biphosphatases; F6P, fructose-6-phosphate; F1,6BP, fructose-1,6-bisphosphate; F2,6BP, fructose-2,6-bisphosphate; 1,3BPG, 1,3 bisphosphoglycerate; 3PG/2PG, 3/2 phosphoglycerate; PGAM, phosphoglycerate mutase; 6PGDH, 6-phosphogluconate dehydrogenase; 6PG, 6-phosphogluconate; R5P, ribulose-5-phosphate; PRPP, phosphoribosyl pyrophosphate; PHGDH, phosphoglycerate dehydrogenase; PEP, phosphoenolpyruvate; PKM2, pyruvate kinase variant M2; PDH, pyruvate dehydrogenase; PC, pyruvate carboxylase; FH, fumarate hydratase; SDH, succinate dehydrogenase; IDH, isocitrate dehydrogenase; IDH*, mutant IDH; aKG, alpha-ketoglutarate; 2HG, 2 hydroxyglutarate; ACLY, ATP-citrate lyase; ACC, acetyl-CoA carboxylase; FASN, fatty acid synthase; FA, fatty acid, SCD, stearoyl-CoA desaturase.

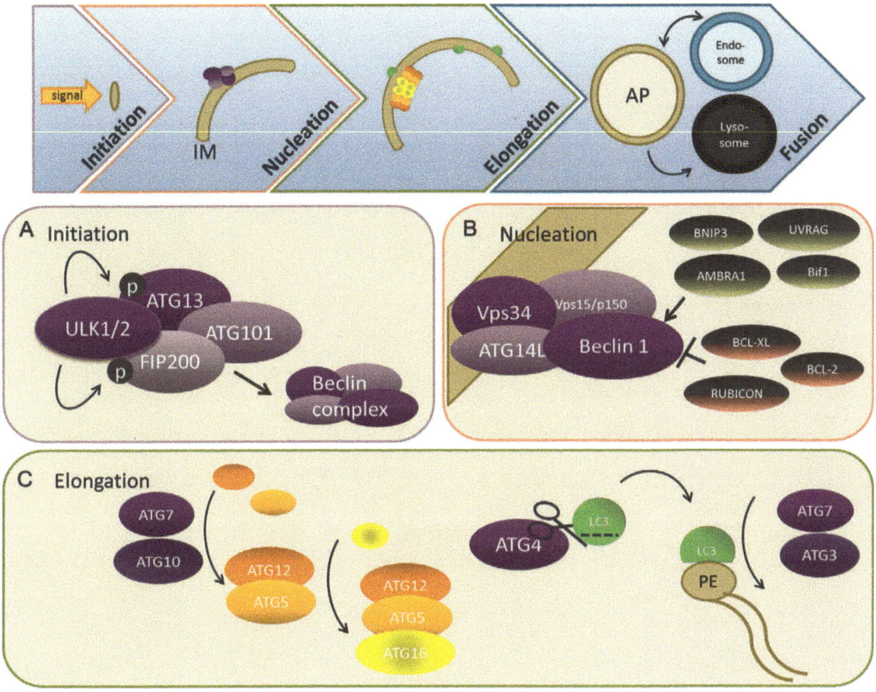

Juliet Goldsmith *et al.*, Figure 2.1 *Overview of the autophagy trafficking process.* The process of macroautophagy occurs in a series of distinct steps: (1) initiation of the isolation membrane (IM); (2) nucleation; (3) elongation of the double-membrane structure to form the autophagosome (AP); and (4) fusion to endosomes and lysosomes ultimately results in the formation of an autolysosome, in which the contents are degraded. (A) Initiation is mediated by the ULK complex. Activation of ULK activity leads to the phosphorylation of FIP200 and ATG13 and initiates nucleation via interaction with the Beclin 1 complex. (B) Beclin 1/VPS34/ATG14/PIK3R4 (p150) complex interacts with multiple interacting partners that positively and negatively regulate Beclin 1/VPS34 lipid kinase activity, resulting in the fine-tuning of autophagosome nucleation. (C) Elongation requires two ubiquitin-like conjugation pathways that form the ATG12–ATG5/ATG16 complex and phosphatidylethanolamine (PE)-conjugated LC3.

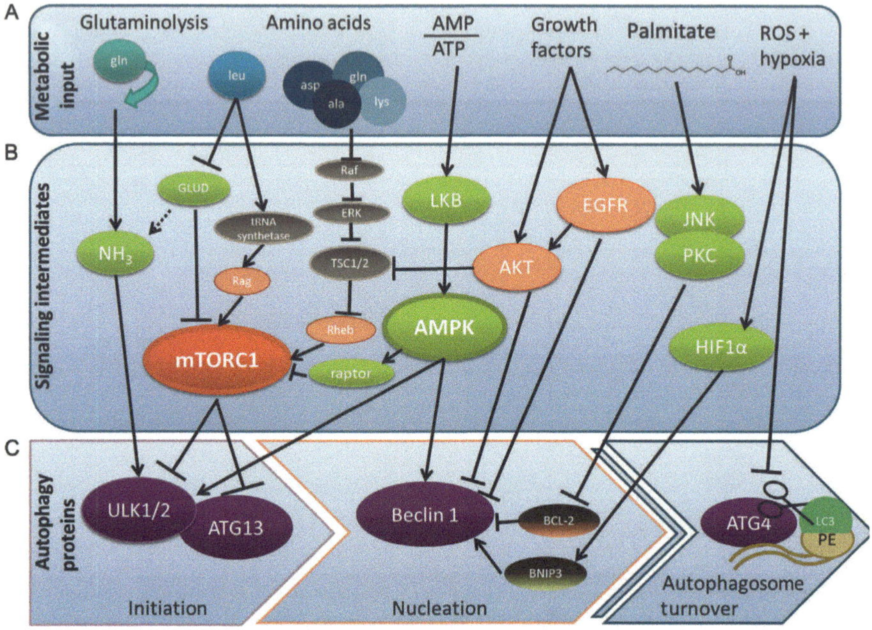

Juliet Goldsmith *et al.*, Figure 2.3 *AMPK and mTORC1 as metabolic regulators of autophagy*. The control of autophagosome formation and turnover is tightly controlled by many upstream metabolic stimuli. Metabolic input (A), such as concentrations of ammonia, general and specific amino acid levels, ATP to ADP ratio, and signals of growth and stress such as growth factors, reactive oxygen species, and palmitate, signals to initiate autophagosome formation and inhibits turnover either through signaling intermediates (B) or by directly inhibiting or activating key autophagy-related proteins (C). AMPK1 and mTOR are the main signaling integrators and modulators of autophagy—they sense glucose and amino acid levels and act on ULK1/2, ATG13, and Beclin 1 to inhibit autophagy in times of plenty and promote autophagy under energy-lean circumstances.

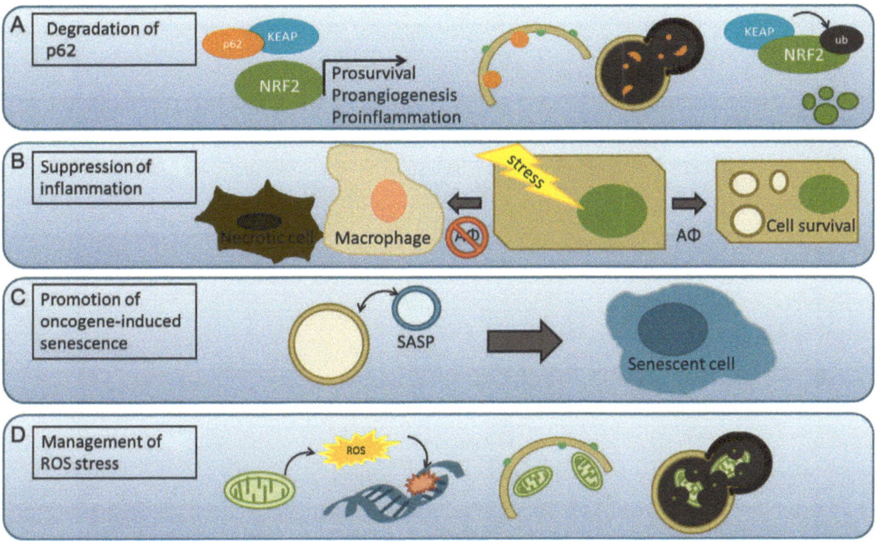

Juliet Goldsmith *et al.*, Figure 2.4 *Tumor-suppressive roles of autophagy.* (A) Autophagy prevents p62/NRF2 pathway prosurvival, proangiogenic, and proinflammatory signaling. During hypoxia, p62 binds to and sequesters KEAP, thereby preventing ubiquitination of NRF2. NRF2 can then promote transcription of prosurvival, proangiogenic, and proinflammatory genes that enhance tumor growth. When autophagy is active, p62 is degraded by sequestration within the autophagolysosome, thereby allowing ubiquitination and degradation of NRF2. (B) Autophagy prevents necrosis and inflammation by promoting survival in stressed cells. Tumor cells with decreased autophagy (AΦ) are more prone to necrosis following stress, which recruits macrophages, promotes inflammation, and fuels tumor growth. (C) Autophagy promotes oncogene-induced senescence by enhancing the senescence-associated secretory phenotype (SASP). (D) Autophagy suppresses reactive oxygen species (ROS) accumulation and genomic damage, which helps prevent genomic instability—an important driver of tumorigenesis. The mitochondria produce ROS under normal metabolic conditions and may increase the production of ROS when damaged. Mitophagy is upregulated in response to ROS and clears excess and damaged mitochondria, which then mitigates ROS production.

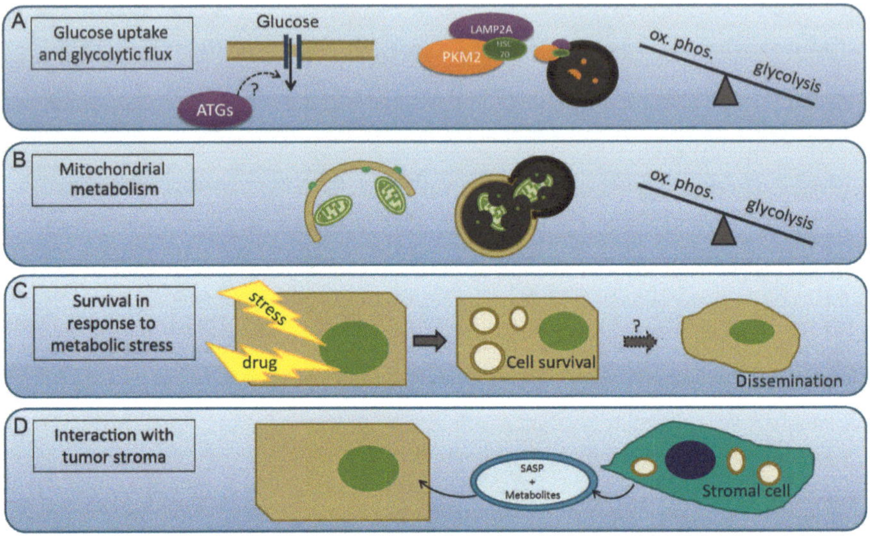

Juliet Goldsmith *et al.*, Figure 2.5 *Tumor-promoting roles of autophagy.* (A) Autophagy promotes glucose uptake and glycolytic flux. Autophagy has been shown to promote glucose uptake in cancer cells, although the mechanism remains to be elucidated. Additionally, increased chaperone-mediated autophagy (CMA) promotes degradation of PKM2, a rate-limiting glycolytic enzyme. Thus, CMA can control the rate of flux through glycolysis and whether glycolytic intermediates are used for energy production or anaplerosis. (B) Autophagy selectively degrades the mitochondria and therefore the machinery necessary for fatty acid oxidation and oxidative phosphorylation. This enhances the shift to glycolysis, which is characteristic of cancer cells. (C) Autophagy promotes survival in response to metabolic stress such as growth factor deprivation, acidic environment, and ER stress by recycling cytoplasmic material in order to maintain the basal energy state and clear damaged, misfolded proteins. This process may also be important for survival during tumor dissemination and metastasis. (D) Autophagy in stromal cells induced by the hypoxic and acidic tumor microenvironment promotes the secretion of metabolites and growth signals via senescence-associated secretion phenotype (SASP) that enhances tumor cell growth.

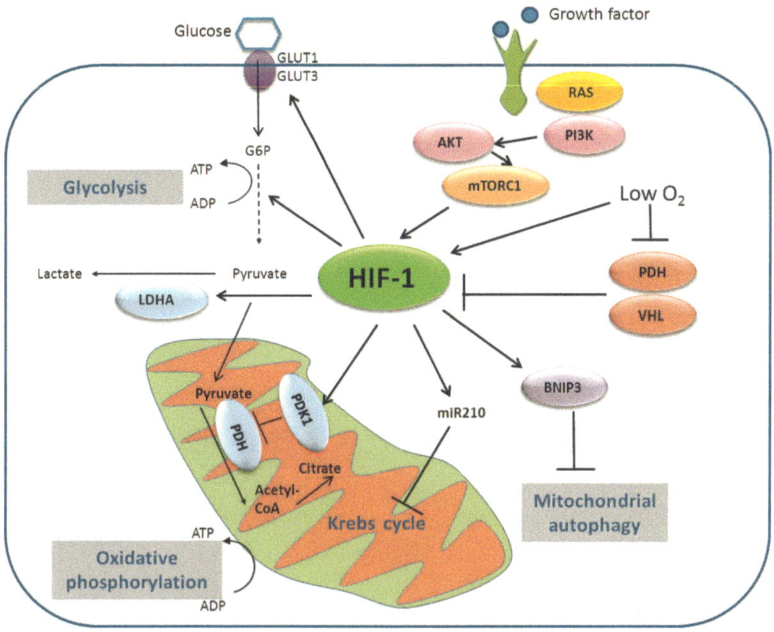

Raffaella Iurlaro *et al.*, Figure 3.1 *Regulation of cancer metabolism by HIF-1.* HIF-1 switches metabolism from oxidative respiration to anaerobic glycolysis. Hypoxia induces HIF-1 by blocking its inhibitors prolyl-4-hydroxylase domain proteins (PHDs) and von Hippel–Lindau (VHL) protein that need O_2 to exert their functions. Once activated, HIF-1 upregulates the glucose transporters GLUT1 and GLUT3, thus enhancing glucose uptake. HIF-1 induces the expression of almost every enzyme of the glycolytic pathway and lactate dehydrogenase A (LDHA), thus resulting in lactate production. Importantly, HIF-1 induces the pyruvate dehydrogenase kinase 1 (PDK1) that phosphorylates pyruvate dehydrogenase (PDH) blocking the entry of pyruvate into the mitochondria. HIF-1 also induces the expression of miR210, inhibiting important enzymes of Krebs cycle, and upregulates the protein BNIP3 that promotes mitochondrial autophagy.

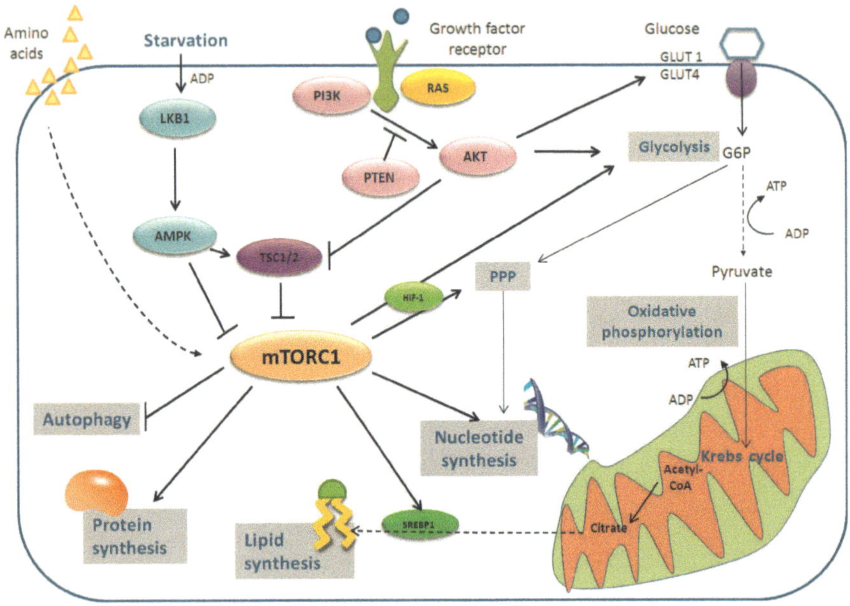

Raffaella Iurlaro *et al.*, Figure 3.2 *Regulation of cancer metabolism by the PI3K–AKT–PTEN and LKB1–AMPK–mTORC1 pathways.* Growth factor receptors activate Ras and phosphatidylinositol 3-kinase (PI3K) leading to the activation of AKT. Once activated, AKT induces glycolysis by regulating glycolytic enzymes and glucose transporters. These effects are counteracted by the phosphatase and tensin homologue (PTEN). AKT can indirectly activate the mTORC1 pathway that promotes lipid, protein, and nucleotide synthesis, contributing to the building of bioblocks necessary for tumor proliferation. Under stress conditions, the AMP-activated protein kinase (AMPK) activation through the liver kinase B1 (LKB1), opposes glycolytic metabolism in part by inhibiting mTORC1. PPP, pentose phosphate pathway.

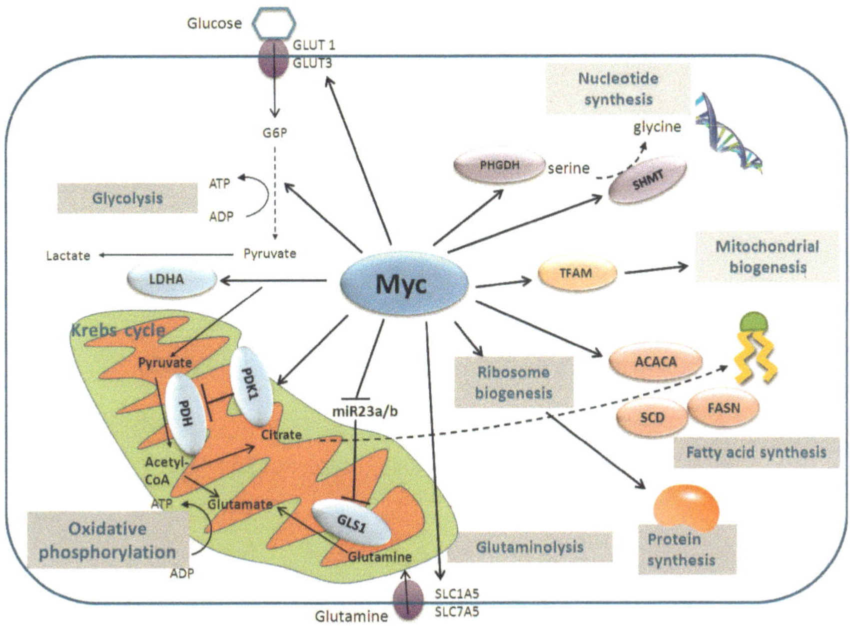

Raffaella Iurlaro *et al.*, Figure 3.3 *Myc regulates cancer metabolism*. Myc promotes cancer cell metabolism at several levels. Myc upregulates the glucose transporters GLUT1 and GLUT3 increasing glucose uptake. It induces several glycolytic enzymes such as the lactate dehydrogenase A (LDHA) resulting in lactate production. Like HIF-1, Myc induces pyruvate dehydrogenase kinase 1 (PDK1) expression, which prevents pyruvate entry into the mitochondria. Myc also regulates glutaminolysis: it upregulates glutamine transporters SLC1A5 and SLC7A5 and induces glutaminase 1 (GLS1) expression. Myc also promotes biomass accumulation essential for proliferating tumor cells. It regulates ribosome biogenesis, mitochondrial biogenesis, and several enzymes involved in fatty acids synthesis such as acetyl-CoA carboxylase (ACACA), fatty acid synthetase (FASN), and stearoyl-CoA desaturase (SCD). Additionally, Myc regulates enzymes involved in nucleotide synthesis such as phosphoglycerate dehydrogenase (PHGDH) and serine hydroxymethyltransferase (SHMT).

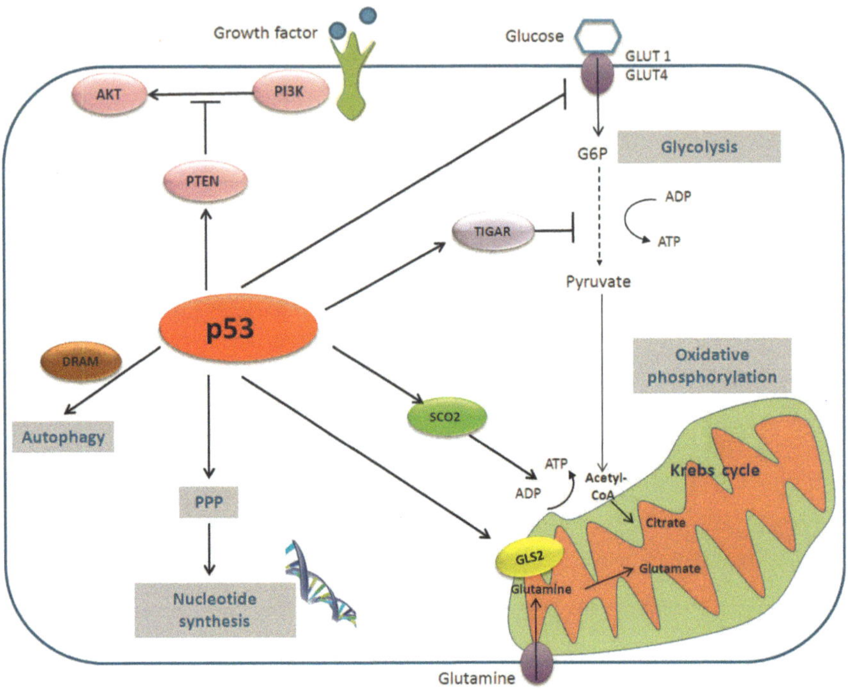

Raffaella Iurlaro *et al.*, Figure 3.4 *p53 regulates multiple metabolic pathways*. p53 responds to metabolic stress and it can inhibit the tumorigenic metabolic switch by suppressing glycolysis and activating the phosphatase and tensin homologue (PTEN). p53 inhibits the transcription of GLUT1 and GLUT4 reducing glucose uptake and it upregulates the TP53 (tumor protein 53)-induced glycolysis and apoptosis regulator (TIGAR), which results in glycolysis inhibition. p53 increases the mitochondrial metabolism by activation of the synthesis of cytochrome c oxidase 2 (SCO2), thus promoting oxidative phosphorylation. p53 can also induce, contradictorily, prosurvival responses in cancer cells, for instance when it increases the flux through the pentose phosphate pathway (PPP) or glutamine utilization. P53 can regulate positively autophagy by increasing the expression of DRAM.

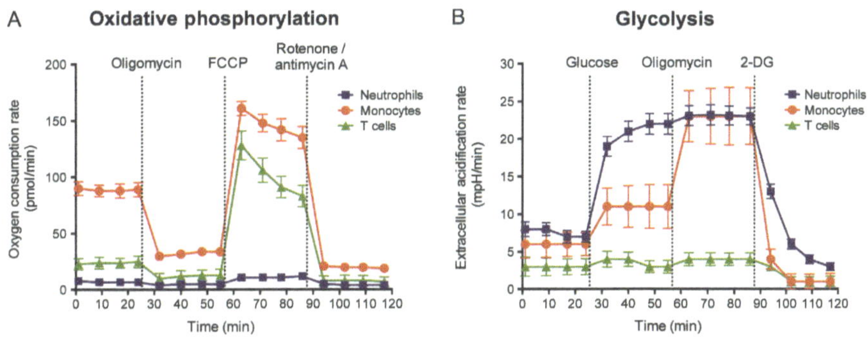

Martin Pelletier *et al.*, Figure 7.4 *Bioenergetic profiles of circulating immune cell subsets.* Human PBMCs and granulocytes were isolated from the blood of healthy donors by density gradient centrifugation using Ficoll-Paque™. Negative magnetic separation was used to further purify monocytes and $CD4^+$ T cells from the PBMC fraction by negative magnetic separation, while neutrophils were purified from the granulocyte fraction. Cells (400,000 cells/well) were seeded in warm XF assay medium on a Cell-Tak-treated XF96 cell culture plate, centrifuged, and incubated at 37 °C without CO_2 for 45 min. Data were acquired with a XF96 Extracellular Flux Analyzer. Cell number was normalized using CyQUANT reagent according to manufacturer's instructions. (A) Oxygen consumption rates were measured using sequential addition of oligomycin (5 μ*M*), FCCP (1.5 μ*M*), and rotenone + antimycin A (1 μ*M* and 100 n*M*, respectively). (B) Extracellular acidification rates were measured in real time using sequential addition of glucose (25 m*M*), oligomycin (5 μ*M*), and 2-deoxyglucose (2-DG; 100 m*M*).

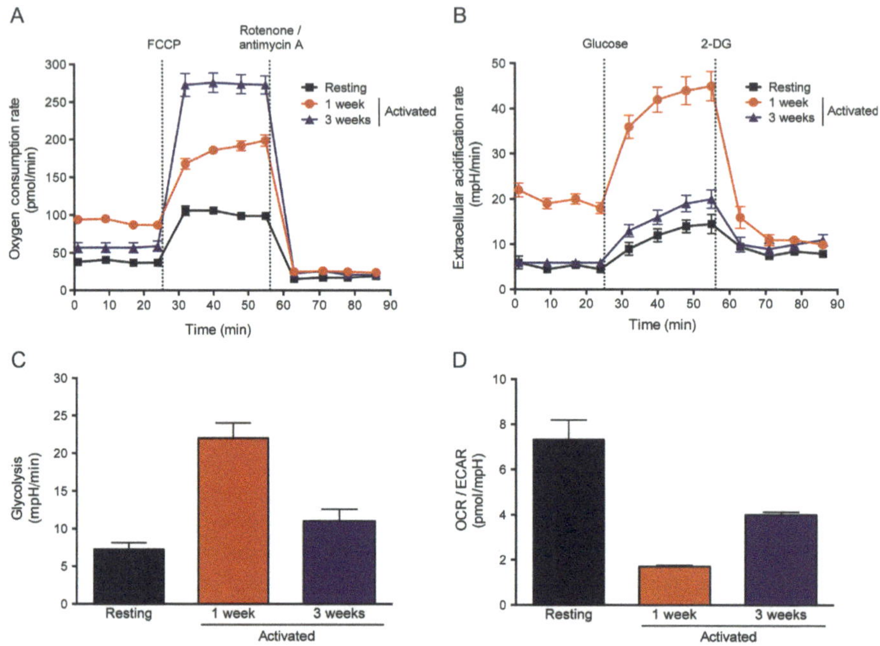

Martin Pelletier *et al.*, Figure 7.5 *Activation-induced glycolytic switch of CD8⁺ T cells.* Total human CD8⁺ T cells were purified using a negative selection kit from a normal donor apheresis sample (resting). The purified CD8⁺ T cells were stimulated for 48 h with anti-CD3/CD28 and then expanded in media with IL-2 for either 5 days (1 week activated) or 3 weeks (3 week activated). (A) Oxygen consumption rates were determined at baseline or after FCCP addition following the mitochondrial stress test protocol. (B) Extracellular acidification rates were evaluated after sequential addition of glucose and 2-deoxyglucose (2-DG) following the glycolytic stress test protocol. (C) Glycolysis was determined by subtracting the extracellular acidification rates following the addition of glucose from nonglycolytic acidification from the data shown in (B). (D) OCR/ECAR ratios representing the relative contributions of basal respiration versus glycolysis are shown for each cell type. Derived from the data in (A) and (B).

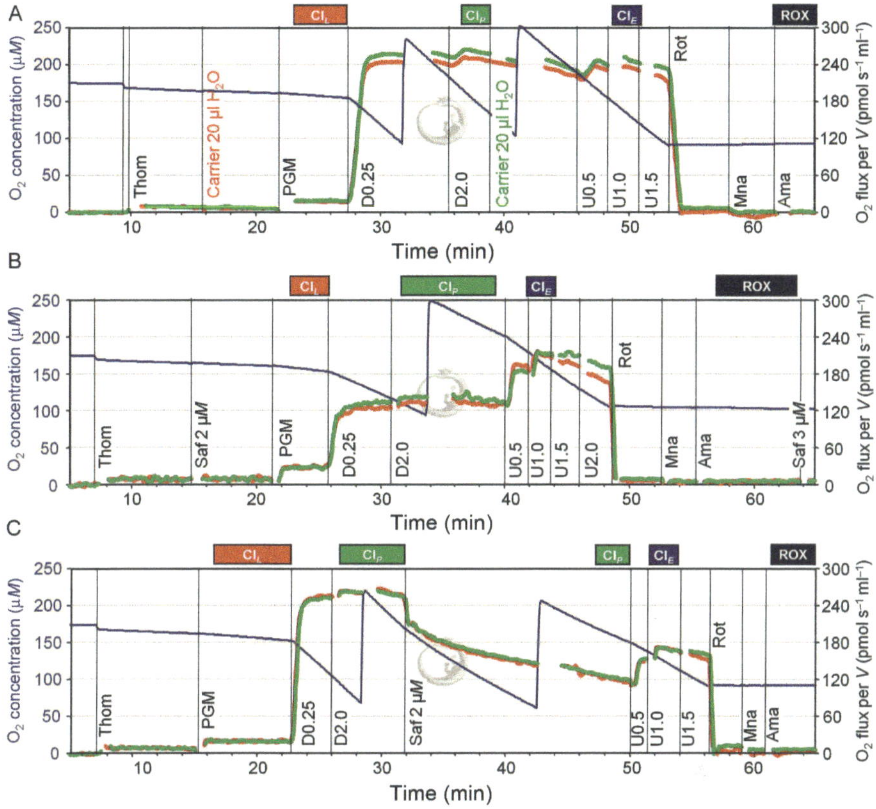

Gerhard Krumschnabel *et al.*, Figure 9.2 *High-resolution respirometry with mouse brain homogenate and substrate–uncoupler–inhibitor titration (SUIT) protocol with concentrations as described in Fig. 9.1.* Oxygen concentration (blue lines; left axes [μ*M*]) is shown for one chamber, whereas plots of oxygen flux per chamber volume (red and green lines, right axes [pmol s^{-1} ml^{-1}]) are superimposed for the two chambers of the O2k operated simultaneously. The plots of oxygen flux are interrupted by disturbances due to titrations or reoxygenations (3 μl of a 200 m*M* stock solution of H_2O_2). (A) Control experiment without safranin, with titration of carrier (20 μl H_2O) at different time points for the two chambers (red or green). (B) Addition of 2 μ*M* safranin into both chambers before substrate titration. (C) Addition of 2 μ*M* safranin into both chambers after inducing Complex I-linked OXPHOS with PGM and 2 m*M* ADP. Thom denotes addition of tissue homogenate (first spin), and bars above each panel denote CI-linked LEAK state (CI_L), OXPHOS state (CI_P), ETS state (CI_E), and residual oxygen consumption (ROX). Exp. 2013-02-13_KL-02, 2013-02-13_AB-02, 2013-02-13_CD-02.

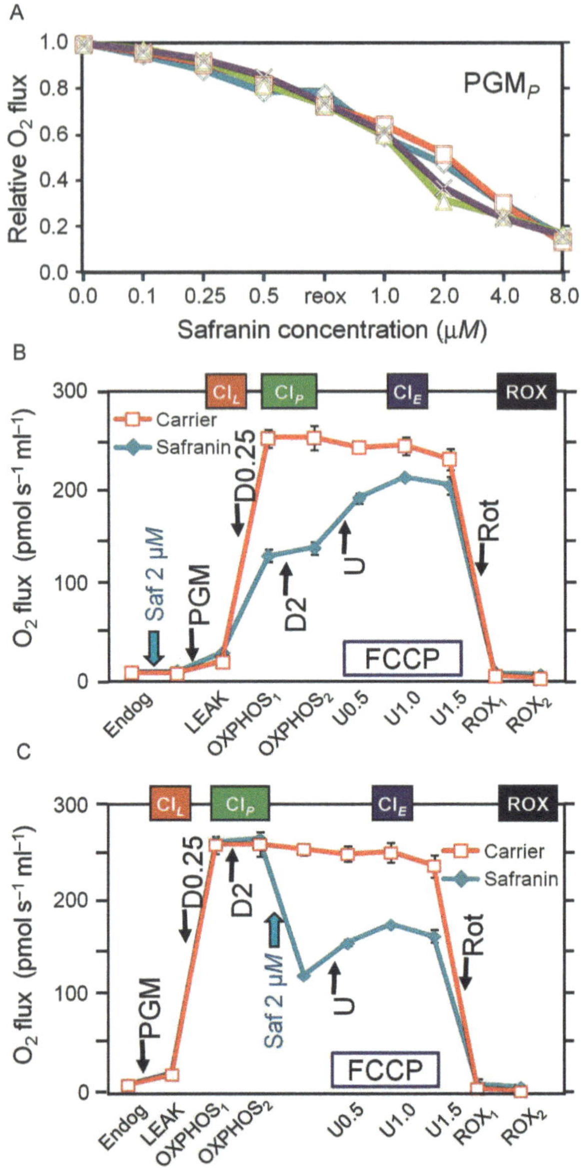

Gerhard Krumschnabel *et al.*, Figure 9.3 *Inhibition of CI-linked respiration by safranin (PGM, pyruvate + glutamate + malate) in mouse brain homogenate.* (A) Oxygen consumption rate as a function of safranin concentration (four assays), normalized to maximum OXPHOS (PGM$_P$). (B, C) Addition of safranin (Saf 2 μ*M*) before or after addition of substrates and ADP. Endog, respiration supported by endogenous substrates; LEAK, respiration supported by PGM without adenylates; $OXPHOS_1$ and $OXPHOS_2$, *P* at 0.25 and 2 m*M* ADP; U0.5, U1.0, U1.5, uncoupled respiration at 0.5, 1, and 1.5 μ*M* FCCP, without effect on carrier controls, but stimulation in the presence of safranin, indicating that the phosphorylation system was more strongly inhibited than ETS; ROX, residual oxygen consumption after inhibition by rotenone (Rot; ROX_1) and malonic acid and antimycin A (ROX_2). Data are means ± SD of four measurements on samples from two independent preparations.

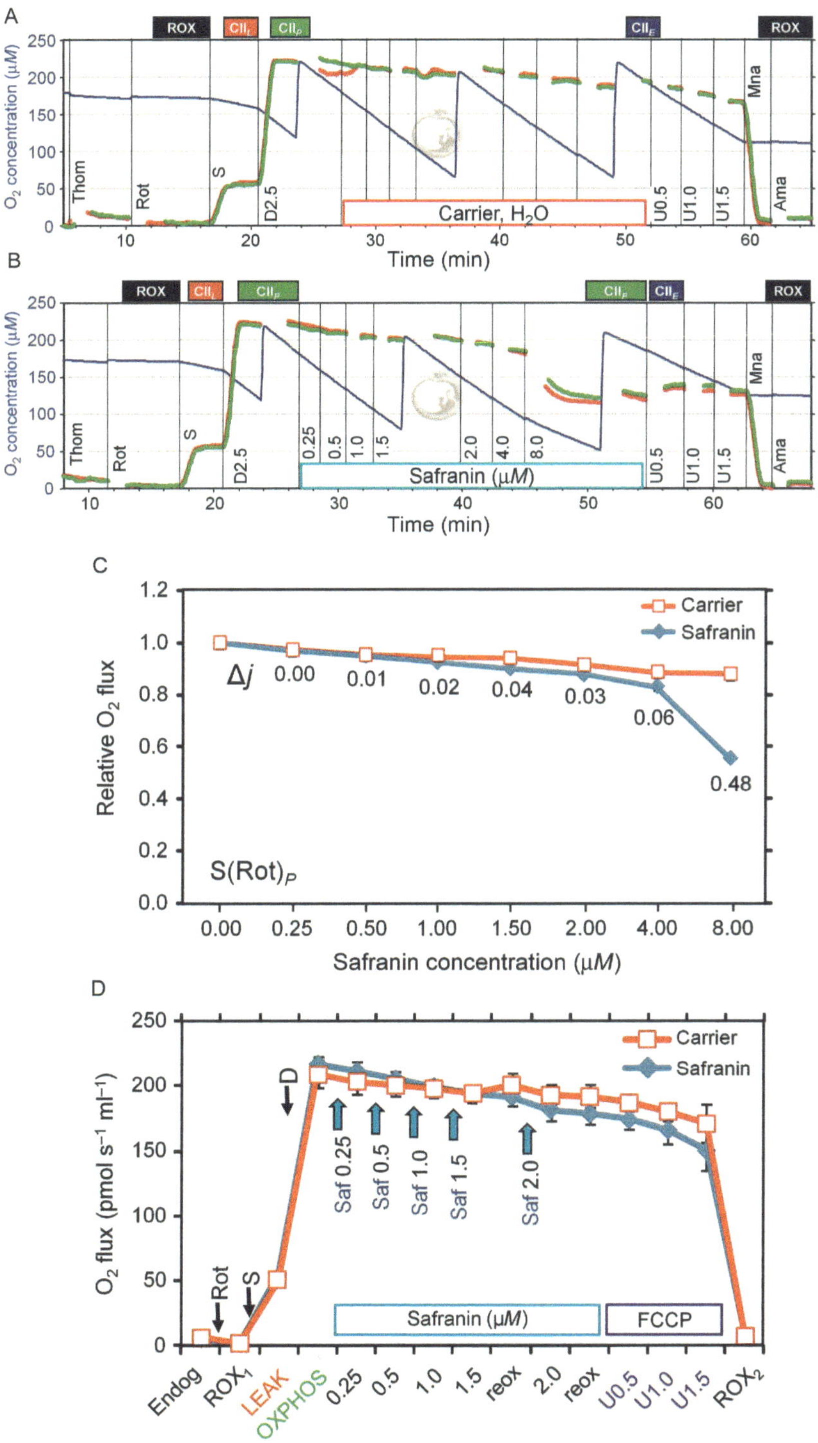

Figure 9.4 See figure caption on next page.

Gerhard Krumschnabel *et al.*, Figure 9.4 *Inhibition of Complex II-linked respiration by safranin in mouse brain homogenate.* Safranin or carrier was added in the OXPHOS state, CII_P. (A, B) Oxygen concentration (blue lines, one chamber) and plots of oxygen flux superimposed for two chambers, with titration of carrier or safranin from 0.25 to 8 μ*M*. Thom, first spin tissue homogenate; Rot, rotenone; S, succinate; D2.5, 2.5 m*M* ADP. Bars above each panel denote CII-linked LEAK state (CII_L), OXPHOS state (CII_P), ETS state (CII_E), and residual oxygen consumption (ROX). (C) Respiration rate normalized to the OXPHOS state; the inhibitory effect is shown as $\Delta j =$ (Carrier–Saf)/Carrier, becoming significant at 4 μ*M* safranin. Data are means ± SD of 10 measurements on samples from four independent preparations. (D) Respiratory flux and safranin titration from 0.25 to 2.0 μ*M*; ROX_1, LEAK, and OXPHOS denote oxygen consumption after addition of rotenone, succinate, and 2.5 m*M* ADP, respectively. Data are means ± SD of three measurements on samples from two independent preparations. Exp. 2013-06-25_MN-02 (A) and 2013-06-25_OP-02 (B).

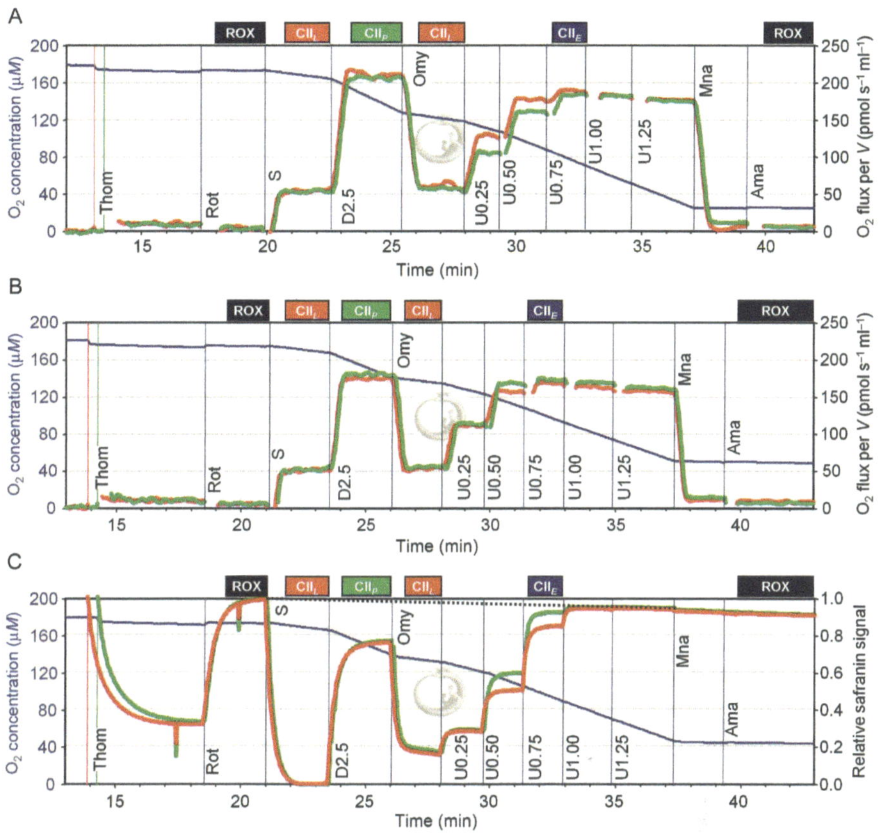

Gerhard Krumschnabel *et al.*, Figure 9.6 *Combined determination of oxygen consumption and mitochondrial membrane potential in the O2k high-resolution respirometer equipped with the O2k-Fluorescence LED2 module.* Oxygen concentration (blue plots, one chamber) and superimposed plots of oxygen flux or fluorescence (two chambers). Safranin was titrated before addition of first spin mouse brain homogenate (Thom). (A) Respiration without safranin. (B, C) Respiration and fluorescence changes with 2 μM safranin. Rot, rotenone inhibiting endogenous respiration to the level of residual oxygen consumption at minimum mtMP (ROX; maximum fluorescence signal); S, succinate-stimulating LEAK respiration and maximizing mtMP in the absence of adenylates (CII_L); D2.5, 2.5 mM ADP-activating OXPHOS and reducing mtMP (CII_P); Omy, oligomycin inducing a LEAK state by inhibition of ATP synthase (CII_L); U0.25–U1.25, uncoupler titration from 0.25 to 1.25 μM FCCP; optimum FCCP concentration for ETS capacity at 0.75 μM (CII_E); Mna and Ama, malonic acid and antimycin A inhibit CII and CIII and reduce respiration to the level of residual oxygen consumption (ROX). The fluorescence signal shows an apparent drift in the uncoupled and inhibited state, which is extrapolated to the initial ROX state (dotted line). Exp. 2013-06-25_MN-04, 2013-06-25_OP-04.

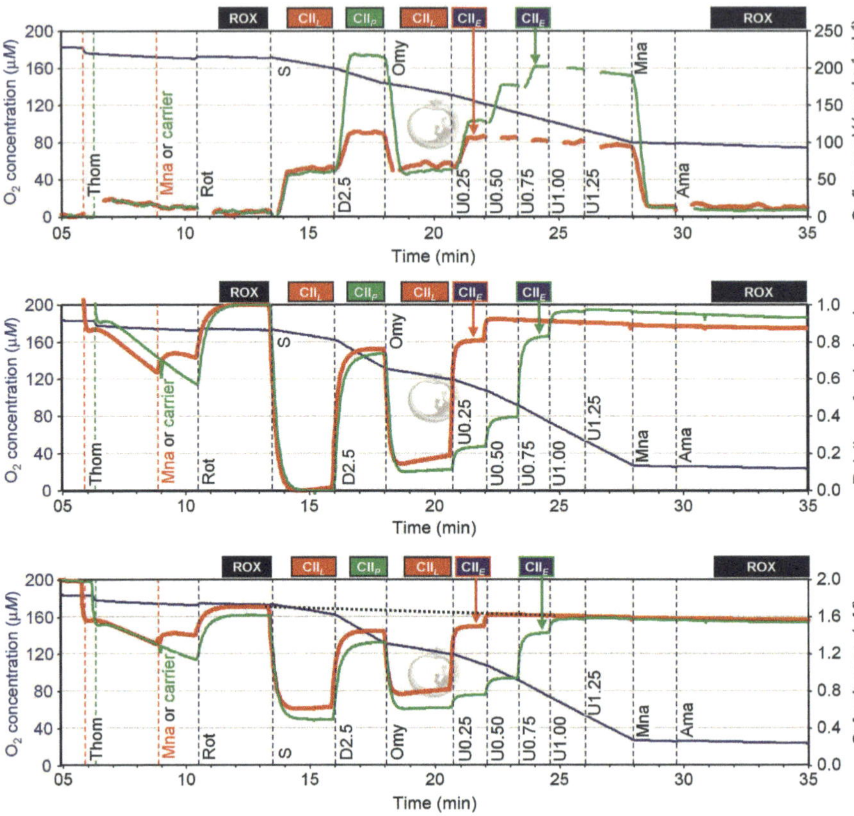

Gerhard Krumschnabel *et al.*, Figure 9.7 *Combined HRR and O2k-fluorometry in a model for a pathological condition compared to controls.* Oxygen concentration (blue plots, one chamber) and superimposed plots of oxygen flux or fluorescence (two chambers). Safranin was titrated before addition of first spin mouse brain homogenate (Thom). Carrier (H_2O) or 0.2 m*M* Mna (inhibiting CII-linked OXPHOS by 50%) was added before starting the SUIT protocol (see Fig. 9.6 for details). (A) Superimposed traces of oxygen concentration and fluxes of controls and Mna-treated cells. (B) Normalized fluorescence signals in controls and Mna-treated cells in the range between minimum and maximum values observed in the ROX and CII_L state. (C) Safranin concentration calibrated before addition of tissue homogenate. The fluorescence signal shows an apparent drift in the uncoupled and inhibited state, which is extrapolated to the initial ROX state (dotted line). Exp. 2013-08-09 MN-05, 2013-08-09 OP-04.

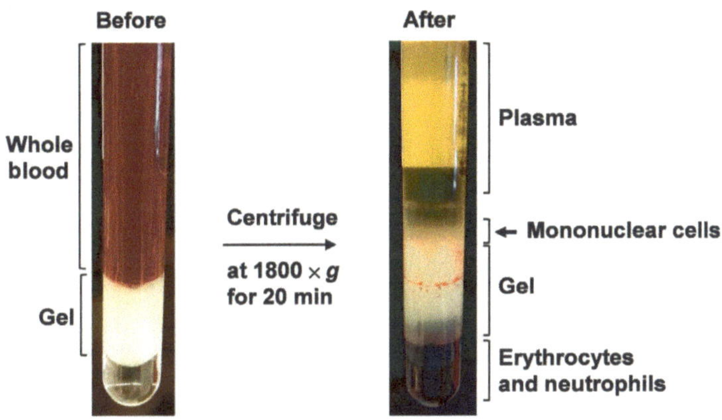

Ju-Gyeong Kang *et al.*, Figure 11.1 *Separation of lymphocytes from whole blood.* The appearance of a CPT filled with blood before and after centrifugation is shown. Density centrifugation allows the separation of peripheral mononuclear cells from other blood cell types and plasma. The mononuclear cell layer is gently transferred with a pipette for further processing and culturing.

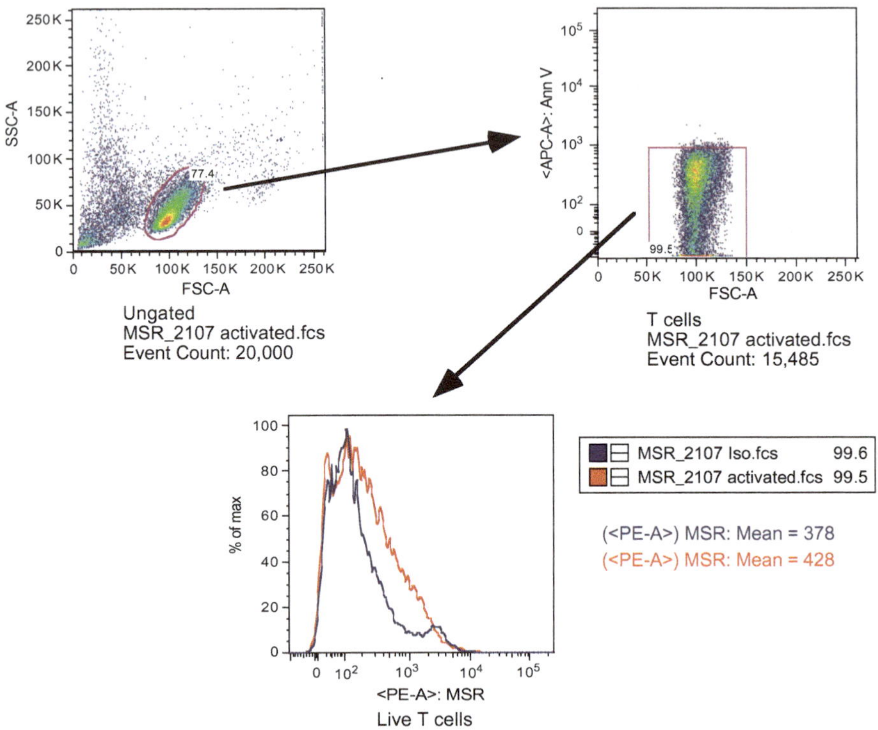

Jing Chen and Clayton E. Mathews, Figure 12.1 *Gating strategy for the cytofluorometic detection of the mitochondrial ROS in human T cells.* Samples are enriched human T cells as mentioned in the text (step 3.2.5). Lower T cell percentage is due to the fact that red blood cells are not lysed in this experiment. The top left panel is a dot plot for the T cell population that was gated on side-scatter and forward-scatter. Live T cells were gated by exclusion of Annexin-V positive cells (top right panel). Lower panel shows comparison of MitoSox Red histogram between activated T cells (red line) and isotype antibody treated controls (blue line).

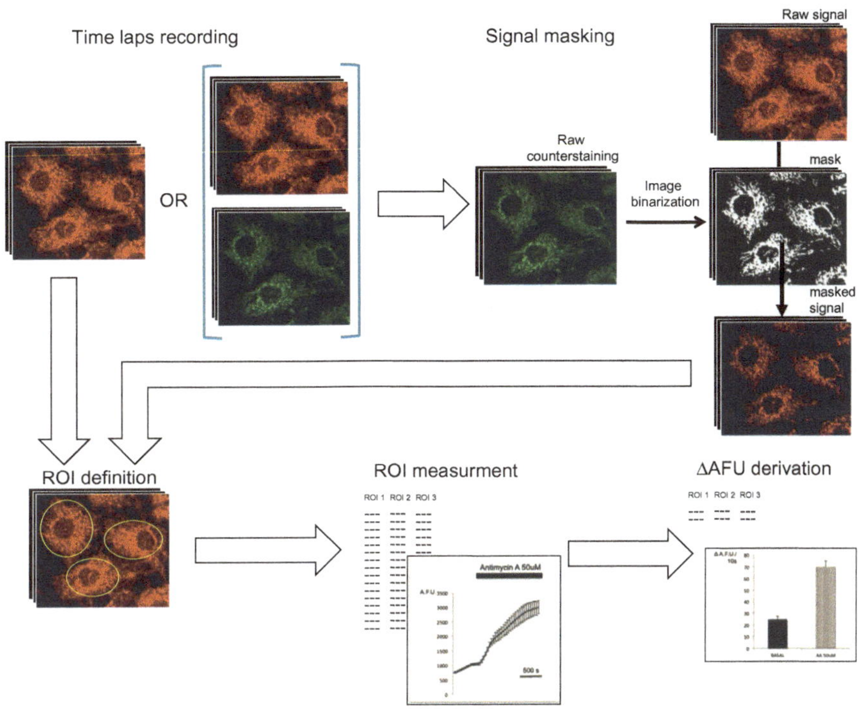

Aleksandra Wojtala *et al.*, Figure 13.3 *Overview of ROS imaging and analysis workflow.* A time-lapse series is acquired with the preferred dye alone or in combination with counterstaining for the specific subcellular regions (e.g., mitochondria) or with other conditions (e.g., co-transfection with an oligonucleotide or cDNA). For counterstaining, the raw counterstaining images should be converted into a mask by image binarization. Such a mask is used to filter the raw signal images for the ROS probe, thereby allowing only the overlapping regions to be considered in the analysis. Regions of interest (ROIs) should then be defined in the raw or filtered images and the average fluorescence intensity calculated for each ROI at each time point; such data will be collected in a spreadsheet for further analysis. The delta value is then determined by calculating the difference between the intensity at each time point and the intensity of the preceding time point. Finally, the average delta intensity, expressed in arbitrary fluorescence units (ΔAFU), of each ROI has to be calculated for each experimental condition.

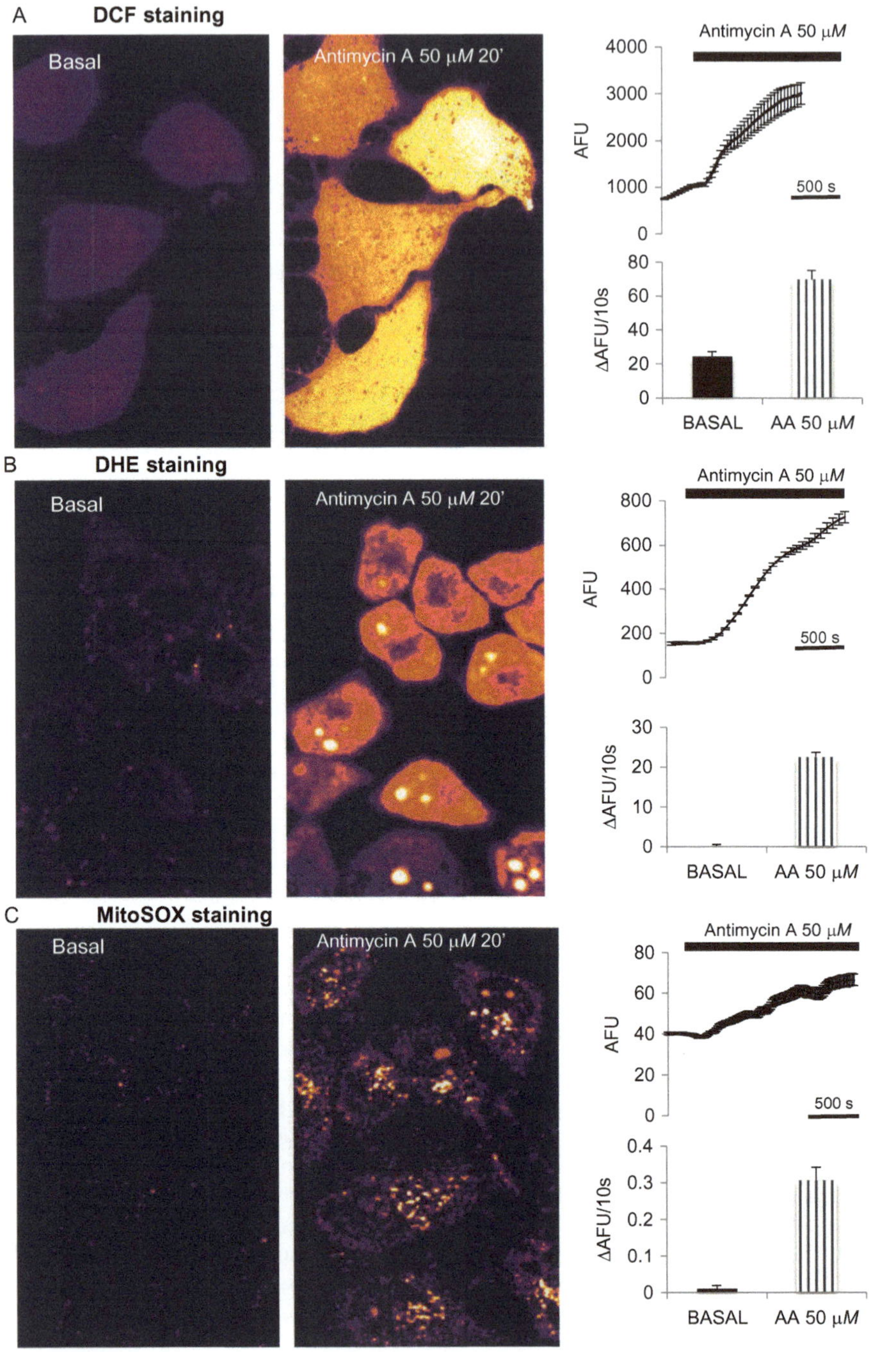

Aleksandra Wojtala *et al.*, Figure 13.4 *Image-based measurement of ROS production after stimulation*. Representative measurements of cytosolic ROS production by CM-H_2DCFDA (A), cytosolic ROS by DHE (B), and mitochondrial ROS by MitoSOX Red (C) before and after stimulation with 50 μ*M* antimycin A (AA) in HeLa cells by laser scanning confocal microscopy. For each group, representative false-color images are displayed, along with graphs of the kinetics and ROS production rates.

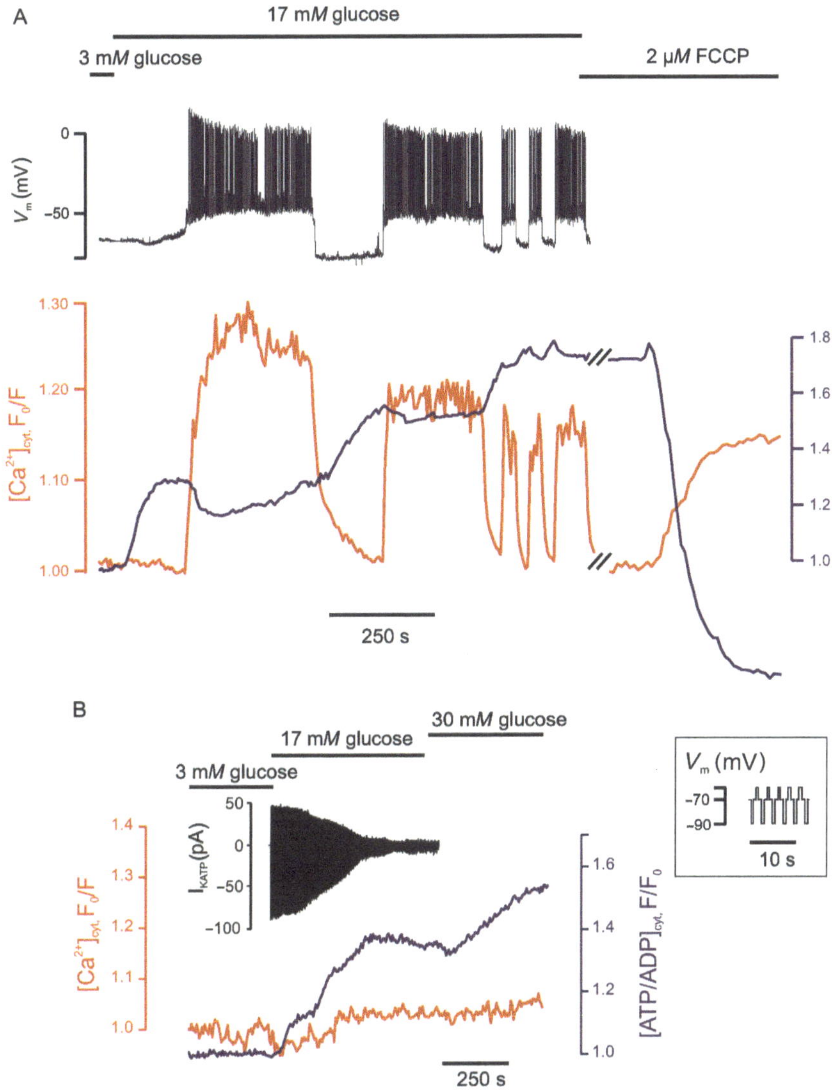

Andrei I. Tarasov and Guy A. Rutter, Figure 15.2 *Changes in [ATP/ADP]$_{cyt}$, [Ca^{2+}]$_{cyt}$, and V_m induced by glucose in pancreatic β-cells.* (A) Typical response in [ATP/ADP]$_{cyt}$, [Ca^{2+}]$_{cyt}$, and V_m to the glucose step from 3 to 17 m*M* recorded in a single pancreatic β-cell. The cell expressing Perceval was loaded with Fura-Red and studied using the setup in current-clamp mode (Fig. 15.1). Note the transient decreases in [ATP/ADP]$_{cyt}$ following the [Ca^{2+}]$_{cyt}$ increases induced by the bursts of electrical activity. FCCP was applied at the end of the recording as a negative control. (B) Voltage clamping of the cell at −70 mV keeps the voltage-gated Ca^{2+} channels shut, thereby preventing Ca^{2+} entry into the cytosol. Small depolarizations (inset) applied to the voltage-clamp cell induced K$^+$ current, mainly through K$_{ATP}$ channels. Note a close correlation of the K$_{ATP}$ channel current ($I_{K_{ATP}}$) with [ATP/ADP]$_{cyt}$. Only a fragment of the $I_{K_{ATP}}$ trace, including the whole range of $I_{K_{ATP}}$ variation, is shown for convenience. The step from 3 to 17 m*M* glucose induces an elevation on [ATP/ADP]$_{cyt}$ and corresponding decrease in $I_{K_{ATP}}$. No Ca^{2+}-induced potentiation of [ATP/ADP]$_{cyt}$ is observed. Interestingly, further elevation of the substrate concentration to 30 m*M* prompts a respective increase in [ATP/ADP]$_{cyt}$.

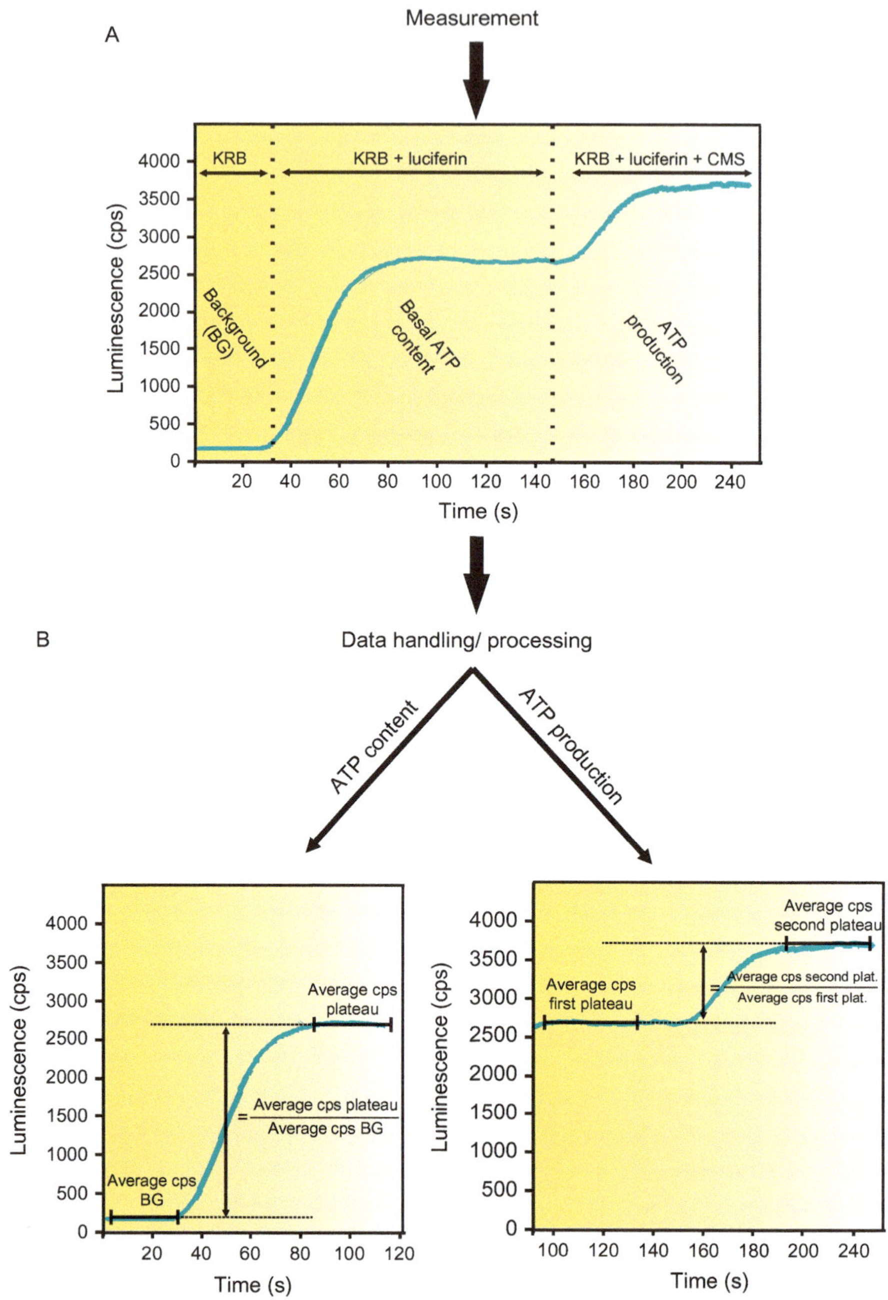

Simone Patergnani *et al.*, Figure 16.1 *Calibration and measurement of intracellular ATP levels using a luminometer.* (A) A representative luminescence trace employed to measure free [ATP] in intracellular compartments using a specifically targeted chimera of the ATP-sensitive photoprotein luciferase. The resting luminescence (background, BG) was recorded before the addition of luciferin to assess the basal ATP levels. Next, the cells were perfused with a Ca^{2+}-mobilizing solution (CMS) to determine the amount of ATP production. (B) The data were processed to estimate the basal ATP content and the levels of ATP produced. KRB, modified Krebs–Ringer solution; cps, counts per second.

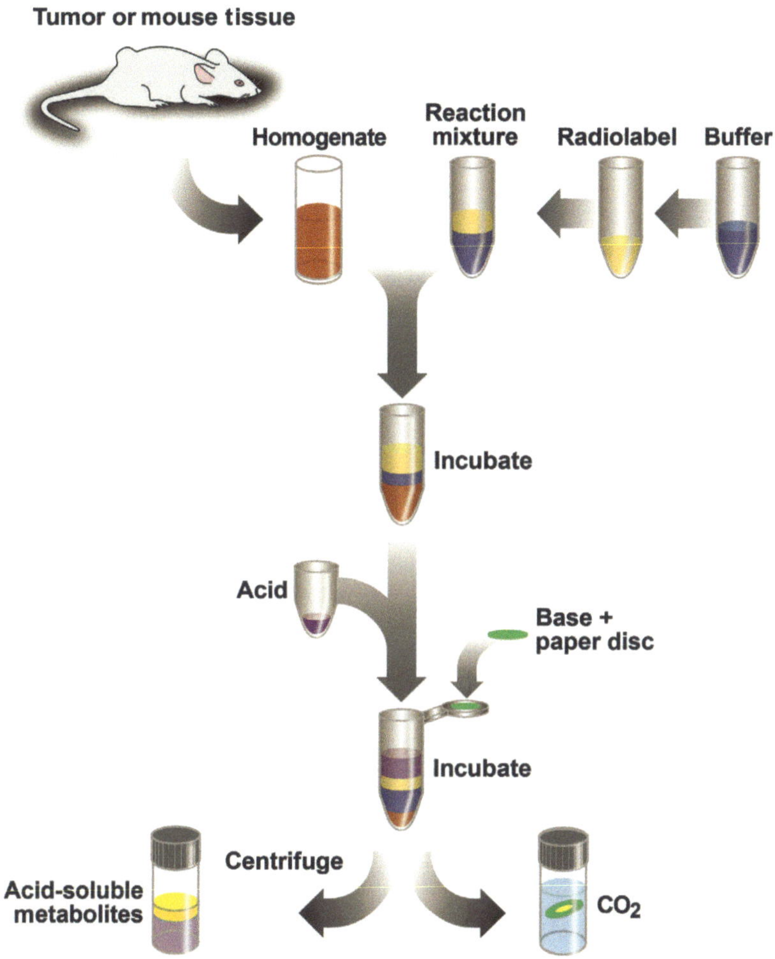

Frank K. Huynh *et al.*, Figure 20.2 *Schematic of the protocol for tissue homogenates.* The tissue or tumor of interest is homogenized with a Dounce homogenizer to lyse cells, while keeping the mitochondria intact. The tissue homogenate is then incubated with a reaction mixture containing radiolabeled palmitate. A strong acid is added to stop the reaction. Unoxidized bound palmitate will precipitate out of solution upon addition of acid. Acetyl-CoA, acetylcarnitines, ketone bodies, and tricarboxylic acid (TCA) cycle intermediates (acid-soluble metabolites) will remain soluble after the addition of the strong acid. Acetyl-CoA that enters the TCA cycle can be oxidized to CO_2 and this will be captured by the strong base in the Whatman paper disc. The amount of ^{14}C in the acid-soluble metabolite fraction and the paper disc CO_2 fraction can each be measured by scintillation counting.

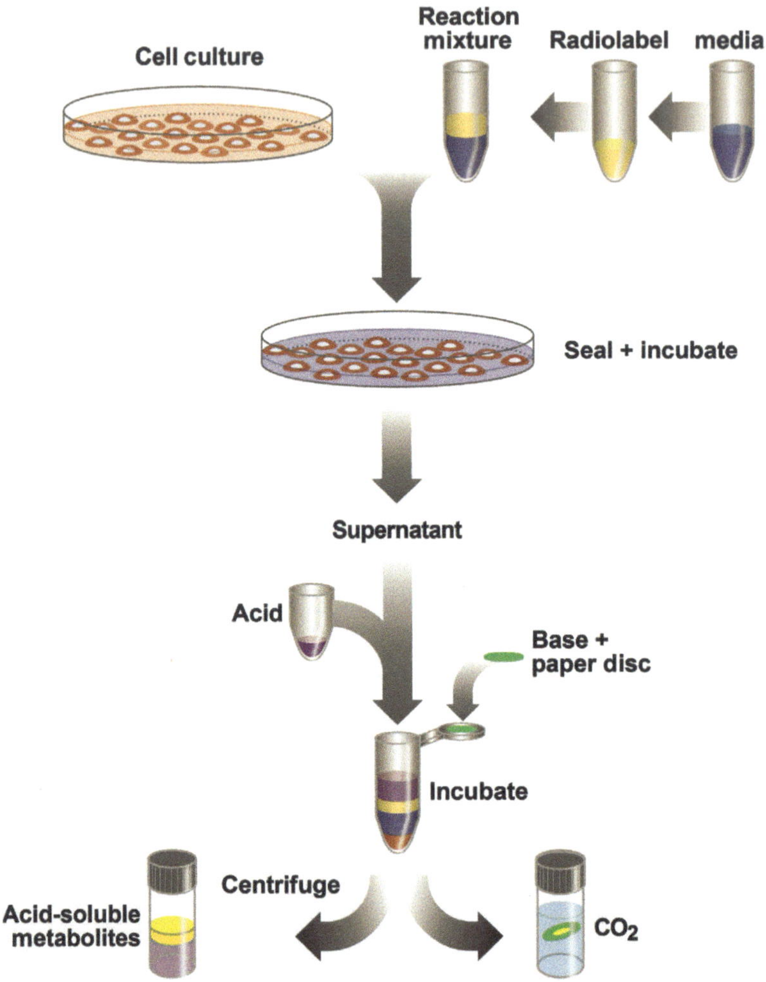

Frank K. Huynh *et al.*, Figure 20.3 *Schematic of the protocol for cultured cells.* In a similar manner to tissue oxidation, the cells of interest are incubated in media containing ^{14}C-palmitate. The supernatant/cells are then added to a strong acid to stop the reaction and the amount of ^{14}C in the acid-soluble metabolite fraction and the paper disc CO_2 fraction can each be measured by scintillation counting.

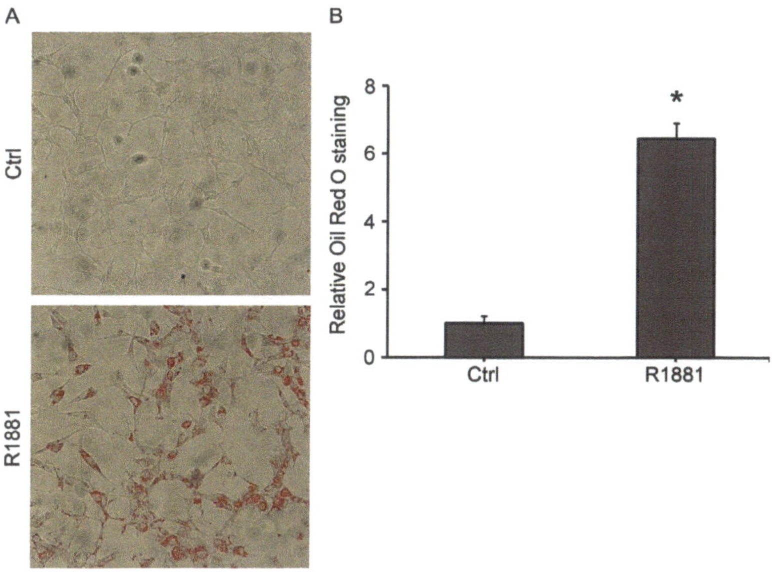

Jørgen Sikkeland *et al.*, Figure 21.2 *Increased lipid accumulation in LNCaP cells in response to androgen treatment detected by Oil Red O.* LNCaP cells were cultured in medium containing 5% CT-FBS for 48 h and treated with 10^{-8} *M* R1881 or vehicle (Ctrl) for 96 h, fixed with PFA, and stained with Oil Red O. (A) Representative bright-field images of stained cells. (B) Quantified Oil Red O staining after isopropyl alcohol extraction. $^{*}p < 0.05$.

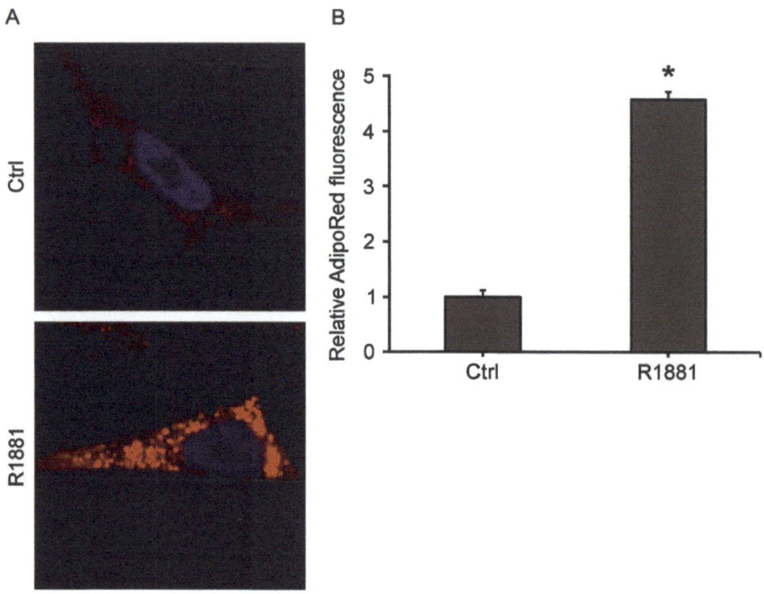

Jørgen Sikkeland *et al.*, Figure 21.3 *Increased lipid accumulation in LNCaP cells in response to androgen treatment detected by HCS LipidTOX Red and AdipoRed.* LNCaP cells were cultured in medium containing 5% CT-FBS for 48 h and treated with 10^{-8} *M* R1881 or vehicle (Ctrl) for 96 h. (A) Representative fluorescence confocal images of cells stained with HCS LipidTOX Red (red) and DAPI (blue). (B) Relative fluorimetric detection using the AdipoRed assay. $^{*}p < 0.05$.

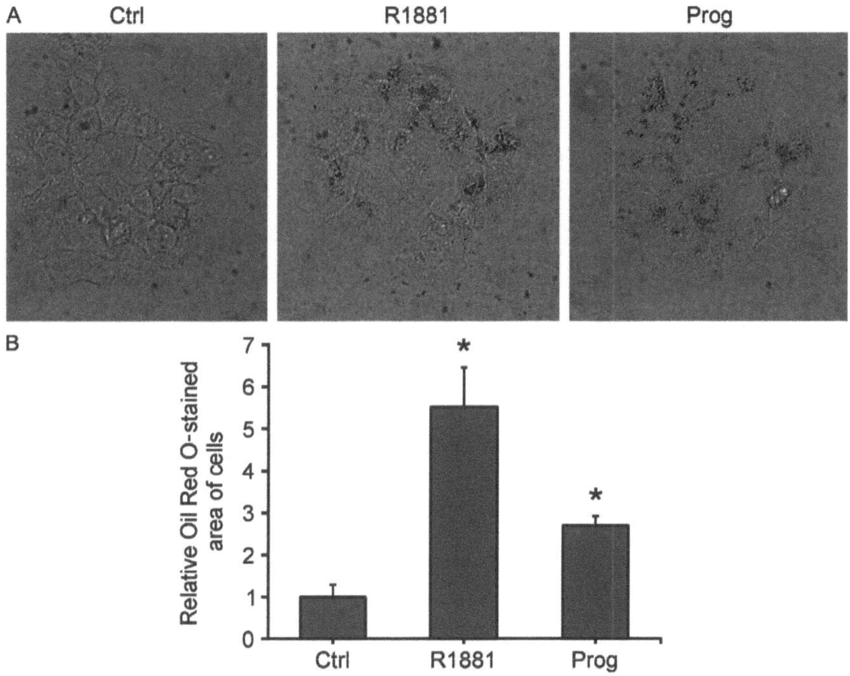

Jørgen Sikkeland *et al.*, Figure 21.4 *Increased lipid accumulation in T47D cells in response to steroid treatment detected by Oil Red O.* T47D cells were cultured in medium containing 5% CT-FBS for 48 h and treated with 10^{-8} *M* R1881, 10^{-8} *M* progesterone (Prog), or vehicle (Ctrl) for 96 h, fixed with PFA, and stained with Oil Red O. (A) Representative bright-field images of stained cells. (B) Relative Oil Red O-stained area of the cells in (A) quantified using ImageJ. $^{*}p<0.05$.

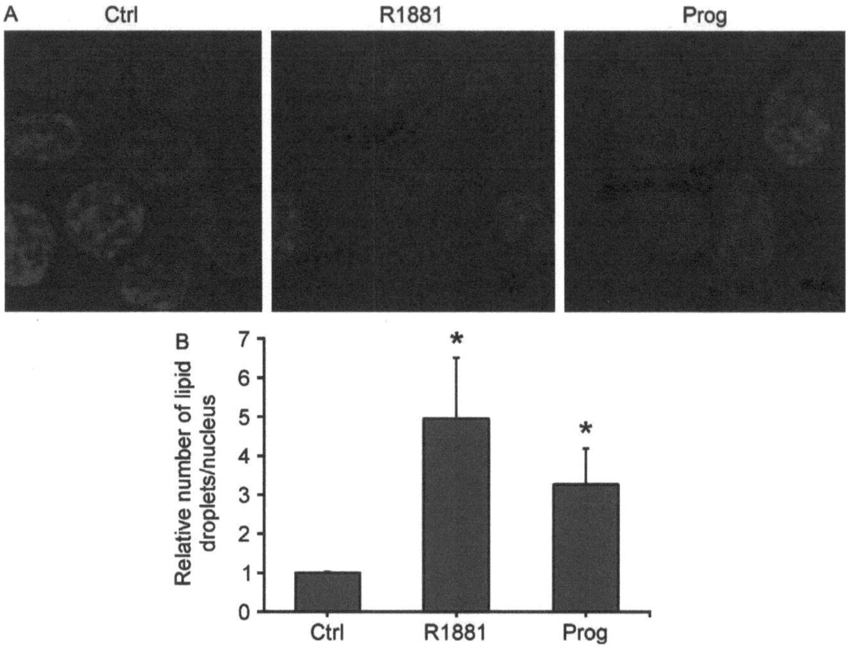

Jørgen Sikkeland *et al.*, Figure 21.5 *Increased lipid accumulation in T47D cells in response to steroid treatment detected by HCS LipidTOX Red.* T47D cells were cultured in medium containing 5% CT-FBS for 48 h and treated with 10^{-8} *M* R1881, 10^{-8} *M* progesterone (Prog), or vehicle (Ctrl) for 96 h. (A) Representative fluorescence confocal images of cells stained with HCS LipidTOX Red (red) and DAPI (blue). (B) Relative number of lipid droplets per nucleus from (A) quantified using ImageJ. $^{*}p<0.05$.

FSC
www.fsc.org
MIX
Papier | Fördert gute Waldnutzung
FSC® C083411

Kolibri 360 GmbH
Ferdinand-Jühlke-Straße 7
99095 Erfurt, Deutschland
produktsicherheit@kolibri360.de